Readings for

Technical Communication

Readings for Technical Communication

Edited by

Jennifer MacLennan

Ron and Jane Graham Centre
for the Study of Communication
College of Engineering
University of Saskatchewan

OXFORD
UNIVERSITY PRESS

OXFORD
UNIVERSITY PRESS

70 Wynford Drive, Don Mills, Ontario M3C 1J9
www.oup.com/ca

Oxford University Press is a department of the University of Oxford.
It furthers the University's objective of excellence in research, scholarship,
and education by publishing worldwide in

Oxford New York
Auckland Cape Town Dar es Salaam Hong Kong Karachi
Kuala Lumpur Madrid Melbourne Mexico City Nairobi
New Delhi Shanghai Taipei Toronto

With offices in
Argentina Austria Brazil Chile Czech Republic France Greece
Guatemala Hungary Italy Japan Poland Portugal Singapore
South Korea Switzerland Thailand Turkey Ukraine Vietnam

Published in Canada
by Oxford University Press

First published 2008

Library and Archives Canada Cataloguing in Publication

Readings for technical communication / edited by Jennifer MacLennan.

Includes bibliograpical references.
ISBN 978-0-19-542322-8

1. Technical writing—Textbooks. 2. Communication of technical information—Textbooks.
I. MacLennan, Jennifer

T11.R37 2007 808'.0666 C2007-903633-3

1 2 3 4 – 11 10 09 08

Cover design: Sherill Chapman

Cover image: Ali Mazrale Shadi/iStockPhoto

This book is printed on permanent (acid-free) paper ♾.
Printed in Canada.

Contents

Introduction

Jennifer M. MacLennan

The teaching of technical and professional communication has too often been approached as a matter of stuffing facts into one of a limited number of standard formats—memos, reports, letters, logbooks, email messages. Apart from admonitions to 'focus on your purpose' and 'remember your reader', such courses have traditionally said little about the political, interpersonal, or ethical demands of writing in a professional workplace.

However, recent growth in technical communication as a scholarly discipline has started to influence what undergraduate education is expected to achieve. Increasingly, professional programs in both colleges and universities are redesigning their introductory offerings and adding additional communication courses to augment the rudimentary skills-based approach that was previously considered sufficient. In this new breed of communication courses, students are encouraged to develop a deeper understanding of the rhetoric and politics of interaction in a professional workplace, and to explore the impact of such issues as language use, style, situation, power dynamics, face, ethics, leadership, and technology on professional communication. Even at the introductory level, students are expected to read with more critical awareness, to become more sensitive to interpersonal and social dimensions of professional interaction, and to exercise more thoughtful judgement.

Readings for Technical Communication is intended to help students do just these things. A collection of thought-provoking essays by both theorists and practitioners, it encourages students to see professional and technical communication as an engaged human process that is shaped and constrained not only by considerations of content and format, but also by such factors as personal credibility, interpersonal sensitivities, relational history, ethical challenges, organizational expectations, and political manoeuvring. Although some of the readings are classics, many have never been anthologized before, and just under a third of the selections were written expressly for this collection. And, although the readings include representation from an international body of scholarship, I am proud to say, as well, that just under half (twenty-one of the fifty-four readings in the book) are Canadian.

The selections are arranged thematically, but may of course be discussed in any order suitable to the instructor's methods. I have tried to encompass a wide variety of issues in the study of professional communication, from the nature of interaction to the impact of technology on our communication, from style and editing to accommodating the needs of an audience. The readings vary in formality, in scholarly and stylistic complexity, and in focus; some were written specifically for the student reader, while others are addressed to a more theoretically advanced audience. I have deliberately included a few selections that were originally published in academic journals, not only because they deal with important theoretical issues, but because they provide valuable lessons in audience adaptation. Many of our students—particularly those in technical programs—have never read anything written for an audience other than themselves. As a result, they often find it difficult to adjust their own style to accommodate different audiences—in part because they have no idea what such adjustments would sound like. Reading articles for whom they are not the original audience challenges them to develop a more sophisticated ear for style, as well as a deeper understanding of the important relationship between audience and rhetorical strategy.

Each selection is accompanied by discussion questions geared to its individual features, which prompt student readers to consider the essays themselves as rhetorical acts. Using these questions as a starting point, instructors can encourage students to develop the habit of reading to understand not only content and argument but also context, and to recognize that style and structure are in part methods for managing the complex of events, objects, people, and relations that make up the technical writer's rhetorical situation. In this way, selections develop students' critical faculties, challenging them to understand what is said, how it is said, and why.

The questions may also be used as the foundation for longer writing assignments, particularly in cases where students are prompted to compare and contrast the current selection with one or more of the other essays in the book, either on the basis of content and argument or on the basis of style, language, footing, audibility, and tone. Suggestions for using the questions in this way can be found in the accompanying online Instructor's Manual (http://www.oup.com/ca/he/companion/maclennanreadings).

In the end, an anthology of readings in any field has as its goal the development of understanding through the process that Margaret Atwood once described as the 'laying on of minds' (1987: 61): the enrichment of students' understanding through exposure to the insights and thoughtful judgement of those who have preceded them. This is the goal of all humanistic education: to train students in the art of *reading*, not just texts but also people and situations, with sensitivity and insight. This essentially rhetorical purpose is what has given shape to this reader, as it has given shape to my efforts as a teacher for the past twenty-five years, and as it will continue to shape my career for the remainder of my working life.

References

Atwood, Margaret. 1987. *The Mission of the University: A Symposium in Six Parts*. Kingston, ON: Queen's University.

Acknowledgements

The completion of any major project brings, along with a sense of culmination, the opportunity to acknowledge those whose support and assistance have made the work possible.

My thanks go first to my contributors, who have so graciously allowed me to include their work in this book; I am grateful to all of them, but most particularly to those who wrote new pieces especially for this collection. I am also grateful to the anonymous reviewers who took time out of their busy lives to read the manuscript and make helpful suggestions.

One of the most onerous tasks involved in producing a reader is organizing and confirming permissions. I am immensely grateful to Burton Urquhart of the Ron and Jane Graham Centre, who so capably and generously assisted me in that process and with other organizational tasks for this and another book project. His help has been indispensable.

I have been glad, too, of the support of the editorial staff of Oxford University Press, especially Roberta Osborne, Dina Theleritis, Jessica Coffey, and Eric Sinkins, all of whom are deserving of praise for their patient good humour. Cliff Newman has been a constant support and a source of good counsel, which I have treasured. It would be difficult for me to estimate the contribution to my writing life of David Stover, whose good judgement has always provided an example of what a publisher should be. Our professional association has now lasted twenty years, and I'm preparing for another twenty—if we both make it that far!

Closer to home, I also want to acknowledge the exceptional faculty of the Ron and Jane Graham Centre for the Study of Communication: Jeanie Wills, Deb Rolfes, Burton Urquhart, and Rebekah Bennetch. Their dedication, collegiality, and hard work are what have made our Centre the vibrant and nurturing intellectual community it is, and I want them to know just how much their efforts are appreciated.

I have also relied on the seemingly boundless talents of Randy Hickson, Faculty Support in the Engineering Computer Centre, who twice rescued my files when my office computer did its best to consign them to oblivion. I am thankful, as well, for the amazing Gweneth Jones of the Engineering Dean's Office, who goes so far out of her way for us on a daily basis, and for the patient good will of Colleen Teague, Coordinator of Personnel and Facilities, who successfully steered us through the extensive office renovations that took place in the middle of the term—and in the middle of this project.

I remain, as always, grateful for the support of D.K. Seaman, who has consistently believed in the value of what I am doing here, and who generously approved and encouraged our efforts to establish a Canadian home for the study of human communication. My indebtedness to Ron and Jane Graham is, of course, profound. We are all gratified by their unquenchable interest in the Centre and its activities, and their enthusiastic commitment to making it everything we hope it will be.

As remarkable as it has been to see our vision turned into the reality of the Graham Centre, there would be no such place without the hard work and investment of many extraordinary people. Chief among these are Claude Laguë, former Dean of Engineering, and Lynn Danbrook, College of Engineering Development Officer. I would like also to acknowledge the kindnesses of numerous other supporters and friends who have contributed along the way, among them Pam Janzen, Garry Wacker, Bob Gander, Ernie Barber, and the wonder-working Franco Berruti, whose belief in my work has made all the rest of this possible. Franco, you were right: the sky is the limit, and we're flying higher than any of us imagined.

I cannot begin to express my personal gratitude, and the gratitude of the rest of the Graham Centre faculty, to our master strategist and constant support, the matchless Malcolm Reeves, Assistant Dean of Undergraduate Programs in the College of Engineering. Malcolm, we would quite literally be nowhere without you—and only partly because we wouldn't have any offices.

My time in the College of Engineering has been especially blessed and enriched by three people who have served as my friends and mentors: David Male, Emeritus Professor of Mechanical Engineering; John Thompson, President Emeritus of St Thomas More College; and Richard Burton, Professor of Mechanical Engineering. The constancy of their support has been remarkable, and the wisdom of their advice incalculable. I am more grateful to them than they will ever know.

In the past year or so I have been deeply saddened by the loss of two people whose influence on my life and my teaching has been profound. Without them, the world is a lesser place. Hubert Spekkens of St Francis Xavier University provided the model of teaching that I have followed all my life. W.B. Lambert, Emeritus of the University of Lethbridge, was a treasured and loyal friend whose decency and goodness made him very nearly unique in that environment. I will miss him always.

It goes without saying that I would be lost—geographically and otherwise—without the constant support of David Cowan, who provides me with the greatest source of contentment and of joy that I have ever known. I would like to dedicate this book, as well as every other accomplishment of my life, to him.

To Dave, for the daily gift of laughter.

PART I

Thinking about Communication

REPEATEDLY, research shows a strong correlation between communicative ability and professional success, life satisfaction, happiness, and even good health. In all these areas of life our ability to function effectively depends on our ability to establish appropriate relationships with those around us.

New graduates in the technical professions are typically surprised to discover how much of their working time must be spent in communicating with others, and how complex that task often is. As we will learn from the experts who have contributed essays to this book, technical communication is rarely a matter of simply marshalling facts and assembling them into a grammatically correct, logically ordered format. Instead, it is frequently beset by interpersonal, cultural, political, and ethical complications—complications for which technical specialists are often ill-prepared.

The single most important communicative skill for any professional is the ability to relate effectively to others. At first glance, the creation and nurturing of relationships may appear to have little to do with technical writing or professional communication. But as the essays in this book reveal, an inability to connect appropriately to an audience is the number one reason that workplace messages fail. Reports, proposals, promotional documents, log books, minutes, email communication, technical presentations, resumes, interviews, even casual exchanges over the water cooler—all of these present situations in which we must get our ideas across, establish our credibility and professionalism, and make ourselves heard and understood by those around us. All of these goals depend on how well we understand others, how effectively we identify their needs and expectations, how genuinely we respect their values and attitudes, how deftly we design our messages to accommodate their interests and priorities. No matter how technically expert we may be, if we can't do these things, we are unlikely to find success or satisfaction in our professional lives.

The essays in this Part are intended to kick-start your thinking about the nature of communication, as well as its central role in your life as a professional. We begin with a consideration of the pragmatic functions of communication, and then move to more theoretical observations that will serve as underpinnings for what is to come in later chapters.

CHAPTER 1

Communicate Well and Prosper: Poor Interaction Costs Companies More Than They Realize—Or Can Afford

Helen Wilkie

How much has poor communication cost your company in the past 12 months? Chances are, you have no idea. Chances are even better it's a lot more than you can afford. But you won't find the numbers in the financial statements or year-end departmental reports. Nothing shows up saying 'lost productivity due to miserable meetings' or 'missed business opportunity through sorry selling skills' or 'the employee quit because there's no communication around here.' Why? Because most people aren't sure what communication really is.

Consider this: when companies conduct internal needs assessments, communication nearly always surfaces near the top of the list. But if you ask 10 people who put it on the list exactly what they meant, you'll get 10 different answers. People often can't pinpoint the problem—they just have a vague feeling communication isn't happening. Unfortunately, this vagueness relegates communication to the bottom of the action list.

Truth is, communication isn't some warm and fuzzy 'nice to have'—it's nothing less than the lifeblood of your organization. If blood doesn't circulate at just the right pressure and speed to all parts and extremities of the human body, that body sickens and eventually dies. So too, does an organization in which communication doesn't flow freely.

Communication isn't limited to vision and mission statements from the top. It's not just news nor internal or client newsletters, annual reports or videotaped messages to the troops. These are all important but they form just a fraction of the communication and miscommunication that takes place every day in the workplace.

What I call applied communication is written, spoken, and non-verbal interaction among people in order to get things done. It takes cooperation to create a product. It takes collaboration to approach a new market. It takes teamwork to implement a strategy. It takes this applied communication to oil and run the machinery of business. And if that machinery breaks down, a great deal of money is lost. It's in this area that you need to look for the financial drain.

Loss of Time

What does your time cost your company for each hour you're at work? A good rule of thumb in calculating hourly cost is: annual salary divided by 2,000 (based on fifty 40-hour work weeks). When you know this figure for your time, you can begin to calculate the cost of applied communication at work not just for you, but for all the staff.

Meetings

Regardless of its purpose, a meeting is an exercise in applied communication. You speak, you listen, you interact. I've never met anyone in business who *has not* complained about meetings: too many, too long, too boring. I would add to that: too expensive.

Consider meetings that are supposed to last an hour but somehow expand to use up most of the afternoon. Calculate the hourly cost of total participant time and multiply by the length of the meeting; and keep in mind that the higher the level of participants, the more expensive the time. The result may not sound too alarming until you consider how many of those meetings take place in your organization every day, every week, every year. Do the arithmetic.

Correspondence

Letters, reports, memos, and now the ubiquitous email—written communication is an integral part of doing business. Unfortunately, statistics show that corporate employees spend altogether too much time writing, and badly at that, so that those on the receiving end spend too much time reading. If a $40,000-a-year employee spends just two hours a day reading, writing, and managing email, that's a $9,000 annual cost. Judging from what people tell me about their work habits, two hours a day is a conservative estimate. And what about those at much higher salary levels who spend much longer writing every day? Do the arithmetic.

Presentations

People at all levels present information in a variety of settings in the workplace every day. These presentations not only consume many hours in the creation and preparation, but also the expensive time of those who must listen to them. Unfortunately, poor presentation skills often result in a futile exercise that communicates less than a simple written report. How many useless presentations take place in a major corporation every day? Do the arithmetic.

Loss of Business

Too many salespeople know their 'pitch' so well that they totally ignore any input that a prospective customer might give them. They barely shake hands and sit down before they start talking. They blithely prescribe their product or service as the cure for a problem, without even finding out whether such a problem exists.

But an effective sales process is, in fact, a conversation, a two-way exercise in applied communication. Done poorly, it can result in lost sales and missed opportunity for continuing business relationships. Customer loss doesn't happen only in the sales process, but can also be spurred by an inept 'customer service' exchange.

When someone calls to complain, the client relationship is at a fragile point. It can be repaired through the right message well delivered, or broken beyond repair by poor communication.

When we consider the lifetime value of a customer relationship, we can truly appreciate the real dollar cost of poor communication.

Loss of People

Whatever people tell their bosses about their reasons for leaving the company, exit interviews often tell a different story. One of the most common reasons cited is that they don't feel anyone listened to them. Day after day in the workplace, millions of people go through the motions of talking with each other in person and on the phone, constantly connected through technology, and never truly communicating with one another.

Study after study tells us that recognition and respect are more motivating than money, and one of the best ways to show people they are valued is to listen to them. Sadly, listening is probably the most underused of all the communication skills.

So people leave. How much then does it cost to replace them? Studies give a wide range, from a low of 25 per cent of salary plus benefits, to a whopping 150 per cent. Employee replacement represents yet another huge cost that can at least sometimes be charged to poor communication.

By improving the way people (and I mean people at all levels) interact to get things done, you can improve morale and increase productivity and the bottom line.

QUESTIONS FOR CRITICAL THOUGHT

1. What does Wilkie mean by 'applied communication'? How, exactly, does she define it? Is her definition broad or focused? Why might she define communication in such a way?
2. Read MacLennan's 'Why Communication Matters' and identity at least three similarities between Wilkie's article and MacLennan's. In what respects are the two different?
3. What is the primary reason, according to Wilkie, that communication matters in a professional context?
4. How does Wilkie suggest that the impact of poor communication may be quantified? What is the purpose of this quantification?
5. This article appeared in the 'Careers' section of the *Globe and Mail*. Who is Wilkie's intended audience? How do you know? What evidence is there in the article itself to indicate its probable readers?
6. Persuasive appeals may be addressed to one or more of the audience's reason, ethical sense, imagination, or emotion. Does Wilkie appeal primarily to reason, to ethics, or to emotion? Provide examples from the reading to justify your answer.
7. What is the biggest challenge, according to Wilkie, of trying to improve communication in a workplace? Why?
8. Wilkie's article appears to be devoted to diagnosing the problem of ineffective communication in the workplace. Does her diagnosis lead to any concrete suggestions or solutions? Explain.

CHAPTER 2

Why Communication Matters

Jennifer M. MacLennan

In a recent article in the *Globe and Mail*, Wallace Immen reports that search committees for executive positions are increasingly 'placing heavier emphasis on "soft skills", like self-awareness, empathy, a sense of purpose, team-building and communication skills' (Immen, 2006: R1). In other words, employers are 'spending much more time probing personalities' in hope of identifying those who fill a wish list of personal characteristics that go well beyond the task-focused skills traditionally associated with management. Overwhelmingly, the qualities on the list are skills in communication—an ability to connect with others, understand their concerns, forge effective relationships, and motivate them to achieve their best. What is revealed by Immen's report, and others like it,[1] is the increasing acknowledgement by employers that effective communication is an essential ingredient in professional success (Aldo, 2006).

Despite the current brouhaha over communication skills, such understanding is not new, especially in the professions. As long ago as 1984, James McAlister identified poor communication skills as one of the foremost contributors to career failure among engineers—more significant even than a failure of technical expertise. McAlister points to poor relationships on the job, an inability to work effectively with other members of a team, an unwillingness to consult, and general ineptness in communication as chief among the career problems encountered by technical specialists (McAlister, 1984). And long before McAlister, Elmer Lindseth, himself an engineer and president of the Cleveland Electric Company, observed that, 'although we may have experts in automation, in engineering, and in production design, their work can be completely offset by a breakdown in human relations and communication' (Lindseth, 1955). Small wonder that McAlister reckoned skill in communication to be 'more important than technical skills' in establishing a young engineer as 'someone with potential' in the eyes of his supervisors.

It's not just in engineering that communication matters; in fact, no matter what the industry or profession, those responsible for hiring decisions have always considered communication skills to be essential (Carnevale, 1990). In a recent survey of managers from 171 companies (NFI Research, 2004), 94.2 per cent of the executives surveyed identified communication as the most fundamental skill an employee can bring to the job; 73.7 per cent of these same people regarded the ability to collaborate with others—actually another communication skill—as equally desirable. In one review of research in the field, communication scholars found that not only are communication skills highly desired by employers, but they are also recognized as 'essential in specific professional careers', including accounting, auditing, banking, counselling, engineering, industrial hygiene, information science, public relations, and sales (Morreale, et al., 2000). Communication has even been argued to be 'the key to successful software development' (Palmer, 2002). And as one researcher in business concluded, 'the skills most valued in the contemporary job-entry market are communication skills (including oral communication, listening, and written communication)' (Curtis, 1988).

Interestingly, although skills like the ones listed are still regarded as essential for nearly any job outside of hamburger jockey, it is the skills of interpersonal relations that are particularly prized in the contemporary workplace, and not only for supervisory and managerial staff. Yet another recent survey showed hiring people with solid interpersonal skills is the single most important factor in creating an effective workplace environment. For example, managers who are good communicators aid staff retention by helping employees to understand how their work contributes to the organization's goals, thereby increasing the employees' sense of value and their commitment to the organization (Business Wire, 2000). Similarly, front-line workers who are good communicators build effective relationships with clients and other personnel, and contribute to the positive climate that keeps job satisfaction high. Since recruiting and training new employees is much more expensive than retaining those already on the job, hiring people who communicate effectively is a sound investment.

The importance that more and more companies impute to communication is in part a result of how much of our time is devoted to it. Some experts estimate that as much as 75 per cent of the average person's day is spent communicating in some way; for students, the proportion is slightly higher (Barker, et al., 1980). Another study estimates that we spend a good 25 per cent of our entire lives communicating with others—a close second to the roughly 30 per cent we spend sleeping (Jones, 2004).

Those in scientific professions are reported to spend anywhere from 50–90 per cent or more (Björk and Turk, 2000; Red Point Management Services, 2006) of their work time engaged in some form of communication—and this is true even at the entry level. One of the measures of the value employers place on communication is its frequent appearance in job postings for a whole range of positions. My own search of the Monster.ca website using the key word 'communication' produced more than a 1,000 hits! Interestingly, few of the job ads that turned up were for positions in the communication field; instead, postings of all kinds listed communication among the required skills for everything from sales to maintenance engineering, from project management to software engineering, from accountancy to office management.

For example, one posting for an Information Systems Project Manager included among its requirements the following communication abilities:

- Exceptional communication skills: listening, verbal [that is, oral], written, public speaking and presentation.
- Expert problem-solving and troubleshooting skills with the ability to think on your feet and to exercise sound judgment and decision-making.
- Committed to professionalism and excellence in customer service.
- Exceptional writing skills to transform a vision into winning Presentations, Proposals, Statement of Work documents, Requirements Documents, etc.
- A high degree of creativity, strong organization skills; ability to plan, prioritize, coordinate and monitor a significant number of simultaneous tasks in a fast-paced, ever-changing environment.
- Ability to work with a minimum of direct supervision or guidance.
- Committed to team goals and success through cooperation. (Monster.ca, 2006b)

Another advertisement for a highly technical position, this one a Production Engineer, asks applicants to demonstrate 'excellent analytic, presentation, written and verbal [that is, oral] communication and interpersonal skills and ability to lead/train by doing with a hands-on style' (Monster.ca, 2006a). Despite the compactness of this list, it names several higher-order communication skills, including writing, public speaking, analysis, leadership, and training. Finally, a position as Payroll Administrator in a technical workplace calls for advanced relational and audience-analysis skills: 'Interpersonal communication skills are essential as the position requires daily interaction with Rig Managers, responding to employee inquiries and maintaining open lines of communication with other departments within the company, as well as information gathering from external sources' (Monster.ca, 2006c).

To round out our consideration of the prevalence of communication in the wish list of requirements for job advertisements, consider the following paragraph, which I have seen featured in numerous job postings for a variety of technical and other positions over the past five years:[2]

> Ability to read, analyze, and interpret common scientific and technical journals, financial reports, and legal documents. Ability to respond to common inquiries or complaints from customers, regulatory agencies, or members of the business community. Ability to write speeches and articles for publication that conform to prescribed style and format. Ability to effectively present information to top management, public groups, and/or boards of directors. Ability to define problems, collect

data, establish facts, and draw valid conclusions. Ability to interpret an extensive variety of technical instructions in mathematical or diagram form and deal with several abstract and concrete variables.

Although it has also been used for a variety of senior posts, I originally saw this formulation in an advertisement for an entry-level engineering position. It struck me then as quite a remarkable list, in that it calls for several sophisticated communication skills: analytical reading and interpretive skills, speech-writing, persuasion, and adapting messages for complex and varied audiences. Few technical programmes adequately prepare their graduates for such demanding communication tasks;[3] in fact, even experienced technical specialists might find it a challenge to measure up to the complete list. However, the fact that this skill set was included in a job ad targeted to recent engineering graduates not only points to the employer's scale of values, but suggests that any applicant who could demonstrate mastery of these skills would have a distinct advantage over the competition.

A skill set like this one, which has been included in recruiting ads for a whole range of professional positions, confirms that, in our highly complex and demanding workplaces, the ability to communicate is regarded as essential no matter what the job is. These employers recognize the value of skills like leadership, cooperation, flexibility of attitude, and responsibility; they are well aware of the importance of being able to interact effectively with people whose priorities, values, and experiences are different from our own; they know, too, that writing clearly and speaking persuasively are just as important as technical expertise to any job. And finally, they know all too well that people whose communication is unclear, incomplete, or discourteous will be ineffective and possibly costly to the company in time, money, and customer relations.

Another indication of how important communication is to the professions is the professional development phenomenon known as 'continuing competence', which has become the standard across a range of professional occupations, from acupuncture to architecture, from audiology to chiropractic, from occupational therapy and dietetics to pharmacy and engineering, to name only a few of those identified by the New York State Education Department in its report on the subject.[4] In Canada, most professions (for example, nursing, occupational therapy, pharmacy, forestry, and engineering,[5] to name only a few) have developed explicit continuing competence policies and programming. Most include communication as one important area in which professionals are encouraged to develop their skills. For instance, a survey by the Association of Professional Engineers and Geoscientists of Saskatchewan (Hein, 1998) asked practising engineers to identify important areas for career development under the guidelines for 'continuing competence' provided by the Canadian Council of Professional Engineers (Day, 1998). In response, nearly all the engineers surveyed identified communication skill development as a main priority.

The survey also asked respondents to list qualities they would use to judge each other's professional competence; several of the categories identified by the engineers depend on the professional's ability to communicate effectively. For instance, 'experience' and 'knowledge' represent competence only if they can be effectively communicated to others; 'approach' and 'attitude' are clearly measures of the professional's interpersonal competence. A separate study by the US Department of Labor showed similar results, identifying 16 qualities for high job performance in any field. Ten are communication skills: listening, speaking, creative thinking, decision making, problem solving, reasoning, self-esteem, sociability, self-management, and integrity/honesty (US Department of Labor, 1991).

It seems clear that any new graduate who can demonstrate a mastery of these desirable communication skills will automatically be considered more attractive to an employer. Some studies even suggest that the ability to communicate effectively and establish a positive relationship in the interview may be regarded by the employer as more important than technical training or experience.[6] The ability to communicate effectively, not only in writing or in public presentations, but interpersonally and one-on-one, is a skill set that every graduating student needs, not only in order to perform well in job interviews, but to ensure a greater likelihood of long-term career success.

But exactly how are graduates measuring up? One answer can be found in a study conducted by the University of North Carolina. Researchers asked the employers of recent university graduates to assess their new employees in both technical preparedness and communication skill, and then to rate the importance of both kinds of skills to the job being performed. The results of the study are interesting from a communication perspective. They showed that in specialized and technical skills, the new graduates consistently *exceeded* employer expectations and job requirements—that is, the grads were more technically skilled than employers expected them to be, and in fact more technically skilled than was specifically required by their jobs.

However, the study showed that their skills in communication were not so well developed. When employers were asked to rank their new employees in the four areas of writing, reading, listening, and public speaking, they consistently rated the actual ability of the students significantly *lower* than the level of skill required by their jobs (Hoey, 2001). The employees' abilities in interpersonal relations and their leadership skills were not specifically addressed in this study, but it is unlikely that their performance would be dramatically better in these areas than in the other 'soft skills' that were surveyed, since few professional programmes have moved to include these in the course work required of their students.

The pervasiveness of communication in nearly every profession and position leaves us with several implications. First, communication skills are both highly valued and highly sought-after by employers, and the wish list of desirable qualities is getting longer. Second, although the traditional communication skills of report-writing, analytical reading, listening, and oral presentation remain as important as ever, other 'soft skills' of communication such as team-building, empathy, self-awareness, ethics, and cooperation are taking on equal or greater value in the eyes of employers. Third, studies like the one carried out at the University of North Carolina suggest that employers believe most new graduates cannot adequately perform the communication tasks expected of them in their positions. Fourth, professional programmes have been slow to react to this expansion in emphasis on communication skills. As a result, they typically provide students with courses that cover only the traditional skills of writing and speaking, but fail to address the other communication skills that are essential to professional success.

Understanding the long-term expectations of their future profession as well as the demands of the current job market is critical for those who are just beginning their careers. Whatever the current buzz among human resources professionals, employers have always needed, and will continue to need, those who can handle the 'soft' skills of working collegially and cooperatively with others. Effective communication is not simply a dispensable luxury; it is an essential ingredient for a successful and fulfilling professional life.

Notes

1. Such as Helen Wilkie's 'Communicate Well and Prosper,' on page 2 in this volume. See also Peter Morton, 'Communication Key to Success in Engineering Careers', *Polytechnic Online* 127. Available online: http://www.poly.rpi.edu/article_view.php3?view=599&part=1 (posted

28 March 2001); A. Terjanian, 'Adequate Communication is Key to Success' (PowerPoint Presentation). Available online: http://www.amcham.am/Adequate.ppt (accessed 18 September 2006).

2. Including, for instance, a Senior Drilling Engineer, Ultra Deep Water, posted by SGF Global for an unidentified oil company in Houston, Texas (http://www.oilcareers.com/content/jobsearch/job_advert.asp?jobadid=28036); Senior Software Developer for First Quadrant (http://www.firstquadrant.com/careers.html); Project Manager for Cardinal Health (http://stlouis.craigslist.org/hea/204630640.html); Inventory Control Manager for Jabil Circuit (http://www.jabil.com/careers/jobDescription.asp?id=2557); Senior Librarian for the Arizona Library (http://www.lib.az.us/jobs/joblisting.cfm?jobID=489), and numerous others.
3. An exception is the University of Saskatchewan's College of Engineering, which offers its Engineering Communication Option, a specialization in communication that can be taken as an add-on to the engineering degree. It is the only program of its kind in Canada. See http://grahamcentre.usask.ca.
4. Including Acupuncture, Architecture, Audiology, Certified Public Accountant, Chiropractic, Dentistry, Dietetics-Nutrition, Medicine, Occupational Therapy, Occupational Therapy Assistant, Optometry, Pharmacy, Physical Therapist Assistant, Physical Therapy, Podiatry, Professional Engineering, Psychology, Registered Professional Nurse, Social Work, Speech-Language Pathology, Veterinary Medicine, Veterinary Technology. See New York State Education Department, Office of the Professions 'Continuing Competence Report' (June 2000). Available online: http://www.op.nysed.gov/contcomp.htm.
5. Canadian Nurses Association, 'A National Framework for Continuing Competence Programs for Registered Nurses'. Available online: http://www.cna-nurses.ca/CNA/nursing/regulation/competence/default_e.aspx (accessed 21 September 2006); Canadian Association of Occupational Therapists, 'CAOT Position Statement: Continuing Professional Education (2006)'. Available online: http://www.caot.ca/default.asp?pageID=153 (accessed 21 September 2006); National Association of Pharmacy Regulatory Authorities, 'A Model Continuing Competence Program Framework for Canadian Pharmacists'. Available online: http://www.napra.org/docs/0/96/191.asp (accessed 21 September 2006); College of Alberta Professional Foresters, 'Continuing Competence Program'. Available online: http://www.professionalforesters.ab.ca/continuingcompetence.htm (accessed 21 September 2006); Canadian Council of Professional Engineers, 'Continuing Competence'. Available online: http://www.ccpe.ca/e/files/guidelinecompetency2004.pdf (accessed 21 September 2006).
6. University of Wisconsin Madison Engineering Career Services, 'The Interview: An Overview'. Available online: http://ecs.engr.wisc.edu/student/interviewing.cfm (accessed 21 September 2006); Jobsite.co.uk, 'Your Attitude is Reflected in Your CV'. Available online: http://www.jobsite.co.uk/articles/candidate/c1/s11/a1181.html (accessed 21 September 2006).

References

Aldo, Paul. 2006. 'Clear Communications: A Key to Success', *The Atlanta Journal-Constitution*. Available online: http://jobnews.ajcjobs.com/news/content/careercenter/features/advice_develop1.html (accessed 18 September 2006).

Barker, L., R. Edwards, C. Gaines, K. Gladney, and F. Holley. 1980. 'An Investigation of Proportional Time Spent in Various Communication Activities by College Students', *Journal of Applied Communication Research* 8: 101–9.

Björk, Bo-christer, and Ziga Turk. 2000. 'How Scientists Retrieve Publications: An Empirical Study of How the Internet Is Overtaking Paper Media', *Journal of Electronic Publishing* 6. Available online: http://www.press.umich.edu/jep/06-02/bjork.html.

Business Wire (press release). 2000. 'Three Out of Four Say Better Communication Equals Greater Employee Retention', *KnowledgePoint* 8 December.

Carnevale, A.P. 1990. 'Workplace Basics: The Essential Skills Employers Want', *ASTD Best Practices Series: Training for a Changing Work Force*. San Francisco, CA: Jossey-Bass Management Series.

Curtis, D.B. 1988. 'A Survey of Business Preferences for College Grads'. Paper presented at the Annual Meeting of the Speech Communication Association, New Orleans (November 1988).

Day, Sir Graham. 1998. 'The Engineering Profession: Some Thoughts on the Implications of Globalization and the Need for Continuing Education'. *Proceedings of the Canadian Conference on Engineering Education*, pp. 2–8. Ottawa, ON: Canadian Council of Professional Engineers.

Hein, Larry. 1998. 'Are We Competent?', *Professional Edge* 54 (February–March): 8–9.

Hoey, Joseph. 1997. 'Employer Satisfaction with Alumni Professional Preparation'. North Carolina State University: University Planning and Analysis. Available online: http://www2.acs.ncsu.edu/UPA/survey/reports/employer/employ.htm (accessed 14 June 2001).

Immen, Wallace. 2006. 'Fancy the Top Job? Get Ready to Be Grilled', *Globe and Mail* 11 September: R1.

Jones, Beth. 2004. 'Conversation "Dying" Even Though We Spend 25% of Our Lives Communicating', *British Gas News* 20 May. Available online: http://www.britishgasnews.co.uk/index.asp?PageID=19&Year=2004&NewsID=567.

Lindseth, Elmer L. 1955. 'The Meaning and Importance of Communication', Publication No. L56-21 (Lecture, 14 September). Washington, DC: Industrial College of the Armed Forces.

McAlister, James. 1984. 'Why Engineers Fail', *Machine Design* (23 February): 47–9.

Monster.ca. 2006a (20 Sept.). 'Brunel Multec's "Industrial Engineer" Job Advertisement', *Monster.ca*.

———. 2006b (20 Sept.). 'CDIT Inc.'s "Project Manager" Job Advertisement', *Monster.ca*.

———. 2006c (20 Sept.). 'Unidentified Employer, "Payroll Administrator" Job Advertisement', *Monster.ca*.

Morreale, Sherwyn P., Michael M. Osborn, and Judy C. Pearson. 2000. 'Why Communication is Important: A Rationale for the Centrality of the Study of Communication', *Journal of the Association for Communication Administration* 29: 1–25. Available online: http://www.natcom.org/nca/files/ccLibraryFiles/FILENAME/000000000158/Why%20Comm%20is%20Important.pdf.

NFI Research. 2004. 'Survey of 171 Managers from Companies with Fewer than 5,000 to More than 10,000 Employees'. North Hampton, NH: NFI Research. Available online: http://www.nfiresearch.com.

Palmer, Stephen R. 2002. 'Communication: the Key to Successful Software Development', *InformIt.com* 22 March. Available online: http://www.informit.com/articles/article.asp?p=26048&rl=1.

Red Point Management Services. 2006. 'Project Minutes', Project Management Software. Available online: http://www.projectminutes.com (accessed 18 September 2006).

US Department of Labor, 1991. *What Work Requires of Schools: A SCANS Report for America 2000*. Washington DC: Secretary's Commission on Achieving Necessary Skills, US Department of Labor.

QUESTIONS FOR CRITICAL THOUGHT

1. According to MacLennan, 'effective communication is an essential ingredient in professional success.' What evidence, and what *kind* of evidence, does she advance in support of this claim? How convincing is this evidence?
2. What is the purpose of MacLennan's article? For whom is it written? Provide some evidence from the article itself in supporting your assertions.
3. Outline MacLennan's argument. How is the article structured?

4. MacLennan reasserts her central thesis in much stronger terms at the end of the article. Why does she do so? Why is the final statement so much more forceful than the initial thesis?
5. MacLennan points out that the importance of communication to employers can be measured in at least three ways. What are they?
6. MacLennan wraps up her article with four implications; what are these? Is the article in any sense a call to action? Why or why not?
7. What is the source of MacLennan's authority? In other words, what gives her the appropriate footing to give the advice she provides?

CHAPTER 3

A Whole New Mind for a Flat World

Richard M. Felder

Interviewer: Good morning, Mr Allen. I'm Angela Macher—project engineering and human services at Consolidated Industries.

Senior: Good morning, Ms Macher—nice to meet you.

Interviewer: So, I understand you're getting ready to graduate in May and you're looking for a position with Consolidated . . . and I also see you've got a 3.75 GPA coming into this semester—very impressive. What kind of position did you have in mind?

Senior: Well, I liked most of my engineering courses but especially the ones with lots of math and computer applications—I've gotten pretty good at Excel and Matlab and I also know some Visual Basic. I was thinking about control systems or design.

Interviewer: I see. To be honest, we have very few openings in those areas—we've moved most of our manufacturing and design work to China and Romania and most of our programming to India. Got any foreign languages?

Senior: Um, a couple of years of Spanish in high school but I couldn't take any more in college—no room in the curriculum.

Interviewer: How would you feel about taking an intensive language course for a few months and moving to one of our overseas facilities? If you do well you could be on a fast track to management.

Senior: Uh . . . I was really hoping I could stay in the States. Aren't any positions left over here?

Interviewer: Sure, but not like ten years ago, and you need different skills to get them. Let me ask you a couple of questions to see if we can find a fit. First, what do you think your strengths are outside of math and computers?

Senior: Well, I've always been good in physics.

Interviewer: How about social sciences and humanities?

Senior: I did all right in those courses—mostly As—but I can't honestly say I enjoy that stuff.

Interviewer: Right. And would you describe yourself as a people person?

Senior: Um . . . I get along with most people, but I guess I'm kind of introverted.

Interviewer: I see . . . (stands up). OK, Mr Allen—thanks. I'll forward your application to our central headquarters, and if we find any slots that might work we'll be in touch. Have a nice day.

This hypothetical interview is not all that hypothetical. The American job market is changing, and to get and keep jobs future graduates will need skills beyond those that used to be sufficient. This message is brought home by two recent books—Thomas Friedman's *The World is Flat* and Daniel Pink's *A Whole New Mind*—that I believe should be required reading for every engineering professor and administrator. The books come from different perspectives—the first economic, the second cognitive—but make almost identical points about current global trends that have profound implications for education.

An implication for engineering education is that we're teaching the wrong stuff. Since the 1960s, we have concentrated almost exclusively on equipping students with analytical (left-brain) problem-solving skills. Both Friedman and Pink argue convincingly that most jobs calling for those skills can now be done better and/or cheaper by either computers or skilled foreign workers—and if they can be, they will be.[1] They also predict that American workers with certain different (right-brain) skills will continue to find jobs in the new economy:

- creative researchers, developers, and entrepreneurs who can help their companies stay ahead of the technology development curve;
- designers capable of creating products that are attractive as well as functional;
- holistic, multidisciplinary thinkers who can recognize complex patterns and opportunities in the global economy and formulate strategies to capitalize on them;
- people with strong interpersonal skills that equip them to establish and maintain good relationships with current and potential customers and commercial partners;
- people with the language skills and cultural awareness needed to build bridges between companies and workers in developing nations (where many manufacturing facilities and jobs are migrating) and developed nations (where many customers and consumers will continue to be located);
- self-directed learners, who can keep acquiring the new knowledge and skills they need to stay abreast of rapidly changing technological and economic conditions.

Those are the attributes our students will need to be employable in the coming American engineering job market. The question is: are we helping them to develop those attributes? With isolated exceptions, the answer is no. We still spend most of our time and effort teaching them to 'Derive an equation relating A to B' and 'Calculate Z from specified values of X and Y.' We also offer them one or two lab courses that call on them to apply well-defined procedures to well-designed experiments, and we give them a capstone design course that may require a little creativity but mostly calls for the same calculations that occupy the rest of the curriculum. Nowhere in most engineering curricula do we provide systematic training in the abilities that most graduates will need to get jobs—the skills to think innovatively and holistically and entrepreneurially, to design for aesthetics as well as function, to communicate persuasively, to bridge cultural gaps, and to periodically re-engineer themselves to adjust to changing market conditions.

Why don't we? It's because people, as a rule, don't want to leave their comfort zones, and engineering professors are as subject to that rule as anyone else. We are all comfortable deriving and solving equations for well-structured single-discipline systems, but most of us are not so sure about our ability to handle ill-defined open-ended multidisciplinary problems or to teach creative thinking or entrepreneurship. So, despite a crescendo of headlines and best sellers about the growing exodus of traditional skilled jobs to developing countries (including high-level research and development jobs, which are increasingly moving to India and China[2]), many engineering faculty members vigorously resist suggestions to make room in the curriculum for multidisciplinary courses and projects or anything that might be labelled 'soft'. Even though most of our alumni in industry—95 per cent? 99 per cent?—assure us (as they have done for decades) that they haven't seen a derivative or integral since they graduated, the traditionalists still insist that we can only produce competent engineers by devoting almost every course in the curriculum to deriving and solving equations, analytically and with Matlab. The same professors are no less resistant to efforts to move them away from the traditional 'I talk, you listen' pedagogy toward the active, cooperative, problem-based approaches that have been repeatedly shown to equip students with the skills Friedman and Pink are talking about (see Suggested Resources, below.)

So far we've gotten away with it, although sharply declining engineering enrollments in recent years should be a red flag. We can't count on getting away with it much longer, however. The relentless movement of industry to computer-based design and operation and off-shoring of skilled functions and entire manufacturing operations is not about to go away. On the contrary, as computer chips get faster and developing countries acquire greater expertise and better infrastructure, the movement will inevitably accelerate. The American engineering schools that respond by shifting toward more multidisciplinary problem- and project-based instruction—the way Olin, Rowan, Rose-Hulman, the Colorado School of Mines, and a number of others have already started to do—will survive. The schools that try to stick with business as usual may not.

Suggested Resources

EFFECTIVE TEACHING METHODS AND THE RESEARCH THAT SUPPORTS THEM

Bransford, J.D., A.L. Brown, and R.R. Cocking, eds. 2000. *How People Learn: Brain, Mind, Experience, and School*. Washington, DC: National Academy Press. Available online: http://www.nap.edu/html/howpeople1.

Felder, R.M., D.R. Woods, J.E. Stice, and A. Rugarcia. 2000. 'The Future of Engineering Education: 2 Teaching Methods that Work', *Chemical Engineering Education* 34,1: 26–39. Available online: http://www.ncsu.edu/felder-public/Papers/Quartet2pdf.

Woods, D.R., R.M. Felder, A. Rugarcia, and J.E. Stice. 2000. 'The Future of Engineering Education: 3 Developing Critical Skills', *Chemical Engineering Education* 34, 2: 108–17. Available online: http://www.ncsu.edu/ftlder-public/Papers/Quartet3pdf.

ACTIVE LEARNING

Felder, R.M., 'Random Thoughts' columns in *Chemical Engineering Education*:

(a) 'Learning by Doing': http://www.ncsu.edu/ftlder-public/Columns/Active.pdf

(b) 'How About a Quick One?': http://www.ncsu.edu/jelder-public/Columns/Quickone.html

(c) 'It Goes Without Saying': http://www.ncsu.edu/felder-public/Columns/WithoutSaying.pdf

See also: http://www.ncsu.edu/ftlder-public/Cooperative_Learning.html.

Prince, M. 2004. 'Does Active Learning Work? A Review of the Research', *Journal of Engineering Education* 93, 3: 223–31.

COOPERATIVE LEARNING

Felder, R.M., and R. Brent, *Cooperative Learning in Technical Courses: Procedures, Pitfalls, and Payoffs*. Available online: http://www.ncsu.edu/felder-public/Papers/Coopreport.html. See also http://www.ncsu.edu/felder-public/Cooperative_Learning.html.

Two meta-analyses of research on cooperative learning vs. traditional instruction can be found at http://www.co-operation.org (University of Minnesota) and http://www.wcer.wisc.edu/nise/c/ll/CL/resource/R2.htm (University of Wisconsin).

A website with links to CL-related papers and to what must be every cooperative learning site in existence is Ted Panitz's site: http://home.capecod.net/~tpanitz/.

PROBLEM-BASED LEARNING

Prince, M.J., and R.M. Felder. 2006. 'Inductive Teaching and Learning Methods: Definitions, Comparisons, and Research Bases', Journal of Engineering Education 95, 2: 123–38. Available online: http://www.ncsu.edu/ftlder-public/Papers/InductiveTeaching.pdf.

Duch, B.J., S.E. Groh, and D.E. Allen. 2001. *The Power of Problem-Based Learning*. Sterling, VA: Stylus.

University of Delaware Problem-Based Learning Clearinghouse. Available online: https://chico.nss.udel.edu/pbl.

Ted Panitz's site (http://home.capecod.net/~tpanitz/), and Deliberations, a site managed by London Metropolitan University (http://www.londonmet.ac.ukldeliberations/problem-based-learning/) are good sources of both information about PBL and links to other PBL-related sites.

Notes

1. If you don't think this is already happening in engineering, check out a 2005 NAE Report called 'Offshoring and the Future of US Engineering: An Overview', available at http://www.nae.edu/NAE/ibridgecom.nsf/weblinks/MKEZ6G6R4D?OpenDocument.
2. S. Lohr, 'Outsourcing is Climbing Skills Ladder', *New York Times*, 16 February 2006. This article reports that of 200 multinational corporations surveyed, 38 per cent said they planned to 'change substantially' the worldwide distribution of their R&D work in the next three years . . . and this particular trend is still in its infancy.

References

Friedman, T.A. 2005. *The World is Flat*. New York: Farrar, Straus & Giroux.

Pink, D.H. 2005. *A Whole New Mind*. New York: Riverhead Books.

QUESTIONS FOR CRITICAL THOUGHT

1. What is the purpose of this essay? Can you identify the original intended audience?
2. Why do you think Felder opens his article with the hypothetical interview scenario? What is its intended effect? What does it add to the impact of Felder's argument?

3. Do you think the applicant in this interview is likely to get the job he has applied for? Why or why not? Are you meant to see the interview exchange from his point of view or from the interviewer's? How do you know?
4. Are engineering schools 'teaching the wrong stuff', as Felder contends? What, then, is the 'right stuff'?
5. What are 'soft' skills? Make a list of those Felder identifies. Why are they so labelled? What, then, would be 'hard' skills? What do these terms imply?
6. Felder diagnoses a problem with engineering education, and he even describes some of the causes; what does he implicitly propose as a solution? Why doesn't he provide a specific prescription of what needs to be done?
7. Felder suggests some reasons that faculty resist changes to the traditional engineering curriculum. Are his characterizations accurate?
8. What, according to Felder, will happen to schools that fail to incorporate the new skills into their curriculum?
9. What is the source of Felder's credibility? How does he establish authority, sincerity, and respect for his readers? Does he retain these throughout the essay?
10. Felder provides a bibliography that is nearly as long as the article itself. The bibliography provides some sources that a reader could consult, but apart from its function as information, what other purpose might it have?
11. How would you describe the tone of Felder's article? Why might he have chosen such a tone?
12. Read MacLennan's 'Why Communication Matters', and compare the arguments made by each author. What commonalities do you see? What are the differences?

CHAPTER 4

First Flight

Andrea McKenzie

Imagine getting your first job in a developing country—one without a computer infrastructure or a modern transportation system. You'd find yourself racing to the bank before it closed to get cash for the weekend, because bank machines don't exist. (In fact, you'd probably need those good old standbys, traveller's cheques, for your first few weeks of expenses.) You wouldn't be able to use a webcam to talk to your family, and you wouldn't be able to instant message your friends or send emails—you'd have to depend on the vagaries of regular mail, air-lifted if you're lucky, or more probably trucked by roads. You'd have a landline telephone (no wireless available), but long-distance calls, especially overseas, would be prohibitively expensive. You might have a fax machine, but again, you'd have to think twice about sending long messages. If your job requires research, you wouldn't have the Internet; you'd have to depend on hunting through books and print journals at the local library, and possibly having them send for what you can't find. And you wouldn't have a word processor to write multiple drafts of documents and print them out; you'd find yourself, instead, banging away at the keys of an electric (or perhaps manual) typewriter and reaching (and cursing) for the white-out whenever you make a mistake—or furiously scribbling corrections for the typing pool whenever they make mistakes on your precious documents. Factory automation, using sophisticated robotics, wouldn't exist; manual labour would.

Welcome to the North America of twenty-five or thirty years ago, the era your parents probably grew up in and remember. For that matter, welcome to the world of that time. In the short span of your life, these are just some of the changes technology has made. It's become trite and clichéd to think about them because they're supposedly obvious; in fact, I've received so many essays beginning with, 'In the last few decades, technology has changed our world' that I've forbidden students to use that sentence. Of course, it happens to be true. The problem is, students write it without really thinking about what it means. History is usually measured by large events like wars; my father, for instance, was born in the year Lindbergh made his solo flight across the Atlantic, grew up during the Great Depression of the 1930s, and saw his older brother leave to fight in World War II. The war is vividly etched in his memory, especially because of its personal impact, but the Great Depression isn't. It was simply the background to his childhood, something he took for granted. Every day events slip by imperceptibly as we're living them, until history books label them as markers of our times.

So it is with technology. Those of us who learned to use slide rules in Grade 9 high school science didn't realize that the electronic calculators and electric typewriters we were using by Grade 12 were the harbingers of a revolution that had already begun. The punch cards, the programming languages like Fortran and Cobol, the character-based operating systems giving way to 'WYSIWYG' graphical systems, the local area networks, the wide area networks, IP addressing, microchips, wireless and satellite communication, the Internet; those of us who lived with these changes, worked with them, wrote about them, helped design them, struggled to learn and create the language so that others could learn about them, did not necessarily believe we were a tiny part of an enormous revolution that would change the way we lived, worked, wrote, designed, learned, created, and communicated. Like my father, who's defined as a participant in the Great Depression because he happened to be there, those of us who happened to work our way through the changes as technical writers often didn't stop to reflect about the changes that were happening. Yet we participated in, witnessed first-hand, and often contributed to those changes, especially those in how we communicate.

Consider that supposedly 'developing' country I described earlier. Communications in an organization would be decidedly local: personal meetings and telephone conversations would be dominant means of communication. You'd learn the foibles of your own small culture, and your colleagues would understand your vocabulary, your references, and your jokes. Global communications would be difficult and expensive, and you'd probably only see distant management occasionally. Open the country up with computer infrastructure, and suddenly management is demanding instant answers to questions via email. They're from another place, so they don't understand your vocabulary, your humour, or the local culture. You can't mediate your writing with voice or personal persuasion; suddenly, you are what you write.

Each term, I show my first-year writing class one of Harold Edgerton's photographs, and ask them this simple question: 'What is it?' Invariably, most of them are puzzled; the closest guess is usually that it's an ice sculpture. Of course, it isn't: it's a stream of water flowing from an ordinary laboratory tap ('Water from a Faucet', 1932). But it looks, to the unknowing eye, like a stunningly beautiful crystalline sculpture. Edgerton's goal was to 'stop time' with a camera; before he developed the technology to do so, we could not see the beauty or the pattern that an ordinary flow of water makes.

The instruments that we use to observe our world change our worldview, the way that we think about and know that world. Galileo looked through a telescope and changed how we think of our physical world; Isaac Newton looked up at an apple tree and saw 'that the same force of gravity, which reaches to the top of the tree, might go on reaching out beyond the earth and its air, endlessly into space' (Bronowski, 1965: 15); the developers who work on PCs, Macintoshes, and other computers can open our world as much, or limit it. We are changed, though we may not know it.

The shift to inexpensive email and electronic documents emphasized writing 'well' as a skill that companies needed; workshops in 'good business writing' abounded to try to rid employees of jargon, wordy structures, and poor grammar. Secretaries, shorthand, and typing pools disappeared; even managers had their own computers, so had to bear the brunt of their own writing instead of dictating to a person or machine. About the same time, colleges and universities realized the need for better writing skills in their graduates, so gave more time and emphasis to writing.

As computers became more accessible, both in their use and their price, we dealt with incompatibility, not just of the machines and their languages, but of the cultures we were using them in. We learned to write gender-neutral, culture-free documents, because we were now working on multi-cultural development teams: our colleagues from other countries, whom we rarely (or never) saw, couldn't understand our local jargon and examples, and we, in turn, struggled with theirs.

We turned to a combination of visuals and text to lessen translation costs. And because electronic slides and images (think PowerPoint) became so easy to create, formal presentations became another emphasized skill: the ability to design and orally present ideas for increasingly complex audiences. It wasn't that they hadn't been done before; it was just that most employees were now expected to be able to do them well. Easier communication made teamwork seem more prevalent (it had been done before, too), so interpersonal communications became more important.

Then came the real irony: cell phones and text messaging. We were, and are, largely accessible at all times: at home via email, in or out of the office via cell phones and Blackberries. We have nowhere to hide except in a technology blackout.

Welcome to the twenty-first century. And understand, as your parents didn't have to, that communications is the key to power in this still-changing world.

Gravity and the Paradox of Flight

You've made it through the interview, you've accepted the offer, and you're triumphantly starting your first day on the new job. You're introduced to a bewildering number of people—you try to grasp their names and invariably can't—a new space to get lost in (where are those restrooms?), and responsibilities that won't become clear for several days or weeks. You're about to become immersed in the organization's culture—the 'local' culture—and the impression you make in the next few weeks is crucial to the way the company views you.

Dorothy Winsor followed four engineering students through their undergraduate careers, studying their gradual immersion into their discipline's writing. One of her findings was that 'the students seemed to experience school and work as two parallel and only loosely related universes' (1996: 97), especially with regard to writing. Not surprisingly, the students gave priority to workplace practices. Winsor's study, however, took place in the 1990s; today, academic institutions pay much more attention to bridging the communication gap between workplace and classroom.

In 1997, the University of Waterloo's Faculty of Mathematics implemented a course specifically tailored to improve their students' oral and written communication skills. Waterloo, deemed the top comprehensive university in Canada by *Maclean's*, began the course because of a recommendation from a committee of industry leaders: students needed to be able to make a better transition from the classroom to the workplace and back. A secondary goal was to create a sense of community in the communication classrooms. Within 12 months, 500 students per year were taking the course, and 97 per cent recommended it to incoming students for three years running. Workshop-based, with a final team project that called for creative solutions to actual on-campus problems, the course bridged the classroom–workplace gap; it became popular because students who had taken it and then gone on co-op employment work-terms returned to say that it actively helped them in the oral and written communication needed on the job. On a larger scale, in 2002, the Massachusetts Institute of Technology changed its writing requirement to a communication requirement: all students must take one communication-intensive course in each year of their degrees. The first two courses may be general or technical writing courses; the last two must be subject-matter courses in their disciplines. All communication-intensive courses must include a specific amount of revised writing, plus at least one practiced oral presentation. Many other universities have had such programs in place for years, and others are implementing them rapidly.

What these academic changes mean is that students graduating from such programs are better prepared to meet the challenges of complex, multiple forms of communication in the workplace. The standards for communication in new employees are definitely higher: among the top ten qualities that organizations look for are teamwork and negotiating skills, leadership, and innovation—all of which either are good communication skills, or require them.

Back to that first day on the job. If you begin work with the idea that you've written your last essay, you're probably right. If you begin work with the idea that writing is auxiliary to your 'real' job, then you've just cut your own lines of communication. If you're a technical writer and believe that your job is to simply to translate other people's ideas, then you've just cut your own power cord and crashed before you've begun.

Winsor's engineering students were imbued with the idea of 'purity' of data; that engineers recorded data, but didn't really 'write'. That was left to managers, who did 'persuasive' writing. Reading between the lines, we can see that these young engineers thought that persuasive writing somehow besmirched the purity of the data they had gathered. Similarly, I see many young technical writers who also believe that their job is data collection: to gather 'facts' about a product and to record them for the user. Both cases remind me of my young science and psychology students, who begin by believing that science is factually based, and that the results of the experiment are all-important. They will eventually learn that both fields are based on questions designed to garner new knowledge, and that the meaning they make by analyzing the data is infinitely more important than merely gathering the results. (Daryl J. Bem, speaking of writing in Psychology, wryly says: 'There are two possible articles you can write: (a) the article you planned to write when you designed your study or (b) the article that makes the most sense now that you have seen the results. They are rarely the same, and the correct answer is (b).')

The myth of objective writing protecting the purity of data has long since been exploded: research in professional communication and scientific discourse has shown us that each discipline develops its own persuasive rhetorical strategies and traditions to present the image considered appropriate for its writers. Charles Bazerman, for instance, explored the traditions and rhetoric of scientific writing to demonstrate that it does use specific persuasive techniques to convince its audience. One of those techniques is objectivity, or the suppression of overt personality and persuasion in the writing. Tradition still holds: as Peter Lennie, Dean of Science at New York University, explained in 2003:

> Science is about establishing the truth of ideas, and scientific methods, both experimental and theoretical, are geared to the critical assessment of ideas. Scientific writing is about evaluating evidence for and against ideas, and the goal is to make it easy for the reader to establish or refute the truth of assertions. . . . Good scientific writing is simple and terse; its plot is always straightforward. . . . [Y]our aim is to convey often complex ideas unambiguously and intelligibly.

The seeming simplicity of organization and sentence structure in engineering, scientific, and technical writing presents students with a seeming conundrum: published experimental studies, like technical documents, give no indication of the revisions the writer has done, the number of failures that occurred, the changes caused by peer reviews and questions, or the time spent analyzing and re-analyzing data to explain the sometimes unexpected results.

What Winsor's students grasped were the rhetorical strategies that led to acceptable engineering writing within the workplace: the ethics of seemingly straightforward data recording. What they failed to understand, however, was that their writing, however straightforward it may have seemed, was persuasive according to scientific standards because it adhered to the conventions of their particular community; that their writing work was part of a complex system shot through with history and laden with cultural significance for a specific discourse community; or that interpretation of results, not the data itself, is the crux of research and idea-work.

Two of the qualities organizations look for in new employees are innovation and initiative: imaginative new ideas that will add to the organization's stock of knowledge, result in better customer relations, or make production more efficient. The ways in which employees can be innovative is endless, but they do not include mere data reporting for either young engineers or young technical writers. To enter an organization with these beliefs is to place yourself in an auxiliary position, on the margins of power instead of at the centre.

Jacob Bronowski takes an example from Karl Popper to explain the difference between data collection and scientific creativity:

> Suppose that [a scientist] . . . sat down, pencil in hand, and for the next twenty, thirty, forty years recorded in notebook after notebook everything that [s]he could observe. . . . and, dying in the calm certainty of a life well spent, [s]he would of course leave his notebooks to the Royal Society. Would the Royal Society thank him for the treasure of a lifetime of observation? It would not. . . . It would refuse to open them at all because it would know without looking that the notebooks contain only a jumble of disorderly and meaningless items. (1965: 14)

In essence, we can spend our lives collecting data, but unless we analyze it, make meaning from it, the data remains useless. As Bronowski goes on to point out, 'order does not display itself of itself; if it can be said to be there at all, it is not there for the mere looking. There is no way of pointing a finger or camera at it; order must be discovered and, in a deep sense, it must be created. What we see, as we see it, is mere disorder' (13–14). Young engineers may spend their time running data and reporting it, but they must present it in some way to their fellow engineers and managers; they must create order of some sort when they write it down to communicate it to their fellows. Order may, of course, be unconsciously created through the rhetorical strategies that have been developed through long tradition in the discipline, or similarly, by presentation strategies that the organization itself has developed, strategies and conventions that are specific to the organization or to a department or group within it.

Technical writers whose main role is to create documentation for the organization and/or its clients about policies, procedures, or products will find themselves in a similar situation. Experienced writers know that information often comes from myriad sources and may or may not be written down in an intelligible form: collecting 'data' to learn about a product, for instance, may come from an engineer or developer's head, specification documents, technical support or customer service reports, software, a mock-up or beta-version of the product, salespeople, and so on. The writer's work becomes much more than data collection: as Carol Wiest, a project manager and senior writer in Toronto comments:

> I like to think of myself as a communication facilitator who helps the organization identify needs and then design, generate, disseminate, and preserve information. The information in turn helps users to do their jobs: to analyze data, identify problems and opportunities, present results to others, and so on. I'm also a model and mentor . . . for written communication and for inter- and intra-departmental communication. (2006)

Wiest demonstrates the cohesive role the technical writer can play within teams and across an organization: not just a 'service provider' or recorder, but an active player whose negotiating skills and expertise can enhance both product and formal and informal communication channels. Technical writers, then, must create order out of a chaos of data (Wiest), keeping audience in mind. Writers don't just sort data; they must educate their readers—the different layers of an organization, not just the end user—by using creative methods of presenting new concepts, products, and technologies. They create synergy, where fusing their own creativity with the organization's store of knowledge results in innovative ways of communicating, both with an organization's clients and with their colleagues in the organization. An innovative writer can work to improve product design, customer relations, and organizational procedures, but only if he or she believes it can and will happen.

Perhaps the most telling (and hilarious) anecdote of writers influencing product design and enhancing an organization's image with its customers, comes from Edward Devereux Sheffe, a corporate communication specialist in New York City:

> When I found myself writing, 'Press OFF to turn this feature ON,' followed by 'Press ON to turn this feature OFF,' I tactfully suggested changes to the firmware. That product had to be released without changes, but my client flew me to Europe to work directly with the developers on re-designing subsequent products to be more user-friendly and intuitive. Never again did users have to press OFF to turn a feature ON. (2006)

Experienced writers such as Sheffe and Wiest recognize that often, designing product documentation reveals product design flaws. When we design, we must think through the logic of a product or system to understand how to educate our readers sufficiently so that they can creatively use that product or system. 'Documenting' the product becomes a means of thought that reveals flaws in that logic. Just as a first draft of a document may allow the writer (but not usually the reader) to understand, learn, or discover a new idea, so the alpha version of any product may be designed to test or discover a new concept instead of reaching out to its users. The author of a 'writer's' draft must re-design the document for its reading audience; a design team must create a beta version that accounts for the end users' level of technical knowledge and understanding. The users, in turn, should be sufficiently educated (through ease of use and documentation) in the system's or product's underlying logic to be able to safely and effectively adapt it for their own purposes.

Consider the role of writing and presentation in creating order out of 'data'. One telling example is that of John Snow, a doctor in London in the mid-nineteenth century. At the time, most people, including doctors and public officials, believed that breathing in bad air caused cholera. Snow, in contrast, theorized that drinking bad water was the cause because patients' first symptoms were severe pains in the gut (stomach) area. During an especially bad outbreak, Snow came to believe that a specific public water pump was the cause of the outbreak, but he couldn't make the public officials close it down, because he had no convincing evidence. Snow then collected the addresses of all those who had died to date in the outbreak, and marked a map with the addresses of the deceased. If you look at the map, you can clearly see the clustering of deaths in one specific area, right around the Broad Street water pump, with deaths becoming fewer as distance from that pump increased. This innovative map finally convinced the public officials to close down the Broad Street pump, which effectively ended the epidemic. Snow's method of creating order out of seemingly random data saved lives in London. Later, his papers on cholera, including his innovative statistical maps, convinced the medical community that cholera was caused by drinking contaminated water, not breathing bad air; Snow thus became responsible for new medical knowledge that saved thousands of lives worldwide.

Snow did many years ago what technical writers and engineers must do now: create new ways of communicating complex information that increasingly varied audiences can understand. His genius lay not only in his analysis of the evidence, but also in the way he presented it: through an innovative technique whose unique, transparent design allowed its readers to examine the evidence, test it, and question their own previous theories. As Edward R. Tufte says,

> What is to be sought in designs for the display of information is the clear portrayal of complexity. Not the complication of the simple; rather the task of the designer is to give visual access to the subtle and the difficult—that is, the revelation of the complex. (1983: 191)

Like Lennie, Tufte believes in the kind of design transparency that will draw attention to the substance. Engineers, scientists and writers must always create order for the audience—must educate them, in essence—to lead them through complexity.

Aerodynamics

Snow's innovations, like Galileo's new theories and Edgerton's photographs, gave us another perspective on the world. When you begin work with a new organization or institution, you enter a unique and specific culture, and like any field or discipline, part of your work consists of learning to read and emulate the communication of that culture: to collect data, if you will, about the organization's values and philosophies, and learn to 'make meaning' from the data so that you, too, can participate. Envision your new environment as a linguistic marketplace, a metaphor taken from Pierre Bourdieu's (1991) theory of language and power. In this marketplace, the authority to speak and be heard is derived from three elements: the ability to use what is recognized as the 'legitimate' language (7); the ability to use that linguistic competence appropriately in context, in a specific situation (37); and the 'recognition, institutionalized or not', that the speaker 'receive[s] from a group' (72). To gain power in an organization, for Bourdieu, means obtaining all three abilities. In a new discipline or workplace, therefore, we must work to learn the mechanisms, written and unwritten, that form the 'legitimate' language of the organization, use them appropriately in specific situations, and gain recognition and power through the credibility we gain.

It sounds fairly simple to gain this capital in an organization: you listen, you learn, you practice, you gain the ear of people who count by emulating the vocabulary, the practices, and the philosophy. But as Bourdieu theorizes, this concept is much more complex: those who hold power like to keep it; those who create the standard or 'legitimate' language try to ensure it's reproduced, so that they keep their power. Ironically, Bourdieu uses an example from education to illustrate his methodologies: educational institutions reinforce and reproduce standard languages by evaluating students on their ability (regardless of cultural or social background) to reproduce them. Those who succeed gain power in the wider community because of the standards thus set and disseminated. Paradoxically, to succeed within an organization, we must demonstrate our abilities to reproduce that organization's language and mechanisms of power, whether or not we agree with them.

Technology has complicated this issue still more. Languages change with time, as have multiple methods of communication. Companies and institutions work within a wider world; the company's ability to lead or keep up with its field determines its power and economic force. This is true for companies who *deal* in technology and for companies who *use* technology. New technologies create new vocabularies and new communication methods: those who adapt stay with the leading edge of power; those who fall behind give themselves away and lose power. Consider how many computer systems and software packages are now obsolete, and how trends in, for instance, text messaging vocabulary, signal whether a user is naive or practiced.

But how and why do organizations develop their own communication systems, their own vocabularies, and their own system of power based on communication? Bourdieu's notion of *habitus* is relevant to understanding organizational communities and the practices generated and consciously or unconsciously learned there. John B. Thompson explains Bourdieu's notion of *habitus* as 'a set of *dispositions* which incline [people] to act and react in certain ways' (12). These 'dispositions . . . become second nature [and] are also structured in the sense that they unavoidably reflect the social conditions within which they were acquired' (12). Thompson goes on to say that 'dispositions' are both 'durable' and not subject to 'conscious reflection and modification' (13). In essence, when we immerse ourselves in a particular community, such as engineering or a specific workplace, we absorb its language, its writing styles, and its means of communication; we also absorb the cultural and social values denoted by that specific community without necessarily being conscious of their meaning or import. For instance, a profession dominated by one gender often unconsciously produces documentation slanted towards that gender. Such documentation reproduces and upholds the social condition of gender by closing the profession to the other gender in its writing.

When we learn such standards—such a system, in essence—uncritically, learning the surface writing and communication strategies without understanding why they are in place, we risk passing on such unconscious assumptions; we become collaborators in upholding them. As Bourdieu says, 'the language of authority never governs without the collaboration of those it governs' (1991: 113). For entry-level employees, such collaboration may seem troubling because we do not seem to have the power to change it, but we must remember that assumptions can be either liberating or confining, depending on the specific community and its attitudes. I mentioned earlier that as a result of globalization, we now write documentation that is culture-fair, a trend that has become standard practice. Conversely, Internet browsers and software are still designed using Western culture's left-to-right, top-down reading habits. These assumptions predicate specific images of expected end-users.

Freefall

If our organizations expect us to be innovative, though, how do we go about it, and how do we create positive change? What exactly is creativity? If we look at all the realms and domains, we find that often, those deemed 'creative'—those who innovate—can't or won't describe the process of discovery, though some have tried. It seems to come as a flash; as an almost unconscious coming together of previous knowledge and resolution of a problem, of 'seeing' in a new way. Alan Lightman, renowned novelist and physicist, describes that moment as 'one of the deepest and most beautiful of human experiences,' but he goes on to say that '[l]ike a timid forest animal, it quickly darts behind a tree when you stare at it' (2002: 68). Lightman argues that this moment provides 'the basis for a powerful understanding between the scientist and the artist', and perceives his creativity in both writing and physics as being grounded in that same sensation (2002: 68). You've probably felt that sensation yourself, whether you're an engineer, a technical writer, or an artist: that deep feeling of satisfaction when a concept or idea you've struggled with is suddenly, without warning, illuminated.

New knowledge is not obtained through collecting data. Copernicus's thesis, for instance, was that 'the earth moves around the sun' (Bronowski, 1965: 11). Previous theories placed the earth at the centre of the universe. In arguing about the nature of creativity, Bronowski says that Copernicus did not create his radical thesis by 'routine calculation. His first step was a leap of imagination—to lift himself from the earth and put himself wildly, speculatively, into the sun' (11–12). Different fields and disciplines define creativity in unique ways. The Humanities, for instance, expects first year students to make the transition from summarizing others' work to analyzing that work and finding a new question to ask and answer. Mathematics, in contrast, considers its first year students must make the leap from routine calculation or computer programming to creative problem solving (Lewis), a perspective that engineering also adheres to.

It's one thing to solve a problem, though, and quite another to convince our colleagues that our solution is credible. From Copernicus to Galileo to Einstein, innovative solutions go through the same process: write down the solution, undergo peer critiques, find weaknesses, solve the weaknesses, and perhaps become accepted. Radical changes in thinking, as Copernicus and Galileo demonstrate, often aren't accepted if they buck the current trends too far.

Notably, innovations are usually publicized within communities through writing. In academics, innovation is published in journals, and prestige depends on acceptance by the best and most rigorous peer-reviewed journals. A scientist published in *Nature*, for instance, has already gained an amount of acceptance within the scientific community, though the 'new' knowledge will be subjected to vigorous questioning: testing by the wider community is part of the scientific method. Technical writers and engineers recognize this method: the writer sends documentation to a panel of experts, who comment on and question the content, accuracy, organization, formatting, and style; the writer must resolve conflicts in the comments, revise the document, and send it out again for review until it's deemed 'approved'.

The same holds true for innovation within any organization, though less formal processes and indirect routes may be possible (Wiest, 2006). I said earlier that innovation and initiative are qualities organizations look for in new hires, along with good communication skills. The three are integrated, for common sense reasons: any change to an established procedure and any new solution to a problem must be tested. It doesn't matter whether the change is to the way documentation is published or to an engineering process: the consequences of the change must be thoroughly assessed before it's implemented. What is the cost of the change in resources, customer relations, economics, and training? How radical is the change? Does it work against established thinking, as Copernicus' imaginative 'leap' into the sun did (Bronowski, 1965: 12), or is it a small change that's easily implemented?

The easiest way for colleagues and management to assess a change is to examine the evidence. And the easiest way to examine the evidence is when it's written down, so that the information can be exchanged, reviewed, sent back with questions, and revised—locally or globally. Earlier, we saw that Bourdieu's linguistic marketplace called for three types of communications 'authority': knowing the appropriate language, using it in a recognized appropriate manner, and having the 'authority' to be heard. Junior employees gain the third by demonstrating adeptness in the first two: a new idea is more likely to be accepted if it is couched in the organization's language and uses the appropriate conventions, from the accepted document genre to the 'right' persuasive strategies.

But that's a complex task. Winsor's students wrote for their fellow engineers, recognizing that to gain credibility with that community, they had to learn acceptable communication methods that lacked overt persuasive techniques. However, as one of the students discovered when he presented his ideas to management, he'd failed to communicate those ideas appropriately: his approach and language were too technical. Yet management makes the decisions about new ideas; if they can't understand the innovation, it won't get implemented. This student lost the opportunity to gain credibility and will have to work doubly hard to regain it when he presents again.

When we write in organizations, we often write for complex audiences, especially when we attempt to create change: our immediate colleagues, management, sales personnel, and marketing personnel. The greater the change we propose, the more complex our audience is going to be, because we must justify all of the costs of innovation. Unlike Galileo, we won't suffer arrest, but we may lose credibility instead of gaining it if we don't provide appropriate evidence couched in appropriate ways that different levels of audiences can understand. Winsor's student, for instance, should have spent time considering strategies for presenting his idea and data to management, finding creative ways to convey complex technical information in ways that his audience could understand. Often, such audiences respond well to visual strategies and metaphors, reinforced with definitions of any technical language used.

A complication that Bourdieu does not mention is that multiple 'standard' languages may compete within the same organization. I said earlier that each discipline maintained its own vocabulary, traditions, and philosophy; these become part of the discipline's 'legitimate' language. Logically, then, organizations that include members of different disciplines will have different, and often competing, 'legitimate' languages on site; these different versions must still adhere to the organization's overall authoritative language. Winsor's student discovered the cost of failing to recognize the conflict. Subject matter experts, such as engineers and developers, may want to use technical terms to describe a product or system, because an inherent value of their fields is technical accuracy. Management needs to understand the potential of a product or change for strategic planning purposes, so may not need to know the technical aspects. The goal of marketing personnel is to have the organization's clients understand the unique qualities of the product and why they should buy it. Different philosophies and values conflict; frequently, as Wiest (2006), notes, the technical communicator becomes the team player who negotiates these minefields, re-envisioning the product or system in terms that resolve conflicts. This work requires imaginative creativity and the ability to persuade others to 'see' the product in a new way.

Finally, we must recognize the conduits of power in the organization. In the 1980s, Jean-Francois Lyotard predicted that our concept of 'knowledge' would change in the computer era, with access to information being a defining factor. I know of no university or college course that teaches its students how to obtain critical information from a recalcitrant developer or manager, but learning how to do so diplomatically is an essential part of any working life. Power is often defined by who has access to information: depending on the organization's ethos and defining philosophies, that information can be jealously guarded and exceptionally hard to obtain. Controversies over patents and copyrights, from file sharing in music to biotechnology, have made these issues even more prevalent. Who owns information? Who has the right to use it? Who should have access to it?

What is evident on a global scale in these issues is also true on a local scale, within an organization. The identity cards we use to access a site, the non-disclosure agreements we sign, the combinations we press to open locks, the passwords we enter to gain access to a system or database: these are all physical reminders of more subtle traces of power. What we must learn is the underlying system of power as it is embodied in an organization's structure, especially its communications strategies. Who can give us access to the information we need? Can we approach the source of information—often a higher level colleague—directly, or are we expected to follow a particular path—our own manager, who will contact a higher level manager, and so on—to get the information? What is the etiquette that governs these negotiations and transactions? What is the best strategy to follow? To follow the appropriate channels, or to bypass the lower echelons and go straight to the top to apply pressure? What will the consequences be in the future? A gain in credibility and respect, or a gain in resentment and a lack of future cooperation?

We must recognize that to gain power ourselves—through demonstrating creativity and innovation—we must learn to both read and put into play the power structures inherent to any organization. We must also recognize that to do so, our communication skills are integral, vital means to that power. The engineer or writer who believes that writing consists of sentences and paragraphs that merely record data works at a severe disadvantage. Only by breaking open the definition and learning that writing is communication, and that we 'write' ourselves and our abilities each time we communicate, will we gain access to the eyes and ears and minds of those in power.

'Writing', then, becomes the document in many mediums that we write; the visual methods we use on the web and in presentations to convey complexity in terms our audiences can understand; the words and strategies we choose when we speak to our colleagues; the knowledge we've gained from 'reading' an organization, its structure and its people put into play in our writing; the negotiations we make to gain knowledge and information. All of these leave traces of ourselves and our abilities for others to 'read', in turn. And because change constantly occurs in ways small and large, writing becomes life-long learning. We cannot turn to formulas and supposedly proven methods, because to do so is to become static and fixed; we fall behind the wave of innovation, and lose credibility once again.

Change takes place as the result of adding to previous knowledge: paradoxically, once conventions do become entrenched, they're often difficult to shift. Causing change requires a fine balance:

you must understand the given conventions—the previous theory or knowledge—so well that you demonstrate your expertise in them. Only then can you break convention, because you've gained the credibility that enables you to persuade others that your change is genuinely for the better; that the benefits outweigh the costs. People new to an organization often blindly follow conventions, especially in communication; only when we learn why a process, a style, a method, is used, can we persuade others to change it. The engineer who records data without understanding how it contributes to a larger whole cannot be innovative; the technical writer who documents a product without understanding genre, audience, and style cannot tactfully recommend changes to the document or product design. Similarly, senior and experienced employees who unquestioningly repeat processes merely because they're traditional have lost their creative edge. We must understand first, then question, then create, then persuade through justification.

If we hone our communication skills constantly, if we consider them an integral part of knowledge creation, then we will ourselves gain in persuasive power. Through communication, we have the power to change our world, whether that world is small and local or built on a global scale. Whether we know it or not, our words matter: we hinder or create change even as we write. The person who writes 'he or she' instead of 'he' chooses a more open view of community; the writer who thoughtfully puts 'developing country' instead of 'third-world country' substitutes a positive perspective for a negative hierarchy; the designer who creates websites readable by people with disabilities makes others socially aware. By such small decisions, we can change our own small domain, and hopefully, the larger worlds beyond.

When we enter an organization—any organization—we hope for that exhilarating challenge, that exciting leap, like Galileo, into a new perspective, new learning, new knowledge. But to get there without, like Icharus flying with waxen wings, failing and falling, we must recognize that when we write—when we communicate—we write ourselves and our beliefs and our awareness of the world beyond ourselves. We write our own histories, and paradoxically, our futures, as we live our present moments.

The Courage of the Early Morning[1]

Forty years after my Dad left his Depression-era neighbourhood, I moved in. When he lived there, it was a working class neighbourhood; when I lived there, it was in transition. Chain stores were crowding out family-run shops; old cottages were dwarfed by new condominiums. Individual character was slowly being crowded out. I was young and thoughtless, and never thought about what it must have been like for him to see those changes in his old haunts. Now, when I go back, I like to imagine him getting up early in the morning and paying a nickel to ride the same streetcar line that I did, the streetcar that would take him away from the Depression and the war to a university degree and a wider world, some 20 years after Lindbergh's *Spirit of St Louis* flew the ocean and changed our notions of flight.

I'll have to ask Dad about it. By email. Tomorrow.

Notes

My thanks go out to Carol Wiest (Toronto, Canada) and Ro Sheffe (New York City, USA) for sharing their expertise, their anecdotes, and their thoughts throughout this project.

1. The title of Canadian World War I flying ace Billy Bishop's memoirs.

References

Bazerman, Charles. 1988. *Shaping Written Knowledge: The Genre and Activity of the Experimental Article in Science*. Madison, WI: University of Wisconsin Press.

BBC Online. N.d. 'Dr. John Snow.' *Historical Figures*. Available at http://www.bbc.co.uk/history/historic_figures/snow_john.html (accessed 5 April 2005).

Bem, Daryl J. 2003. 'Writing the Empirical Journal Article', in John M. Darley, Mark P. Zanna, & Henry L. Roediger III, eds, *The Compleat Academic: A Career Guide*, 2nd ed., pp. 185–219. Washington, DC: American Psychological Association.

Bourdieu, Pierre. 1991. *Language and Symbolic Power*, John B. Thompson, ed., Gino Raymond and Matthew Adamson, trans. Cambridge, MA: Harvard University Press.

Bronowski, J. 1965. 'The Creative Mind', in *Science and Human Values*, rev. ed. New York: Harper & Row.

Edgerton, Harold. 2000. 'Water from a Faucet, 1932' in Gus Kayafas, ed., *Stopping Time: The Photographs of Harold Edgerton*. New York: Harry N. Abrams, Inc.

Lennie, Peter. 2003. 'Clear and Simple Scientific Writing', *How We See* lecture handout. New York University. October.

Lewis, Adrian. 1997. Personal interview with the Director of First Year Studies, Faculty of Mathematics, University of Waterloo. October.

Lightman, Alan. 2002. 'The Art of Science', *New Scientist* 21/28: 68–71.

Lyotard, Jean-Francois. 1984. *The Postmodern Condition: A Report on Knowledge*. Geoff Bennington and Brian Massumi, trans. Minneapolis, MN: University of Minnesota Press.

Sheffe, Edward Devereux. 2006. Personal interview. 6 February.

Summers, Judith. 1989. *Soho—A History of London's Most Colourful Neighborhood*. UCLA: Department of Epidemiology, School of Public Health. Available at http://www.ph.ucla.edu/epi/snow/broadstreetpump.html (accessed 3 February 2006).

Tufte, Edward R. 1983. *The Visual Display of Quantitative Information*. Cheshire, CO: Graphics Press.

Wiest, Carol. 2006. Personal correspondence (email). 6 January.

Winsor, Dorothy A. 1996. *Writing Like an Engineer: A Rhetorical Education*. Mahwah, NJ: Lawrence Erlbaum.

QUESTIONS FOR CRITICAL THOUGHT

1. Why does McKenzie open with the developing country scenario?
2. McKenzie describes technological advances as a 'revolution'. What did they revolutionize? What changed?
3. Why does technology mean 'you are what you write'? What does communication technology do to context? How important is shared context to a writer and reader?
4. 'The instruments we use to observe our world change our worldview.' Comment on the extent to which this statement echoes the arguments presented in McLuhan's 'Motor Car', Grant's 'Thinking about Technology', or Lorinc's 'Driven to Distraction' essays.
5. What effects did the advent of electronic communication (email, word processors, the Internet) have on how most workers communicate, and are expected to

communicate, in the workplace? How does McKenzie link those changes to the need for effective writing skills?

6. Is McKenzie right that communication is the key to power in a technological world? How is this so?
7. What does writing have to do with helping students make the transition from school to the workplace?
8. McKenzie describes changes that have taken place in university programs to offer better foundations in communication skill. To what extent do such changes respond to the problems identified by MacLennan in 'Why Communication Matters' or Felder in 'A Whole New Mind for a Flat World'?
9. McKenzie, like Charles Campbell ('*Ethos*: Character and Ethics in Technical Writing'), Mary Fran Buehler ('Situational Editing'), Cheryl Forbes ('Getting the Story, Telling the Story: The Science of Narrative, the Narrative of Science') and others, challenges what she refers to as 'the myth of objective writing'. In what sense is this objectivity a myth?
10. According to McKenzie, objectivity in scientific writing is a rhetorical strategy rather than an indication of the 'purity' of scientific observation. In what sense is this kind of writing nevertheless persuasive, according to McKenzie?
11. Is science a culture, as McKenzie contends? What is 'scientific creativity'?
12. Read Dombrowski's 'Can Ethics Be Technologized?' and recall his assertion that 'raw technical information does not signify its own ethicality.' Similarly, McKenzie points out that data are meaningless until given some order by the writer, whose job is to 'create order out of the chaos' of information. Can you find a parallel in your own writing experience?
13. Why does McKenzie quote from so many sources?
14. Why does McKenzie consider facility with writing a matter of power?
15. In what sense is an organization a kind of culture? What effect would such a culture have on the writing that takes place within it?
16. Explain what McKenzie means by the statement 'New knowledge is not obtained through collecting data.' How, then, is new knowledge obtained?
17. According to McKenzie, innovation, initiative, and communication skills are interlinked. How? To what extent does this assertion echo John Lorinc's views in 'Driven to Distraction'?
18. What is the meaning of the title of McKenzie's essay? Why does she end the piece on such a personal note?

CHAPTER 5

Functional Communication: A Situational Perspective

Lloyd F. Bitzer

. . . This essay considers rhetoric as a functional, or pragmatic, communication and thus a critical mode of functional interaction in which the chief interacting grounds are persons on the one hand and the environment on the other. The decision to underscore the process of functional interaction and to regard persons and environment as interacting grounds tends to generate a view of rhetoric which may be called situational.[1] The essay asks, What are the conditions, characteristics, and factors operative in that process of communication through which human beings achieve harmonious adjustment with the environment?

The Process of Functional Interaction

The situational view of rhetoric takes as its starting point the observable fact that human beings interact functionally with their environment. This is not an inevitable starting point. Typically, stylistic rhetorics commence with the relation between the nature and resources of language on the one hand and the intentions or meanings of a speaker on the other. The scientific rhetorics of the eighteenth century commenced with the relation between natural psychological processes and communicative intentions and activities. These and similar approaches either dismiss the relation of persons and their messages to environment, or regard this relation as secondary. The situational view, however, seeks to discover the fundamental conditions of rhetoric—of pragmatic communication—in the interaction of man with environment. . . .

To say that human beings are involved in functional interaction with the environment is, of course, a truism. Nevertheless it is necessary to explain briefly what this means in order to underscore the radical nature of this interaction. Upon reflection all of us know that in seeing, touching, and other sense acts we participate with what we take to be real. Our habits and patterns are keyed to natural pulsing for days and seasons, and our daily work involves our hands, bodies, and thoughts with objects and relations presented by the world. We build roads and homes to suit the terrain and modify the terrain to suit our needs. When the physical environment jars us with unexpected, sometimes cataclysmic events, we respond by adjusting ourselves or the environment to assure safety or contentment. In sum our existence and well-being depend upon skillful and ongoing adjustments to, or modifications of, our environment. We cannot live at all, let alone well, in a totally inhospitable world. Our somewhat harmonious relation to the environment is achieved through striving: human societies and individuals do not merely rest, like rocks, upon the earth; they are actively engaged in adjusting, responding, overcoming, planning, labouring, making, and acting. These comments are meant not to overstate or falsely dramatize, but to make clear that human interaction with the physical environment, marked by striving, is a fundamental and pervasive condition of life.

Interaction with the mental environment is no less fundamental and pervasive. Mental environment includes the whole field of ideas, images, meanings, symbols, laws, rules, conventions, attitudes, feelings, interests, and aspirations that constitute the mental world. We live and act as sentient beings in a universe stocked with mental as well as physical entities, and our interaction with the environment often involves striving with respect to mental entities. For example, we seek to resolve a conflict between an attitude and a principle; our present desire competes with a long-term interest; components of an ideology are contradictory; one idea calls into question another. The mental environment acquired from parents, culture, education, and the whole of experience is often discomforting, sometimes hazardous, and presents challenges we strive to resolve by various adjustments. While the environment sometimes offers challenges that are exclusively mental (mathematical or philosophical puzzles, for example), the total environment within which we live and constantly interact is clearly a massive and complex mix of the physical and mental. It is perhaps sobering to note that we alone give existence to much of this total environment and are ultimately responsible for many of its constituents and qualities.

A necessary condition of normal and deliberate pragmatic interaction is the presence of certain environmental constituents, which form a structure. What minimal constituents must be present? From among the mass of details comprising the total environment something is recognized as other than it should be, that is, an exigence; and something else is recognized as a means of remedy or modification. These two constituents—exigence and remedy—are essential; in a genuine pragmatic situation both are real. The presence of the person who interacts introduces other important conditions of interaction. After all, the person must recognize the exigence and see the connection between the instrument or remedy and the exigence; often the person will estimate accurately the propriety of his adjustment.

Rhetorical or pragmatic communication is a special kind of pragmatics, and involves additional constituents in the environment: the necessary constituents form a structure I call a rhetorical situation. What constituents must be present in the situation as a condition of successful pragmatic communication? First, there must be an exigence—a problem or defect, something other than it should be. Second, there must be an audience capable of being constrained in thought or action in order to effect positive modification of the exigence. Third, there must be a set of constraints capable of influencing the rhetor and an audience.

The presence of an audience is necessary because only by means of its mediating influence can the exigence be modified. Since the audience must be capable of modifying the exigence positively, it follows that listeners incapable of this modifying influence will not count as a rhetorical or functional audience. As individuals we address ourselves pragmatically; thus the self is sometimes rhetor and audience. The exigence is sometimes in the audience addressed—for example, its false beliefs.

Every rhetorical situation contains some constraints such as persons, events, objects, relations, rules, principles, facts, laws, images, interests, emotions, arguments, and conventions. Having the power to influence decision and action needed to modify the exigence, these constraints are parts of the situation and influence both rhetor and audience. The rhetor's central creative task is to discover and make use of proper constraints in his message in order that his response, in conjunction with other constraints operative in the situation, will influence the audience. Professor Carroll Arnold has remarked that constraints may be regarded as opportunities and limitations 'within which and through which any rhetor must work to evoke change. These opportunities and limitations consist of (1) everything at the disposal of the rhetor for purposes of inducing change and (2) demands and readinesses that exist in the situation in any given moment' (Arnold, 1974: 28–9).

A rhetorical situation may be defined as a complex of persons, events, objects, and relations which presents an exigence that can be completely or partially removed if discourse—introduced into the situation—can influence audience thought or action so as to bring about positive modification of the exigence.

Genuine rhetorical situations, as opposed to sophistical and spurious ones, are real in the sense that the situational constituents—exigence, audience, and constraints—are present in the historic environment and available for scrutiny by persons prepared to see them. This does not mean that all constraints and exigences are pure objective facts. Some constraints are pure facts, however, and every genuine exigence has an observable factual component. Other constraints exist only as elements of mental life, and every exigence has a component consisting of an interest. The rhetorical situation is real and objective, however, in the sense that an observer, possessing appropriate knowledge and interests, usually can see its parts and appreciate its force. For example, if an earthquake is imminent, the exigence (from a human perspective) is the danger to human life and property; the constraints consist of the information, motives, and arguments that will be sufficient to influence the audience and persuade it to alter its conditions and secure safety. Inasmuch as this situation is easily scrutinized, it can be determined to be genuine.

Situations are not fixed or unchanging. Some of their elements, after all, are physical objects and events caught up in streams of history. For this reason exigences, constraints, audiences—and the relations among them—will be altered by historical processes and causal networks. Some situational elements are mental entities—principles and motives, for example—which may be altered by numerous things such as changes in perspective, shifting interests, and recognition of new arguments or maxims. Thus situations may be expected to change because of forces and tendencies in the environment, whether or not anyone apprehends or acts to alter them; unquestionably there have been many situations that came into existence and ultimately disintegrated, unrecognized as situations. Nor are situations isolated from one another; within a single frame of time and place, they may overlap and implicate one another. Any attempt to resolve one may well generate another. These observations underscore the extent to which situations are located in the environment and governed by forces outside our control. The situational perspective does not deny the influence of the individual's creativity in the apprehension of situations and in the efforts to modify them through creation and presentation of messages; nor does it ignore the degree to which thought constitutes the environment. Instead the situational perspective considers thoughts as well as things to be parts of historic reality, and it calls attention to the close relation of pragmatic communication to actual historic conditions.

In summary: (1) Functional interaction with the environment exhibits purposeful striving to achieve adjustment, balance, and harmony. (2) The situational view regards rhetorical communication as essentially functional. (3) Functional communication involves two distinct grounds—persons, on the one hand, and the environment, consisting of physical and mental entities, on the other. (4) The environment presents real constituents as necessary conditions of any functional activities, and it presents situations, consisting of exigences, audiences, and constraints, as necessary conditions of any functional communications. (5) Situational rhetoric thus commences not with attention to speaker intention and artistry, nor with focus on language resources, the argumentation process, or natural psychological processes; rather it commences with the critical relation between persons and environment and the process of interaction leading to harmonious adjustment.

The process of pragmatic communication begins with apprehension of situational constituents. The exigence is pivotal among these because human beings respond to situations in proportion to their perception of matters that are other than they should be. An examination of the notion of exigence and an account of rhetorical, or functional, response by means of pragmatic messages will clarify this.

Exigences

Let us suppose that our total environment presents no problems—no poverty, injustice, or war; no personal illness or tragedy; no petty squabbles; no questions inviting answers; no controversy needing resolution; no object or idea awaiting discovery or invention; no condition of any sort inviting us to adjust ourselves or the environment. Would there by any pragmatic communication in this 'best of all possible worlds'? If our environment were perfect, would we seek to effect change through discourse or other means? In the absence of perplexity or exigence of any kind, there would be no need for adjustment; no need for rhetorical inquiry, advocacy, persuasion, debate, or mediation; no reason to effect change through pragmatic thought, action, or communication. There would be no work for teachers, scientists, philosophers, and news reporters because in the ideal world there are no lacunas in that which we want to know, should know, or could know. In the absence of exigences rhetorical communication would be unnecessary.

However, between the best of all possible worlds and the world we live in—between the ideal and the real—falls a shadow. From the human perspective the world is afflicted by imperfections. Immense exigences face us: many of the world's people are hungry and poor; the diseased and homeless need care; many persons have yet to enjoy basic human rights; world peace and community elude us. Recognition of these and similar problems gives rise to a fundamental proposition grounded in individual and collective experience: that *the environment and persons invite change*. That is to say, all of us recognize that we rightly seek to alter those aspects of the environment and of ourselves that are thought to be other than they should be.

An exigence is an imperfection marked by some degree of urgency; it is a defect, an obstacle, something to be corrected. It is necessarily related to interests and valuations. John Dewey remarked: 'Valuation takes place only when there is something the matter; when there is some trouble to be done away with, some need, lack, or privation to be made good, some conflict of tendencies to be resolved by means of changing existing conditions' (1939: 34). Exigence is the necessary condition of a rhetorical situation. If there were no exigence, there would be nothing to require or invite change in the audience or in the world—hence there would be nothing to require or invite the creation and presentation of pragmatic messages. When perceived the exigence provides motive.

Many exigences cannot be modified by human effort; one cannot change the inevitable, such as earthquakes. Other exigences, although not inevitable, come into existence and resist modification; for example, persons permanently crippled by disease or accident suffer from exigences that cannot be positively modified.

Clearly many exigences can be modified without the creation and presentation of messages to mediating audiences. Some yield to perception, exploration, and experiment; others to physical labour or use of tools; some to the invention of mechanical devices; others to thoughts and discourses by which things, events, emotions, and other matters are merely represented. Such exigences invite response, but not response that needs or requires messages addressed to mediating audiences. I fear that a book is not in my briefcase; I need only look. My door is locked; I need a key. The bolt is loose; a wrench is needed to tighten it. My garden tomatoes are not healthy; I experiment, using first one remedy then another. My friend composes a poem that gives expression to his insight; his poem, in its character as an expression of his discernment, needs no mediating audience. My colleague in physics who discovers a principle and composes a report about his discovery needs no mediating audience. I seek to express my views on the nature of rhetoric; my verbal representation of my thoughts does not need to engage a mediating audience. In these and similar instances exigences are not rhetorical.

Exigences are clearly rhetorical when they are capable of positive modification and when such modification requires or invites messages that engage audiences who can modify those exigences through their mediating thought, judgement, or action. Thus in the course of his work, a labourer may need to instruct, encourage, or persuade a fellow worker. Because I know little about the afflictions of tomatoes, I ask advice of an expert; he serves as rhetor, and I am his audience. My poet friend, his poem created, seeks an audience because he believes others should experience the feeling and insight evoked by his poem. My colleague in physics, his discovery now described in his essay, feels duty bound to announce his findings and thus alter the conceptions of his scientific community. My own essay on rhetoric is intentionally rhetorical to the extent that I seek to remove objections and misunderstandings in the minds of critics and to win the agreement of my colleagues toward a perspective I believe correct and useful. The proposition seems undeniable: not only does the world invite change, but *there is a class of exigences whose modification requires or invites the assistance of messages addressed to mediating audiences*. Such exigences are clearly rhetorical.

It is not always easy to know whether exigences are rhetorical. Some are indeterminate: we cannot know with confidence whether discourse could modify them. For example, when a child is near death even parents who believe God does not confer personal favours may pray on the chance that He does. A candidate for political office may expect defeat but continue to speak because he might be elected in spite of the odds. In some instances the candidate speaks simply to finish out the game; this implies a different exigence. In ordinary instances of indeterminate exigences the speaker's well-considered decision to speak is based mainly upon the urgency of the exigence and the probability that it could be positively modified.

The person who speaks in order to modify a nonrhetorical exigence—one which discourse cannot possibly change—is engaged in a futile venture. The opposite poles of futile rhetoric, however, are not without significance. At the one extreme the rhetor is comic or foolish: for example, a man alone in a boat and adrift at sea shouts for help although he knows his words will be unheard. At the other extreme the rhetor is heroic: a man or woman pushed toward the death chamber hurls invective at Nazi guards; or a person speaks a truth in need of utterance even though his audience closes its mind to it. A speaker who responds to a momentous exigence, even though its modification is absolutely futile, is engaged in heroic effort. In such situations discourse usually turns inward to become lamentation or soliloquy.

A rhetorical exigence consists of a factual condition plus a relation to some interest. Literally billions of factual conditions in the environment are not perceived as being for or against anyone's interest, and no one seeks to alter such neutral conditions. Speakers are motivated to create messages when they perceive factual conditions related to felt interests. In other words they seek to change by means of discourse something which matters, something about which they are not indifferent. A wholly neutral factual condition does not solicit a response; and an interest unrelated to a factual condition has no object. An exigence exists when a factual condition and an interest are joined.

By a factual condition I mean any set of things, events, relations, ideas, meanings—anything physical or mental—whose existence is (or is thought to be) independent of one's personal subjectivity. By an interest I mean any appreciation, need, desire, or aspiration which, when related to factual conditions, accounts for the emergence of motives and purposes. The desk now before me, the objects on it, and the relations among the objects—all amount to a factual condition. The red maple displaying autumn colours and the workman about to fell the tree—a factual condition. The principles of constitutional government, the laws of the land, and the actions of persons involved in the Watergate affair—a factual condition. The addition of interest makes each an exigence—something other than it should be, a defect, a matter to be altered. The desk presents work to be done by me, and in duty I should not delay. The red maple ought to survive, I believe, because I appreciate its colour and shade. Those responsible for the Watergate burglary and cover-up violated principles and laws we hold to be just and essential; consequently, they ought to be punished. Factual conditions plus related interests thus comprise specific exigences we can modify if positioned to do so and if the total situation issues a strong invitation.

Consider an elementary situation. At some distance ahead we see an acquaintance who is deaf and blind. A few steps in front of him lies a deep excavation where sections of sidewalk have been removed. If he continues walking he will fall and probably be injured seriously, perhaps fatally. Furthermore there are no signs he knows of his danger. This is the factual condition, consisting of objects and relations immediately perceived, plus other assumed and inferred facts. Is this factual condition related to an interest? Of course it is. We fear for the man's safety and do not want to see him injured. The condition plus the related interest comprise the exigence, which we will modify if we can. To complete the situation, suppose a woman, reading a book, is seated in a lawnchair a few yards from the blind man. She is our audience—she and only she is in a position to receive our message and act to modify the exigence. Perceiving all this we respond by creating a message: 'Hey! You with the book! That man is deaf and blind! Stop him!' She hears, understands, and is constrained by the information in the message as well as by the exigence she now perceives. She runs to the man and guides him to safety. Both the exigence, which has been modified perfectly, and the rhetorical situation cease to exist. If she understood our message but did not act, we would suppose that she failed to see the factual condition or did not experience the requisite interest. . . .

Responsiveness to Exigences

Speakers and audience, when perceiving exigencies, are more or less responsive to them. The following factors seem to be of primary important.

DEGREE OF INTEREST

Other things being equal, responsiveness depends on the degree of interest experienced in the apprehension of an exigence. Among the factors influencing interest six appear obvious in the light of common experience; most of them are identified by rhetoricians in one context or another—for example, as factors of passion in Campbell's *Philosophy of Rhetoric*, book 1, chapter 7.

1. An exigence will generate a degree of interest in proportion to the probability of the factual component. Other things being equal, a zero-probability factual condition will produce the least interest, a certain factual condition the maximum interest. A factual condition thought improbable or unbelievable ordinarily does not solicit interest sufficient to motivate a pragmatic response.
2. Possessing knowledge about a factual condition is different from knowing the condition. That millions of Jews were murdered by the Nazis is a fact which appreciably affects us; but in another sense to know that same phenomenon is to apprehend it directly and sensibly or to receive from a communicator a vivid representation of it. Knowing in this second sense has a more powerful effect. Jonathan Edward's *Sinners*

in the Hands of an Angry God systematically argued toward the certainty of the factual condition of the unsaved and, in addition, through vivid representation the sermon dangled the unsaved over the fires of hell. Generally interest will increase insofar as the factual condition is known directly and sensibly, or through vivid representation.

3. A third factor is proximity in place and time. An exigence near in time and place generates more interest than a distant one. People are more interested in exigencies which are near their own places than in distant places, and more interested in those which are imminent than those occurring ten years past or in the distant future.
4. The degree of interest depends in part upon the magnitude of the factual condition or some aspect of it. An exigence involving people unknown and few in number will be less affecting than one involving many people who have public reputations. An exigence involving an action or event extraordinarily large will be more affecting than one involving an action or event of common proportions. An exigence will generate more interest if its likely consequences are numerous and of great significance, less interest if consequences are few and insignificant.
5. Sometimes speaker and audience perceive an exigence in which they themselves have little or no personal involvement. On the other hand, people frequently have a strong relationship with the exigence and in one way or another are personally involved: they feel somehow responsible for it or understand that the exigence or its modification will bring about good or evil for them, their families, or friends. An exigence that involves speaker or audience personally will generate more interest than one in which they are not directly involved.
6. The quality of the specific interest is a factor. The urgency of a particular exigence depends on the intrinsic quality of the specific interest involved; pain and fear, for example, are normally more powerful than curiosity and self-esteem. Such generalizations regarding the intrinsic quality of interests are risky, however, since much depends upon a person's physiological and psychological state at a particular moment in time.

MODIFICATION CAPABILITY

Someone contemplating whether to respond to an exigence asks, with reason, Could my effort have the effect of modifying it? A person who believes his response could not in any way modify it is not likely to respond, even though the exigence is urgent. If he believes his response could modify the exigence, then he is likely to respond. As modification capability increases, readiness to respond increases.

Modification capability will increase to the extent that there is little rather than much difficulty involved in positive modification due to the nature of the exigence; an audience is well positioned with respect to speaker and audience and is competent to function as a mediating agent; needed constraints are readily available in the situation and can easily influence the audience; there are channels through which the speaker can effectively address the audience, along with message-genres familiar to speaker and audience; the speaker possesses knowledge and skills needed in the analysis of situations and in the creation and presentation of pragmatic messages.

RISK

Speaker and audience may also ask, What is the risk if I respond? What will my response cost—in time and energy, prestige, danger, personal commitment, and money? And what will be my satisfaction, gain, or reward? Readiness to respond will vary according to the degree of risk: other things being equal, a person who stands to lose much and gain little is less inclined to respond than one who stands to gain much and lose little.

OBLIGATION AND EXPECTATION

A person's responsiveness may be increased by recognizing that only he or she is capable of modifying the exigence and by the knowledge that other people expect a response, or that duty or obligation requires it.

FAMILIARITY AND CONFIDENCE

Responsiveness is facilitated by familiarity and confidence. The speaker and audience who are familiar and practiced in responding to certain situations will be more ready; if the speaker is fearful of speaking or the audience fearful of deciding and acting, each will be less inclined to respond than if confident and experienced.

IMMEDIACY

Speaker and audience are also more likely to respond if they believe their efforts must occur now or never—that there is no opportunity for postponement.

When these factors are 'strong', an observer can often predict accurately that messages of a certain kind will occur. For example, if news reporters witness a tragic event—a California earthquake causing destruction of the highest magnitude—they will respond to a situation whose exigence and constraints are so powerful that we can predict with practical certainty the central themes and topics of their news reports. Is it possible for the reporters to ignore the situation, dismiss the exigence, free themselves from situational constraints? Of course it is. But should they do so, we would think they suffered some mental lapse or failed their responsibilities. These observations do not imply a deterministic relation among situations, rhetors, and responses. They do indicate that rhetors sometimes are involved in complicated ways with situations, roles, motives, causal forces, genre, and other factors, which make their communication behaviour highly predictable. There is no inherent contradiction between predictable communication behaviour, on the one hand, and the presence of freedom and creativity on the other. . . .

The Fitting Response

The fitting response is sometimes easy to judge: we apprehend situations which clearly invite a particular response—a word or phrase, a command, an argument, a speech; sometimes we need only say what we have said previously in similar contexts. Other situations permit alternative fitting responses, and we may be unable to anticipate beforehand, or judge later, which of several responses would fit best. The 'Cooper Union' situation prescribed alternative responses that we as critics might have sketched out in terms of purposes and topics. But could any of us have specified Lincoln's address at Cooper Union? Critics often see the fit between situation and response only in retrospect.

Speakers sometimes misconceive the situation because they make mistakes about the exigence, constraints, or audience and believe falsely that their artistry will be adequate to the task. Richard M. Nixon made two speeches upon his resignation, one to the nation and another to the White House staff. Both situations invited fitting responses, but he misconceived them. His most important error was this: he mistakenly thought that the central exigence was his own suffering of what he regarded to be an unmerited punishment. Given this error his messages were likely to fail—and they did: neither speech was fitting as a corrective, and neither was fitting in the sense that the situation enabled it to be well received. The basic flaw in Mr Nixon's perception of his 'resignation situation' was his failure to recognize, or his refusal to admit, that the situation was epideictic and that his fitting response must blame himself or praise the very persons and institutions responsible for his fall.

This essay has examined conditions and factors involved in functional or pragmatic communication and in so doing extended the argument for the situational perspective. In particular I have emphasized persons and their environment as the fundamental interacting grounds, and argued that the rhetorical situation is a necessary condition of communication by which persons strive to achieve balance and harmony with the environment. The narrow scope of this essay precluded detailed attention to many aspects of the art of rhetoric. However, this omission does not imply that interest is absent or that problems inhibit accommodation of the rhetorical art to the situational view.

In retrospect I am aware that the examples of pragmatic communication cited in this essay are, by and large, clear and straightforward instances of functional communication; moreover they are single, purposeful, and managed efforts to modify identifiable exigences. Someone who notices this fact might ask, But do not all or nearly all kinds of communicated messages function—somehow, ultimately, and perhaps upon some larger scale—to adjust man to environment and environment to man? Thus is there not a great deal more to be said?

A complete account of how messages of many kinds function in the interaction of man and environment is a large and complicated subject. Surely kinds of messages other than those treated in this study contribute to the harmonious adjustment we seek to maintain. Over many years we engage the environment in a rhetoric of adjustment; through poetry, drama, ritual, education, philosophy, religion, and through patterns of response beyond those we have treated and of which we are seldom conscious, we come to terms with the environment and with ourselves, thus maintaining a delicate balance.

Notes

1. A view of situational rhetoric was published by the author in an essay titled 'The Rhetorical Situation', *Philosophy and Rhetoric* 1 (Winter 1968): 1–14. I am indebted to numerous students and colleagues whose valuable criticisms, presented in discussions and publications, have led me to make some modifications and extensions of my view.

References

Arnold, Carroll C. 1974. *Criticism of Oral Rhetoric*. Columbus, Ohio: Charles E. Merrill.

Dewey, John. 1939. *Theory of Valuation*. Chicago: University of Chicago Press

QUESTIONS FOR CRITICAL THOUGHT

1. Bitzer's article was originally written for an audience different from the audience of this book. What challenges does this fact present to you as a reader?

2. According to Bitzer, 'our existence and well-being depend upon skilful and ongoing adjustments to, or modifications of, our environment.' In what sense is it true that our messages modify or adjust our environment?
3. Following Bitzer's analogy, what does rhetorical communication have in common with a discipline like engineering? Why do you suppose Bitzer has introduced this comparison?
4. Bitzer distinguishes between the physical environment and the mental environment. What, exactly, is the mental environment?
5. What is a rhetorical situation, according to Bitzer? State his definition.
6. Bitzer argues that the constraints we face in a given situation can be understood as both 'limitations and opportunities' for the creation of messages. In what sense are they both?
7. How does Bitzer define an exigence? What makes an exigence rhetorical? What does it mean to say that an exigence is made up of both 'facts' and 'interest'?
8. What, according to Bitzer, are the factors influencing an audience's willingness to respond to someone else's message?
9. If all rhetorical messages are situated, as Bitzer argues, is his own essay also a product of its context? Can you point to any evidence that this is so?
10. What is the value of reading a selection that was originally written for a completely different audience? What can be learned from this exercise?
11. What implications does Bitzer's theory have for the practice of technical or professional communication?
12. Read Felder's essay 'A Whole New Mind for a Flat World', and show how Bitzer's theories might be used to explain what went wrong in the new graduate's interview.
13. How formal is the writing style in this article? Is it more or less formal than others in this book? Why might this be so? To what extent is its formality, or lack of formality, determined by its context?
14. What is the purpose of Bitzer's article? Are his intended readers generalists, or are they specialists? How can you tell?
15. How much authorial distance does Bitzer maintain? How personal or impersonal is the writing here? Is the style appropriate to its purpose and context?

16. How does Bitzer establish his credibility? To what extent is credibility a function of appropriate audience adaptation?

PART II

Communicating Science

THE work of scientists and engineers has created extraordinary benefits for society, not only in essential fields such as medical research and technology, but also in leisure, travel, and entertainment. In part as a result of that work Canadians enjoy one of the world's best standards of living. But as technology becomes more complex and scientific discoveries advance beyond the understanding of a lay public, these same scientists and engineers face increasing scrutiny from those concerned about the social, ethical, and environmental consequences of their activities.

Unfortunately for many technical specialists, the skills needed to persuade non-specialists of the meaning and value of scientific and technical advances are different from what is required to communicate those same findings to other scientists and engineers, who typically share the same values about what is important and what constitutes proof. In order to be successful, an engineer's communication, like that of any other professional, must establish and maintain credibility and authority with those who may be unfamiliar with technical subjects.

Most people recognize that an engineer's ability to communicate highly technical subjects in a clear and understandable way is important, but they may not realize that good communication involves more than simple information transfer. Instead, it requires careful attention not only to the clarity of the message itself, but also to satisfying the needs of the audience and to establishing the speaker's credibility. This is a greater challenge than it may appear, since most of those with whom the engineer regularly communicates are non-specialists whose technical expertise varies widely. Unfortunately, scientific training does little to prepare scientists for this task.

Scientists and engineers need to be able to communicate about their work in a way that non-specialists can understand, if for no other reason than that it is often non-scientists—clients, managers, administrators—who make policy, procedural, and funding decisions for technical and scientific projects. Because such lay readers cannot directly judge technical skill, they will instead rely on the clarity and confidence of a professional's communication as a basis for judging technical competence. Thus, skill in communicating specialized information often becomes the measure of an engineer's competence, irrespective of his or her actual technical expertise.

For whatever reason, as the essays in this Part show, science and engineering communication is often done poorly, especially when directed to a non-specialized audience. These selections detail the challenges inherent in communicating science, both to the general population and to anyone outside the writer's field of specialization. Some of these selections offer insight into why scientific communication often fails; others present some suggestions for improving practice. All, however, recognize the fundamental importance of communication to the practice of science and engineering.

CHAPTER 6

Communicating Science

J.S.C. McKee

Looking back in history, it is not long since the presentation of scientific facts and figures to the public was often regarded as an activity neither advisable nor appropriate. When I was a young tenured lecturer at Birmingham University in England, it was considered highly inappropriate to attempt to inform the public of recent advances in scientific discovery. The reasons for this were twofold. Firstly, the detailed understanding of new phenomena was the forté of a few people in a given field, and to try to communicate at a low intellectual level the significance of new work was not only to undervalue the effort that many scientists had invested over a long period of time but to trivialize the essential complexity of what had been accomplished. It was also regarded as totally unreasonable to expect a listener or member of the public, having neither an undergraduate science degree or even a high school diploma behind them, to appreciate—at any level—the import or significance of a new discovery. Things are different now, but the transition to wholehearted communication of science to the public has been a difficult one and is an activity in which too few of us participate with any enthusiasm.

While reflecting on this problem recently, I recalled to mind an interesting column in the *San Francisco Examiner* of 1967, in which an apocryphal tale of two assistant professors seeking tenure at the University of California at Berkeley was presented. The article announced that a tenured position at the University had become open for competition and that the short list for the appointment comprised two scholars with widely different backgrounds. The first, an expert in Sanskrit, had been employed on a continuing appointment for several years. His work was highly esoteric and when he gave an annual public lecture, as he was obliged to do by the by-laws of the university, it was given in his own rooms and attended by an audience of one, namely his graduate student who had worked with him for several years. The second candidate, on the other hand, gave a first year lecture course in Comparative Religions that was among the most popular in the whole university. His enrolment of 2,000 students was limited only by the capacity of the auditorium in which he presented his material. This class was enthusiastic both about his presentation and his content and eager students were known to form a line two hours in advance of each presentation.

When the Board of Regents finally came up with its decision, it was that the professor of Sanskrit be confirmed in a tenured appointment. On being questioned by the press and public as to the nature of this bizarre decision, the Chair of the Board of Regents commented, in relation to the qualifications of the loser, that 'the University of California at Berkeley is one of the great storehouses of knowledge in the western world and this young whipper-snapper is continually giving the stuff away.'

As we approach the end of the twentieth century, one fact of life that has become crystal clear is that all able and practising scientists are in the business of 'giving the stuff away'. The 'stuff' being scientific knowledge and the ability to apply it to a myriad of both academic and technical situations. The research scientist is now only too well aware of his or her need to inform and woo the taxpaying public, if funding for fundamental and applied research is to be forthcoming. In fact, if the scientist is unable to persuade others of the value of the work proposed, then chances of gaining the opportunity to proceed with it will be minuscule at best.

It is not only enlightened self-interest, however, that necessitates the communication between the scientist and non-scientist. There exists in the community at large a hunger for increased scientific knowledge and the ability to function in an increasingly technological environment. Some 12 years ago, a study carried out in Europe by OECD indicated that over 60 per cent of the community would welcome or needed further scientific education than they had acquired and more scientific information than was readily available to them. In Canada, we find a burgeoning of scientific journals aimed at the non-scientist or layperson, which partly meet the need for scientific information and illustrate effectively the significance of new discoveries as they come along. The fact that such journals sell with alacrity on most newsstands indicates that the need is there. The continuing popularity of television programs such as *The Nature of Things* in Canada and *Nova* in the United States, plus the initial acceptance of the Discovery Channel, also serve to indicate that there is a healthy hunger in Canadian society for new knowledge and new skills.

The advent of new style museums where interactive exhibits dominate the scene now enables people who are otherwise uncomfortable in a scientific or technological environment to educate themselves through button-pressing response mechanisms that enable them to perform more effectively at their workplace or in the home.

The reasons as to why a need for science in society has suddenly arisen are many and varied. Clearly, the thirst for scientific knowledge was not confined to the ancient Greeks and there are many today who find areas of scientific endeavour, such as space research, to be of continuing fascination. There are also people with a modest background in scientific knowledge who continually educate themselves through literature of varying degrees of factual correctness and difficulty. There are housepersons who find the home environment a more challenging place than it has ever been in the past. The advent of the home computer, the microwave oven, the garburetor, the television set with its VCR attachment, the electronic dust precipitator, and the year-round air conditioning system all require new, if modest, technical skills on behalf of the user if they are to be operated effectively. In addition, there are always a number of people in the community who do not trust scientists further than they could throw them, and eagerly acquire such scientific information as is available in order to keep, as it were, a check on the scientists and what they are doing.

The evolution of a scientific culture in Canada is becoming essential to the generation of a highly skilled work force and a vibrant economy. The traditional idea that there exist two separate cultures within society—a scientific culture and a humanistic culture is becoming passé. If a culture is indeed the sum total of the beliefs and experiences of a society, then the interaction between science and the society it serves has to be ongoing and develop to the benefit of all. The ubiquitous nature of science requires that scientific knowledge and training be made available to any and all that wish it. Public awareness is but one side of the coin. Science education and technical training is the other. Effective communication is essential to both.

QUESTIONS FOR CRITICAL THOUGHT

1. This opinion piece was originally written as an editorial for the journal *Physics in Canada*. Given that context, who might McKee's intended audience have been? What appears to be the purpose of the article?
2. What evidence does McKee offer to support his view of the necessity 'to inform and woo the taxpaying public' regarding advancements in science? You should be able to find at least four separate reasons.
3. Why does McKee tell the story of the two young professors seeking tenure? What does its inclusion add to his argument?
4. Why, according to McKee, has the communication of science become a more pressing issue for the lay public? To what extent do their interests coincide with those of the scientists themselves?
5. How formal in style is this essay? Point out some specific features that mark it as either formal or informal. What are the qualities of formal writing?
6. Is McKee's article personal or impersonal? Is its level of personal disclosure consistent with its level of formality? To what extent are the two features independent?
7. McKee makes several assertions that are inherently critical of scientists and scientific writing. How does he establish the necessary authority to do so? What is the source of his credibility?
8. McKee points out that communicating science to a lay public is challenging; does he offer any concrete advice for doing so? If so, what is it? If not, why not?
9. Compare McKee's article with the selections by Stephen Strauss ('Avoid the Technical Talk, Scientists Told. Use Clear Language') and Cheryl Forbes ('Getting the Story, Telling the Story: The Science of Narrative, the Narrative of Science'). What commonalities do you see in their arguments? What differences do you see? Can any of these be accounted for by a consideration of the audience for whom each is writing?

CHAPTER 7

Avoid the Technical Talk, Scientists Told. Use Clear Language

Stephen Strauss

Scientists who want research dollars from Canada's largest granting agency will have to spell out what their projects are about in everyday language.

And if the scientists are not sure that their popularization skills are up to snuff, they are advised to try out their straight talk on non-scientific friends and relations.

The new directive comes from the Natural Sciences and Engineering Research Council that dispensed $258-million in grants to about 7,300 Canadian scientist and engineers last year.

In 1995, NSERC was attacked by Randy White, Reform MP for Fraser Valley West in British Columbia, for funding what he deemed frivolous research. He was particularly incensed that this was happening while the federal liberals were making cuts in social programs.

Among the projects Mr White singled was research on the paternity of squirrels, the energetics of hummingbirds, 'information processing in pigeons' and 'cubitus interruptus locus in Drosophila'.

Initially, the council had difficulties responding to the charges because, while experts in the field decide who should get the awards, the staff in Ottawa was not skilled enough to parse out the precise meanings of winning grant proposals.

'At the time, all we could release under the Access to Information Law was the title of the research. For each individual researcher, we had to go back and ask what they were doing and could we release that information,' said Arnet Sheppard, an NSERC spokesman.

The new policy is a two-pronged attempt to address exactly what the research is and what it is good for.

'I feel very strongly in a time when public funds are at a premium that researchers' accountability takes the form of explaining to the public what they are doing,' said Tom Brzustowski, NSERC president. 'I think the public has the ability to understand its [research] importance once the veil of jargon has been lifted.'

He said he thought Mr White's initial criticism was a rhetorical device 'to call for public accountability in the expense of government funds. And as far as I am concerned, he is right on.'

Calls to many scientists across the country who received NSERC grants revealed that while there is a general belief that the popular summary is a good idea, there are going to be communication problems in some fields.

Mathematics, in particular, is likely to contain domains where simplification is impossible. 'I don't think I really can explain things on that level,' said Yuri Billing, a mathematician at the University of New Brunswick in Fredericton, about the new guidelines. 'I can explain what I do to graduate students, but I have real trouble in explaining it to undergraduates,' he said.

He received a $12,000 grant to study 'Kac-Moody algebras, groups and their applications'.

Henry Reisweig, a professor of biology at McGill's Redpath Museum, received $18,900 to study 'systematics, biogeography and ultrastructure of hexactinellid sponges'.

One translation is that he has become the absolute world expert on the species, distribution, and structure of a class of deep-water sponges. It is not a crowded intellectual domain. 'There has not been another expert in this group since 1921,' he said.

One of the things that happens to him is scientists from all over the world send him samples of the sponges from their collections and ask him to decide if they are new genuses or species. One out of every three or four has never been described in scientific literature. After deciding they are truly new, the McGill professor gets to classify (name) them.

Who cares about how many kinds of a certain class of sponge there are? People who are interested in the evolution of life. Sponges were present when animals appeared on earth. 'They [the sponges] are really strange, they don't have cells, they are one big cell,' Professor Reisweig said.

However, for those who demand an economic payoff to justify research, there simply will not be one. 'They are of no economic interest that we know of,' Reisgweig said.

But in some cases, very applied, if basic, research has hidden under a veil of jargon that only scientists could love.

Christopher Bender, a chemist at the University of Lethbridge, received $10,300 to study 'the photochemical and thermal reactivity of 2,3-benzobicyclo (4.2.0) octa-2,7-dienes'.

Professor Bender's group is trying to use heat or light to develop alternative or shorter routes to making taxol. Taxol is a recently approved cancer drug that originally came from the yew tree but is now being synthesized in laboratories.

Among the most enthusiastic proponents of simplification are Mr White and one of the scientists whose work he attacked.

'It's a wonderful idea,' said Clifton Gass, a University of British Columbia zoologist. He received $17,000 to study hummingbird bioenergetics, an area that he roughly defines as 'How do hummingbirds get enough energy and how do they keep from spending more energy than they bring in?'

He believes that while the public will learn what it is that scientists do, so too will people on the grants committees that fund them. The granting committee members 'complain all the time that they don't know what people are talking about [in their applications] because they speak in jargon all the time,' he said.

What's the purpose of his work? It has shown that hummingbirds are very smart when it comes to figuring out how to get the most energy bang for their flying buck. 'We can tell stories about hummingbirds' energy strategies which force not just scientists but everyone to change their ideas about animal intelligence.'

Mr White could not help but crow a little about the new policy. 'It's nice to know that the Reform Party is doing things which are making the system change to be a little more accountable, although this should have been 10 or 15 years ago,' he said from his BC riding office.

'I look forward to reading [the new explanations] with pleasure, but keep in mind if I read them and I don't like what I see, I will be speaking my mind.'

The simplified explanations will be displayed on the NSERC home page on the Internet. Good ones will be highlighted as examples to other scientists.

QUESTIONS FOR CRITICAL THOUGHT

1. Strauss reports on a directive from the National Science and Engineering Research Council of Canada about improving the clarity and readability of scientific grant applications. What problem is this directive, and indirectly Strauss's article, intended to address?

2. The article is a primarily informative report on the new NSERC policy. To what extent is it also persuasive? What strategies does Strauss use to persuade his audience?
3. Strauss uses the term 'veil of jargon', borrowed from one of the sources he quotes. Is the metaphor effective? What does it imply? Why does Strauss choose to use it?
4. In several places in this piece, Strauss describes a research project in highly technical language, and follows up with a plain-talk version of the same project. What is the rhetorical intent of this approach? Does it advance his purpose? What is its emotional impact on the reader?
5. This article shows elements of journalistic style, such as a 'hook' (a catchy opening) and the strategic use of quotations, to name just two. What is the function of these strategies?
6. A good piece of journalism, even a report like this one, tells a story. Outline the plot of the 'story' told by Strauss.
7. How informal or formal is this passage? How personal or impersonal is it? What is the correlation between these two qualities (formality and personality)?
8. Are Strauss's sympathies in this report with the scientists who must find a way to write plainly about complex scientific research, or with the lay public who must read their prose? How do you know?
9. How readable is Strauss's article? Does he himself write in the plain, clear style advocated in his report? Provide some examples to support your answer.

CHAPTER 8

Getting the Story, Telling the Story: The Science of Narrative, the Narrative of Science

Cheryl Forbes

Darwin knew how to tell a good tale, and nineteenth-century readers responded just as we do today. Poincaré was no mean raconteur, either, mathematically speaking. Booksellers in France couldn't keep his books in stock. Primo Levi brought the elements to life in *The Periodical Table*, which my ninth grade science teacher could have used to excite us about the properties of copper, zinc, and titanium. Book reviewers raved.

What made the books of these scientists best-sellers? What made these scientists good writers? Each had a sense of narrative or story, which is important in fiction, memoir, and biography—but in science, in professional communication? Certainly when we professors prepare students to write scientific reports, we don't usually emphasize narrative; if anything, we forbid it, perhaps because, like second-hand smoke on hair and clothes, the odours of subjectivity, invention, and triviality, and 'dumbing-down' cling to the word *narrative*. We want our science presented objectively, with the appropriate authorial distance, sobriety, factuality; or so we instruct students.

Yet the style we teach makes a subtle epistemological argument. When we insist that students write objectively, dispassionately, in the third person, using the passive voice, avoiding metaphors, similes, analogies—in other words, in prose devoid of voice or humanity—we are arguing that science is not what human beings make, but something human beings merely report as having been made or discovered. Science is the way to find what is Out There, objective, eternal, unchanging. It gives readers an inaccurate perception of the messy, chaotic, contingent, accidental, disputatious nature of scientific inquiry and discovery—which makes science so exciting. The objective style keeps readers at arm's length and reifies science, or, as Alan G. Gross claims, such a style 'creates a sense that science is describing a reality independent of its linguistic formations' (1990: 17).

That so few rhetoricians of science tackle narrative discourse or the discourse of popular science indicates how reified the objective style is. For example, though Gross considers analogy in scientific discourse and analyzes Watson's *The Double Helix* as a fairy tale (1980: 58ff), he does not put his discussion in the large context of narrative itself (later in this essay I return to Watson and Crick's paper on DNA). Or consider Rom Harré's article, 'Some Narrative Conventions of Scientific Discourse' (1990), in which the discussion of narrative focuses on the 'story' in academic prose: the scientist as hero, the presentation of results as a linear, logical progression from a set of hypotheses, and the final inductive work. Harré's point is that the scientific article presents a fiction, a fairy tale, because the reality of scientific research is nothing like that described in the literature (Greg Myers makes a similar argument in 'Making a Discovery: Narratives of Split Genes' [1990]).

Although I agree, this is not the kind of narrative I am talking about, as I make clear below. Nor am I (primarily) concerned with the discourse of the academic scientific journal article, even though I would like to urge that most, if not all, scientific discourse forego an objective style in favour of the narrative style I explore here. I would like to read such discourse in articles in disciplinary journals and not just in *Harper's* or *The New Yorker*. I would like scientists to tell me stories, but though my heart lies with this suggestion, my head does not (and of course it would merely replace one kind of discourse requirement for another, when what we need are multiple kinds of discourse). For busy scientists who need to read succinct, highly coded discourse, narration could well conflict—and the typographic and formatting requirements of many disciplinary journals are designed to make articles quickly understandable (for instance, the abstract, subheadings that outline contents, the conclusion, tables, and figures).

Nevertheless, and keeping this caveat in mind, a scientist could introduce narrative elements into an objective style (through choice of verbs, for instance, or through metaphor), without sacrificing succinctness or shifting the focus from the scientific information to the scientist herself. However, my purpose here is not to discuss the academic scientific journal article and its narrowly focused readership. Rather, I am concerned with professional communication written by scientists or science journalists for a general, 'liberally' educated audience. Why this focus, when most recent work on scientific prose has centred on the academic article? (In addition to those already cited above, see, for instance, *Understanding Scientific Prose*, edited by Jack Selzer [1993]; *Textual Dynamics of the Professions: Historical and Contemporary Studies of Writing in Professional Communities*, edited by Charles Bazerman and James Paradis [1991]; Bazennan's *Shaping Written Knowledge: The Genre and Activity of the Experimental Article in Science* [1988]; and *The Rhetorical Turn: Invention and Persuasion in the Conduct of Inquiry*, edited by Herbert W. Simons [1990]).

I focus on scientific discourse for a general, educated audience because, like Selzer (1993), I believe that scientific prose is a central discourse in our culture (this goes beyond Selzer's 'especially important'), but I disagree with Selzer's claim that 'scientific discourse today is typically carried out not in public but in more private communities . . .' (7). On the contrary, the scientific discourse that is 'especially important' in and for our culture is precisely that carried out in public: through newspapers, news magazines, scientific periodicals, and 'public service' brochures and advertisements, whose readership is the generally educated public, and over the air and radio waves (though I limit myself here to scientific discourse in print, I could well and fruitfully apply my argument to broadcast scientific discourse).

To return, then, to the scientists cited at the outset, we need to understand that their discourse is effective because of the narrative elements. We need to answer our students who read Darwin or Poincaré (often at our request) and then ask why those scientists write one way when they have been told to write in another. And, finally, we need to graduate more scientists who can write for a broad audience, precisely because science and its discourse is so important in and for our culture. Stephen Jay Gould himself calls for this in the introduction to *Bully for Brontosaurus* (1992).

So in thinking about stories and science and whether Darwin, Poincaré, and Levi are exceptions and exceptional among scientists or writers of science, I conducted an experiment. I pulled from my bookshelves several volumes of contemporary scientific dicourse, most of them marketed to a general, educated audience, as were Darwin's, Poincaré's, and Levi's. (These volumes were not chosen with forethought or because I thought they would support my position; they simply happened to be the science books on my shelf; in this sense, I chose them randomly.)

I opened the introduction or preface and the first chapter of each book to find out how the author starts out because the beginning of any discourse sets the tone, establishes the style, and prepares readers for what follows. For scientific discourse, the lead, to use a journalistic word, is crucial, and for a general audience, the more technical the subject, like astrophysics, quantum mechanics, or DNA research, the more attention a writer must pay to the opening words in order to keep readers reading, to engage them at the outset in the adventures they are about to participate in. In other words, from the opening sentences, a writer wants readers to participate in the journey, to be part of the discovery, to catch the excitement of the adventure, as if they had been in on things from the beginning. Thus, I read the introductory matter as well as opening chapters. I studied the pronouns each writer used, what verb forms, what adverbials, all of which are significant discourse markers for narrative. I wondered how (or whether) writers establish a sense of plot and character (pronouns), of drama and comedy (passive or active verb forms), of place and time (the function of adverbials). Here, then, are seven samples, with a few brief comments after each to began the discussion of narrative in 'popular' scientific discourse; each book was written for a general audience (broadly defined, sometimes very broadly):

- Dennett, *Darwin's Dangerous Idea: Evolution and the Meanings of Life* (1995). 'We used to sing a lot when I was a child, around the campfire at summer camp, at school and Sunday school, or gathered around the piano at home. One of my favorite songs was "Tell Me Why"' (17). Dennett says in the preface that he is writing about science, not writing science per se, so it might be unfair to point to his use of the first persons singular and plural or the sense of place and time he establishes. But *Consciousness Explained* (1991) is science, and there he begins his preface with another personal story and the first chapter with a 'supposition': 'Suppose evil scientists removed your brain from your body while you slept, and set it up in a life-support system in a vat' (3). A supposition by any other name is a story, and he uses the second person to engage our attention, to involve us in his story.
- Sacks, *An Anthropologist on Mars* (1995; Sacks, a neurophysiologist, uses the word anthropologist metaphorically). 'I am writing this book with my left hand, although I am strongly right-handed' (xv). There it is again, the first person singular. His first chapter begins, 'Early in March 1986 I received the following letter' (3). Although Sacks could have begun this sentence also with the first person singular, he chooses to begin with an adverb of time (undoubtedly for a smoother transition into the letter he is about to quote). But the emphasis remains on the first person.
- Johnson, *Fire in the Mind* (1995). 'Several years ago, on a visit home to New Mexico from my self-imposed exile in New York City, I was driving through the predominantly Catholic village of Truchas, on the high road from Santa Fe to Taos, when I rounded a corner and was startled to see . . .' (1); this sentence, the first of the Introduction, continues for four more lines. The first chapter begins, 'In the evening, just as their planet is about to complete another revolution, small bands of earthlings . . .' (11). He tells a story from his point of view, and though this sentence doesn't contain the first person singular (he does use it in the course of the chapter), Johnson nevertheless makes it clear to readers that we are hearing the story from his perspective. (Here is a good example of the way writers can present a first person perspective without actually using 'I'.)
- Gould, *Dinosaur in a Haystack* (1995). 'I have always seen myself as a meat-and-potatoes man. You can take your ravioli stuffed with quail and . . . well, stuff it somewhere. (I am also quite capable of releasing my own ground pepper from a shaker.)' (ix). In the first sentence of the first essay, Gould writes that 'Galileo described the universe in his most famous line: "This grand book is written in the language of mathematics, and its characters are triangles, circles, and other geometric figures"' (3). No first person singular, to be sure, but we do find narrative: in 29 words, Gould cleverly packs a story within a story.
- McPhee, companion books *In Suspect Terrain* (1983) and *Assembling California* (1993). 'The paragraph that follows is an encapsulated history of the eastern United States, according to plate tectonic theory and glacial geology' (1983: 3) and 'You go down through the Ocean View district of San Francisco to the first freeway exit after Daly City, where you describe, in effect, a hairpin turn to head north past a McDonald's to a dead end in a local dump' (1993: 3). Although neither sentence includes the first person singular, like the second passage cited from Gould above, we do have narrative. We have stories. We have 'encapsulated history', and history is by definition a narrative. McPhee is promising to tell us the geological story of one part of

our country, and throughout the rest of the book, he tells us the story as narrated by scientists who hold or dispute plate tectonic theory (thus also narrating and explicating the theory itself). In the second quotation, McPhee uses the even-more-frowned-upon and forbidden second person singular (think of the many admonitions professors give students about using 'you'), which I noted in Dennett and found again in Gould. 'You' automatically implies 'I'.

- Angier, *The Beauty of the Beastly* (1995). 'When I was a girl'—more first person singular—'I had a terror of cockroaches that bordered on pathological' (ix). What about the beginning of the first chapter? 'Ah, Romance. Can any sight be as sweet as a pair of mallard ducks gliding gracefully across a pond' (3). Angier does not use the first person here or throughout the chapter. Nevertheless, who could deny that she establishes a strong narrative voice or that the first person singular is lurking suspiciously close to the surface of her discourse? Does she not invite readers to stand shoulder to shoulder with her and watch those mallards 'glide gracefully' as she tells us their story?
- Nuland, *How We Die* (1994). 'Every life is different from any that has gone before it, and so is every death. The uniqueness of each of us extends even to the way we die' (3). Not what I'd call a scintillating start for a story, I may have found my one real exception, but then the second paragraph (and the acknowledgments, of all things) changes my initial response: 'The first time in my professional career that I saw death's remorseless eyes, they were fixed on a fifty-two-year-old man, lying in seeming comfort between the crisp sheets of a freshly made-up bed, in a private room at a large university teaching hospital' (4). Death is no longer an abstraction, but a character in a narrative. As with Gould's opening line about Galileo, here is a story within a story. In Chapter 2, Nuland gives readers the medical story of heart disease, the story that lies underneath the first two he tells.

Let me repeat. I am putting forward the case that good science writing, particularly (though not exclusively) for a general, liberally educated audience, depends on the conventions of narrative—place, time, characters, dialogue, dramatic tension, humorous incongruity, suspense, mystery, intrigue, plot, and a believable, reliable narrator.

Sometimes the story reads like a whodunit, sometimes like an Aesop's fable or folktale, sometimes like slapstick comedy, but regardless of the (sub)genre, at the heart of good science writing for an educated audience—an audience not narrowly specialized—what drives the writing and the writer and what engages readers, what makes the text succeed, is narrative. And without narrative—or with mishandled, inconsistent narrative—the text fails to fulfill the expectations of readers and the goals of the writer.

Let's look a little more closely at the few sentences I have quoted in my sample list of books, before moving to an extended analysis of the narrative elements in two scientific essays. Although not every quotation contains every item in my narrative catalogue (place, time, characters, dialogue, dramatic tension, suspense, mystery, intrigue, plot, humorous incongruity, and a believable, reliable narrator), we do find several in most of them. Take place, for example. Johnson and McPhee and Nuland give us place, as does Dennett. Gould, Sacks, Angier, and Johnson emphasize time, also introducing characters and dialogue, even if they do it by citing letters or other writing. McPhee subtly includes dialogue by paraphrasing a scientific theory, which makes his first book on plate tectonics, *In Suspect Terrain* (1983), a conversation with that theory (and the scientists who hold or reject it); in the second book, he establishes a conversation with the reader by his emphatic use of the second person singular.

Although dramatic tension, suspense, mystery, and intrigue can all be subsumed under plot and seen as synonymous, I list them separately to emphasize their importance, for every sentence cited above evidences these characteristics: Something is about to happen to someone, and we'd better keep our eyes open.

Because such discourse has tension, contains mystery, and suspends or delays the main point, readers ask questions, our curiosity is peaked, and we continue to read. Humorous incongruity can work similarly, and though not all the writers cited above use humorous incongruity, several do, among them Gould and Angier. Welcome to the seriously playful side of science.

The narrator in science writing deserves a separate paragraph (if not a separate article). Every writer, even McPhee, establishes a distinctive narrative voice, usually with the first person singular (McPhee creates a narrator by his use of 'you' and the implied 'I'), which affects verb choice. The result is an active, not passive, voice. For of all the characteristics of narrative, the credible, personable, reliable, conversational presence of the teller of the tale is the most important. We might do without an unambiguous sense of place or humorous incongruity (many narratives have neither), but we cannot do without the narrator, who provides presence and perspective and establishes empathy and connection with the reader (and not, coincidentally, with the subject). So although the narrator comes last in the list, she initiates the rest, for without a narrator we cannot have a narrative: cause and effect.

Given the above list of characteristics of narrative and to further develop the theory that there is a narrative of science and a 'science' (or system) of narrative, I want to look closely at two quite different but well-known writers of science, a journalist who specializes in science and a scientist who writes journalism. Both kinds of writers direct their discourse at a general, educated audience, and this audience (I am included in it) needs both kinds of writers for a rich understanding of science and scientific inquiry. The biological essays of journalist Angier have been praised as 'science writing at its best', and she has been compared to writers from Aristotle to Lewis Thomas. Gould's collections on evolution, culled from his monthly column for *Natural History* magazine, have consistently landed on the bestseller lists. I have chosen them, therefore, because they are highly respected, widely read, and representative of the two types of writers who publish in the field today.

Angier's 'The Scarab, Peerless Recycler' in *The Beauty of the Beastly* (1995), takes an unlikely subject and through narrative changes our presuppositions. Gould's 'Creation Myths of Cooperstown' from *Bully for Brontosaurus* (1992), the collection Gould himself considered his best to date, interrogates the nature of story itself to ask fundamental epistemological questions: How do we know what we know; What counts as or constitutes science? His essay (and indeed the whole book), therefore, exposes the heart of writing science as narrative.

Angier

Recycling, Superfund cleanup, landfills, supertankers trolling the waters with garbage and no port or harbour willing to accommodate the trash: Such issues form the backdrop for Angier's Aesop-like essay, 'The Scarab, Peerless Recycler' (1995). Implicitly she asks what humans might learn (about ourselves?) from studying the scarab. She also implicitly asks us to reconsider our value system—a kind of 'least of these' approach to understanding nature.

Therefore, from her opening lines, Angier wants to jar us. We might think that we are superior to the Egyptians, who might have worshiped the dung beetle, but perhaps they recognized how important the dung beetle was, unlike us. We would rather avoid beetles altogether—just as we would rather keep our dung out of sight and out of mind.

So Angier says to readers, 'Don't think you're wiser than the Egyptians. Listen to this story.' She uses phrases like 'stamp of nobility', compares a beetle head to a diadem and a beetle body to 'glittering mail of bronze or emerald or cobalt blue', and catalogues a beetle's symbolic significance: 'rebirth, good fortune, the triumph of sun over darkness' (1995: 109). All this for a creature that feeds on feces. How many women sporting an emerald or ruby scarab brooch know the real story? Angier's essay reminds me of Primo Levi's tale about lipstick, for which manufacturers need, among other ingredients, a goodly supply of chicken shit (the AngloSaxon, which is his, has delighted students every time I include it on a syllabus); as Levi discovers, it isn't all that easy to come by.

Angier (1995) tells a story at once medieval, magical, and masculine, for her words conjure up images of jousting tournaments, damsels in distress, the courtly love tradition. The adjective *majestic* and the adverb *romantically* reinforce such images so that readers write a particular kind of narrative as they read, not one normally associated with science in general or with dung beetles in particular. Angier's metaphors, almost a requirement for narration in science because of the implied story inherent in any metaphor, create a singular (that is, unique) voice, even though she does not use the first person singular. They also create the humorous incongruity of dung beetles as knights in shining armour doing battle against the forces of evil: They 'assiduously clear away millions of tons of droppings, the great bulk of it from messy mammals like cows, horses, elephants, monkeys, and humans'. Yes, Angier assures us readers, dung beetles do us a favour, work nobly for our good, and, she implies, it's about time we give credit where it is due. (Angier sneaks in the revelation that we are among the messiest of mammals—and by listing humans last she implies that we are, actually, the messiest. A climax comes at the end of the sentence; we know and are not surprised by cows and elephants being messy, but humans? She nails us by surprising us.)

In other words, what a story this is.

Every sentence, every paragraph in Angier's (1995) succinct essay contributes to the story, at the same time that each is a story in miniature, making the whole far greater than the sum of its parts. In terms of narrative, what do we find?

- A protagonist—the quintessential youngest son, the lowly dung beetle
- A problem—too much waste material on our overcrowded planet
- Dramatic tension—thousands and thousands of dung beetles descending on a large pat of dung and eating it sometimes within minutes; or the story within the story about how Australia in the 1960s began to solve its dung problem by introducing several varieties of beetles
- Romance—'singles bars', 'courtship dances', dung beetles mating and forming families
- Artistry—those dung beetles that shape their food into beautiful geometric forms
- Exotic locations—savannahs of Africa, the deserts of India, the Himalayan meadows, the Panamanian jungles, the redwood forests
- Time—early bird beetles to late night revelers
- Costumes—some beetles don disguises, for instance, looking like sticks; and
- Competition and crime—fighting for the first bite.

About the only narrative element Angier lacks is dialogue, for though she unabashedly and unapologetically loves to anthropomorphize, she stops short of writing explicit beetle conversations the way a writer of beast fables might (where beasts have human language and even human characteristics). Nevertheless, in what she does narrate we can write the dialogue she only implies, as in the following: 'Robber beetles sneak in and try to steal balls painstakingly shaped by others. Joining the fray are many species of dung-eating flies. The scene resembles a fast food outlet at lunchtime, with all the patrons grabbing something to bring back to their desk' (1995: 112). Because she uses an analogy from readers' everyday experience, it is easy to imagine the dialogue:

> 'Your money or your life.'
> 'Stick 'em up.'
> 'I'll take a Big Mac and a large fries.'
> 'Hey you—whaddya mean, butting into line like that?'
> 'Jeez, it's late. I've gotta get back to work or the boss'll *kill* me.'

We can do this because we know what happens at a fast-food franchise during a frantic American worker's lunch hour. And we now also have a pretty good idea of the feeding frenzy that hits dung beetles when yet another cow drops a pat.

Angier writes economically, tells the story vividly, and helps us participate enthusiastically. Will we agree with her that beetles are 'key organisms in the environment?' (1995: 113). Absolutely, thanks to her skill as a storyteller. Will we think of beetles the same way after reading her story? Never.

Gould

When we think of scientists who write today, Gould is the first name to come to mind. Yet he stands in a contradictory place when it comes to science, to writing, and to scientific discourse, and for this very reason he is worth studying. He brings a great deal to the table. Like Angier, Gould takes writing science for a general audience as a high calling, almost a priestly vocation. He deplores the equation of popular writing with pap and distortion: 'Such a designation imposes a crushing professional burden on scientists (particularly young scientists without tenure) who might like to try their hand at this expansive style' (1992: 11).

Yet Gould also dislikes journalism that gives 'instant fact and no analysis . . . for the dumbdowners tell us that average Americans can't assimilate anything more complex or pay attention to anything longer' (1992: 91; Angier, who never provides facts without analysis, does not fall within Gould's critical scope). Nevertheless Gould boldly urges young scientists to imitate him, and practice and perfect good science writing for average Americans: 'We must all pledge ourselves to recovering accessible science as an honorable intellectual tradition' (12).

Whom does Gould mean to include with the first person plural? Scientists? Science journalists? Professors of science? Professors of scientific and technical writing? He doesn't say, other than that 'several of us are pursuing this style of writing in America today. And we enjoy success if we do it well' (1992: 12). I would, therefore, include all the writers mentioned above, and many, many more; and I would make the case that Gould wants teachers to instruct all students of science in the same 'style of writing' he and his cohorts are practicing (which does not imply that it should be the only style students learn). It's about time, he implies, to put this 'honorable intellectual tradition' back in the curriculum for the benefit of students, professors, and the public.

Although Gould does not define his style of writing as narrative, nevertheless he hints at two narrative strands in the history of science writing, which he labels the Franciscan and the Galilean. Angier (1995) lies in the Fransciscan tradition, the poetic, exultant celebration of nature's glory and quirkiness—as her comparisons, analogies, similes, metaphors, and whimsy in 'Scarab, Peerless Recycler' demonstrate. Although Gould loves a good Franciscan narrative, he refuses to tread that trail because he claims he would fall on his face. Instead, as he documents the human effort to understand nature, he follows the Galilean tradition. Using words like history, Gould explains that he loves 'the puzzles and intellectual delights' nature provides (1992: 13). It is not stretching narrative to the breaking point to say that a 'puzzle' is just another name for a 'mystery' or that an 'intellectual delight' implies a search for the telling incongruity.

If we readers don't find a fully expressed theory of scientific narrative in the opening pages of *Bully for Brontosaurus* (1992), by studying Gould's discourse we can infer one, for instance in the third chapter, 'The Creation Myths of Cooperstown'. This essay is remarkable on two grounds: It provides another example of narrative, and its subject is, ultimately, narrative itself—the narratives that compete for our attention. Thus, he implicitly addresses why scientists should write in this way.

Gould juxtaposes the scientific hoax of the Cardiff Giant (which he compares to Piltdown Man) and its story of human origins to the Abner Doubleday hoax about the founding of baseball, which is immortalized in Cooperstown (itself the repository of numerous stories, town as book, so to speak) and to the story of evolution, the only story that Gould credits with telling the scientific truth. It doesn't give the plot away to quote from the end of Gould's intricate tale:

> And why do we prefer creation myths to evolutionary stories? . . . Yes, heroes and shrines are all very well, but is there not grandeur in the sweep of continuity? Shall we revel in a story . . . that may include the sacred ball courts of the Aztecs . . . Or shall we halt beside the mythical Abner Doubleday . . . thereby violating truth and, perhaps even worse, extinguishing both thought and wonder? (1992: 58)

Because Gould understands the pull of stories, as he admits here, he wants evolutionary stories to hold their own with other stories, which he calls myths. And how can they hold their own in the popular imagination—to return to Gould's first chapter—without writers who know the story and how to tell it well, Gould asks? I would add, where will these informed, skillful writers come from, unless students learn how to write narrative?

Because opening lines tell us so much, as I pointed out above, consider the opening lines of 'Cooperstown'; 'You may either look upon the bright side and say that hope springs eternal, or, taking the cynic's part, you may mark P.T. Barnum as an astute psychologist for his proclamation that suckers are born every minute' (1992: 42). Gould's second person singular establishes a narrative voice and, indeed, a narrator, who says, in effect, Let me tell you about . . ., or, Do you recollect the time that . . .? Any time a writer directly addresses the audience, readers find a story about to commence, if not already in progress, which is the impact of Gould's opening words. We readers feel as if we have arrived in the middle of a story, and though we don't know what the narrator has already said, we sense that it won't take us long to figure out what's going on. To help us, Gould uses many of the same narrative elements as Angier:

- Mystery and dramatic tension—how the Cardiff Giant succeeded in fooling people
- A problem—why origins of anything fascinate people and why we like to believe in 'myths'
- A villain—George Hun
- A hero—Gould himself?
- Romance—our love of baseball and Cooperstown, and
- Stories within stories within stories—from Abner Doubleday to A.G. Spaulding to Henry Chadwick to the Civil War to the Victoria and Albert Museum in London and then back to Lower Manhattan and Alexander Joy Cartwright.

Unlike Angier, Gould does use the first person. He wants readers to come with him as he stalks his prey, as he unravels the ultimate story, as he tries to discover why people love myths. And here occur some hints at a theory: 'Creation myths . . . identify heroes and sacred places, while evolutionary stories provide no palpable, particular object as a symbol for reverence, worship, or patriotism' (1992: 57).

Can scientists compete with mythic tales? Are evolutionary stories deficient as stories because they lack heroes whom leaders can turn into symbols, objects to revere? (Stories, as I noted above, do have heroes and villains.) Does Gould's claim undercut the very argument I have been making? Or is Gould in this passage being disingenuous or subtle or modest?

Let me answer my questions by calling Gould himself as witness in my defense and against himself:

- 'We care deeply about Darwin's encounter. . . . The details do not merely embellish an abstract tale moving in an inexorable way. The details are the story itself' (1992: 29).
- 'We are bombarded with too much. . . . If we cannot sort the trivial from the profound, we are lost in terminal overload. The criteria for sorting must involve context [the story] and theory—the larger perspective that a good education provides' (91).
- 'The full story . . . contains lessons' (212).

And most tellingly and directly in the final essay of the collection, 'The Horn of the Triton', Gould rejects his earlier declaration about science lacking heroes and sacred places:

- 'This essential tension . . . has been well appreciated by historians, but remains foreign to the thoughts and procedures of most scientists. We often define science (far too narrowly, I shall argue) as the study of nature's laws and their consequences' (500).
- 'We are truly historians by practice, and we demonstrate the futility of disciplinary barriers between science and the humanities' (501).
- 'They [the planets] are objects in the domain of a grand enterprise—natural history—that unites both styles of science in its ancient and still felicitous name' (508).

Science and story, heroes and heroines, good guys and bad guys, events and the places where they occur—on earth, inside the earth, above the earth—how can Gould assert on the one hand that scientists have no story to tell, that scientists shouldn't surrender 'science to the domain of narrative' (506), and then write so persuasively that science is story?

Gould narrates the story that is involved in the scientific fact, demonstrating theory and revealing assumptions by writing about particular people in particular places doing particular things at particular times. Scientific discovery, as I said at the outset, does not come fully formed from on high. Often the best way to understand a fact or theory is to know how it came to be known—in other words, what people did.

Gould also writes with one of the strongest narrative voices in the science-writing business. For instance, he directly engages readers through his use of parenthetical asides, which are ubiquitous (I cite one example, above), and by telling us directly of his passions, his curiosities, his questions: 'I confess that I have always viewed . . .' (1992: 331); 'I read Stewart's letter and sat bolt upright with attention and smiles' (389); 'I find something enormously ironical in this old battle' (412); 'I need hardly remind everybody' (181); 'My odd juxtapositions sometimes cause consternation; some readers might view this particular comparison as outright sacrilege' (122).

If we are to take Gould at his word that he practices the kind of science writing that he wants others to practice as well, and if we find in his discourse every element that constitutes narrative, as we do, then we can only conclude that, despite the few paradoxical passages that appear to reject the very style of discourse he is writing and recommending, he nevertheless believes that successful science writing means narrative. We don't need less narrative, we need more. We don't need fewer narrators of science, we need more of them.

To return to the point I made at the beginning, my focus is on scientific discourse for a general, educated audience. However, as I also indicated at the beginning, I think it possible for scientists writing in disciplinary journals to use some of the narrative elements I identify. A case in point is one of this century's seminal scholarly papers in biology, which announced the discovery by Watson and Crick of the double helix DNA.

Not only do Watson and Crick use the first person, but they use it throughout their two reports. They also use the active voice and strong verbs. For instance, here is their opening paragraph: 'We wish to suggest a structure for the salt of deoxyribose nucleic acid (DNA). This structure has novel features which are of considerable biological interest' (1980: 237). Later they write, 'We wish to put forward a radically different structure . . . we have made the usual chemical assumptions' (238). And near the end of their short, tight paper, we find, 'It has not escaped our notice that . . .' (240).

Although I cannot claim that Watson and Crick completely fit the stylistic model demonstrated here, their reports do reflect a sense of mystery and intrigue intended to pique readers' interest and curiosity; they do use some narrative elements: 'In our opinion, this structure is unsatisfactory for two reasons' (1980: 237), which leads them into their 'radically different structure'. They are saying, in effect, that many people have been working to unravel a mystery, some good attempts have been made, but they have fallen short, and we now know why.

Here is another example from Watson and Crick's second paper, which also incorporates a sense of mystery and uses the first person: 'Despite these uncertainties we feel that our proposed structure . . .' (1980: 246). Therefore, there may well be a place for narrative, even in articles for disciplinary journals. In fact, last year the editors of *Science* announced that they wanted to make their articles more readable, more jargon-free, less arcane—and, their examples indicated, more narrative-like.

Narrative, then, isn't simply a way to dress up, entertain, or sweeten science for nonscientists—not the Mary Poppins approach to broad, ongoing education. Rather, in Gould's words, 'The details [and he means the narrative details] are the story itself': people, places, conversation, serendipitous mistakes (who did what and why). We professors need to teach our science students to imitate Gould, Angier, McPhee, Dennett, Johnson—and Darwin, Einstein, Poincaré, and others. For as Berthoff, a teacher and researcher of composition theory has put it, 'how we construe is how we construct' (1983: 166). How we conceive of a discipline—our epistemology—is defined or revealed by the discourse we write.

For Further Reading

The following list, by no means exhaustive, is intended as a supplement to the works cited in the text. With one exception, I have not included other titles by authors named there; certainly any reading list should contain additional books by Gould, McPhee, and Sacks. In putting the list together, I wanted breadth in terms of scientific discipline and writing style.

W.J. Broad, *The Universe Below: Discovering the Secrets of the Deep Sea*. New York: Touchstone, 1998.

G. Grice, *The Red Hourglass: Lives of the Predators*. New York: Delacorte, 1998.

S. Hubbell, *Broadsides From the Other Orders: A Book of Bugs*. New York: Random House, 1993.

B. Lopez, *Arctic Dreams: Imagination and Desire in a Northern Landscape*. New York: Scribner's, 1983.

S.N. Nuland, *The Wisdom of the Body*. New York: Knopf, 1997.
D. Quammen, *The Song of the Dodo: Island Biography in an Age of Extinctions*. New York: Touchstone, 1997.
D.R. Wallace, *Idle Weeds: The Life of an Ohio Sandstone Ridge*. Columbus, OH: Ohio State University Press, 1980.

References

Angier, N. 1995. *The Beauty of the Beastly: New Views on the Nature of Life*. Boston: Houghton Mifflin.

Bazerman, C. 1988. *Shaping Written Knowledge: The Genre and Activity of the Experimental Article in Science*. Madison, WI: University of Wisconsin Press.

Bazennan, C., and J. Paradis, eds. 1991. *Textual Dynamics of the Professions: Historical and Contemporary Studies of Writing in Professional Communities*. Madison, WI: University of Wisconsin Press.

Berthoff, A.E. 1983. 'How We Construe Is How We Construct', in P.L. Stock, ed., *fforum: Essays on Theory and Practice in the Teaching of Writing*, pp. 166–70. Upper Montclair, NJ: Boynton/Cook.

Dennett, D.C. 1991. *Consciousness Explained*. Boston: Little, Brown.

———. 1995. *Darwin's Dangerous Idea: Evolution and the Meaning of Life*. New York: Simon & Schuster.

Gould, S.J. 1992. *Bully for Brontosaurus: Reflections in Natural History*. New York: W.W. Norton.

———. 1995. *Dinosaur in a Haystack: Reflections in Natural History*. New York: Harmony Books.

Gross, A.G. 1990. *The Rhetoric of Science*. Cambridge, MA: Harvard University Press.

Harre, R. 1990. 'Some Narrative Conventions of Scientific Discourse', in C. Nash, ed., *Narrative in Culture: The Uses of Storytelling in the Sciences, Philosophy, and Literature*, pp. 81–101. London: Routledge.

Johnson, G. 1995. *Fire in the Mind: Science, Faith, and the Search for Order*. New York: Vintage.

McPhee, J. 1983. *In Suspect Terrain*. New York: Farrar, Straus and Giroux.

McPhee, J. 1983. *Assembling California*. New York: Farrar, Straus and Giroux.

Myers, G. 1990. 'Making a Discovery: Narratives of Split Genes', in C. Nash, ed., *Narrative in Culture: The Uses of Storytelling in the Sciences, Philosophy, and Literature*, pp. 102–26. London Routledge.

Nuland, S.B. 1994. *How We Die: Reflecting on Life's Final Chapter*. New York: Vintage.

Sacks, O. 1995. *An Anthropologist on Mars: Seven Paradoxical Tales*. New York: Vintage.

Selzer, J., ed. 1993. *Understanding Scientific Prose*. Madison, WI: University of Wisconsin Press.

Simons, H.W., ed. 1990. *The Rhetorical Turn: Invention and Persuasion in the Conduct of Inquiry*. Chicago: University of Chicago Press.

Watson, J.D. 1980. *The Double Helix: A Personal Account of the Discovery of the structure of DNA*, G.S. Stent, ed. New York: W.W. Norton.

QUESTIONS FOR CRITICAL THOUGHT

1. Who are Darwin, Poincaré, Primo Levi, and Stephen Jay Gould? Why does Forbes mention them? Why doesn't she tell us who they are?
2. Forbes argues that it is narrative that makes good scientific writing. What, exactly, does she mean by narrative?
3. To what extent can Forbes's article be considered to provide a solution to the problem described by Stephen Strauss ('Avoid the Technical Talk, Scientists Told') and J.S.C. McKee ('Communicating Science')?

4. Forbes suggests that scientific style typically privileges such features as 'objectivity, authorial distance, sobriety, and factuality', among others. In your experience, is she correct? To what extent do the readings in this book conform to these expectations?
5. What is the purpose of Forbes's article? Who are her intended readers? What does she want them to do?
6. Why might the style of scientific journals be different from the prose of popular magazines such as those named by Forbes (*Harper's* or the *New Yorker*)?
7. Much of Forbes's argument flows from her assertion that scientific prose is a 'central discourse in our culture'. What, according to her reasoning, are the implications of this fact?
8. According to Forbes, the beginning of any work sets the tone, establishes the writer's style, and prepares the reader for what is to come. Does Forbes's own work conform to this expectation? Pick any other essay in this book; does the work of your choice conform to this expectation?
9. Forbes argues that, to uncover the narrative features of good scientific writing, a critic must focus, among other things, on pronoun usage (which determines plot and character), the use of active or passive voice (which heightens or deadens the drama), and adverbials (which establish a setting in place or time). Try this kind of analysis with Forbes's own article; what did you discover? Then turn to any other selection in the book and do the same thing. How useful is this method for understanding how the readings work?
10. Is Forbes's audience made up of scientists, science writers, teachers of technical writing, or student writers like yourselves? How do you know? Point to some specific features of the text that support your opinion.
11. Forbes asserts that 'good science writing depends on the conventions of narrative.' What, exactly, are these conventions as she conceives of them? To what extent are they the features of all good writing, in science or otherwise? To what extent are these features typical of the selections in this book?
12. How formal or informal, personal or impersonal is Forbes's writing? Why do you think she has made the rhetorical choices she has?
13. What is Forbes's attitude to her readers? How do you know?

CHAPTER 9

Advancing Science Communication: A Survey of Science Communicators

Debbie Treise and Michael F. Weigold

The writings of science communication scholars suggest two dominant themes about science communication: it is important and it is not done well (Ziman, 1992; Nelkin, 1995; Hartz and Chappell, 1997). This article explores the opinions of science communication practitioners with respect to the second of these themes, specifically, why science communication is often done poorly and how it can be improved. The opinions of these practitioners are important because science communicators serve as a crucial link between the activities of scientists and the public that supports such activities. To introduce our study, we first review opinions as to why science communication is important. We then examine the literature dealing with how well science communication is practiced.

The Important of Science Communication

Science communication is typically thought of as the activities of professional communicators (journalists, public information officers, scientists themselves). Effective science reporting is perhaps the only mechanism for most people to learn about fast-breaking events and exciting developments that affect everyone. Beyond informing people about what is happening in science, science communication also places scientific activity within a broader context. Thus, it can provide the public with information essential to forming opinions about public policy and about the costs and benefits of governmental expenditures on science. Hartz and Chappell (1997) suggested that 'the populous needs as much information as possible to act wisely and intelligently—whether it is about high technology or garbage collection' (117). This idea, that science knowledge permits the public to make effective decisions about science policy, is a common theme in the science communication literature. An educated public should be better equipped to choose from among competing technical arguments on topics such as energy conservation, solid waste disposal, pesticide risk, and social welfare policy.

Nelkin (1995) offered an additional claim for the benefits of science knowledge. Not only can effective reporting 'enhance the public's ability to evaluate science policy issues;' but it also can aid 'the individual's ability to make rational personal choices' (2). Thus, the communication of scientific findings may aid people in making better decisions about their own lives, health, and happiness. An educated public may resist associating responsible science with irresponsible science and may better discriminate the activities of scientists from those of 'pseudo' scientists (psychics, astrologers, etc.) (Shortland and Gregory, 1991).

For scientists and for others who work in scientific organizations, effective communication can be conceived of as returning a debt created by public support. It can create favourable attitudes toward science and science funding among policy makers and the broader public by making clear the benefits that scientific activity offers to society. Daniel S. Greenberg, founder and former editor and publisher of *Science & Government Report*, suggests that a public ignorant of science might resist efforts at science funding. This, in turn, could create an economic crisis since American prosperity is sustained in large part by the great scientific and technological achievements of the past (cited in Hartz and Chappell, 1997). Communicating about science also allows scientists to share insights from their work about the nature of the world and those who occupy it. Greater understanding of the world may be important for its own sake, and it can help to generate excitement among young people who might otherwise not consider scientific careers.

Historically, the free press has been the most important way that nonscientists learn about new advances in science. For most people, after completion of their formal education, exposure to science tends to occur through chance encounters with news reporting on science. As Nelkin (1995) suggested:

> For most people, the reality of science is what they read in the press. They understand science less through direct experience or past education than through the filter of journalistic language and imagery. The media are their only contact with what is going on in rapidly changing scientific and technical fields, as well as a major source of information about the implications of these changes for their lives. (2)

In summary, many scholars argue that communicating science information to the public is important. In addition, the primary responsibility for science communication appears to fall to the news media. Thus, it is possible to ask, How well is science communicated to the public? The consistent answer to this question seems to be, not very well.

The Effectiveness of Science Communication

At one time, the news media were seen as having the potential to create a country of science-literate citizens. Seventy years ago scientists believed science journalism would advance an awareness of science that would elevate public understanding plus foster appreciation, literacy, and tax-supported dollars for research (Tobey, 1971). However, many contemporary scholars believe that science is not communicated effectively to the general public. One piece of evidence for this is that much of the public appears to be scientifically 'illiterate' (Hartz and Chappell, 1997; Paisley, 1998; Maienschein and Students, 1999). Those involved in the science communication process also appear to find fault with science communication. For example, scientists are frequently disappointed or angry about media coverage of their research, their fields, or science generally (Shortland and Gregory, 1991; Hartz and Chappell, 1997). Journalists report frustration with the difficulties of describing and understanding important scientific findings (Hartz and Chappell, 1997) and with the low levels of support provided by their news organizations for reporting on science news.

There is no shortage of reasons given for the presumed poor quality of science communication. Some problems may originate with reporters and their methods for covering news (Trumbo, Dunwoody, and Griffin, 1998). For example, research suggests that those who cover science frequently lack any but the most cursory backgrounds in the sciences and mathematics. 'Most journalism graduates will be exposed to science journalism issues, if at all, in passing during basic and advanced reporting courses' (Palen, 1994: 607). In a Canadian study, fewer than one in three editors had taken a single science course in college (Dubas and Martel, 1975). Ismach and Dennis (1978) found that more than 60 per cent of newspaper and television reporters major in journalism, while only 21 per cent major in the social sciences, and a mere 3 per cent major in science or engineering. This lack of expertise may contribute to widespread error in reporting on science (Ankney, Heilman, and Kolff, 1996).

News-gathering norms also may hinder effective science communication. Valenti (1999) argued that scientists and journalists are governed by different norms with respect to valuing knowledge that is quantitative, technical, certain, theoretical, cumulative, complete, and restricted. Friedman (1986) suggested that the values of journalism create a short-term focus for reporters. A reporter who breaks a story is given far more credit than one who follows up with detail. In addition, news values require that issues that have received coverage in the past be approached from 'a new angle'. Editors and reporters tend to value stories that contain drama, human interest, relevance, or application to the reader (Friedman, 1986), criteria that do not always map easily onto scientific importance.

Editorial pressures may create problems for reporting on science. Dubas and Martel (1975) noted that city editors prefer stories with a sensational angle or an element of conflict. Not surprisingly then, many science writers are unhappy with the priorities of their editors (Dennis and McCartney, 1979), believing they like to scare readers, ignore continuing stories, and waste space and air time on junk. Editors often write story headlines (Friedman, 1986) and control story revisions, with the consequence that science reporters sometimes write for editors rather than for the public. Editors also have different news priorities than lay readers (Dubas and Martel, 1975) or reporters (Dunwoody, 1986).

Problems in science communication also have been attributed to scientists. Neal Lane, former director of the National Science Foundation, claimed that:

> With the exception of a few people . . . we don't know how to communicate with the public. We don't understand our audience well enough—we have not taken the time to put ourselves in the shoes of a neighbor, the brother-in-law, the person who handles our investments—to understand why it's difficult for them to hear us speak. We don't know the language and we haven't practiced it enough. (cited in Hartz and Chappell, 1997: 38)

The failure of scientists to communicate with the public is not merely a matter of inadequate skill. Journalists have accused scientists of being uncooperative or unwilling to describe their work in easily understood ways. However, the scientist wishing to communicate directly with the public faces several important hurdles. Perhaps the most basic of these is language. According to one source, as recently as 1920, the language used in a journal such as *Nature* would be comprehensible to literate audiences and would not sound dramatically different from other forms of literature. However, 'scientific language has diverged from the mainstream of literary language and divided into a large number of small, winding tributaries' (Shortland and Gregory, 1991: 12). In addition, scientists who wish to communicate directly to the public are rarely rewarded and often face organizational and professional impediments (Gascoigne and Metcalfe, 1997).

Paradoxically, scientists are sometimes criticized for saying too much, when, in the eyes of their critics, they become spokespersons on issues that fall outside their areas of professional competence. For example, Carl Sagan voiced views on issues as diverse as abortion and nuclear winter, which are areas that some suggested went well beyond his expertise.

A final impediment to science communication may be the public itself. Media consumers differ with respect to their knowledge about, and interest in, science and technology news (Krieghbaum, 1967). Despite the assumptions of scientists and science communicators about the importance of science and technology information, many adults show little interest in science (Miller, 1986). Prewitt (1982) and Miller (1986) have both argued that only a fraction of adults are concerned enough about science to attend to, and process, science communication. Patterson (1982) noted that young adults generally support science but also fall into attitudinal types ('true believers', 'anxious dissenters', and 'seasoned supporters'). O'Keefe (1970) found that doctors generally consider medical information from mass media to be reliable but judge it to be of little personal importance.

There is even disagreement concerning what public understanding of science means (Lewenstein, 1992; Logan, Zengjun, and Wilson, 1998). For example, does it refer to the knowledge of nonscientists about developments of science? Does it refer to science literacy (i.e., basic understanding of accepted scientific facts and theories)? Is it an appreciation of the methods of science? Does it refer to familiarity with new technologies or to sophistication about the implications of scientific findings? Is it perhaps all of these things, or some combination? There is no clear consensus that emerges from the literature on this issue.

To summarize, a dominant theme that runs through current scholarship on science communication is that while the communication of science is important, it is not being done very well. Surprisingly, almost no recent large-scale studies exist that explore this problem from the perspective of the science communication practitioners. According to one estimate, there are 5,000 science writers, defined as people who 'spend all or most of their time writing about subjects in the bodies of highly organized knowledge known loosely as the sciences' (Burkett, 1986: 3). A broad range of topics can be seen as relevant to science writing, including the physical sciences, the social sciences, applied fields such as engineering and medicine, and areas of technological development.

Science reporters tend to have somewhat different values than regular reporters. A 1973 survey found that science reporters tend to favour alternative formats to hard news, feeling that these alternative formats allow more effective communication about science issues (Friedman, 1986). A 1978 survey of science writers found the most common formats of science journalism were science-writing features (38 per cent) and spot or breaking news (32 per cent). By comparison, the hard news format is still the most prevalent in journalism. This poses challenges for science writers because science stories often lack a peg, and scientific discoveries do not happen overnight (Friedman, 1986).

The purpose of this exploratory study is to examine the beliefs of a specific constituency in the communication of science: science writers, editors, and science communication researchers. This important group of practitioners and scholars are the gatekeepers and message crafters of news about science and technology. We approached this group to understand their perspectives on current problems with science communication and to determine the needed changes that could advance science communication in the near future.

Method

To answer our overall research question, an open-ended survey was mailed to a diverse group: a random sample of 800 members of the National Association of Science Writers (NASW), a purposive sample of 30 editors of news sections and science publications (who were not members of NASW) listed in Bacon's Guide, and 20 mass communication scholars conducting research in science communication. The questionnaire asked, What are the most important questions, opportunities, or unresolved issues in science communication that, when addressed, would advance the current state of the field?

A more qualitative approach was chosen because the purpose of the study was to identify and understand the issues from the perspective of those who are closest to the scene. In other words, the emic view was sought from those who are in the business of science communication (Lindlof, 1995). As with other qualitative research, our interest is in 'their logic and the kinds of evidence they considered worthwhile and relevant. We suspend our own notions of what is right' (Lindlof, 1995: 57). We were seeking to define the boundaries of the phenomenon in question, namely, the parameters of the concerns regarding science communication. Finally, our decision to use qualitative research was consistent with the suggestion of Press (1991) that the need for qualitative research often arises from 'considerations of important issues framed as "problems": circumstances that involve the unsatisfactory state of affairs for some person or group' (423).

In sum, qualitative research was chosen because the research question is exploratory and 'stresses the importance of subjects' frame of reference' (Marshall and Rossman, 1989: 46). In addition, an open-ended survey was selected over in-depth interviews as a data collection method after a pretest suggested that respondents wanted time to reflect. Also, given the number of desired responses, in-depth interviews were deemed unworkable.

Of the 850 questionnaires that were mailed out, 497 were returned for a response rate of approximately 58 per cent. (Specifics on number of respondents in each category are not provided as respondents had the option of identifying their field.)

The responses were read and coded by three independent coders trained in qualitative content analysis. Because in many cases respondents provided more than one idea, a total of 1,249 issues were identified. These responses were collapsed into five independent categories, concerns, or opportunities: science literacy (n = 290), the process of science learning (n = 283), ethics (n = 269), technology (n = 261), and scientist and journalist training (n = 146).

Results

SCIENCE LITERACY

Based on the sheer number of concerns raised, our participants believed that science literacy was the most pressing issue for science communicators and researchers to address. Many participants felt that the current method of assessing science literacy, while providing a benchmark, was not comprehensive enough. For example, many respondents suggested that extensive knowledge of our solar system or the rotation of the Earth does not measure science literacy accurately. One researcher raised the following question:

> Perhaps I may not be familiar with Earth systems science but I may be very knowledgeable about physiology or anatomy. Would I not then be science literate? Earth systems science may not be my area of specialty, but that does not exclude me from being science literate.

In a related vein, participants raised the issue discussed in an earlier section about what people should learn to be able to read a publication such as *Science News*. Should people have the basic levels of science knowledge to be able to understand each and every article published in *Science News*? Other participants asked whether people need to know such diverse information or to what extent the 'general population' wants to spend the time accumulating such information.

Other respondents felt that more accurate measures of science literacy would need to assess, for example, the extent to which words such as HTML (Hypertext Markup Language) and genetic altering are now a part of the vocabulary of science literates. In sum, our participants were less troubled by the traditional concerns of whether Americans are becoming less science literate and were more concerned with the accuracy of current science literacy measures.

THE PROCESS OF SCIENCE LEARNING

Closely tied to the issue of science literacy is the question of just how science learning takes place. Many of our participants felt that after years of research, science communicators still do not understand how, or if, readers use mass-mediated science information to form understanding or to acquire in-depth knowledge. For example, one science writer said, 'I don't believe we understand how readers make sense of news about science. I don't think we know what they get from science news stories.'

Another writer asked more fundamental questions:

> Do we know what is science news to readers? Do we know how they pick which stories to read first? Not to read at all? In other words, do we know what their expectations are about science news? I don't think we know; otherwise, we'd frame many of our stories to meet those expectations.

Similarly, another respondent asked about the process of reading about science: 'As a writer, it would help me to know if readers' use of science stories varies by topic.'

Other participants felt it is important to know how actual science story mechanics and writing affect the process of science learning. For instance, one researcher said:

> It is vital for us to know if science stories are being written in such a way as to facilitate learning. Are we helping to educate people to evaluate evidence or are we hindering them? Do the *details* provided in stories make the information more understandable or more confusing?

Similarly, another researcher said, 'It is imperative that we know how strongly the availability of pictures affects comprehension, enjoyment and inclination to read.' In other words, our participants were concerned that science communicators themselves are unsure of the science-learning process and effects of their efforts on readers.

DISASTERS, HYPE, ETHICS

Many of our science writers felt that the current formula for successfully communicating science involves hype and publicity that carries with it inherent ethical issues. For example, from the perspective of producing science information, many scientists who rely on government or foundation funding are faced with increasingly shrinking research dollars. This situation has caused a serious competition among scientists and has led to burgeoning efforts to 'self promote and shape the presentation of results'. Likewise, another practitioner felt end results might suffer. That person said, 'We need to determine the effects of sensationalism and hype on science news and reporting before we collectively shoot ourselves in the feet. Plus, this brings to bear serious ethical breaches.'

This concern was reflected in comments about the framing of potential disasters as well. Many of our respondents cited the recent imminent 'meteor crashes to Earth killing hundreds of thousands'. One editor said it was the result of a 'media machine hell-bent on entertaining with hype and sensationalism' and 'using legitimate science and legitimate institutions [in this case, NASA] to scare the . . . out of readers'.

As a result, many of our respondents across all disciplines mentioned that we need to conduct research to assess the effect of premature release of this type of information. In addition, one writer suggested that research would illuminate 'how the issues are defined, how they are reported, and how journalists use science sources to frame the risks'.

TECHNOLOGY

While technologies such as the Internet have provided a boon to science and its products, many of our participants expressed concerns about how it will continue to change the landscape of science communication. At a basic level, journalists, editors, and researchers alike see the web as a viable outlet to disseminate science news; however, all anticipate serious problems. To date, most journalists have relied on peer-reviewed articles to judge the authenticity, validity, and reliability of potential science news. However, much of the research now appearing on independent science websites has not reached this crucial stage. The question then becomes, as one editor asked, 'How do we filter, judge, and sort science information? How will reporters recognize "key" versus "minor" science stories?'

Other participants were concerned with how the public, who already may be ambivalent—or even critical of past conflicting coverage of heath issues, for example—will react to the speed and proliferation of science news on the web. As one writer suggested, 'This will affect the whole process of science communication. We need to know now how, or if, this amount of information is affecting public cynicism and confusion.'

SCIENTIST AND JOURNALIST TRAINING

The topic of the 'journalist versus the scientist' and its many ramifications has been researched extensively. The fact that so many of our participants continued to voice concerns about this issue suggests that the academic research has not been properly or extensively disseminated. For our participants, however, the issue has been distilled to one of lack of scientist and journalist training. For example, our participants feel that it is important for research to assess how many science writers have received basic statistical training and how that affects their reporting. One researcher said, 'We need to know if science writers are trained to point out the relative risks and limitations of the studies they are reporting about. For example, have these writers been trained to judge the veracity of health risk assertions?' Likewise, another editor suggested that we need to uncover whether 'science writers or their editors are trained to understand the nature of science and its uncertainties or the nature of scientific concepts in general'.

Another writer suggested that research should see if there is 'a correlation between how much untrained science writers rely on "predigested" press releases to obtain information because they aren't in a position to judge the credibility of raw information.' Many of our respondents were concerned about scientist training as well. One participant stated, 'We need to assess how much we're helping the scientist to improve the science communication process. Are we giving them support in the form of access to training opportunities or information about how we work?'

Still another respondent raised a more basic issue:

> I feel we need to define, once and for all, the responsibility of the journalist and the scientist—and not just for science news. Is it our responsibility to educate, to inform, or to promote critical thinking? Or is it none of the above?

Discussion

Our participants identified many key issues that stress the need for improving science communication. The largest concern was that of the public's science literacy. Various studies have documented that the public lacks basic knowledge of many key areas of science. However, our respondents went one step further in noting the absence of solid metrics for science literacy. The development of science literacy measures can help scientists and journalists alike more accurately gauge what people do and can understand about science.

Respondents also expressed concern about their own lack of knowledge as to how people learn about science. This concern has direct application to how stories can be crafted and what learning aids might best advance public understanding. This issue also dovetails with another concern, the training of science writers. Science writers are at the critical intersection of the practice of science and the public understanding of science. This calls for extraordinary skill, as the writer must be comfortable understanding scientific findings from diverse fields of science and also must be sophisticated in the translation of these findings into language and images accessible to ordinary people. Perhaps in response to this need, science journalism programs have been springing up at research universities across the United States. It is hoped that part of the curriculum in these programs will help writers deal with another concern mentioned among our respondents, the ethics of science reporting of 'hype' and 'catastrophe'.

In sum, the responses from our sample of science writers confirmed that there is considerable consensus between scholars of science communication and practitioners from the field of science reporting. Many of the issues mentioned by the writers correspond to issues noted in recent years by scholars such as Chappell, Dunwoody, Miller, Prewitt, Logan, and Lewenstein. Several strands of communication research during the next several years can play a vital role in advancing solutions to these problems. Scholars should better identify components of scientific literacy and determine the relationships between literacy, appreciation, and political support for science. New programs for training science journalists will be laboratories for determining the optimal skills of science communicators, but only if such programs institute comprehensive benchmarks and measures of these skills.

References

Ankney, R., P. Heilman, and J. Kolff. 1996. 'Newspaper Coverage of the Coronary Artery Bypass Grafting Report', *Science Communication* 18: 153–64.

Burkett, W. 1986. *News Reporting: Science, Medicine, and High Technology*. Ames: Iowa State University Press.

Dennis, E., and J. McCartney. 1979. 'Science Journalists on Metropolitan Dailies', *Journal of Environmental Education* 10: 10–11.

Dubas, O., and L. Martel. 1975. *Media Impact: A Research Study on Science Communication*. Ottawa: Ministry of Public Affairs.

Dunwoody, S. 1986. 'When Science Writers Cover the Social Sciences', in J.H. Goldstein, ed., *Reporting Science: The Case of Aggression*, pp. 67–81. Hillsdale, NJ: Lawrence Erlbaum.

Friedman, S. 1986. 'The Journalist's World', in S.M. Friedman, S. Dunwoody, and C.L. Rogers, eds, *Scientists and Journalists: Reporting Science as News*, pp. 17–41. New York: Free Press.

Gascoigne, T., and J. Metcalfe. 1997. 'Incentives and Impediments to Scientists Communicating through the Media', *Science Communication* 18: 265–82.

Hartz, J., and R. Chappell. 1997. *Worlds Apart: How the Distance between Science and Journalism Threatens America's Future*. Nashville, TN: First Amendment Center.

Ismach, A., and E. Dennis. 1978. 'A Profile of Newspaper and Television Reporters in a Metropolitan Setting', *Journalism Quarterly* 55: 741.

Krieghbaum, H. 1967. *Science and the Mass Media*. New York: New York University Press.

Lewenstein, B. 1992. 'Introduction', in B. Lewenstein, ed., *When Science Meets the Public*, pp. ix–xvi. Washington, DC: American Association for the Advancement of Science.

Lindlof, T. 1995. *Qualitative Communication Research Methods*. Thousand Oaks, CA: Sage.

Logan, R., P. Zengjun, and N. Wilson. 1998. 'Evaluating Assertions about Science Writing, Reporting, and News Selection: A Content Analysis of the *Washington Post* and *Los Angeles Times*'. Paper presented at the annual meeting of the Association for Education in Journalism and Mass Communication, August, Baltimore, MD.

Maienschein, J., and Students. 1999. 'Commentary; To the Future—Arguments for Scientific Literacy', *Science Communication* 21: 75–87.

Marshall, C., and G. Rossman. 1989. *Designing Qualitative Research*. Newbury Park, CA: Sage.

Miller, J. 1986. 'Reaching the Attentive and Interested Publics for Science', in S.M. Friedman, S.

Dunwoody, and C.L. Rogers, eds, *Scientists and Journalists: Reporting Science as News*, pp. 55–69. New York: Free Press.

Nelkin, D. 1995. *Selling Science: How the Press Covers Science and Technology*, rev. ed. New York: Freeman.

O'Keefe, T. 1970. 'The Mass Media as Sources of Medical Information for Doctors', *Journalism Quarterly* 47: 95–100.

Paisley, W. 1998. 'Scientific Literacy and the Competition for Public Attention and Understanding', *Science Communication* 20: 70–80.

Palen, J. 1994. 'A Map for Science Reporters: Science, Technology, and Society Studies Concepts in Basic Reporting and News Writing Textbooks', *Michigan Academician* 26: 507–19.

Patterson, J. 1982. 'A Q Study of Attitudes of Young Adults about Science News', *Journalism Quarterly* 59: 406–13.

Press, A. 1991. 'The Impact of Television on Modes of Reasoning', *Critical Studies in Mass Communication* 8: 421–41.

Prewitt, K. 1982. 'The Public and Science Policy', *Science, Technology, & Human Values* 36: 5–14.

Shortland, M., and J. Gregory. 1991. *Communicating Science: A Handbook*. New York: Longman.

Tobey, R. 1971. *The American Ideology of Natural Science*. Pittsburgh, PA: University of Pittsburgh Press.

Trumbo, C., S. Dunwoody, and R. Griffin. 1998. 'Journalists, Cognition, and the Presentation of an Epidemiological Study', *Science Communication* 19: 238–65.

Valenti, J. 1999. 'Commentary: How Well Do Scientists Communicate to Media?', *Science Communication* 21: 172–8.

Ziman. J. 1992. 'Not Knowing, Needing to Know, and Wanting to Know', in B. Lewenstein, ed., *When Science Meets the Public*, 13–20. Washington, DC: American Association for the Advancement of Science.

QUESTIONS FOR CRITICAL THOUGHT

1. According to Treise and Weigold, most experts in the subject agree that communicating scientific information is important, but that it is nevertheless not done well. Does your reading in this book so far confirm or refute this assertion?
2. What, according to the authors, makes science writing important? How do they support this assertion? Do you think they would agree with Cheryl Forbes ('Getting the Story, Telling the Story: The Science of Narrative, the Narrative of Science') that scientific prose is a 'central discourse in our culture'? Why or why not?
3. According to Treise and Weigold, most lay people get their information about science from science journalists rather than from scientists themselves. Why might this be? What problems does this fact present?
4. List at least three of the reasons, as outlined by Treise and Weigold, that science writing is largely ineffective.
5. This article is a research report. What elements of its structure make this fact evident? Compare it with some of the other research essays in the book; what similarities do you see? What differences? (Some comparables include Bitzer's 'Functional Communication', MacLennan's 'Disciplinarity, Identity, and the "Profession"

of Rhetoric', Halloran's 'Classical Rhetoric for the Engineering Student', or Talbott's 'The Deceiving Virtues of Technology'.)

6. Relative to other readings in the book, would you describe this article as more or less formal? More or less personal? What elements of its context might account for these differences?
7. In her article 'Getting the Story, Telling the Story', Cheryl Forbes argues that much scientific writing strives to appear 'devoid of voice or humanity'. Is this charge true of Weise and Treigold's article? Use evidence from the text to support your reply.
8. If good scientific writing tells a story, as Forbes maintains, what is the story told by Treise and Weigold's article?
9. How readable is this article? How engaging is it? Why or why not?

CHAPTER 10

Communicating Science in the 'Digital Age': Issues and Prospects for Public Engagement

Richard Holliman

Introduction

These are both exciting and challenging times to be communicating science. Developments in technology are providing opportunities for greater levels of interactivity with traditional mass media, such as digital television and online newspapers, providing new ways for media professionals and audiences to communicate. In this way the traditional dislocation between producers and receivers (see Thompson, 1999 for discussion) is being eroded. The introduction of affordable and accessible technologies that facilitate the production of user-generated content are also providing alternative spaces for citizen-consumers to communicate outside of mainstream media and editorial control, therefore potentially democratizing these aspects of citizen engagement. Thus, 'citizen media' (for a discussion of citizen media, see McAfee, 2005), is providing a new critical voice in public debates—particularly with respect to political news—in the form of a citizen-led 'fifth estate' (Gibson, 2005a). This democratizing influence has even led some to predict the demise of the 'media baron' (Gibson, 2006).

Meanwhile, moves to further deregulate the UK media marketplace have seen a proliferation in the number of digital television and radio channels, with audience share for existing analogue TV channels reducing dramatically in recent years, as (hard copy) newspaper sales also continue to fall. To counter these trends, traditional broadcast and print media have also explored new media platforms. For example, newspapers have introduced online editions, and broadcast media are developing digital 'on-demand' services, allowing viewers to personalize broadcast schedules, choosing what to watch, where, when, and on which device (Thompson, 2006). As a result, professional communicators are increasingly expected to work across traditional and new media platforms.

Taken together, the impact of technology and deregulation means that audiences for traditional broadcast media have fragmented as choice increases, and there are many more ways for citizen-consumers (including scientists and professional communicators) to communicate (read: produce and consume) scientific (and other) information in the digital age. The challenge for these actors is therefore:

- To develop the necessary skills to access and navigate through this plethora of sources and to identify credible relevant and useful information.
- When motivated to do so, to acquire the skills required to produce and share user-generated-content.

On a similar timescale to these developments in technology and deregulation, the legacy of previous high-profile science-based issues in the UK has also led to policy changes with respect to science and society. The bovine spongiform encephalopathy/variant Creutzfeldt-Jakob disease (BSE/vCJD) episode is an important case in point. The resulting context for science communication and public engagement is one that has seen significant changes in recent years. In this short essay I will briefly explore how technology, deregulation, and the legacy of recent high-profile science-based issues have converged to produce new opportunities for dialogue, engagement, and deliberation about science-based issues.

The Role of New Technologies

New media, such as those you can access through home computers, mobile phones, and handheld personal digital assistants (PDAs) are providing opportunities for a wide range of citizen-consumers to engage interactively with developments in science, the latter two allowing users to access information whilst on the move (e.g., using wireless networks). These media also provide opportunities for citizen-consumers to learn about a range of high profile (e.g., climate change) and emerging (e.g., nanotechnology) science-based issues that may have relevance to their everyday lives. This issue of relevance is of particular importance when it comes to biomedical issues. The emergence of the term 'expert patients' illustrates the demand for biomedical information; an expert patient is someone who chooses to access and learn specialist information about illness and disease (e.g., in relation to conditions that they, or relatives and friends, are suffering from). Anyone who has accessed biomedical information before visiting a doctor (e.g., in the UK through NHS Direct Online [2006]), which can be accessed through the Internet, a national phone network, or digital television—or bought therapeutic treatments over the Internet—is an example of someone who is motivated to engage with biomedical knowledge that is useful and relevant to their lives. This presents a potential challenge for health professionals and expert patients who need to (re)negotiate their relationship and reconcile possible disjunctures between professional and specialist knowledge.

It is not surprising that these new media are of particular interest to young people as they are more likely to become 'early adopters', those who enthusiastically embrace new technologies as they enter the marketplace. A recent survey commissioned by the *Guardian* newspaper and conducted by the pollster ICM illustrates this trend. It showed that, of those people between 14 and 21 who were online, over a third had produced their own website or weblog (or blog as it is more commonly known) (Gibson, 2005b).

A blog is a form of online interactive diary, where the user-generated content, often a combination of text and images, is (usually) listed in reverse chronological order. (More recently, video logs—or vlogs—have also emerged, where audiovisual materials are the primary content.) Produced and updated regularly by anyone who has access to a networked computer and the skills necessary to use blogging software, this form of communication allows, indeed encourages, feedback and comments. In this sense, blog postings move from being strictly linear—the audience receiving information, as is the case when you read a book—to being more interactive and dynamic. The communication therefore becomes a dialogue between *users*, as they exchange information that is relevant and useful to the activity/phenomena they are discussing; in effect, producing a form of collaborative authorship. Success, of course, requires commitment by the users (the producer of the blog and the readership). Indeed, most blogs fail either because the producer of the blog fails to provide sufficient interesting material and/or the readers stop posting sufficient, interesting comments. If successful, however, blogs have been known to cross over into more traditional media, such as books, or 'blooks' as they have become known. Indeed, some blogs have been created with this specific purpose in mind, for example, to illustrate the authors' interest in publishing a science magazine for women (Gosline, Law, and Casselman, 2005).

There are numerous examples of blogs currently online in what has become known as the 'blogosphere'—you may even have your own blog or one that you regularly read and send comments to—and they are produced for a range of purposes. With respect to science and science-based issues, a wide range of actors (e.g., scientific journals) have produced blogs as a way engaging users with newly published science (Nature Publishing Group, 2006). Indeed, recent evidence suggests that scientists and scientific institutions are well aware of the instrumental benefits of using new media platforms, such as blogs (Luft, 2006) and podcasts (Editorial, 2006), unconstrained as these artefacts are by limitations of space and broadcast schedules, respectively.

Alternatively, blogs can be used to enhance learning opportunities, extending the boundaries of the classroom or seminar (Day, et al., 2006), thus blurring the lines between informal and formal learning. Blogs are also produced to promote and discuss the interests of networked (scientific) citizen-consumers, who, in effect, create or access pre-existing online communities. The power of this form of communication is well illustrated by the example of Laurie Pycroft, a 16-year-old blogger who helped to form a pro-vivisection campaign called Pro-Test[1] through his blog (for an outline of this story, see Booth, 2006) to campaign for the building of a new laboratory in Oxford where animal experiments would be conducted.

It could be argued that science blogs produced by citizens have blurred the boundaries of expertise, beyond those who might usually be defined as a scientist (for a more detailed discussion of expertise, see Collins and Evans, 2002). It is just as likely, however, that these blogs have made this already blurred distinction more visible. Indeed, those citizen-consumers with expert or specialist knowledge, but no formal qualification or membership of a professional society—so-called 'pro-ams' (Leadbetter and Miller, 2004)—have been enthusing about their particular interests for many years, just not on the Internet. Now new technologies allow users to communicate more effectively through blogs, podcasts, online conferences, discussion lists, and email, debating and consulting about a wide range of (scientific) information. This information can be shared, but also personalized, creating new networks of information exchange, regardless of temporal or spatial boundaries. In this way, networked (scientific) citizen-consumers are becoming more visible, illustrated by (sometimes transitory) membership of online networks.

Table 10.1 Extracts from UK television audience share (%) for individual viewers[2]

Year	BBC1	BBC2	ITV	Channel 4	Channel 5	Others
1981	39	12	49	–	–	–
1983	37	11	48	4	–	–
1991	34	10	42	10	–	4
1997	30.8	11.6	32.9	10.6	2.3	11.8
2005	23.3	9.4	21.5	9.7	6.4	29.6

Source: Broadcasters' Audience Research Board Ltd., 2006.

The Influence of Deregulation

Technology then has an important role to play in influencing the ways in which citizen-consumers communicate. But it would be wrong to assume that technology is the only factor in providing new opportunities for dialogue, engagement, and deliberation about science-based issues. Deregulation, increasing the influence of free market economics in the broadcast (e.g., television and radio) media marketplaces—print and online media have been working in a largely deregulated marketplace since their inception—has also played an important role. In practice, deregulation is likely to have two key interrelated effects on traditional broadcast media:

- First, multi-channel, on-demand, digitally-enabled audiences are fragmenting; these viewers are increasingly switching from mainstream channels (e.g., BBC1 and ITV1) to recently launched channels (see Table 10.1).
- Second, the influence of public sector broadcasting—encapsulated in the phrase 'inform, educate, entertain'—may diminish.

These effects have important implications for science communication, because deregulation is likely to bring greater commercial influence to the broadcast media marketplace, meaning that science will have to compete to maintain its current position in broadcast schedules and in viewers' programme choices.

At the same time, however, the increasing take-up of digital television in the UK can offer real-time audience deliberation as viewers watch programmes—you may have even pressed that red button yourself—providing opportunities to engage with complex science-based issues (Holliman, 2006). A recent example of this was illustrated by part of the BBC's *DoNation* season,[3] which promoted issues related to organ donation. It was a joint venture between two popular BBC TV medical dramas, resulting in *Casualty@HolbyCity*.[4] Introduced by Professor Robert Winston, this program offered viewers the chance to vote on which of two endings of the episode they wished to view. With social and ethical issues revisited throughout the programme, the storyline featured two patients who were desperately ill and in need of organ transplants, but there was only one donor. In effect, the viewers were invited to choose which patient received the organs of another patient who was a recently deceased organ donor. Of course, one of the key aims of the programme was to highlight the difficult decisions that health professionals regularly make due to the lack of suitable organ donors. Hence, viewers were asked not only to deliberate by choosing who should receive the donated organs in this fictional scenario, but also to be part of the solution by volunteering to become organ donors themselves through the associated *DoNation* season website; an example of real-time citizen engagement with a complex science-based issue (Holliman, 2006).

National newspapers too are responding to the competitive media marketplace and declining newspaper sales, which are a long-term trend. In recent years, the elite newspapers the *Times* and the *Independent* have moved to compact (read tabloid-sized) formats and, more recently still, the *Guardian* adopted the Berliner format, purchasing new printing presses that allow full-colour editions, including a full colour double-page centrefold (Douglas, 2005). More prosaically, you may have noticed the introduction of email addresses of journalists and URLs—links to websites—at the bottom of newspaper articles in recent years. Some science articles now also include 'click through' glossaries where technical terms are listed and underlined in blue in the body of an article—as per an online hyperlink—'linking' the Internet-literate reader to a small glossary on the side of the page. In this way, newspapers are adopting some of the language and practices of the web, even in their hard copy editions. These changes have followed the success of online versions of newspapers, which were initially introduced for the 1997 UK general election campaign. *Guardian Unlimited*[5]—effectively, the online edition of the *Guardian*—proved to be so popular that it was retained and expanded after 1997. There are now plans to increase convergence between the printed copy and online editions with the introduction of 'web first'; where reports destined for the hard copy edition of the newspaper are published on *Guardian Unlimited* first (Mayes, 2006).

Now one of the most popular online newspapers in the world—another effect of publishing on the Internet—*Guardian Unlimited* regularly includes science coverage and a weekly podcast of science news. The site also runs online chats, again in real-time, where readers get to ask questions of experts (e.g., with well-known scientists). Although not a novel idea with respect to public engagement with science—for example, the sociologist Steve Fuller ran a global online conference to discuss aspects of the public understanding of science in 1998 (see Fuller, 1998)—this format does extend the debate beyond academia, allowing media professionals and invited experts to see what is important to those willing and motivated enough to post a question; another example of real-time citizen engagement. However, as the recent attempt by the *Los Angeles Times* to introduce a 'wikitorial'—effectively a collaboratively authored editorial—to its online edition indicates, controversial and complex issues may not lend themselves easily to this form of interaction (see Glaister, 2005 for an outline of the story). Structured smaller-scale face-to-face attempts at public engagement, such as citizen juries and consensus conferences, may prove more effective when debating highly charged topics where there is unlikely to be a single policy solution.

The Legacy of High Profile Science-based Issues

So far I have briefly considered how developments in technology and the influence of deregulation are providing new opportunities and challenges for science communication and public engagement. There is another important factor to consider: the legacy of previous high-profile science-based issues, in particular the way they were governed and portrayed in the public sphere. The BSE/vCJD episode provides a useful illustrative example of such an issue.

Science communication, particularly in the form of news media reporting, played an important role in informing members of the public about developments in the BSE/vCJD episode (Adam, 2000), a situation that is the case for many science-based issues (Holliman, 2004). Research has consistently shown that once citizen-consumers have left formal education, media coverage is a key informant about developments in science. It should come as no surprise then that, following a previous campaign of reassurance that promoted UK-produced beef as entirely safe for human consumption, sales of this product reduced significantly in the immediate aftermath of official announcements in March 1996 that suggested there was a link between BSE and vCJD.

Research suggests that the wider legacy of the BSE/vCJD episode can be characterized by a 'crisis in trust' between official sources, scientists, the food industry, and citizen-consumers (see Reilly, 1999). Recommendations from two reports (House of Lords Select Committee on Science and Technology, 2000; The Phillips Report, 2000) were subsequently adopted in an attempt to increase public confidence in risk governance. For example, requirements for openness and transparency in science communication, in part facilitated by new media—through posting materials and reports on the Internet—are now a central aspect of government departments and agencies, and scientific institution communications strategies.

The Food Standards Agency (FSA), set up as an independent government department in the aftermath of a number of high-profile food scares illustrates these new policies. The FSA places great emphasis on the ideals of openness, transparency, dialogue, and consultation. In practice this involves a number of pragmatic measures, such as producing an up-to-date website,[6] distributing podcasts of open board meetings, holding regular public consultations, and providing advice on clearer labelling of foods with the aim of informing consumer choice. Still a relatively new department, time will tell whether FSA measures are effective in increasing public trust.

On a similar timescale, government agencies and scientific institutions have responded to the House of Lords and Phillips reports—alongside critiques from social researchers (e.g., see Irwin and Michael, 2003; Stilgoe, Wilsdon, and Wynne, 2005)—promoting new opportunities for dialogue and consultation with citizen-consumers as a means of increasing confidence in decision-making about complex science-based issues. To this end, the feasibility of 'upstream engagement'—beginning dialogue and deliberation as complex science-based issues begin to emerge as areas of public debate, and then continuing those activities at regular points downstream—is being discussed by the scientific community, social researchers, and policy makers (see Willis and Wilsdon, 2004 for detailed discussion). In particular, the discussions focus on which issues are worthy of upstream engagement, who should be involved, and when and how to conduct these activities.

The changing culture of public engagement with science-based issues has also seen a growing interest in citizen-led dialogue and online deliberative exercises. As the success of *Café Scientifique*[7] illustrates (Editorial, 2004), innovative ways of engaging a range of actors in discussions about complex science-based issues where citizen-consumers set the agenda for discussion do not have to be confined to new media platforms; whether ICT literate or not, engaging face-to-face is still a viable option.

Issues and Prospects for Public Engagement: The Challenge of Communicating Science in the 'Digital Age'

Although this paper has only scratched the surface of these issues, I hope to have demonstrated that science communication and public engagement in the UK are undergoing a period of significant change, in part due to the development of technology, the deregulation of the broadcast media marketplace and declining print media sales, and the legacy of high-profile science-based issues. Traditional broadcast and print media outlets are exploring new media platforms, while the social-networking phenomenon ensures user-generated content continues to flow. The ever greater visibility of networked (scientific) citizen-consumers that such communication facilitates suggests that this trend will continue to develop. The overall result is an expansion in the number of ways to communicate science and engage with citizen-consumers, including alternative voices, shifting linear one-way communication to more fluid, dynamic, democratized communication. The potential for these forms of communication to deal with complex and often controversial science-based issues remains to be seen, however.

The growth of personalized information and collaboratively authored information, largely outside the constraints of space (e.g., blogs, wikis), time (e.g., podcasts), and editorial control provides interesting opportunities for citizen-consumers (including scientists and professional communicators) to engage with science-based issues. These developments suggest that the interface between science and society could become more blurred in terms of the distinction between experts and non-experts. Of course, citizen-consumers have always had local knowledge that is of value when engaging with science-based issues. The real challenge is to ensure that engagement activities are genuinely inclusive and meaningful to those affected by the outcomes. If citizen-consumers cannot access these debates or their views are overlooked, or simplistically dismissed, then the 'crisis of trust' is likely to remain.

Notes

1. You can access the Pro-Test website at http://www.pro-test.org.uk/. All the websites listed in subsequent notes were last accessed on 14 August 2006.
2. These figures include viewers who chose to watch BBC1, BBC2, ITV1, Channel 4, and Channel 5 via 'digital' (satellite, cable, and freeview) and analogue signals. The category 'others' includes channels that are *only* available via 'digital' signals.
3. This site is available online at http://www.bbc.co.uk/health/donation/.
4. This site is available online at http://www.bbc.co.uk/drama/casualty/donation/.
5. This site is available online at http://www.guardian.co.uk/. Other UK national newspapers also run online editions include the *Telegraph* (http://www.telegraph.co.uk/), the *Independent* (http://www.independent.co.uk/), and the *Times* (http://www.timesonline.co.uk), as does *BBC News* (http://news.bbc.co.uk/).
6. This site is available online at http://www.food.gov.uk/.
7. You can access the *Café Scientifique* website at http://www.cafescientifique.org/.

References

Adam, B. 2000. 'The Media Timescapes of BSE News', in S. Allan, B. Adam, and C. Carter, eds, *Environmental Risks and the Media*. London and New York: Routledge.

Booth, R. 2006. 'Bedroom Blogger, 16, Takes on Animal Rights Protestors', *Guardian* 25 February. Available online: http://www.guardian.co.uk/animalrights/story/0,,1717618,00.html (accessed 14 August 2006).

Broadcaster's Audience Research Board Ltd. 2006. 'Annual % Shares of Viewing (Individuals) (1981–2005). Available online: http://www.barb.co.uk/ (accessed 14 August 2006).

Collins, H., and R. Evans. 2002. 'The Third Wave of Science Studies', *Social Studies of Science*, 32: 235–96.

Day, G. et al. 2006. 'Medical Humanities—A Conversation about the Intersection between Medicine and the Arts'. Available online: http://www.medhum.blogspot.com/ (accessed 14 August 2006).

Douglas, T. 2005. 'Is New *Guardian* Too Little, Too Late?', *BBC News Online* 12 September. Available online: http://news.bbc.co.uk/1/hi/uk/4236846.stm (accessed 14 August 2006).

Editorial. 2006. 'Sound Science', *Nature* 439: 2.

Editorial. 2004. 'The Rise of Café Culture', *Nature* 429: 327.

Fuller, S. 1998. 'The First Global Cyberconference on Public Understanding of Science', *Public Understanding of Science* 7: 329–41.

Glaister, D. 2005. '*LA Times* "wikitorial" Gives Editors Red Faces', *Guardian* 22 June. Available online: http://www.guardian.co.uk/international/story/0,,1511745,00.html (accessed 14 August 2006).

Gibson, O. 2006. 'Internet Means End for Media Barons, says Murdoch', *Guardian* 14 March. Available online: http://media.guardian.co.uk/newmedia/story/0,,1730539,00.html (accessed 14 August 2006).

———. 2005a. 'The Bloggers Have All the Best News', *Guardian* 6 June. Available online: http://media.guardian.co.uk/mediaguardian/story/0,,1499801,00.html (accessed 14 August 2006).

———. 2005b. 'Young Blog Their Way to a Publishing Revolution', *Guardian* 7 October. Available online: http://www.guardian.co.uk/frontpage/story/0,,1587081,00.html (accessed 14 August 2006).

Gosline, A., K. Law, and A. Casselman. 2005. 'The Higher Purpose of Inky Circus (I'm a poet and I do know it)', *Inky Circus—Life in the Girl Nerd World* 4 October. Available online: http://www.inkycircus.com/jargon/2005/10/the_higher_purp.html (accessed 14 August 2006).

Holliman, R. 2006 (in press). 'Representing Science Through Multiple-channel Digital Television: Opportunities for Dialogue, Engagement and Deliberation', in the Proceedings of the *2005 Communicating European Research*, Springer.

———. 2004. 'Media Coverage of Cloning: A Study of Media Content, Production and Reception', *Public Understanding of Science* 13: 107–30.

House of Lords Select Committee on Science and Technology. 2000. *Science and Society* (Third Report). London: HMSO. Available online: http://www.publications.parliament.uk/pa/ld199900/ldselect/ldsctech/38/3801.htm (accessed 14 August 2006).

Irwin, A., and M. Michael. 2003. *Science, Social Theory and Public Knowledge*. Maidenhead: Open University Press.

Leadbetter, C. and P. Miller. 2004. *The Pro-am Revolution: How Enthusiasts are Changing Our Economy and Society*. London: DEMOS. Available online: http://www.demos.co.uk/catalogue/proameconomy/ (accessed 14 August 2006).

Luft, O. 2006. 'Nature Compiles List of 50 Top Science Bloggers', *Online Journalism News* 11 July. Available online: http://www.journalism.co.uk/news/story1928.shtml (accessed 14 August 2006).

Mayes, I. 2006. 'Open Door', *Guardian*. Available online: http://www.guardian.co.uk/Columnists/Column/0,,1811430,00.html (accessed 14 August 2006).

McAfee, N. 2005. 'Insights for the Future of Public Media: A Report on the Global Voices Summit', *Centre for Social Media*. Available online: http://www.centerforsocialmedia.org/future/docs/mcafeegv05report.pdf (accessed 14 August 2006).

Nature Publishing Group. 2006. 'Nature Blogs'. Available online: http://www.nature.com/blogs/index.html (accessed 14 August 2006).

NHS Direct Online. 2006. 'Welcome to NHS Direct Online'. Available online: http://www.nhsdirect.nhs.uk/index.aspx (accessed 14 August 2006).

Reilly, J. 1999. '"Just Another Food Scare?" Public Understanding and the BSE Crisis', in G. Philo, ed., *Message Received—Glasgow Media Group Research 1993–1998*. Harlow: Longman.

Stilgoe, J., J. Wilsdon, and B. Wynne. 2005. *The Public Value of Science: Or How to Ensure that Science Really Matters*. London: DEMOS. Available online: http://www.demos.co.uk/catalogue/publicvalueofscience/ (accessed 14 August 2006).

The Phillips Report. 2000. *The BSE Inquiry*. London: HMSO. Available online: http://www.bseinquiry.gov.uk/ (accessed 14 August 2006).

Thompson, J. 1999. 'The Media and Modernity', in H. Mackay and T. O'Sullivan, eds, *The Media Reader: Continuity and Transformation*. London: Sage.

Thompson, M. 2006. 'BBC 2.0: Why on Demand Changes Everything', Royal Television Society Baird Lecture, 22 March. Available online: http://www.bbc.co.uk/print/pressoffice/speeches/stories/thompson_baird.shtml (accessed 14 August 2006).

Willis, R., and J. Wilsdon. 2004. *See Through Science: Why Public Engagement Needs to Move Upstream*. London: DEMOS. Available online: http://www.demos.co.uk/catalogue/paddlingupstream/ (accessed 14 August 2006).

QUESTIONS FOR CRITICAL THOUGHT

1. Who are the likely readers of Holliman's article? How do you know?
2. What is his primary purpose? Identify his thesis.
3. What are 'citizen media', according to Holliman? How central is this concept to his main argument?
4. Holliman surveys several forms of 'new media' that he claims are 'democratizing' public or mass communication, and which in the process are presenting challenges to scientific experts. What are the features that make such media more 'democratic'? You should be able to identify at least three.
5. Holliman is interested in 'dialogue, engagement, and deliberation about science-based issues'. According to him, what is the central challenge of communicating science in the digital age?
6. How readable is Holliman's article? Is it good science writing, according to the standards that have emerged from your reading so far? Why or why not?
7. How personal or impersonal is Holliman's writing? How formal or informal is it? To support your assessment, point to some specific strategies he uses.
8. Consider the prevailing mood of Holliman's article: is he generally optimistic about developments in the communication of science, or is he generally pessimistic? What aspects of his text provide you with evidence for one or the other? Compare his general attitude to that of any of the other writers featured in this book. What differences, if any, do you see?
9. Using a reliable dictionary, look up the words 'citizen' and 'consumer'. Record and compare their meanings. What does Holliman mean to suggest by yoking the two words in the unlikely configuration of 'citizen-consumers'?
10. What is the crisis of trust of which Holliman speaks? How did it originate? What has it got to do with the main purpose of Holliman's essay?

CHAPTER 11

Negotiating Organizational Constraints: Tactics for Technical Communicators

Marjorie Rush Hovde

> You gotta know when to hold 'em
> Know when to fold 'em
> Know when to walk away
> Know when to run.
> —Don Schlitz, 'The Gambler'

Imagine the following scenario: A person is working on an undergraduate degree in technical communication and has begun an internship of writing computer manuals for a four-person software company as its first technical communicator. No standardized procedures guide the writing and revising of documentation. Salespeople wrote existing manuals in their spare time. The president is also the programmer, and he has openly stated that he hates documentation. He never records on paper the upgrades he has made to the software. The company's *sole* salesperson, who also serves as the trainer, knows a great deal about the software and the users, but he is seldom in the office because he trains clients at their sites. The only other employee is an office manager who also handles customers' questions about the software when they call in. Where does the new technical communicator start? How does she gather the information she needs about the software and the users? On what basis does she make decisions about the format and design of the manuals? Most importantly, how does she relate to the people with whom she works in order to produce good quality documentation?

This scenario, describing one of the situations in this study, supports Sopensky and Modrey's (1995) claim that in addition to technical communication abilities, technical communicators need procedural or 'how-to' knowledge of how to interact socially within their organizations in order to improve the quality of their written products. Indeed Van Wicklen (2001) estimates that technical communicators can spend as much as one-third of their time interacting with co-workers in their organizations, facing obstacles such as 'difficulty obtaining information, reticent or uncooperative engineers, canceled projects, unreasonable or unclear deadlines . . . and office politics' (8).

Thus while the job of a technical communicator is usually not as fraught with risk as that of the gambler mentioned in the epigram, both gamblers and technical communicators do need procedural knowledge in their contexts. In particular, apprentices (Lave and Wenger, 1991) in the field may lack judgment about when to hold, fold, walk away—or run. Consequently, awareness and use of a range of intra-organizational communication tactics may help student and apprentice technical communicators gain influence over their work.

To gain a greater understanding of such tactics, I conducted observational case-study research in two different work sites. During my data-gathering, I noticed that apprentice technical communicators often had good ideas for designing documentation but did not know how to address corporate constraints that prevented them from carrying out those ideas. However, more experienced technical communicators who had a variety of tactics for dealing with their contextual situations, and who knew when to use appropriate social-interactional tactics, were frequently more influential than technical communicators who did not.

Below I discuss the literature in technical communication on intra-organizational communication dynamics, the design of the present study, its findings, and implications for theory, pedagogy, and practice. The study takes as its point of departure two central questions: What tactics does a technical communicator need to know how to employ in dealing with others within the organization? How are the uses of these tactics related to the influence that technical communicators exhibit within their organizations?

Technical Communicators and Inter-Organizational Communication Practices

For technical communicators, the ability to communicate with co-workers in organizations is vital. In entry- and mid-level positions that demand intense work with co-workers, technical communicators make decisions in situations where people may disagree and where resources are limited (Pfeffer, 1995). Therefore, technical communicators need to understand how influence works and how they can work within the constraints of organizational dynamics to exercise a measure of control over their work.

The relatively powerless positions of many technical communicators within their organizations may lead to unhappiness with their jobs, Reportedly, technical communicators express dissatisfaction with co-workers on a greater level than in other occupations, especially regarding the time and resources allocated for their work, noting that documentation is frequently not a high priority in their organizations. Many technical communicators also indicate that they would like to have more responsibility in making decisions about their writing (Philbin, Ryan, and Friedel, 1995). A repertoire of intra-organizational communication options may help such technical communicators feel empowered to deal with workplace constraints, thus reducing their dissatisfaction.

This organizational challenge is particularly acute for apprentice technical communicators whose social-interactional tactics have been shaped largely by academic experiences. They face a daunting transition involving a high degree of 'uncertainty' (Miller and Jablin, 1991) when they move into professional roles largely because they do not know how to understand and deal with internal organizational politics within their organizations (Thomas, 1995). New employees need to learn how to develop working relationships with co-workers who can assist them in their work, while not alienating others (Freedman and Adam, 1996). Furthermore, because typical political practices are often not overtly acknowledged (Thomas, 1995; Smart, 2000), apprentices' struggles to work effectively with colleagues may be made even more difficult.

The silence about power relationships within organizations also extends to silence in recent technical communication scholarship about how to deal with those power relationships (Thomas, 1995). The few authors who have tried to describe the complex expertise needed by technical communicators have tended to overlook organizational communication abilities (Green and Nolan, 1984; Dobrin, 1997). (Ironically, Dobrin mentions that one of the reasons he left technical communication was that 'given the constraints' [105], he was not doing the work he thought he should do.) Fortunately, however, a few authors do provide practical advice for technical communicators on taking organizational power relationships into account when communicating with reviewers of their documents (Hackos, 1994; Smart, 2000) or mention technical communicators' intra-organizational communication abilities tangentially (Heneghan, 1992; Raven, 1997). However, we need a rich understanding of effective tactics in many different situations.

When it is found, advice for technical communicators on communicating within organizations offers only general suggestions that one should try to effect changes largely through personal initiative (Grove, Lundgren, and Hays, 1992; Sopensky and Modrey, 1995; Barker, 1998). Van Wicklen (2001) is somewhat more aware of the complexities of organizational power relationships, but even so, she still encourages a great deal of personal initiative on the part of the technical communicators. Apprentice technical communicators certainly need personal initiative, but they also need to develop a repertoire of specific tactics for dealing with complex organizational constraints and expectations.

Two qualitative, empirical studies point to certain tactics that technical communicators need. Heneghan (1987) noted that the apprentice technical communicators in her workplace study needed intra-organizational communication abilities such as writing collaboratively and interviewing for information. In Raven's study (1992), the technical *communicators* observed needed to know how to negotiate the contradictions that arose when different reviewers disagreed about changes in documentation. Raven's technical communicators' strategies included: negotiating (checking with an approver, or getting the approver to conclude the change is necessary); capitulating; waiting; and escalating (making an executive decision, negotiating with other managers, verifying technical information, or seeking an arbitrator). The tactics Raven observed deal only with how to resolve differences of opinion. However, because of the dual-directional nature of how individuals interact with a system, I wished to look also at tactics that initiated action or reacted to non-conflict situations. Additionally, I wished to learn if technical communicators in other organizations practiced her four tactics.

Important social-interactional abilities needed by technical communicators in complex organizational situations have not been thoroughly explored. The social-interactional tactics discerned through this study begin to build an understanding of what those abilities are and how those abilities look in practice. This understanding can provide a range of options to beginners, as well as to more experienced technical communicators, as they attempt to carry out their complex work. In this study, I designed a project that would begin to shape a fine-grained picture of technical communicators' tactics for negotiating organizational constraints.

The Study

The methods and model for this study are located within the understanding of activity theory that human behaviour is best studied in its contexts, as part of an activity system aimed at achieving specific goals and using tools and genres to achieve those goals. In considering the tactics that technical communicators employed, I assumed that they were influenced by and worked within the larger system. The tactics that I observed could be understood as both tools and genres of behaviour that function during the processes of producing documentation. Below, I discuss my research questions, the theoretical framework of the study, and its methodology, research sites, and participants.

RESEARCH QUESTIONS

This study is located in the context of one broad question: What is the nature of procedural knowledge or expertise for technical communicators? On the basis of my reading of research literature (for instance, Green and Nolan, 1984; Grove, et al., 1992; Heneghan, 1987; Raven, 1992; Sopensky and Modrey, 1995) and on my observations of technical communicators at work, I have created a working map of abilities needed by technical communicators (Figure 11.1), indicating the complexity of their expertise. (This map is not intended to be exhaustive or representative of every technical communicator's situation. Rather, it represents abilities that technical communicators may need as they create documentation.)

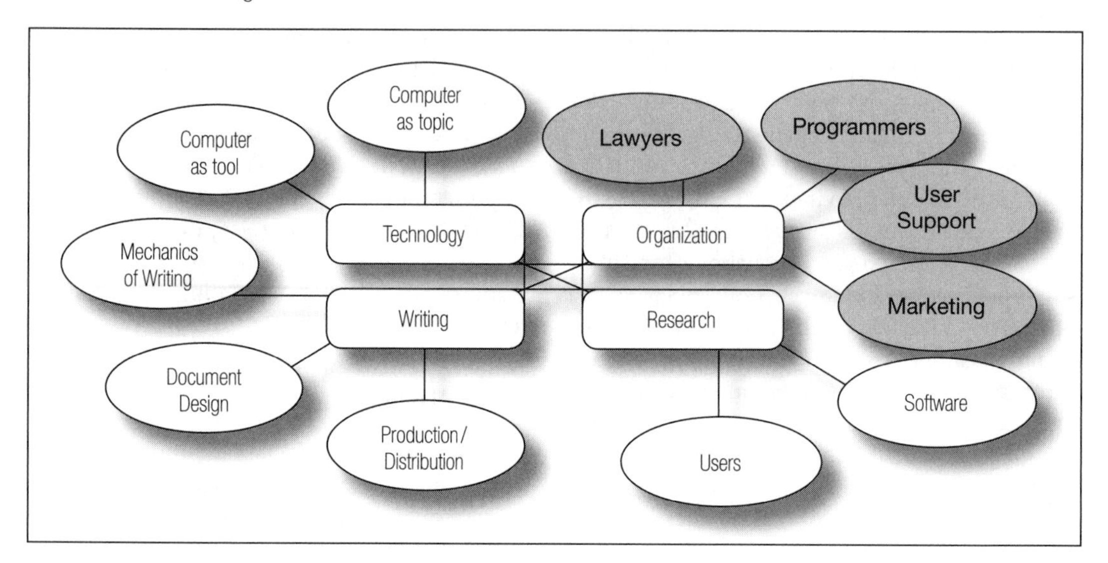

Figure 11.1 Technical communicators need abilities to function in four areas.

In this article, I focus specifically on the upper right shaded quadrant of the map as I explore the questions: What does a professional technical communicator need to know how to do in dealing with others within the organization? What tactics are employed by influential technical communicators and by those with less influence? In addition, I wished to explore how activity theory could illuminate the practices of these technical communicators and how the analysis of their practices could influence the principles of activity theory.

I found these questions especially compelling after I observed the two least experienced and influential technical communicators in this study. Unlike the more experienced technical communicators, both apprentices were products of a professional writing undergraduate major with an emphasis in writing for business and industry. Yet, their specialized education in technical communication did not necessarily teach them tactics they needed for dealing with the realities of working with others in their organization. They both held many valuable ideas about the nature of effective documentation, but neither had mastered the ability to argue effectively within their organizations for major changes. They knew how to write and design documents, but they did not know how to deal with organizational constraints that influenced writing. As Pfeffer (1992) argues, '. . . we need to understand strategies and tactics of using power so that we can consider the range of approaches available to us and use what is likely to be effective' (341). The observations in this study begin to provide such practical knowledge.

THEORETICAL FRAMEWORK

Social-interactional practices are inextricably intertwined with the uses of power and influence. (For my purposes here, I am defining 'power' as having the means to compel certain behaviours from co-workers; 'influence' for me has a milder connotation, one of shaping behaviour without an extensive use of force.) Technical communicators without much knowledge of their organization may experience unproductive encounters because 'Not understanding the degree to which the situation is politicized may cause a person either to use power and influence when it is unnecessary, and thereby violate behavioral norms as well as waste resources, or to underestimate the extent to which power needs to be employed, and fail in the task of implementation' (Pfeffer, 1995: 33). This principle is played out in Raven's (1997) study in which engineers held a higher status than the technical communicators and acted to protect that status, sometimes withholding information from the technical communicators to be sure that not too many people would understand engineers' specialized information. Because previous technical communicators in the organization had lacked technical knowledge and were seen as 'glorified scribes' by some engineers, these engineers were accustomed to directing the development of both the product and the documentation; they resisted the technical communicators' suggestions for how to create documents for users. These technical communicators worked in close physical proximity to engineers, so they may have followed the engineers' suggestions for changes in documentation in order to avoid insulting the engineers and thereby jeopardizing working relationships. This challenging situation is far from unusual, so technical communicators in similarly complex organizations need all the tactics they can find or devise.

One reason that political rules are often not named is that they are complex, situational, and subject to constant, albeit gradual, change as part of an 'activity system'. Recent proponents of activity theory argue that one should understand typical patterns of behaviour in an individual by considering the entire 'activity' that motivates and shapes that behaviour, including the collective use of physical and conceptual tools. As David Russell (1995) explains, 'Activity theory analyzes human behavior and consciousness in terms of *activity systems*: goal-directed, historically situated, cooperative human interactions . . . functional system(s) consisting of a subject (person or persons), an object(ive) (an objective or goal or common task), and the tools (including signs) that mediate the interaction' (53, emphasis in original). Such a system operates as 'a local sphere of goal-directed collaborative endeavour, where thinking, knowing, and intellectual accomplishment are mediated by a matrix of physical settings, symbol systems, analytic methods, technologies, and structured social interaction' (Smart, 2000: 226). This perspective can enrich our understanding of the interaction between behaviour and context because activity theory 'embeds consciousness in a wider activity system and describes a dynamic by which changes in consciousness are directly related to the material and social conditions current in a person's situation' (Nardi, 1996: 13). In understanding the community and its rules, one can see the forces affecting the behaviour of the individuals.

Using activity theory in conjunction with empirical study (Russell, 1995, 1997; Nardi, 1996) offers a researcher an 'elaborated theory of context . . . that embraces objects and motives of collectives and their participants to explain reciprocal interactions among people . . . ' (Russell, 1997: 505). This approach guided this study of individuals' actions as they employed discourse tools within social contexts to accomplish specific ends. Such qualitative empirical study can lead to 'grounded theory' (Berkenkotter and Huckin, 1995: x). As grounded theory is applied to practice, it should prove more useful because it has taken into account the entire social system in which an activity takes place.

Methodology, Research Sites, and Participants

In order to begin to answer the research questions posed above, I conducted observational case-study research in two different organizations that created user documentation for software products produced in-house. The first site, a small organization that I call B&F Programming, employed four people. This organization produced and marketed software to a small, specialized user base. The B&F technical communicator I studied, Sue (all names are pseudonyms), worked there in the summers and part-time during the school year. She was relatively new to the organization and revised paper manuals for their two major software products.

The second site, a larger, not-for-profit organization that I call Money Services, provided financial services to other not-for-profit organizations. Money Services' primary function was to provide these services; the software they sold to clients allowed them to provide the services more efficiently. The key informant at that site, a Technical Writer II whom I call Trish, had about five more years of experience as a technical communicator than Sue. Although Money Services had about 12 Technical Writers at the time of my data collection, I focused on three, Trish, Hanna (a Senior Technical Writer), and Faith (a Technical Writer I), as they created paper and online documentation for a software package that I call QuickCash.

I observed and recorded interactions of these technical communicators at work one or two days a week for a period of 10 months. As an observer, I attempted to have as little effect on the technical communicators' work as possible. The participants also kept a small notebook in which to record significant events when I was not present. I made special note of meetings and interactions during which technical communicators made decisions with co-workers. While on site, I asked the participants to talk aloud about what they were doing, as long as it did not interfere with their work. Extensive field notes, tape recordings of a few meetings, collected documents, and open-ended interviews with technical communicators and their co-workers provided additional data.

In analyzing the data collected at the two sites, I attempted to observe what technical communicators actually did rather than relying on what they might say they did, focusing on their actual 'communication behavior' (Falcione, Sussman, and Herden, 1987). Upon completion of the data collection, I analyzed and categorized the social-interactional tactics that these technical communicators used when dealing with others within the organization. I looked especially at instances in which the technical communicators took actions outside of situations covered by routine procedures. I also analyzed other factors, such as the relative experience and influence of the technical communicators within the organization.

In political situations within their organizations, these technical communicators experienced varying levels of power and influence. At B&F, technical writing had a low priority as evidenced by the fact that Bill, the president and programmer, had not initiated the hiring of a technical communicator; Sue had called him at the beginning of a summer on the chance that he might need one. One of her biggest job challenges was to get information about the software updates from Bill, the sole programmer. She often had to find circuitous ways to complete her work; at one point, I observed her eavesdropping on a conversation Bill was having with someone else explaining his software. Sue did not perceive that she had the power to question him directly for information or to alter the typical practices of the organization. One advantage she experienced was that because there was little oversight of her work, she could determine the direction of many of her tasks and some of her documentation decisions.

Technical Writing at Money Services held a slightly stronger position within the company, even though they were housed in an office building about five miles distant from the Money Services headquarters because it was not large enough to accommodate all of Money Services' employees. (See Pfeffer, 1992, on the relationships between physical settings and power.) Several of the technical communicators commented that their situations at Money Services were better than at other places where they had worked. Several employees in other departments who worked closely with the technical communicators commented to me that they were impressed with what the technical communicators accomplished given limited resources. However, corporate culture posed barriers to ideal circumstances. For instance, the technical communicators had no direct access to external users because Marketing representatives were concerned that too much interaction with users might jeopardize accounts (Hovde, 2000). In addition, the organization's six-month software and documentation revision schedule did not permit the technical communicators to add user research to their already multitudinous tasks. Furthermore, the technical communicators could not decide by themselves how to organize manuals. Over the years, the manuals had been organized according to the architecture of the software because Systems Developers liked them that way, although in the middle of my data gathering, the technical communicators received permission to reorganize the manuals to be structured according to users' typical workflow. The technical communicators also had to be accountable to the Client Support Department, which was largely responsible for the distribution of the manuals.

Findings

The discussion below of these technical communicators' social-interactional tactics begins to create a picture of an important aspect of the expertise of technical communicators. While this picture will need to be tested and supplemented with further studies, nevertheless the findings provide useful insights into options available to technical communicators as they interact with people in their organizations using tactics that perpetuated, resisted, and/or shaped the activity system.

VARYING LEVELS OF INFLUENCE AMONG THE TECHNICAL COMMUNICATORS

In this data gathering and analysis, I observed that the use of several social-interactional tactics was one of the factors that affected the amount of influence that the four technical communicators observed in the study were able to employ. Below I briefly sketch a portrait of the technical communicators and then discuss the uses of each of the tactics they employed.

An influential technical communicator

Hanna, as Senior Technical Writer, exercised a great deal of influence within Money Services. People both inside and outside the Technical Writing Department respected her and sought her advice. Bright, articulate, and busy, she sat on many cross-departmental planning committees for revisions to the software and the documentation. Within Technical Writing, she supervised the work of other technical communicators and served as editor of some of their documentation. (Soon after I completed my data gathering, she was promoted to Department Head when the Supervisor stepped down.)

In her use of social-interactional tactics, Hanna knew how to go along with others' decisions when necessary, even if she did not always agree with them. But she was also able to argue well for changes when needed. She frequently served as a liaison between Technical Writing and the other departments with which they needed to work, bringing information from the members of one department to those of another.

Hanna frequently agreed with the proposals of others, but she also resisted some of them. When she resisted, she usually gave a reason that others might not have thought about. These reasons frequently were based on knowledge gained through her interaction with others on cross-departmental committees. The technical communicators working with her usually agreed with her wishes when she presented such extra-departmental reasons.

A moderately influential technical communicator

Trish, my primary informant at Money Services, had worked there nearly as long as Hanna had, but she did not wield the influence that Hanna did. Although Trish interacted frequently with people from other departments, she did not participate in early design decisions for revised software and documentation. Rather, the meetings she typically attended gave her opportunities only to discuss revised software specifications or participate in technical reviews of the documentation.

Trish had built an image of the documentation expectations of people in other departments and did not seem to resist their wishes very often. She generally complied with requests, especially from those whom she perceived had influence and power at Money Services. As reasons for her writing decisions, Trish frequently stated, 'The Systems Developers want it this way,' or 'They said they wanted us to do it this way. I would change it if it were up to me.' She did, however, resist more frequently within the Technical Writing Department. She did not initiate or make decisions as frequently as Hanna did, and this limited her influence on the work she was doing. In addition, the quality of the documentation sometimes suffered from her too-frequent compliance.

Trish's level of comfort in employing a variety of tactics to effect major changes in the QuickCash manuals evolved gradually through the time of my data gathering. Early in the study, at a planning meeting for the next version of documentation with Faith and Hanna, Trish brought up a list of small changes in formatting conventions that she wanted made in the QuickCash paper manuals. However as the study progressed, she was given permission from colleagues outside Technical Writing to reorganize and streamline the next version; and after that Trish listed many larger aspects of the manuals that she wished to change. Overall, Trish's typical uses of socio-interactional tactics led her to be only moderately influential in communicating with co-workers.

The least influential technical communicators

The two relatively new technical communicators whom I studied, Sue at B&F and Faith at Money Services, exhibited even less influence in their organizations than did Trish. As a novice and the only technical communicator at B&F, Sue needed to create documentation within an organization that did not value it highly. In contrast to Money Services, B&F had few standard procedures for creating documentation so Sue had to devise her own procedures and count on the cooperation of co-workers. Both Sue and Faith experienced initially what Miller and Jablin (1991) characterize as a period of uncertainty in which 'newcomers perceive that they receive less information from those around them than they believe is needed' (92). As Schein (1987) has noted, the stage of being new to an organization can be a 'crucial time of learning . . . often painful and full of surprises' (158). The tactics that these apprentices employed indicate that they were still struggling to maneuver socially within their organizations.

Faith, as a Technical Writer I, suggested several innovations at Money Services, but many of them were never implemented, partly because of her limited ability to argue effectively for them with reasons that addressed organizational constraints. As a newcomer to Money Services, she had been initially assigned to revise documentation for QuickCash. Because she spent a great deal of time learning about the software well enough to document it, she had little opportunity to interact with people from outside her own department. Throughout the course of the data gathering, Faith frequently expressed frustration at how slowly traditions at Money Services changed. After working at Money Services for a year and a half, Faith complied reluctantly with creating system-oriented manuals. Although she would have preferred manuals organized by user tasks rather than by the system, she realized that tradition was strong in the Technical Writing Department and that major projects, such as revising a manual to be user-task oriented, took a great deal of time to plan. This compliance indicated that she was becoming aware of organizational constraints that prevented her from achieving an ideal. (Near the end of my data gathering, she announced that she was looking for another job, possibly one in freelance or magazine journalism.)

As mentioned earlier, both Sue and Faith were products of a professional writing undergraduate major with an emphasis in writing for business and industry. Each had taken a course in writing for the computer industry, and each had more formal technical communication education than the others in this study. Both apprentices held valuable ideas about the nature of effective documentation, but neither had mastered the ability to argue effectively in their organizational contexts for major changes. As apprentices, both felt frustrated at times about not knowing their organizations well enough to be influential, a condition similar to that noted by Miller and Jablin (1991). Both technical communicators typically complied or deferred when decisions had to be made. Nonetheless, on occasion both were able to provide fresh perspectives to the traditions of each organization, as also noted in Freedman and Adam (1996).

USING SOCIAL-INTERACTIONAL TACTICS WITHIN THE ACTIVITY SYSTEMS

Given the assumption of activity theory that individuals and the culture are constantly influencing one another, I have divided these technical communicators' typical social-interactional tactics into those used by individuals to react to the culture, either perpetuating it or resisting it, and those used by individuals to attempt to influence the practices of the culture. (In this discussion, I do not argue that the individuals used the tactics with these intentions, but rather that the tactics had the potential to promote specific effects on the culture.)

Tactics used in reaction to the culture

Within any activity system, participants react to the expectations and demands expressed by the culture. Frequently, these technical communicators reacted to the culture in ways that had the effects of perpetuating the practices and values of the system, but they also found opportunity to resist, as discussed below.

Tactics to perpetuate the culture

The tactics discussed below, all of which had the effect of perpetuating the culture, can be divided into three types.

1. Agreeing. The technical communicators in the study often went along with requests and traditions because they were in agreement with the rationale behind them. They did their best to accommodate the requests of other departments in planning and producing the documentation. Such a high level of agreement is necessary for an organization to function, so this level of agreement is not surprising.

2. Complying. At other times, the technical communicators complied with a request, but the reason behind the compliance was often that they had not thought about reasons for or against the action. They simply did as they were told. For assistance, Trish complied with organizational work processes that contributed to the creation of the manuals. She planned the documentation development processes around the schedules of the Systems Developers and the people in Systems Support who needed the manuals and online help in place before they could begin beta testing of the newest version of the software.

Table 11.1 Social-interactional tactics have the potential to exhibit varying effects within an activity system

Tactics used in reaction to the culture		Tactics used pro-actively to influence the culture
Perpetuating the culture	Resisting the Culture	
Agreeing Complying Acquiescing	Suggesting options Refusing Ignoring Postponing	Making new decisions Proposing/innovating Educating others

The two least experienced technical communicators complied more frequently than the more experienced ones. At B&F, Sue's compliance with her co-workers' requests sometimes reflected her uncertainty about principles of effective document design. For instance, Bill, the president and only programmer at B&F, wanted topics in the manuals arranged as one-page modules, so Sue complied, even though she had to squeeze information to make it fit onto some of the pages. Once, Sue tried a different page layout to make the information fit better, but Bill did not like it, so she returned to the standard layout. In this compliance, Sue was unable to argue successfully for changes that she wished to make, but she did not display resentment when one of her ideas was not adopted.

At times, Faith complied in a way that was to her detriment. After one of Trish's reviews of the manual written largely by Faith, she made only the changes that Trish had marked. Faith did not go through the manual herself to find errors that Trish might have missed. For reasons that were not clear to me, Faith did not go above and beyond compliance in this case, which gave Trish great concern about the overall quality of Faith's manuals.

3. Acquiescing. In acquiescing, these technical communicators reluctantly went along with a request or expectation. Pfeffer (1992) has noted that many people in organizations seem to want to acquiesce, to yield authority to higher-ups. Although he decries this 'passivity', at times it was necessary given the realities of competing organizational claims, especially of forces outside Technical Writing. For instance, Hanna once consented reluctantly to having the divider tabs of the QuickCash manuals printed in-house instead at an external facility that could give them a more professional look. She realized that the decision was made to reduce expenses for the company and that the standards of the Technical Writing Department had to take lower priority.

Trish and Faith also acquiesced to the Money Services' tradition that Technical Writers could not get in touch with end users, even though they believed such contact would have improved the quality of the documentation. (The Marketing Department was cautious about allowing anyone other than Marketing representatives to have contact with clients. Hence, the technical communicators knew little directly about clients and about how they used the software and documentation.) Furthermore, upper management did not budget time for technical communicators to conduct any form of user testing. Trish accepted these conditions as 'given' and never proposed that the technical communicators have more direct contact with users, even though she was sure such contact would help the documentation. Faith, on the other hand, proposed from time to time that the Technical Writers find ways to find out more about the end users. Although some of her more ambitious plans were never implemented, she did redesign the user response card in the paper manuals, an initiative that led to an increased rate of return from users.

This acquiescence, even to co-workers' unwise decisions, should hardly surprise us. As Pfeffer (1992) has noted, 'Authority is . . . obeyed because it is inconceivable not to. The power of leaders and bosses becomes institutionalized, and is thus not questioned or even thought about' (133). Furthermore, Pfeffer notes that 'social proof' is powerful in that people tend to agree with others rather than make independent judgments. Apprentice technical communicators understandably may have chosen this path of least resistance, but the more experienced technical communicators often took it as well. These technical communicators usually realized that the request was not the ideal that they would have liked, but they also realized that factors in the activity system required that they compromise their ideals. They also sometimes realized that they did not have enough influence to carry out the ideal plan. Ironically, the more they knew about the organization's goals and values, the more they were able to acquiesce when they deemed that larger goals and values were more influential than their own. However, this knowledge of the organization also allowed them to resist the culture at times, as discussed below.

Tactics that resist the culture

As Russell (1995) has argued, 'activity systems are dialectical. Change is not unidirectional. It is accomplished through joint activity, whether cooperative or conflictual. . . . The participants . . . *appropriate* (borrow and transform) the tools and object(ive)s and points of view of others, leading to changes in the means of pursuing the object(ive) of the activity system' (55, emphasis in original). In this vein, given the relative lack of influence that these technical communicators held within their organizations, I was surprised at how frequently they employed tactics that resisted the culture, albeit usually in indirect ways. Each of the four tactics described below had the potential to change the directions of their organizations' practices.

1. Suggesting other options. Many times, when these technical communicators did not agree with someone else's proposal, they did not say 'no' directly, but instead proposed other options. This tactic can turn a culture in new directions. This tactic was most often successful when the options were feasible within the typical practices and situations of the culture. Because suggesting other options is a way of saying 'no' indirectly, the technical communicators were usually able to maintain smooth working relations with co-workers while still influencing the practices of the organizations.

When suggesting other options, Hanna and Trish usually gave reasons based in their knowledge of corporate constraints and their knowledge of the documentation. When the technical communicators were planning the 4.2 version of the QuickCash documentation, some people outside the Technical Writing Department proposed that all documentation be placed online. Because Hanna knew that such an approach would take several years to accomplish, she proposed instead that they send only the large manuals to the internal clients until the online documentation was fully implemented. In another instance, a User Support person suggested that material about professional judgement for the users might go into the manual. Trish replied that it would be difficult to include such information in the manuals, but that perhaps such material could be included in the quarterly newsletter that Electronic Services sent to clients. Had she not known about that newsletter, she may have been less able to propose a feasible option.

Although the apprentices proposed other options less frequently than did the more experienced technical communicators, occasionally insights gained from their exposure to other ideas and activity systems worked to an advantage because they were able to challenge typical practices. While participating in the planning of a new version of the QuickCash documentation, Faith suggested they eliminate the field definitions and the quick key access in one of the manuals because it was covered in another book. This proposal was accepted, as was her idea to index the paper manuals by topics rather than by chapter and section headings. Although seemingly small, these changes enhanced the usability of the manuals. The two newcomers were able to question tradition and suggest new ways to meet needs, even if these suggestions were not always implemented.

2. Refusing or resisting. Directly refusing to do what someone else in the organization has requested is probably the riskiest of the tactics that I observed these technical communicators use. This risky behaviour was almost always accompanied by a reason for the refusal. The more influential technical communicators were likely to think of strong reasons to refuse. While accompanying refusals with reasons indicates the relatively powerless status of even the most influential of the technical communicators, the inclusion of a reason communicated to co-workers that the technical communicators were not refusing for arbitrary reasons, but for reasons that fit the value structures of the organizations.

Giving persuasive reasons for resistance seemed to be a part of both Trish's and Hanna's expertise as technical communicators. Simply saying 'no' was not enough. In one instance, a member of another department suggested that Money Services switch from using a large three-ring binder to using a small three-ring binder to hold the manual pages. Hanna resisted this suggestion by pointing out that a smaller binder might actually involve substantial cost. At another time, people from another department suggested distributing information for clients on an electronic bulletin board. Trish cautioned against this, mentioning that clients at a previous job did not like to pay the communication costs for an electronic bulletin board. (This was in pre-Internet days.) Their reasons seemed to be most persuasive when they took the values of the organization into account.

Despite her propensity to comply with requests from outside of Technical Writing, Trish felt more free to resist or refuse within her department. At one time, Sally, the Supervisor of Technical Writing, asked Trish to write a proposal for a new project. Trish refused, stating that she was behind schedule on her regular work because she had spent the last week compiling figures for another new project. She frequently showed her ability to fit her reason to the person, knowing what reasons the other person would find persuasive.

Sometimes Trish's reasons for refusals seemed to be based more on her own standards than on what was best for the users. For instance, at one time, the technical communicators received requests from User Support to include non-software related business procedures in the QuickCash manuals. Trish argued that such knowledge was outside the typical work of a Technical Writer and that they would have to conduct additional research and interviewing in order to learn non-software information. She saw her responsibility as documenting only the software, not necessarily providing the user with non-software information. In this view, she had been influenced by the values of an organization that did not see documentation as meeting the users' non-software knowledge needs as well as training and phone support could.

The least influential technical communicators seldom refused requests directly. Instead they used other more indirect tactics when they wished to resist the culture.

3. Ignoring. This tactic, another risky one, was seldom used. When these technical communicators ignored someone, it was usually in response to a suggestion from someone with little power. Even if these technical communicators might have wanted to ignore a request from someone in greater power, they did not do so in my observations.

The few instances of ignoring that I observed occurred in reaction to Faith, the least experienced technical communicator. For instance, while planning a new version of the QuickCash manual, Faith proposed that they send a survey to end users to see what they would like. Neither Trish nor Hanna responded to this suggestion. I do not know if they had not heard Faith, or if they had heard her and had chosen not to respond. However, conducting such a survey would have been a complex process at Money Services, involving gaining permission from many departments that would have resisted such relatively direct client contact initiated by the Technical Writers. Hanna and Trish may have elected not to take time to explain the complexities to Faith at that moment. Their lack of response may also have indicated Faith's relative lack of power. When impractical suggestions came from Sally, the Supervisor of Technical Writing, Trish and Hanna generally briefly and gently explained why her suggestion could not be carried out.

In most instances, suggestions from co-workers received responses, either supportive or resisting. This lack of use of the tactic of ignoring may reveal the nature of the power relationships between Technical Writing and other departments. Ignoring suggestions would have damaged the credibility the Technical Writers had worked long to build and would have strained relationships with colleagues on whom Technical Writing depended. Contrary to the observations of Lee-Arm Kastman Breuch (2001) who found that her transitional technical communication students often ignored requests from external clients, I observed no instances of Sue or Faith ignoring comments or requests from co-workers. This absence may indicate their relatively powerless position, but also may indicate that they were sufficiently aware of the practices of the workplace activity system to realize that ignoring was not typically a wise tactic.

4. Postponing or deferring. In using this tactic, the technical communicators delayed a decision and/or asked for someone else's input before making a decision. At times they even deferred the decision entirely to a co-worker. Even the least experienced technical communicators used this tactic frequently; it seemed to be one that they learned early in their careers.

Several strategic reasons may have prompted these technical communicators to postpone or defer a decision. One was that deferral allowed for greater thought and collaboration so that wise decisions could be made, especially in complex, problematic situations. As Pfeffer (1992) has argued, 'Delay gives you a chance to learn more about other people's points of view and this knowledge can be employed in formulating tactics that will be more successful' (227). Delay can allow time for people to think of other options and evaluate them.

As a case in point, Trish and Faith once discovered discrepancies between the software specifications and how the software actually worked. It was too late to change the software, but it also would have looked bad if an external client found the documentation to be inaccurate. Trish decided to talk to Hanna before documenting this information because, as she noted, 'We've opened up a can of worms.' When two competing values were in tension, Trish chose not to make the decision by herself.

Another reason for deferral was that sometimes a decision lay outside of the technical communicator's responsibilities. The most experienced and influential technical communicator, Hanna, seemed to be the one who delayed the least. One reason may have been that her job responsibilities did not entitle her to make those decisions. Making a decision that went beyond the boundaries of their job expectation could have landed these technical communicators in trouble.

A third reason to defer was that, at times, the technical communicators did not want to take responsibility for the outcome of a decision or wanted to make sure that accountability lay elsewhere. When Hanna faced a decision that might yield negative results, she made sure that the Technical Writing Supervisor, who would take responsibility, approved it. In addition, Hanna routinely deferred responsibility by asking upper management to sign off on the Technical Writers' planned documentation schedules, so that the technical communicators would not be accountable for missed deadlines caused by a 'curve' thrown into the project by management. Thus Hanna attempted to ensure that responsibility for delayed publications lay elsewhere.

Trish frequently deferred or postponed interdepartmental decisions as a 'tail-covering' tactic, such as when she asked Frank, the Supervisor of User Support, to sign off on printing redesigned manual covers. She explained, 'It's his money, so having him sign off covers us.' In another instance, she wrote a memo letting Sally's superiors know that a discrepancy in the numbers of QuickCash manuals planned and the number actually produced was caused by Frank having changed his mind after approving the plan. Trish wanted to be sure that they understood that the consequent depletion of inventory supplies was not her fault.

A fourth reason to delay or defer a decision was to make a point diplomatically. For instance, Frank, the Supervisor of User Support, once suggested that all software 'cascades' be listed in an appendix of the manuals, almost like a walk-through of the software. Trish did not want to include these because it would create too much duplication to maintain, but she did not want to come across as the 'bad guy'. Consequently, she brought the idea up at an interdepartmental planning meeting (with Frank present) and asked the group to decide on whether or not to implement it. They decided not to.

Not being able to defer or deferring too much could have negative consequences for these technical communicators. For instance, Trish sometimes seemed unable to relinquish control as she delegated work to Faith. At times, Trish asked Faith to take responsibility for a project, but then went to relevant meetings with Faith or took the project back into her own hands if Faith was gone for a few days. Trish worked diligently and produced work of high quality, but was frequently passed over for more supervisory positions, apparently because of her weak managerial and 'people' skills. In addition Faith sometimes asked other Technical Writers to deal with issues raised by people outside of the department. Doing so may have prevented her from gaining valuable experience in working with people across the organization.

Knowing when to defer, although it was sometimes a sign of indecision, also indicated that these technical communicators knew when to wait in order that a decision could be made thoughtfully. Deliberate delay played an important part in these technical communicators' professional knowledge. However, not handling deferral wisely may have made them less influential in the organization at times.

Tactics used pro-actively to influence the culture

Although the majority of the tactics that these technical communicators used were in reaction to the requests of other participants in their respective activity systems, these technical communicators also at times initiated actions that could potentially affect the normal practices of that system. Like the tactics for resisting, the use of pro-active tactics could lead to changes within the typical genres of behaviour of the activity system.

Making new decisions

At times, these technical communicators ran into situations in which no one else had made a decision for them to follow. In those instances, they often made decisions without a great deal of consultation. Usually, these decisions were about small changes. Making decisions also entails taking responsibility for the outcomes, however, and so the least experienced and influential technical communicators often avoided making decisions, especially about major issues.

Hanna made decisions, especially decisions within the Technical Writing Department, more frequently than the others did, based on her knowledge of users and on her knowledge of organizational constraints. Although she often made these decisions collaboratively in discussion with Trish and Faith, this practice indicates Hanna's decisiveness, her ability to create solutions that solved several problems at once, and her influence within the company. At times, Hanna took responsibility for decisions that were potentially expensive. When Trish discovered that 600–700 manual covers were printed with the QuickCash name in a format different from the copyrighted one, Hanna made the decision to discard all of them and to print new covers.

Occasionally, Trish made decisions largely on her own about small matters within Technical Writing and within her responsibilities as a Technical Writer II. For example, when planning a new version of the manuals, she and Faith decided on their own that, in order to make the manuals concise, they would show only the shortest way to do a task. In another instance of a decision that did not involve much risk, Trish altered a routine thank-you memo to the head of the Copy Centre that the Supervisor, Sally, had asked Trish to ghost write after a manual was complete. Trish added her own name to Sally's and turned the thank-you into a planning memo for the next version.

Trish explained that she added her name because Sally did not participate in day-to-day planning, and it would look awkward if it seemed that the time-line was coming from Sally only. Trish thought the head of the Copy Centre needed the time-line. Such instances of Trish's making new decisions happened rarely during my observations, however. She usually followed the lead of others. Avoiding responsibility for decisions about large matters also meant that she did not have opportunities to exert influence on the typical organizational practices.

Although Faith did at times participate in decision-making processes with Trish and Hanna, thus learning how and when to make decisions, it was rare that Faith or Sue made decisions on their own. The apprentices did not appear to want the individual responsibility entailed in making a decision.

Proposing/innovating

In proposing or innovating, these technical communicators were typically not reacting to a situation presented to them, as in the previously discussed tactic. Instead, they seemed to be coming up with a new idea for coworkers to consider. This tactic also differs from that of making a decision in that not all of the courses of action that these technical communicators proposed were implemented. As Pfeffer (1992) has argued, 'Innovation almost invariably threatens the status quo, and consequently innovation is an inherently political activity' (7). Therefore it should not be surprising that Hanna and Trish innovated relatively more frequently than Faith and Sue, the two least experienced and effective technical communicators, especially about matters that carried a great deal of weight.

Hanna seemed to propose more freely than did any of the other technical communicators in this study. She was more inclined to initiate and give persuasive reasons, to act rather than only react. She frequently proposed new practices outside the Technical Writing Department, indicating that she knew how to deal with the power relationships of the larger organization. Although she solicited responses and input from others, she also suggested possibilities for how revised documentation could deal with problems that the others observed. Hanna's greatest source of influence, in addition to her position as Senior Technical Writer, seemed to lie in her ability to propose new ways of working. Innovation may have been less frequently observed in the other Technical Writers because, as Pfeffer (1992) argues, organizational innovation often . . . involves obtaining the power and influence necessary to overcome resistance' (71). Hanna demonstrated such influence by initiating requests for information from people in other departments about QuickCash users and about what should be excluded from the manuals. She proposed new formats for the manuals when the audience for them changed. These proposals provided opportunity for the typical practices of the activity system to change.

While Trish was willing to innovate in some circumstances, her work was characterized by less innovation than that of Hanna. In one instance she proposed that they make page breaks more frequent in the manuals, but such innovations were not typical for her. She worked cautiously within what she understood to be her responsibilities. Although Trish had ideas for possible directions, these ideas were often expressed only in reaction to someone else's comments. As Pfeffer (1992) has argued, initiating can make one vulnerable, and can cause actions to be taken that cannot be easily undone. Trish apparently did not wish to take on such uncertainty and responsibility, although she did find ways to innovate in lower-risk situations.

While Faith and Sue were generally less likely to innovate than Hanna and Trish, nevertheless they were able to propose and implement several changes in the manuals that added to their usability. For instance, Sue reorganized some of the manuals into a more chronological structure, and Bill approved of the change. She also experimented with new page designs and suggested adding a glossary to one of manuals. In these innovations, she demonstrated an emerging sense of users' documentation needs and of what might meet those needs.

Faith also frequently proposed new approaches. Not all were accepted, but her proposing of them indicated that she felt free to innovate (or that she was unaware of the risks she was taking). For instance, at a planning meeting for revising the 4.2 documentation, Faith argued that online help should be brief, task-oriented, and without background information. This suggestion was not implemented, but her others about a new page for each task, the inclusion of a mail-back card for client responses, and an index for the manuals organized by topics rather than by headings from the text all were implemented for the 4.2 QuickCash manuals. During Faith's time at Money Services, she had begun to learn how to generate persuasive reasons for her proposals. When the QuickCash documentation team was planning to revise manuals, Faith argued for starting each new task on a new page. Hanna expressed concern that this change might make the manuals longer, but Faith countered with a prototype chapter of the QuickCash manual that she had created with each task on a new page; the section had not become significantly longer. In doing so, Faith displayed that she had learned that she needed to provide evidence to support her claim to Hanna's satisfaction.

However, Faith had not yet learned tactics for presenting or proposing ideas in interdepartmental settings, a tactic that the more influential technical communicators practiced much more frequently then she did. The apprentices' relative newness seemed to be an asset in thinking of new ways to approach their work—and one could speculate that this was because of their recent experience in other activity systems—their inability to argue effectively for some of their innovations limited the amount of changes that they could actually implement.

Educating others

At times, the technical communicators in the study, especially the more experienced ones, took advantage of opportunities to educate co-workers about the typical practices of their area of their organizations. As could be expected, they educated newer technical communicators, but they also took opportunities to educate people outside of the Technical Writing Department about its needs. In a sense, this tactic could be viewed as one that perpetuates the activity system, but it also can be seen as an attempt to be pro-active rather than reactive.

As Green and Nolan (1984) have argued, educating others is typically one of the tasks of more experienced technical communicators. Educating others often happened during the technical communicators' everyday tasks rather than in structured, formal settings (as in Freedman and Adam, 1996). Hanna's role as liaison gave her opportunity to educate co-workers both inside and outside the Technical Writing area. In one instance, a manager in Systems Support commented in reviewing drafts of new QuickCash Quick Reference manuals that information was not complete. Even though Faith was writing the manuals and receiving reviewer comments, Hanna called this manager and explained to her that Quick Reference manuals were intended only to give basic information. The manager's comments stopped after that call.

Hanna was able to educate and to provide reasons for explaining why a certain action might not be able to be taken. Educating another person, especially a novice to the organization, allows an insider to 'rais[e] to awareness' about what she 'knows but has relegated to . . . her subconscious or has taken for granted' (Sutton and Louis, 1987: 350). Opportunities for education also may 'challenge or affirm the insider's schema' (1987: 350). Hanna's ability to educate others reflected her great knowledge of the organizational climate at Money Services and provided clues to her understanding of its constraints.

Although she did so less frequently than Hanna, Trish also educated co-workers in the course of her daily work. For instance, she was able to use the QuickCash documentation structure to show other Technical Writers how to incorporate information in documentation without capturing prompts from the software, because capturing prompts was difficult and took a great deal of time. In addition, she educated Faith in how to learn about new software by reviewing specifications for a QuickCash variation and writing questions for Faith to cover in the specifications walkthroughs.

In addition, Trish frequently educated Frank, the Supervisor of User Services, who had worked for Money Services for only one-and-a-half years and at times seemed unaware of the complexities of the organization. When Frank commissioned a special Quick Reference manual, Trish educated Frank and others from User Support about the technical communicators' writing cycles, emphasizing that when a document went out for review, it was largely complete. This action seemed necessary to her because several people in Client Support wanted to make major changes in a review draft of a Quick Reference manual late in the document cycle. Her tactic of educating a newcomer helped to solidify her position as an insider (Sutton and Louis, 1987) and may have contributed slightly to her power or influence within the organization.

Faith and Sue were not in positions from which they could educate others, as would be consistent with Green and Nolan's findings (1984). They were still in the position of being educated about the practices of their new activity systems.

Implications for Practice, Theory, and Pedagogy

The socio-interactional tactics discussed above seem consistent with two of Pfeffer's (1992) principles of working with power within an organization: understanding the various interests within the organization, and understanding why others may think the way they do. The technical communicators observed in this study did not seem to use these tactics deliberately, but rather used them spontaneously, contrary to Pfeffer's recommendation that in the process of implementing a project within an organization, one should set goals, study others within the organization, and then select appropriate tactics to get the job done. As the table below indicates, technical communicators with varying levels of influence used these tactics in different ways. How frequently the technical communicators used each tactic, especially at the varying levels of influences that each one held, reveals the practice of one highly significant aspect of their expertise.

EFFECTIVE AND INEFFECTIVE PRACTICES OF INFLUENCE

In this study, the effectiveness of the social-interactional tactics was influenced by factors including the person using the tactic, this person's position in their organization, and the political context in which the tactic was used.

This study indicates that knowing how to interact with co-workers is an important component of procedural knowledge/expertise for technical communicators. The tactics discerned in this study begin to provide a range of options that might help the technical communicator mentioned in the opening paragraph of this article to work within a difficult context. In naming these tactics, I hoped to put names to elements of tacit practical procedural knowledge and dispel some of the mystery about how technical communicators deal with organizational power relationships. The findings also begin to answer the research questions addressed in this study regarding how technical communicators deal with organizational constraints, how these findings relate to activity and discourse theories, and how we can use these insights in pedagogy.

Table 11.2 Tactics and influence variables among the four technical communicators

Tactics	Most influential technical communicator (Hanna)	Moderately influential technical communicator (Trish)	Least influential technical communicators (Sue, Faith)
1 Agreeing/complying/ acquiescing	• Agreed when she had a good reason, especially a user-based reason • Agreed reluctantly sometimes, given corporate constraints	• Agreed more frequently with extra-departmental requests than with in-department requests • Agreed to content and design wishes of people outside of Technical Writing	• Agreed frequently, but often reluctantly • Agreed at times to their detriment • Acquiesced at times because they were not aware of alternative courses of action
2 Suggesting other options	• Felt relatively free to suggest options • Based reasons on corporate needs and users	• Suggested options less frequently than Hanna • Dealt with meeting several organizational needs at once	• Provided fresh perspective • Often did not give persuasive reasons for new ideas
3 Resisting/refusing	• Cited reasons for resistance based on corporate needs and values	• Refused more frequently than one might have expected • Refused indirectly, accompanied by a reason • Resisted use-friendly innovations at times when they made her job harder	• Refused less frequently than other technical communicators
4 Ignoring	• Ignored others rarely, especially outside of the department	• Ignored others rarely, especially outside of the department	• Not observed in these technical communicators
5 Postponing/deferring	• Postponed or deferred in order to shift responsibility or to be certain that a decision was made with sufficient thought	• Postponed or deferred quite frequently • Postponed to avoid risk as well as to make collaborative decisions • Deferred especially on out-of-department matters	• Postponed decisions at times because of a lack of information • Learned early in their careers to postpone or defer
6 Making a new decision	• Made decisions collaboratively • Made more decisions on large matters than did the less influential technical communicators • Made decisions based on her knowledge of organizational constraints	• Made decisions usually only on small matters	• Usually made decisions collaboratively • Made decisions mainly on small matters
7 Proposing/innovating	• Proposed and innovated much more frequently than the other technical communicators	• Proposed or innovated at times, but not frequently • Supported proposals with reasons based on her knowledge of the corporation	• Proposed or innovated when they were given freedom to do so • Could not often support proposals with persuasive reasons
8 Educating	• Educated no only people within Technical Writing, but also outside it	• Educated mainly other technical communicators or newcomers from outside of the department	• Not observed in these technical communicators

What does a professional technical communicator need to know how to do in dealing with others within the organization?

Each social-interactional tactic that these technical communicators practiced carried with it risks and benefits that technical communicators had to assess quickly in situations of decision or change, depending on the role they held or the position held by the person with whom they were communicating. The most influential individuals were able to play various roles and improvise as circumstances arose, consonant with Hanks' (1991) observations of 'mastery'.

Some of these tactics are similar to those that technical communicators used in Raven's (1992) study. Raven's categories of Waiting and Escalating could be subsumed under mine of Postponing/Deferring. Her Capitulating parallels the Acquiescing/Agreeing described in this study. Her Negotiating can be seen as similar to Suggesting other Options. The technical communicators in this study used additional tactics, however, in situations that did not involve conflict resolution, thus giving a richer picture than that in Raven of technical communicators' options for communicating within organizations.

Although I did not do a quantitative analysis of the data, I observed that the more risky tactics such as Ignoring, Educating, Refusing, and Innovating were the least frequently used, especially by the less experienced technical communicators. This avoidance of risk is not surprising given the typical status of the technical communicators in this study. Additionally, many of the tactics were used in reaction to coworkers. Tactics that initiated action were used with less frequency. The relative status of the technical communicators in this study may not have permitted them a great deal of opportunity to initiate.

What tactics are employed by influential technical communicators and by those with less influence?

Influence for an individual within an organization can come from many sources including one's job title, experience, knowledge, ability to persuade, communication tactics, and/or coworkers' perceptions of that individual. It appears that a combination of the title, tactics, and level of knowledge were primary factors affecting the influence a technical communicator had. As Pfeffer (1992) suggests, the job title is only one source of a person's influence and is not all one needs to exercise power. For instance, Money Services' Technical Writing Supervisor, Sally, by virtue of her job title, appeared to be in a position to implement improvements for her department, yet she was frequently ineffectual. I suspect this was because she overused the tactic of compliance. She sometimes initiated actions, but she seldom proposed new options. On the other hand, Faith and Sue, the least experienced and lowest in status of these technical communicators, at times were able to effect changes that, although seemingly small, improved the usability of the documentation. Thus it would appear that job titles alone do not determine technical communicators' influence; the influence may come in part from the socio-interactional tactics they employ while working within their positions.

The technical communicators observed in the study needed to understand well the organizational constraints and culture in order to do their best work. Experienced technical communicators know that practices and reasoning within a corporation are Byzantine and irrational. For all of the technical communicators observed, simply using socio-interactional tactics themselves to accomplish their ends was not sufficient. They also had to know how to provide persuasive reasons for action. Reasons these technical communicators gave typically included (1) user needs and how they are likely to respond to texts, (2) formal conventions, (3) the desires of others within the company, and (4) the technical communicators' own time, energy, and expertise constraints. The more appropriately these technical communicators crafted their reasons to the audience, the more effective they were in arriving at good solutions to problems. The more influential technical communicators frequently adapted their reasons to the audience at hand. At times, a technical communicator gave slightly differing reasons for a single action, depending on the audience, indicating that they had learned how to ascertain what would be persuasive to different audiences. These influential technical communicators had learned to consider the input of co-workers but also to put forward their points of view in order to affect the finished documentation, as did the writers in Smart (2000). Within their activity systems, their actions were influenced by the prevailing conditions, but were not wholly determined by those conditions. The technical communicators were also, at times, able to influence the direction of their systems.

Beginning technical communicators may believe and act as if organizations will respond to arguments for taking an action simply because it appears to be a 'common-sense' good idea. For instance, Faith desperately wanted to have greater contact with the end users of her documentation so that she could create manuals to meet their needs, an idea that seemed reasonable to her. But she was new enough to the corporate culture that she did not know how to work with the people in Marketing nor with tight schedules that prohibited the technical communicators from taking time for more audience research and analysis. Faith had several ideas worthy of implementing, but she did not understand the corporate culture well enough to know how to argue effectively for their implementation.

As a result of their academic backgrounds, Faith and Sue knew a great deal about what is typically thought of as technical communication; they knew how to compose and design text for the benefit of the readers. However, they were unable to employ their knowledge if someone else with more influence in the organization opposed them or if other constraining aspects of organizational culture were in play.

THEORIES OF SOCIAL INTERACTION AND DISCOURSE

What can these findings contribute to activity theory and theories of discourse that explore interactions between the culture and the individual?

This study begins to expand our understanding of how technical communicators negotiate within organizations that may 'condition but not determine' their actions (Russell, 1997). The findings support the claim made in activity theory that all individual needs the resources of an organizational context in order to create documentation, but also that the individual's activities are constrained by the values and practices of that context.

Activity theorists also argue that change can happen within a culture when individuals see contradictions and act to resolve them. While the apprentices in this study were able to see many contradictions, the more experienced technical communicators were likely to ignore some of these contradictions. The study indicates, however, that individuals such as these technical communicators may have only a small impact on the practices of an activity system. At least three factors may stand in the way of change: (1) technical communicators may lack the knowledge to argue effectively for change, (2) large organizations with their multiple and complex activities may be difficult to change, and (3) levels of power may affect who can be an agent of change within a large and complex organization. Further study of technical communicators with more influence within their organizations may indicate how an activity system changes.

In addition, this study supports Kastman Breuch's (2001) contention that students moving from school settings to workplace settings may experience a clash of activity systems. Faith and Sue, as apprentices in transition from one system to another, were learning the typical practices of a new activity system and experiencing dissonance when the two sets of expectations and traditions clashed.

Somewhat contrary to the claims of activity theorists, however, these technical communicators did not seem to use these tactics deliberately with a clear objective in view. Rather, as they faced new situations, they improvised ways to act or react. I seldom observed the technical communicators explaining why they chose a particular social-interactional tactic.

Questions remain about how these tactics fit into activity theory's classifications of the elements of an activity. Are the tactics 'tools' that the technical communicators used to accomplish goals? As Christiansen (1996) has argued, typical tactics may be considered 'tools' because they were created by those who used them, based on their motivations in their work; the tactics then guided their work. If so, these tactics can also be viewed as artifacts created by the technical communicators in context and then used as tools to allow them to function socially.

Or, are the tactics a form of 'genre knowledge' regarding the production and use of documentation within an activity system? Genre knowledge can refer to an understanding of the formal conventions of discourse (Berkenkotter and Huckin, 1995) but also to 'social action arising in response to perceived regularities in situations and exigencies' (Smart, 2000: 226). These genre rules can allow for variation within one's activities. Use of the tactics could be also be seen as knowledge of 'behavioral genres' that suggest typical actions for a variety of situations. These 'behavioral genres' may lie between 'rules' and 'tools' in activity theory models (Clay Spinuzzi, personal communication: 24 March 1999). The socio-interactional tactics discussed here, whether viewed as 'tools' or 'behavioral genres', are worthy of study from an activity theory perspective because they mediate the technical communicators' social functioning and because the technical communicators use them for specific purposes. Understanding the tactics is part of understanding the expert practice of these technical communicators in context. The tactics are 'artifacts' that combine with 'actions', and both are 'interwoven with each other and with the social worlds of the human beings they mediate' (Cole, 1996: 120). The insights of activity theory allow us to observe how the individual and the group interact and change each other, mediated by their tools. However, this study raises issues about how these technical communicators' tactics fit into activity theory models.

PEDAGOGICAL STRATEGIES

The results of this study provide students and apprentice technical communicators with a repertoire of actions and reactions for communicating within an organization. Conscious knowledge can help one employ a variety of tactics and thus enhance one's effectiveness in the organization (Pfeffer, 1992). Without a conscious grasp of these options, apprentices may assume that their only options are either to do what they are asked to do or quit their jobs. However, awareness of these social-interactional tactics is only a starting point. Not all options will be equally effective or appropriate in all situations. Students also need to learn how to read a rhetorical situation to see which tactic may be appropriate (as indicated in the second table) and to learn how to generate persuasive reasons appropriate for that situation. While apprentices may learn some of this ability through trial and error, beginning technical communicators who already possess a sense of options and alternatives from their prior training may feel less trapped into doing only what they are told.

How can beginning technical communicators acquire a repertoire of effective tactics? Although it may seem that '[p]roblem solving skills evolve more from using common sense and learning from past experience than from following a cookbook' (Sopensky and Modrey, 1995: 104), this 'common sense' may be enhanced through deliberate instruction. Smart (2000) argues that 'learning to play a role in an unfamiliar sociorhetorical "game" . . . involves development on various levels, development that can only come from experience' (245). However, the observations from this study provide options that apprentices might learn even before they gain workplace experience. Educators can provide 'experience' in well-supported atmosphere that prepares novice technical communicators for their experience in organizations.

Given the complexity of workplace culture and power relationships, technical communication students need exposure to writing problems and situations that are as 'real-world' and complex as possible. Pedagogical conditions in which students, under the guidance of an experienced instructor, engage in collaborative and authentic tasks relating to a complex 'messy' activity system hold the most promise for preparing students to experiment with the range of social-interactional tactics that they will need in the workplace (see Freedman and Adam, 1996). The pedagogical program that we seek should take into account Hanks' (1991) contention that 'learning a process takes place in a participation framework, not in an individual mind' (15). Involving students in that participation framework may prove valuable to their mastery. In addition, there are indications that 'apprentices' who learn from interacting with each other learn practice more effectively (Lave and Wenger, 1991). This understanding counteracts the advice presented above that individuals just have to 'work smarter' in order to negotiate organizational constraints. In this study, Faith often collaborated with more experienced Technical Writers and thus was exposed to the typical tactics employed by the more experienced colleagues. Sue, on the other hand, often floundered, partly because she had only an office manager to mentor her, not an experienced technical communicator. Educators and mentors may provide guided opportunities for educational interactions to take place.

Academically supported internships and job-shadowing can also provide students with valuable guidance during the time of transition from academia to the workplace. During this transition time, students need to look at all writing tasks in the context of 'power and political situation variables' (Thomas, 1995: 467). In these settings, students can experience learning to analyze a 'newly encountered domain of discourse in order to recognize goals, values, and social relationships that are significantly intermeshed with writing' (Smart, 2000: 246). Rather than learning through trial and error once on the job, students need deliberate training in interpersonal communication skills (Philbin, et al., 1995). Educators may wish to make the students' roles in projects similar to the entry-level positions they are likely to have immediately after graduation. Instructors need to provide guidance in making students aware of the options that are available so that they do not always take the path of least resistance.

Students also need abilities for analyzing and adapting to the culture of their organizations, and they need to know how to effect change within those settings. Beginners who do not learn the culture of the organization, but who see suggestions from others only as individual preferences, will be less effective within an organization (Freedman and Adam, 1996). Additionally, newcomers need to understand the organization's 'social system' in order to find persuasive reasons for change (Pfeffer, 1992). Because the technical communicators worked with many departments at Money Services, they held positions favourable to learning about the organizational culture. The most influential technical communicators needed to know not only writing processes, but also how to give co-workers persuasive reasons for what they wished to do. In addition to an understanding of general communication theory, students also need to know how to learn about the social dimension of an organizational activity system in order to generate reasons appropriate to co-workers.

Future Research

Although the socio-interactional tactics discussed here come from specific cases, they may have broader applications for students and technical communicators. Because this was a preliminary study, the following questions could profitably be answered in future research projects:

1. Do technical communicators in other organizational situations employ the socio-interaction tactics observed in this study? Are other similar tactics also used? How do varying situations affect the use of socio-interactional tactics?
2. How frequently do technical communicators of varying influence employ these tactics? How does use of these tactics lead to influence within an organization? What other sources of influence within an organization might technical communicators have?
3. How do apprentices learn and develop these tactics? How do they learn when to apply which tactic?
4. What reasons do technical communicators provide for their actions or reactions? How persuasive are those reasons with their audiences?
5. How do technical communicators in new organizations learn the organization well enough to begin to generate persuasive reasons for action?

Research attempting to answer these questions can enhance our theories of the practice of technical communication within complex social systems. From such research, we may also be able to design heuristics that help apprentices learn more quickly.

Conclusion

This study provides a basis for beginning to understand how technical communicators use social-interactional tactics within an organizational activity system, enhancing our understandings of a key aspect of the universe of expertise employed by technical communicators, as modelled in Figure 11.1. Awareness of these tactics can help students and beginning technical communicators moving into new situations to gain more influence over their work, to know when to hold, fold, walk away, or run. Such procedural knowledge or 'knowing in practice' (Lave and Wenger, 1991) can be used effectively to contribute to improving the quality of technical documents and the work lives of technical communicators.

References

Barker, T.T. 1998. *Writing Software Documentation: A Task-oriented Approach*. Needham Heights, MA: Allyn & Bacon.

Berkenkotter, C., and T.N. Huckin. 1995. *Genre Knowledge in Disciplinary Communication: Cognition/Culture/Power*. Hillsdale, NJ: Erlbaum.

Breuch, L.M.K. 2001. 'The Overruled Dust Mite: Preparing Technical Communication Students to Interact with Clients', *Technical Communication Quarterly* 10, 2: 193–210.

Christiansen, E. 1996. 'Tamed By A Rose: Computers as Tools in Human Activity', in B. Nardi, ed., *Context and Consciousness*, pp. 175–98. Cambridge, MA: MIT Press.

Cole, M. 1996. *Cultural Psychology: A Once and Future Discipline*. Cambridge, MA: Belknap Press.

Dobrin, D.N. 1997. 'Guest Editorial: Why I Don't', *Journal of Technical Writing and Communication* 27, 2: 105–17.

Falcione, R.L., L. Sussman, and R.P. Herden. 1987. 'Context: Internal and External Environment', in F.M. Jablin, et al., ed., *Handbook of Organizational*

Communication: An Interdisciplinary Perspective, pp. 195–227. Newbury Park, CA: Sage.

Freedman, A., and C. Adam. 1996. 'Learning to Write Professionally: "Situated Learning" and the Transition from University to Professional Discourse', *Journal of Business and Technical Communication* 10, 4: 395–427.

Green, M.M., and T.D. Nolan. 1984. 'A Systematic Analysis of the Technical Communicator's Job: A Guide for Educators', *Technical Communication* 31, 4: 9–12.

Grove, L.D., R.E. Lundgren, and P.C. Hays. 1992. 'Winning Respect Throughout the Organization', *Technical Communication* 39, 3: 384–93.

Hackos, J.T. 1994. *Managing Your Documentation Projects*. New York, NY: Wiley.

Hanks, W.E. 1991. 'Foreword', in J. Lave and E. Wenger, *Situated Learning:Legitimate Peripheral Participation*, pp. 13–24. Cambridge: Cambridge University Press.

Heneghan, B.B. 1987. 'The Composing Processes of Computer Documentation Writers'. Unpublished doctoral dissertation, George Mason University, Virginia.

Hovde, M.R. 1994/5. 'The Knowledgeable Practice of Computer Documentation Writers: Tactics for Constructing User and Software Images and For Negotiating Organizational Boundaries'. Doctoral dissertation, Purdue University, 1994. *Dissertation Abstracts International* 56: 9523365A.

———. 2001. 'Research Tactics for Constructing Perceptions of Subject Matter in Organizational Contexts: An Ethnographic Study of Technical Communicators', *Technical Communication Quarterly* 10, 1: 59–95.

———. 2000. 'Tactics for Building Images of Audience in Organizational Contexts: An Ethnographic Study of Technical Communicators', *Journal of Business and Technical Communication* 14, 4: 395–444.

Kuutti, K. 1996. 'Activity Theory as a Potential Framework for Human–Computer Interaction Research', in B. Nardi, ed., *Context and Consciousness: Activity Theory and Human–Computer Interaction*, pp. 17–44). Cambridge, MA: MIT Press.

Lave, J., and E. Wenger. 1991. *Situated Learning: Legitimate Peripheral Participation*. Cambridge: Cambridge University Press.

Miller, V.D., and F.M. Jablin. 1991. 'Information Seeking during Organizational Entry: Influences, Tactics, and a Model of the Process', *Academy of Management Review* 16: 92–120.

Nardi, B. 1996. *Context and Consciousness: Activity Theory and Human–Computer Interaction*. Cambridge, MA: MIT Press.

Philbin, A.I., A.M. Ryan, and L. Friedel. 1995. 'How Technical Communicators Feel About their Occupation: Facets, Attitudes, and Implications for the Future of the Profession', *Journal of Technical Writing and Communication* 25, 3: 303–20.

Pfeffer, J. 1992. *Managing with Power: Politics and Influence in Organizations*. Boston, MA: Harvard Business School Press.

Raven, M.E. 1992. 'Analyzing and Adapting to Multiple Audiences: A Study of Two Writers in the Computer Industry'. Unpublished doctoral dissertation, Rensselaer Polytechnic Institute, New York.

Russell, D.R. 1997. 'Rethinking Genre in School and Society', *Written Communication* 14: 504–54.

———. 1995. 'Activity Theory and Its Implications for Writing Instruction', in J. Petraglia, ed., *Reconceiving Writing, Rethinking Writing Instruction*, pp. 51–77. Mahwah, NJ: Erlbaum.

Schein, E.H. 1987. 'Individuals and Careers', in J.W. Lorsch, ed., *Handbook of Organizational Behavior*, pp. 155–71. Englewood Cliffs, NJ: Prentice Hall/ Simon & Schuster.

Sopensky, E., and L. Modrey. 1995. 'Survival Skills for Communicators Within Organizations', *Journal of Business and Technical Communication* 9, 1: 103–15.

Smart, G. 2000. 'Reinventing Expertise: Experienced Writers in the Workplace Encounter a New Genre', in P. Dias, A. Pare, and M. Farr, eds, *Transitions: Writing in Academic and Workplace Settings*, pp. 223–52. Cresskill, NJ: Hampton Press.

Sutton, R.I., and M.R. Louis. 1987. 'How Selecting and Socializing Newcomers Influences Insiders', *Human Resource Management* 26: 347–61.

Thomas, S.G. 1995. 'Preparing Business Students More Effectively for Real-World Communication', *Journal of Business and Technical Communication* 9, 4: 461–74.

Van Wicklen, J. 2001. *The Tech Writer's Survival Guide: A Comprehensive Handbook for Aspiring Technical Writers*. New York: Facts on File Checkmark.

QUESTIONS FOR CRITICAL THOUGHT

1. Why does Hovde open her essay with the scenario about the novice technical writer? What is the function of this narrative to the article as a whole?
2. Hovde asserts that technical writers need to know more about communication than simply the specifics of writing technical documents. Instead, she argues, they need to understand 'how to interact socially within their organizations'. To what extent is this also true of other professionals?
3. Does Hovde imply that professional education is failing to teach students these valuable skills? Compare her article with those by MacLennan ('Why Communication Matters'), Felder ('Whole New Mind for a Flat World'), and McKee ('Communicating Science'). To what extent do the views expressed by these various writers coincide? To what extent do they differ? Does the fact that similar observations have been made by multiple observers influence the credibility of the assertions? Why?
4. What is the function of the gambling analogy in the overall rhetorical impact of Hovde's article? In other words, what purpose does it serve for the reader? Does it contribute to Hovde's argument? How?
5. How readable is Hovde's article? What are the features that contribute to its readability?
6. According to Hovde, novice technical writers (and by extension, other new professionals) 'do not know how to understand and deal with internal organizational politics'. What are the reasons for this lack of understanding on the part of novices? Does it constitute a flaw in their education? (You may wish to compare Hovde on this point with Richard Felder's 'A Whole New Mind for a Flat World' or Jennifer MacLennan's 'Why Communication Matters'.)
7. Hovde points out that many of the political and power constraints operating in organizations are not ever made explicit, and suggests that this might be the reason so many new professionals find them difficult to negotiate. Why do you suppose that such constraints are left unstated?
8. Hovde's report presents what is essentially a set of case studies, bolstered by references to other research. How convincing is it? Why?
9. Hovde explains in detail how her data were collected. Why does she do so? Does this information contribute meaningfully to your understanding, or is it a distraction? Present evidence from the article to support your answer.

10. This is a research report, but throughout, Hovde speaks in the first person ('I'). How effective is this stylistic choice? What is its effect? Does her use of the personal pronoun in any way contradict her assertion that she wished to be as unobtrusive as possible while carrying out her observation visits? Why or why not?
11. Hovde never addresses this directly, but her report suggests that there is a relationship between interpersonal dynamics (what she calls 'socio-interactional' tactics) and persuasion or influence. Describe this relationship as fully as you can. What insights does this article provide into interpersonal relations and into persuasion as factors in a professional's career success?

PART III

The Case for Rhetoric

MANY of us are used to hearing the word 'rhetoric' used to refer to double-talk—political or advertising language used to manipulate and confuse an audience, a deliberate substitution of style in place of substance. *The Canadian Oxford English Dictionary* confirms this assumption, defining the word as 'language designed to persuade or impress (often with an implication of insincerity or exaggeration, etc.)'. In light of such a definition, it must seem odd to find a chapter on rhetoric in a book about scientific and technical communication. Surely what matters most to the communication of science is factual accuracy and logical deduction: no room there for the arts of deception so typically ascribed to rhetoric.

But before we throw the baby out with the bathwater, perhaps it would be useful to take a second look at what we're talking about when we use the word 'rhetoric'. That same dictionary offers another possibility when it also defines rhetoric as 'the art of effective or persuasive speaking or writing'. In other words, rhetoric can also be understood to be about how we convince other people to accept our interpretations of the facts, to understand what those facts mean, and to act in response to these findings. All of these are tasks we perform through our technical communication: we report the findings of an assessment visit, we present a design solution to a client, we record in a log book the data that appear to support the conclusion we expected from an experiment. In all these cases, we are hopeful of convincing others to see the facts in the same way that we have seen them, to find them as compelling as we do, and perhaps in response to take this course of action instead of that one. Whenever we do any of these things, we are, whether we realize it or not, employing the persuasive art of rhetoric.

The extent to which we are acting rhetorically when we practice technical communication is one of the central questions of this Part. Here, several writers and researchers of technical communication tackle head-on the question of how rhetorical technical writing really is, and how much the study of rhetoric has to offer to the professional communicator. To answer some of these questions, they turn to the oldest extant book on the subject: Aristotle's *Rhetoric* (circa 330 BCE).

Like the writers in this section, Aristotle discusses both the strategies that people can use to make their communication effective, and the practical reasons why they communicate in the first place. In the course of his discussion, he identifies three primary ways in which a speaker or writer can influence an audience: through a genuine concern for the audience's interests (what Aristotle called 'pathos'), through his or her own credibility and sincerity ('ethos'), and through a sound understanding and well-reasoned argument ('logos'). As several of the essays in this section confirm, this advice remains as sound now as it was when Aristotle taught it over two thousand years ago, and can be employed with just as good effect by the contemporary technical writer.

CHAPTER 12

An Engineer's Rhetorical Journey: Personal Reflections

Richard T. Burton

I want to make it clear at the outset of this essay that I am not an expert in communication. What I am is a professional engineer. As it happens, I am also a university professor who conducts research, supervises graduate students, and teaches many courses. Until very recently, I also did senior administrative work. At all of these tasks, I have been successful, and I always considered myself an able communicator—or at least, I kidded myself that my abilities were acceptable.

Like many practising professionals, I had assumed that communication is something that just happens. I had long ago learned some grammar and composition skills in a basic English course, and I had built on those skills through years of professional activity. As a successful professional engineer and teacher, I was satisfied that I knew how to communicate effectively—after all, engineers are expected to do so every day, both in writing and in oral reports, and you don't get very far in university administration without being able to write acceptably or make your point at meetings.

In fact, I was like most everyone else. Most of us take for granted that we can communicate; after all, we can talk sensibly to our friends and colleagues, present an occasional toast at a wedding, order something over the phone, and express our displeasure when something goes wrong. We are also able, for the most part, to successfully manage the communication demanded by our careers. By this measure, we are successful. Like most people, I was confident in my ability to make myself understood and to engage my audience effectively, and my success in my chosen profession was all the evidence I needed to convince me that I was as good a communicator as I needed or wanted to be.

That is, until relatively recently. Over the past few years, I have had the humbling pleasure of teaching a required course in communication to undergraduate engineering students. The course I was given to teach was unlike anything in my own undergraduate experience, where an introductory literature class or a 'technical writing' class that consisted of naming the parts of a report was considered sufficient to teach us how to communicate. Instead, the course that I was set to teach was one firmly grounded in the principles of rhetoric, a discipline that I had never even heard of until the moment I was assigned the class. It was to be a life-changing experience for me.

In keeping with what I would soon learn about rhetoric, the focus of the course was to teach the students how to 'discover, in any given case, the available means of persuasion' (Aristotle, 1954: 1355b). In other words, our goal in teaching the class was much more than to provide guidelines for writing technical documents; instead, the course was designed to train the students' judgement about how to position a message appropriately for a variety of audiences—to teach them to size up a situation and respond with an appropriate message tailor-made for that audience in that circumstance.

To my own surprise, the experience challenged, and eventually overthrew, many of my assumptions, both about communication in general and about my own communication in particular. Under the guidance of the course designer and other instructors who specialize in the study of rhetoric, I became very conscious of my weaknesses as a professional communicator, and I have learned to recognize and challenge some of the practices that I took for granted. I am aware that I have as yet only 'scratched the surface' of rhetoric as an area of study, and yet I know that the experience has changed—for the better, I believe—my approach to my own writing and speaking, as well as my approach to teaching. In the remainder of this essay, I want to share some of the discoveries I've made as a result of learning some basic rhetorical theory; I'm sure some of these examples will resonate with you. In the process, I hope to query the notion of what it means to communicate well, or to be a 'good communicator', especially within the professions.

Since rhetoric began as the study of public speaking, it is perhaps not surprising that my first rhetorical discovery came in that context. As an engineer, I am thinking in particular about seminars and formal paper presentations in my field. If you are familiar with such presentations, let me ask you this: When was the last time you really enjoyed a talk, or came out with something meaningful? When was the last time you encountered a speaker who seemed genuinely interested in communicating a message clearly rather than simply impressing a select and privileged few in the audience with a flashy visual presentation or a morass of jargon? Finally, when was the last time a speaker made the subject vivid and lively for you, without getting wound up in technical details?

Let me recast that same question from another perspective: If you were the author of such a presentation, did you take the time to really think about who was in the audience and what level of specialized jargon they could understand, or tolerate? Was your first thought about how you might clarify the information for the audience, or did you instead concentrate on a flashy computerized presentation or an impressive level of abstract conceptual language, and just assume that your audience would keep pace? If you thought more about your presentation than you did about making sure your audience understood you, here's some news: you just failed the most basic test of rhetorical effectiveness.

The idea that a speaker or writer has to think about the audience's needs, understanding, or level of interest might seem simplistic to someone who has mastered fundamental rhetorical principles, but it's a potentially revolutionary assumption for many writers and speakers. The sad part is that, though most of us may know we should work to accommodate our audience, most of the time we focus more on content than we do on making that important connection. Almost everyone can tell you that audience matters, but for every writer or speaker who can actually put this awareness of audience into practice, there are a hundred who can't.

Like most professionals, engineers are often guilty of assuming they are better communicators than they actually are. Part of this assumption is based on the fact that, much of the time, their audience consists of other technical professionals who share the same views, use the same jargon, and have similar levels of professional experience. In such an instance, accommodating the audience seems an insignificant point, since the engineers' communication appears to serve them adequately. But let me digress a bit and explain to you about why I no longer take it for granted that engineers and other professionals can communicate as well as they believe they can.

The failure of audience awareness that I see among my fellow professionals is clearest when we have to communicate with those outside of our own professional community, who may not share the same knowledge, assumptions, or attitudes. The question for many of us is not whether we can adapt to the expectations and needs of a different audience and 'shift gears' appropriately; it is more an issue of whether we even know that such a shift is required.

Perhaps an example from my personal experience will help to illustrate what I mean. The area of 'controls engineering' in which I work has become so entrenched in jargon and mathematical complexities that it is almost impossible to describe even the most basic concepts to a non-control engineer. Thus, an important audience—industry—is very cautious about supporting research in this area or implementing the theory in a practical setting, mainly because they cannot understand what is being said. Such failures are costly, both in terms of lost financial support and in terms of delays in implementing some very powerful ideas, and these problems are the direct result of publications and presentations that ignore the primary audience and address only those whose understanding of the jargon and mathematics equals that of the researcher. I must admit that in the past I too have been guilty of overlooking my audience, but now that I have learned to think about their needs, I have made it my own personal objective to change the attitude of my colleagues in my own area of specialization. This hasn't been an easy task.

My personal discovery and appreciation of rhetoric have not been without some repercussions. The clearer your publications and discourses are, the more susceptible you are to critical review, because if the audience can understand your message, they are more apt to challenge or critique your ideas. But as intimidating as it can be to have your ideas debated and challenged, isn't simulating discussion exactly what you want to do? Who knows, the critic's challenge might just draw your attention to some detail or idea that you'd overlooked; in turn, that renewed attention might be the very thing that allows you to preserve your credibility in future publications.

I have always subscribed to this philosophy, but it was only in watching experts practise rhetorical communication that I realized how much more I still could do. By practising what I have learnt from my mentors, I have been able to stimulate—and welcome—more debate about my ideas than I was comfortable with in the past, and I even feel that I'm receiving a more positive reception when I speak in public. And yes, I have sometimes been found wrong in some of my ideas, but this has not been nearly as embarrassing a situation as I had expected.

My growing understanding and appreciation of rhetoric have also affected my work with graduate students, I believe for the better. I now expect them to be able to clearly explain their research concepts in terms that a person outside their area of expertise can readily understand. For some, this has been a very frustrating experience because they want to 'get on with it' and not spend time with this kind of detail. However, I have discovered for myself what rhetoricians already know: that in many cases, a strong reluctance or inability to explain concepts often masks a serious lack of understanding. Once we have identified the lapses in understanding and cleared them up, I feel that most of my graduate students do appreciate the effort to bring clarity to their communication, especially when it comes to oral exam or paper presentation time.

It isn't my intention to be too negative about the abilities of students, engineers, or other professionals; in fact, I think they are often victims of their working environment, which in many cases actually endorses this kind of behaviour. For example, in submitting reports, memos, and theses, or in delivering presentations, an engineer is frequently required to follow 'in-house' standards or recipes for format and delivery. I am not referring here to the use of an outline structure, which can work very well in assisting a writer to organize ideas; instead, I am talking about templates intended to standardize the organization and content of certain routine tasks. These are, in some respects, a necessary evil: they are meant to provide some consistency in the kind and order of material presented, and to ensure that details are not left out. If they are measured purely in terms of efficiency or uniformity, these templates can be successful. However, this same uniformity can also stifle the writer's creativity or personality, rendering the report lifeless or even unreadable.

For all these reasons, I have never liked the template-driven approach, but until I began to learn about rhetoric, I couldn't articulate why I was resisting it. Now I realize the reason for my instinctive response: a template necessarily focuses the writer's attention away from the interaction with the audience and onto a standardized one-size-fits-all cookie-cutter structure. For inexperienced writers, the 'packaging' can even come to take precedence over the purpose of the communication, so that the only things that can be said are those that fit the template. As a result, the format that was intended to assist a writer in organizing and presenting information can actually distort important elements of the message by casting emphasis on unimportant details at the expense of more significant information.

To overcome the limitations of the template, we might do well to ask some questions about its origin and its value to what we're writing. For instance, who developed the report recipe, and why? Was the format developed in-house, specifically to accommodate our own circumstances, or is it some generic package that someone paid big bucks for? Does the template adequately serve our communication needs, or do we find ourselves having to skew our messages to fulfill the demands of the template? A knowledge of rhetoric can help us develop the skill and the confidence to dump the template in favour of a format that more effectively accommodates the needs of the audience, the shape of the message, and our own standards as writers. If you work in an environment that typically uses the template-driven approach, then of course you will want to run your work by your superiors for their comments, but if you can achieve your goal of a clear and effective message, everyone wins.

Another failure of professional culture when it comes to clear communication is the proliferation of jargon and acronyms. Engineers aren't alone in being unable to adjust their language use to the demands of audiences other than their peers. Everyone who works in a specialized field—from the physician who looks after you to the mechanic who fixes your car, from the salesperson who shows you the latest in personal digital assistants to the accountant who helps with your tax return—is steeped in specialized language, and it's a struggle for all of us to think outside of the buzzwords we've been trained to use. But thinking outside those buzzwords is exactly what we need to learn to do if we're going to communicate clearly to the audiences of non-specialists with whom we regularly interact. Think of the last time you purchased a computer, and recall the jargon that the clerk threw at you, when all you wanted to know was how to connect to the Internet. In effect, the rest was gibberish. How much gibberish do we routinely present to our own readers, without even being aware of it? The answer, I'm afraid, is often 'too much'. For example, I have just finished reading a graduate thesis that had over 25 acronyms on a single page. Even with a reference page beside me, I could not decipher the writer's meaning, and I finally gave up. Too bad for me, you might say. But it's too bad for him if his thesis fails because the committee could not understand what he had written.

In the technical world, such examples are all too typical, and unfortunately, we have learned to tolerate them in too much of our communication. Writing to accommodate the reader's knowledge level is viewed by some as 'dumbing down'. Not so: in fact, it is the writer's job to communicate ideas clearly. Frequently, the audience isn't 'dumb'; in fact, often the only limitation they have is a limited knowledge of the subject matter and a lack of knowledge of the jargon imposed on them by a careless writer. Sadly, this fondness for obscurity affects the profession at all levels.

In my experience, one of the worst contributors to bad writing these days is email, which has proliferated to the point that it is becoming a nuisance. Email messages have replaced not only letters and memos, but also phone calls and even personal visits. Indeed, have you ever caught yourself sending an email to the person in the next office instead of simply poking your head around the door?

This email problem is exacerbated when the written message is so obscured by thoughtless errors that we can hardly make out the meaning. Who would think of sending a letter or report that contained no capitals, no punctuation, with every second word abbreviated and not a single verb to be found? Yet many people, including professionals who should know better, think nothing of dashing off an email message in exactly this form. Even if the recipient normally isn't fussy about matters of grammar, spelling, and capitalization, messages that completely ignore these details are actually much harder for the recipient to read and understand. In the end, what you're communicating loud and clear to your colleagues and clients is how little regard you have for their time and how little concern you have for the issue at hand. If the message is important enough to send and email, it's important enough to spend a couple of extra minutes to get it right. If you have trouble catching the errors until after you've sent your message, try sending it first to yourself. You might spare yourself some embarrassment.

Teaching this course has sensitized me to the challenges of accommodating my audience on several levels, and as a result I am acutely aware of their needs and expectations, of the message content, and even of the appropriateness of responding or sending the information in the first place. Indeed, I now spend much more time on the email messages I send out, because I have become more sensitive to what I read from others.

Like many other professionals, I had taken my communication for granted. But now I wonder what makes so many of us believe we are experts in communication, just because have some experience in writing reports and making a few presentations. I don't think of myself as an expert electrician because I can wire my home based on the building code (a kind of recipe); nor do I think of myself as a medical specialist because I can learn to take my pulse, blood pressure, and temperature. Why then are we so quick to think we have nothing more to learn about communication, when it's so clear that the opposite is true? I now believe that engineers and other professionals can benefit immeasurably by admitting that they could use some professional guidance to become better communicators, and that we could all profit from learning the basics of rhetorical judgement: how to understand the audience's expectations, how to adapt our message to those expectations, and how to present ourselves credibly.

The first step to communicating more effectively is to realize that recipes or cookie-cutter procedures are not enough if you want to understand what you're doing and why. Nor is it enough to run your report through a spell-checker and grammar-checker. Though these can help to catch errors, by themselves they won't make your writing effective. Report or memo templates, too, can offer rudimentary guidelines, but becoming an effective communicator, like becoming a skilled and effective engineer, requires much more thoughtful consideration. The second step is to answer in your own mind the following: Who exactly is your audience? What is the point of the message you want to share with them? Finally, what exactly is your expertise in the material being delivered (in essence, what is the source of your credibility)? If we actually take the time to think through the answers to these questions before we create a message, we can improve our communication significantly.

In all fairness, engineers and other professionals are beginning to recognize the importance of becoming better communicators. Those who are self-employed or who deal with the public, for example, come to realize the importance of understanding the audience's needs and of talking to them in language they can understand. If they don't learn these things, they do not survive for long in business. I just recently talked to a self-employed engineering entrepreneur who reported that miscommunication had cost him several contracts, and who observed that it was only after a great deal of soul-searching that he could admit to himself what the problem was. This same realization seems to be growing across the professions, as companies large and small invest in communication training for employees at all levels. What other professional workshops see CEOs and custodial staff learning the same material together?

Of course, if you're after real improvements in your communication skills, the trick is to pick the right course. Getting a new set of recipes is not the answer to improving communication skill—what you need is a better understanding of the dynamics of the interaction between speaker, audience, and message. Before you invest time and money, investigate a bit: Does the workshop include some sound base of principles that will help to develop your skills of assessment and response, or does it simply provide more templates and recipes?

One last discovery that I have made in the last few years is the realization of my deficiencies as a teacher. Don't get me wrong: I have always considered myself a good instructor and have received very positive feedback from my students. However, teaching a communication class quickly becomes very humbling because of the way it shines a spotlight on your own classroom practices. Unlike in other courses, in the communication classroom your delivery and your message are one and the same; truly you must demonstrate what you are saying in your own practices, or your credibility and indeed the message itself can soon evaporate. If you fail, you will not only have lost your own credibility but also have lost your audience for the whole subject. I have come to realize the dangers of having a communication course taught by someone who is not an outstanding practitioner in his or her own right. Fortunately for me, my mentors and teachers of rhetoric were experts, and because of their guidance and my willingness to accept an approach based on sound theory, I received the best student evaluations I have ever had over a span of some 30 years. I have seen for myself that a solid foundation of rhetorical principles works, and as a result I feel that I have made a contribution to these students' communicational well being which will stay with them the rest of their lives.

I feel that I have always applied the basic elements of the 'rhetorical triangle' of speaker–audience–message to all my lectures, papers, and presentations without knowing formally about the triangle. However, it has only been since I understood the interrelationship between the elements and their importance that I have became aware of areas where I could improve my communication. Were my expectations from the students driven by sound reason or by convenience and past practice? Because of my exposure to rhetoric principles, I now ask myself these questions each time I give a lecture, assign a problem, or grade an exam.

I began this discussion with a promise to share some of the discoveries I've made as a communicator, and I would like to end by returning to this idea. In writing this essay, I have in some sense turned the spotlight on my own skills. I am an academic with many years of experience, and have admitted—perhaps even illustrated—my own limitations as a communicator. Writing about communication when I'm still learning to master the fundamentals of rhetoric myself is a task full of risk, and I realize that I may have blown my credibility. But I am hopeful that by taking these risks I have tempted you to learn more about what it takes to be a better communicator, and in so doing, that I have taken the first steps to becoming a better communicator myself.

References

Aristotle. 1954. *Rhetoric and Poetics*, W. Rhys Roberts, trans., Fredrich Solmsen, ed. New York: The Modern Library.

QUESTIONS FOR CRITICAL THOUGHT

1. Why does Burton begin by declaring that he is *not* an expert in rhetoric? How does this admission affect his credibility?
2. What is the purpose of Burton's essay? Can you find its thesis?
3. How does Burton define rhetoric? Is his treatment of the term similar to or different from that provided in Smith's 'What Connection Does Rhetorical Theory Have to Technical and Professional Communication?' Why?
4. Compare the style of Burton's essay with Smith's. What differences do you see? As you develop your answer, consider the vocabulary, purpose, tone, voice, and audience relationships each employs or establishes.
5. What does Burton consider the 'most basic test of rhetorical effectiveness'? Does he himself pass?
6. What is the reason, according to Burton, that engineers' communication often fails? Compare his assertion with the claims made by Urquhart in 'Bridging Gaps, Engineering Audiences: Understanding the Communicative Situation', Felder in 'A Whole New Mind for a Flat World', or McKee in 'Communicating Science'.
7. How formal is this essay? How personal or impersonal is it? Why might Burton have chosen to write in this way for this essay?
8. Who is Burton's intended audience? How do you know?
9. How has his new understanding of rhetoric affected Burton?
10. What's wrong with the 'template-driven approach' to writing that Burton speaks about? How does it differ from the 'heuristic' approach that Smith describes?

11. Burton suggests that poor communication is at least partly the result of engineering culture. In what respect is this so? Have any of the other contributors to this book made a similar point?
12. How much freedom does an individual have to depart from a profession's standards and expectations?
13. What does Burton mean by 'rhetorical judgement'? When such a judgement is rendered, what is it that is being assessed?
14. Does Burton make an effective case for the value of rhetoric to an engineer? Do his views have anything in common with those of Neil Ryder in 'Science and Rhetoric'?

CHAPTER 13

Science and Rhetoric

Neil Ryder

On Language, Metaphor, and the Communication of Science

Ours is a culture overrun by persuasive messages, with science a weapon in the persuaders' armouries. Glance at an ad for hair care products; overhear a politician defending their actions on BSE. These persuasive acts are examples of the art of rhetoric—examples in which science and rhetoric are intimately connected.

'Rhetoric' is a term that sits very uncomfortably with science, especially in empiricist Britain. Yet as well as these examples of explicit persuasion, attempts to use the ideas of science in newspaper articles or radio and television programs are also rhetorical acts. What sense can we make of such a conjunction between science and rhetoric?

There are a number of possible responses to the suggestion that science need be considered alongside rhetoric. For instance, some may feel that the intellectual standards of the one, rhetoric, are incapable of meeting those of the other, science; worse, they may be actually incompatible or even antithetical. But another response is that the very practice of science is governed by rhetoric.

Some forms of communication are clearly typical of science and as such form a rhetoric of science. Peter Medawar pointed to a discrepancy between the way science is formally reported amongst scientists and what scientists actually do. He advocated that the discussion part of the scientific paper, usually relegated to the end of the paper, should be presented right at the top, so that the openness and the tentativeness of the whole debate are foregrounded. Since then linguists have taken a close look at scientists' writing practices and although it appears that there is some variety in the way structures appear across the different scientific disciplines and journals, the rhetorical dimensions remain.

When we turn to the presentation of science to the public some of the forms of presentation that scientists use amongst themselves have to be abandoned. There is considerable work necessary to transform ideas from the scientific sphere to even the semi-popular pages of *Scientific American* or *New Scientist*. These transformations must be wrought in every aspect of a text, from its overall structures, to its grammars and vocabulary. On the whole many of the changes can be represented as simple instructions; putting, for instance, the familiar before the esoteric, introducing living actors instead of inanimate substances, preferring active grammar for passive. We can think of this as a fairly routine translation strategy for science journalism.

But at this point a major choice emerges and the alternative you choose will depend on your view of science. The choice is: are you comfortable with a situation where scientists draw conclusions, make discoveries, and then have them translated somehow into everyday language and leave that as the extent of the intercommunication between the two parts of the culture? Or do you believe that the quality of argument in science itself is no different from that which applies to, say, literary criticism, or political debate?

If you believe the former, then scientific ideas and conclusions stand clear of the accidents of the language and of political struggles in lay society—the society into which we hope they will be received—and you will not have much trouble with the translation strategy. But if you believe that the quality of argument in science itself is no different from that which applies to, say, literary criticism or political debate, then clearly the skills and the status of the scientist have to be different. Scientists have to come down off their pedestals and hustle for a hearing along with the rest of us. The ideas and conclusions a scientist entertains and accepts are the product of processes no purer than any other intellectual activity.

Scientific knowledge has no God-given authority over poetry, politics, or even fine art, and the system of evaluation or selection and approval are analogous, if not identical, in the different fields. Science journalism here is no longer translation. It is the writing of new stories with quite different characters and relationships.

If we now turn to scientific institutions and their policy decisions about language then we find that a distaste for metaphors lies at the heart of modern science. A well-known campaign against the use of metaphor was planted in the origins of the Royal Society.

Spratt, in his manifesto for the young Royal Society, called for scientists to eschew all tropes including metaphor. 'Give me as many things in so many words, give me as many ideas in so many words.' Spratt's argument—his intuition—was to cast aside literary elaboration. Metaphor does elaborate. Metaphor names things 'incorrectly', so in one sense it describes an incorrect referent. Spratt's desire to 'cleanse' the language of science arose from the perceived power of metaphors to inflame passion in one of the most internally poisonous periods of English domestic history. Yet Spratt's ideas about good language have a surprisingly modern resonance. The idea that language should stay close to the speech of the artisan—that the language of Anglo-Saxon origin is close to experience, whereas the Latinate is remote from it—finds sympathetic echoes in F.R. Leavis, I.A. Richards, and many other commentators. But the language of science no longer confirms that experience.

An examination of the way in which scientific ideas gain credence within a scientific community, shows that the role of metaphor is absolutely crucial. At a certain point the scientist will encounter a conceptual problem and in order to try to solve that problem, a leap of the imagination is necessary. All the scientist can do at that stage is try to find something that fits, something necessarily from your existing experience, from some parallel world. Hence Niels Bohr used the idea of the solar system as an analogy for his model of the atom. And Crick and Watson used the idea of code to imagine what is going on when DNA splits up, reformulates, and gets transcribed—how patterns, information, are being passed down the line. In all this metaphors are absolutely crucial to the initial description of scientific models. If metaphor forms the basis of scientific imagination, scientists cannot afford to throw out the whole of rhetoric. They need to accept that some of it must be useful.

It is the consequences of science for people and their physical world that matters. Galileo, that arch-rhetorician, defended himself against the Church's threats of torture with carefully constructed arguments. If his story teaches us anything, it is that Science for People is a struggle against powerful institutions and that, ultimately, institutionally endorsed torture is ineffective against the art of rhetoric.

QUESTIONS FOR CRITICAL THOUGHT

1. In his opening sentence, Ryder describes science as a 'weapon in the . . . armouries' of persuaders whose messages have 'overrun' the culture. What kind of metaphor is he working with here, and what's its effect? What does it contribute to his overall argument?
2. Ryder uses a number of terms that he doesn't explicitly define, such as 'rhetoric', 'literary criticism', and 'science journalism'. Why doesn't he provide definitions for these terms? Is he right to expect that his audience will know what they are?
3. How does Ryder appear to be using the term 'rhetoric'?
4. What does Ryder seem to be talking about when he refers to a 'rhetoric of science'?
5. Ryder declares that 'there is considerable work necessary to transform ideas from the scientific sphere to even the semi-popular pages of *Scientific American* or *New Scientist*.' What sort of transformation is needed, in Ryder's view? Why are such changes necessary?
6. What is the purpose of Ryder's article? Can you find a thesis statement?
7. Who are Ryder's intended readers? How do you know?
8. Ryder outlines two sides of a debate about the extent to which science is rhetorical. What are the two opposing views? Which, if either, does Ryder share? How do you know?
9. How well organized is the article? That is, how strategically has Ryder crafted his argument? Can you see any aspect he might have improved upon?
10. Who are Spratt, Niels Bohr, Crick and Watson, and Galileo? Who are F.R. Leavis and I.A. Richards? Why does Ryder not identify them? Why does he refer to them at all?
11. Ryder closes his article with what might be considered a restatement of a well-known cliché ('the pen is mightier than the sword'). Which of the two versions of this sentiment is the more effective? Why? Does this statement provide an appropriate conclusion to this article?

CHAPTER 14

What Connection Does Rhetorical Theory Have to Technical and Professional Communication?

Tania Smith

Introduction

In this essay I provide an overview of rhetorical theory and theoretical terms in everyday language for the use of working professionals and post-secondary students of communication. Because rhetoric is a vast and complex field, I will focus on the best-known aspects of Western European rhetoric grounded in Classical rhetoric, giving an overview of its key principles and terms and their applications in a variety of modern settings.

Rhetorical theory is worth studying because it can be used to produce communication as well as to analyze or critique communication. It is both a practical skill and a philosophical humanistic study. Communicators who grow accustomed to rhetorical concepts and terms can use them among each other in order to defend their own strategic choices and to improve each other's communicative effectiveness.

Rhetorical theory is probably as old as humankind, but its textual record can be traced back to the rise of democracy in ancient Greece, where it was popularized by its early teachers: Aristotle (*On Rhetoric*), Isocrates (*Antidosis* and *Against the Sophists*), and Plato (*Phaedrus*). Cicero (*On Oratory*) and Quintilian (*Institutes* of *Oratory*) further developed theories of rhetorical education and practice in the ancient Roman empire. Rhetorical theory is culturally adaptable, surviving more than two millennia and forming new branches even as it was transplanted to various nations in Western Europe and, more recently, North America. Other theories of rhetoric have been found in the history of China and other non-European nations. Today more than a hundred graduate (MA and PhD) programs in Rhetoric and/or Professional and Technical Communication rely on rhetorical theory as a foundation for the training of modern scholars and public, professional, and workplace communicators.

In a nutshell, rhetorical theory is a branch of communication theory that is concerned with the production and reception of persuasive discourse. The average person may think of advertisements or political speeches when hearing the terms 'rhetoric' and 'persuasion'. However, modern rhetoricians have broadened the definition of rhetoric, and of persuasion itself. Rhetoric includes all forms of communication intended to have specific effects on a target audience. These effects may include various degrees of informing, entertaining, and persuading. Informing and persuading are the primary end results of communication used in workplaces, used by professionals, and in the communication of science and technology, as well as in communication used for political or commercial purposes.

What is 'Rhetorical' Communication?

Rhetoric encompasses the functional and strategic uses of communication to change the way people think, believe, feel, and act. Basically, rhetoric is occurring whenever a communicator draws on communication principles to develop a strategy using language (including words, symbols, and gestures) to influence a particular audience (or audiences) for a particular purpose (or purposes).

Crafting rhetorical communication involves a conscious process, and presumes an author who intends a specific range of outcomes for his language use directed at others. Because it is conscious and strategic, rather than purely intuitive and inspirational, rhetoric is an art that can be taught and learned, and demands skill with language and close observation of one's social context.

In this category of 'rhetoric' or 'persuasive discourse' can be found many novels, graffiti, and political speeches, but also business memos, reports, meetings, and other genres used in technical and scientific communities and in a variety of professions.

Is every form of technical and professional communication persuasive? The majority of rhetoricians would say yes, because teaching and informing are necessarily persuasive acts. Rhetoricians have argued that even when merely describing an object to someone else, that description includes a unique perspective (personal, cultural, functional) that is not actually representative of the object itself. In even the driest technical description, features of an object are selected and omitted, de-emphasized, and enhanced. This is necessary because most descriptions are structured to be useful to an audience in their social context and must be efficient rather than comprehensive. For example, in a description of the pencil-making process found in a document sent to prospective investors in a pencil factory, we would not see a description of a pencil lead that goes on and on for pages (or verbally, for hours), but one could indeed write one if one wished to be as comprehensive as possible about every detail down to the microscopic and chemical level. Therefore, even simple descriptions are in fact persuasive of the hypothesis that 'this is all you need to know—a faithful and relevant description of this object (or phenomenon)'.

However, not every act of communication is rhetorical, and not all communication theories provide a rhetorical perspective. Imaginative literature and art often have rhetorical functions, yet literary and aesthetic theories emphasize representation and the freedom of aesthetic appreciation and creative interpretation rather than the persuasive or didactic functions of communication. More importantly, rhetoric does not have to do with every grunt and groan or every word we say. Rhetoric also excludes merely 'expressing oneself' in a haphazard way to others in general, or to oneself (however, some have theorized our 'internal rhetorics').

Rhetorical Theory and Education

Rhetorical training includes guided practice and the study of both the production and the reception of persuasive discourse. When we study how to produce rhetoric, *heuristics* are useful. These are the various sets of questions, steps, and tips one often sees in textbooks and 'how-to' instructions addressed to communicators—how to research one's audience, how to structure a certain kind of proposal, how to choose emphatic verbs or metaphors, how to label graphs, etc. These are helpful as we continually teach and learn from each other how to produce effective rhetoric. The Classical canons of invention, arrangement, style, memory, and delivery are a representation of the process of crafting communication from start to finish.

But if we were to focus exclusively on heuristics, rhetors would only learn to answer the question 'how'. Heuristics can be so persuasive through their powerful efficiency and simplicity ('Step 1, Step 2 . . . ') that inexperienced learners find in them little room for questioning, discussion, or debate. Certainly we may gain handy tips that put us to work as communicators, but the more specific the tips are, the more often they can become inflexible guidelines, like the 'five-paragraph essay' students have learned for decades in Canadian high schools. The heuristics often focus on the rhetorical forms or genres produced by professional communicators (media releases, FAQs, usability test reports), and rest on assumptions about the requirements of communication technologies and organizational or social relationships.

To guard ourselves against making the assumption that 'frequent practices' of communication are in fact the 'best practices' in our situation, we must continually ask the essential question 'why'. Why is it better to say something one way rather than another? What difference will it make for the communicator and audience? Who benefits, who is excluded or put at risk, and is it the most efficient way?

That's why rhetoricians infuse their apprenticeship in the production of rhetoric with the study of its reception, which we call *rhetorical criticism*. While study of its production involves *heuristics*, the study of its reception involves *hermeneutics*, or theories of interpretation and audience reception. Rhetorical criticism builds knowledge about how various types of people in a variety of situations will likely interpret and react to other people's rhetoric, and why.

Because all the variables in rhetoric are subject to change, critical studies in rhetoric build the discernment skills and general knowledge needed for a rhetorician to discover and deploy what is most persuasive in a given situation. The three appeals of *ethos*, *logos*, and *pathos* are helpful in analyzing persuasive strategies and their potential effectiveness.

The distinction between production and reception is not absolute—there is a significant degree of overlap between the functions, not merely a symbiotic relationship. Rhetoricians have pointed out that the acts of listening and reading are just as constructive and productive as creating one's own message, because inevitably the audience collaborates with the text and its author to fill in the unstated assumptions and to produce new meanings relevant to the present situation. Also, in order to produce effective rhetoric, we must continually imagine ourselves in the position of the target audience listening or reading to what we are saying.

Rhetoric and the Production of Communication

Rhetoric does indeed encompass the full process of communication, not just the act of composing the *texts* or messages themselves. The reactions of *audiences* are an essential part of evaluating the effectiveness of rhetoric, and hypotheses about the audience's reactions are the continual preoccupation of every good rhetor. Rhetorical theorists also look at the social and physical *contexts* that gave rise to the need (or 'exigency') to communicate for a purpose, for these contexts give form to the communication. They also consider the *actors* (called 'Rhetors') and their cultural and educational backgrounds.

Traditionally rhetoric has named the processes of producing communication through 'The Five Canons': Invention, Arrangement, Style, Memory, and Delivery.

INVENTION

Invention is what you do when you discover what to say. Many people are familiar with 'brainstorming', which would be included here. Brainstorming and interviewing heuristics aid the rhetor to gather information about the communication context, the content of what should be said, and the general argumentative strategies.

In the larger sense, rhetorical invention includes what we understand today as 'invention'—the research and development that goes into any new scientific discovery, consumer product, social system, or technology. Communication is essential to the communal creation of knowledge, as none of us creates 'ex nihilo'. Creativity builds on the communication of previous generations and all the communication of our past and present relationships. Any comprehensive theory of rhetoric or technical communication that pays no attention to invention is sadly incomplete, as it matters greatly *how we discover what we say*, not only how we say it.

ARRANGEMENT/DISPOSITION

Arrangement obviously comprises the processes and principles of organizing the communication product so that the audience can receive it in an optimal way. Communication handbooks subdivide speeches and reports into sections, and communication textbooks and teachers today also name and characterize the various sections or parts of technical reports, sitcoms, novels, and websites.

STYLE

Style can be thought of as arrangement at the microscopic level, with an intangible 'flair' of connotation that is often called 'tone' or 'voice'. The choice of one's words and their arrangement produce various types of humour, levels of formality, and degrees of complexity. Grammar rules and dictionaries explain the rudiments of style, but rhetors/communicators are the ones who give language 'style'. In this phase of composition, rhetors pay attention to copyediting and the rhetorical nuances of sentences and words. Visual rhetoricians decide whether to use a sepia photograph with the appearance of age (cracking or curling), or an image of the same subject made using a camera filter that softens or blurs the image. Documentary filmmakers may coach their interviewers to use a certain phrase when they ask a particular question.

For the last few centuries, rhetoric has often become limited to a matter of style, perhaps because of the scientific interest in language at the microscopic level. A large number of rhetorical terms were developed in the Greek and Roman eras to help explain the varieties of figurative language. In the late Renaissance, style became the primary focus of many forms of rhetoric, making it seem as if rhetoric was only meant for elaborate, formal, ceremonious communication. In the enlightenment, many scientists and authors reacted against this trend and started to theorize a 'plain style'.

MEMORY

Orators (public speakers) were the focus of ancient rhetoricians, and so they had theories and practices in place to help rhetors memorize their speeches before delivering them. All good public speakers rely to a great extent on their memories so that they can give their full attention and eye contact to their audience rather than reading from notes.

Modern rhetoricians have been interested in how memory in general, and even public memory, influences communication. But when the canons are viewed as the rhetor's process, memory is all about ensuring the stabilization of the communication product prior to its delivery. It is not really about the audience remembering the message after it is delivered, or how it relies on an audience's memory of prior events or messages.

Memory has to do with the communicator or writer making a reproducible record of a message that has already been well-crafted. In our modern developed cultures, the canon of memory may be seen as the act of writing something down as the 'final draft' after you are finished researching, organizing, writing, and editing. You have already selected and arranged your text and/or images and sounds. Memory involves recording or finalizing the content and form of the message itself, prior to its adaptation to various media that make it ready for delivery and reception.

Of course, one must keep delivery (mentioned next) very much in mind when considering font sizes, colours, sounds, and time limitations. Memory is analogous to the acts of typesetting and printing the book or magazine, of hanging the art in the gallery prior to opening night, and of uploading the website prior to its public launch. But in some situations, the acts of memory and delivery are inseparable—the sending of the final version of an email message, for example.

DELIVERY

In the ancient context that formulated this set of rhetorical canons, delivery was the last step in the rhetor's job—the performance itself in front of the target audience. Action. Preparation for this part of one's art included voice training, tips on how to wear one's clothes, how to use gesture, facial expressions, and the like.

As mentioned above, in modern print-oriented and new media cultures, 'delivery' is often difficult to disconnect from memory, especially since the acts of memory and delivery are no longer limited to time and space. Memory and delivery can become two perspectives on the same step in the production process. However, one can consider delivery as a more audience-centred part of the process than the act of memory. Delivery is simply that—the way in which we deliver recorded or printed messages to the audience via mail, email, a website, architecture, or telephone.

In complex communication processes, such as those occurring in organizations and in public media, often publishers, media producers, and event planners have more control over this final process of rhetorical production than do the original authors or speakers. The author has already supplied a stable record of the approved message, but the delivery phase is when we adjust the message's clothing, when we frame its images or place the ad on page seventeen, or when we time the communication's appearance appropriately for the situation of its reception.

An important part of delivery is situational psychology: getting an audience's attention and delivering one's message clearly within a certain setting. Delivery considers the role of one's own communication within the local or global ecosystem of communication. When the audience receives the message, will the message be packaged together with other discordant messages? Will the physical setting be dark, and require the message to have lights and colours? Is the audience going to be sitting down listening to your message, or moving through the space where your message is posted? When will the audience need silence, white space, or an opportunity to get refreshments? Will it be delivered at the end of a workday? Will you adapt the introduction or delay the release of certain messages if you hear that a major public disaster has just occurred?

Ethos, Logos, Pathos

Aristotle first developed the three types of rhetorical appeals, *ethos*, *logos*, and *pathos*, which are very versatile and useful terms for rhetorical criticism and production even to this day. Generally speaking, *ethos* includes appeals to the credibility and character of the speaker, *logos* concerns the content and argumentation of the speech, and *pathos* evokes the emotions and values of the audience. All three are factors in almost every step in the process of communication, and almost every passage in a speech or text.

Modern teachers of rhetoric have developed what is known as 'the rhetorical triangle'—a useful visual model that places ethos, logos, and pathos on each tip of the triangle. Often there is a circle around the whole triangle representing 'the rhetorical situation'—the time, place, social context, etc. One drawback of the triangle model is that it makes it seem as if an 'ethos' strategy cannot simultaneously be a 'logos' or 'pathos' strategy. Not so. A strong argument may simultaneously provide evidence for a claim, arouse emotion in the audience, and enhance the credibility of the speaker.

It is not necessarily the case that all three appeals must have equally high frequency or intensity in order to make communication effective. A particular communicative act may be most effective when it emphasizes only one or two of the appeals. For instance, a job application must naturally emphasize ethos. When applying for a job in Canada, however, intense pathos appeals (i.e. evoking pity for one's unemployed status) or lengthy logos appeals (detailed narratives of workplace experiences) would usually be inappropriate.

In addition, *either increasing or decreasing* the emphasis on one or more of the appeals may make communication more effective:

- *Pathos* must be handled with care. Some extreme uses of pathos can be unethical or manipulative, especially when ethos and logos are weak. Yet pathos is a factor in every form of communication even if the use of pathos is nearly absent, for when communication is supposed to be objective and unemotional, even the absence of extreme emotions, or the presence of 'calmness' in the audience, may be a desirable 'pathos effect'. On the other hand, when nonprofit organizations appeal for donations, it would be ineffective *not* to rely on pathos, since powerful emotions and values are needed to inspire or reinforce an audience's behaviour.
- *Ethos* works in a similar way as pathos. A writer or speaker may effectively build her ethos indirectly (by being clear and well organized) or situationally (an ethos formed 'extrinsic' to rhetorical communication, through her prior reputation). At other times, when an audience does not know an organization well or doubts its credibility, it is necessary for public relations communication to directly describe the organization's values and contributions (this is called 'intrinsic' ethos because it is constructed within rhetorical communication). Without a foundation of good ethos, it is difficult to retain an audience's attention and good will. Certain communication faults such as grammar errors, foul language, and disrespect for audience diversity, are likely to harm ethos in the eyes of certain audiences, thereby preventing other appeals from being effective.
- *Logos* is necessarily part of any message, even at the sentence level. Logos, the thought-content of communication conveyed through logic and effective arrangement, is a basic requirement for clarity of the message, and is the 'meat' of any communication. A strong logos is highly ethical because it is usually produced by careful thought and research. Comprehensive logos strategies enable an audience either to be persuaded with good reasons and evidence, or to mount a refutation and resist persuasion. Yet if the arguments are too complex, abstract, or technical, the communication may lose the attention of certain audiences who are not open to an intellectual challenge.

RHETORIC'S ADAPTABILITY THROUGH HISTORY

The ancient rhetorical theorists were primarily concerned with the public oral communication of statesmen, lawyers, and leaders of public opinion, and the education of men in training for these roles. (Their three types, or roles, of oratory were called deliberative, forensic, and epideictic, respectively.) But this model, though useful in its own day, could not remain unchanged in terms of its applications.

As society became more commercialized and education spread, rhetoric became associated with other communication roles and genres. In the middle ages, Augustine applied rhetorical principles to Christian teaching and preaching. In the renaissance, courtesy books described how to establish romantic liaisons or impress a monarch. As more people acquired writing skills and the economic access to pens and paper, rhetoricians taught how to write business letters, love letters, and even poetry. In the period of Enlightenment, as western society transitioned from monarchy to democracy, rhetorics of conversation and everyday conduct cropped up in essays in the daily periodical press, theorizing the communication of the common person and teaching men and women how to be good citizens. Looking into the history of rhetoric, one can see that rhetoric adapted itself to the needs of society at the time.

Rhetoric Today in Technical and Professional Communication

Technical and Professional Communication theorists draw on rhetoric because of its adaptability across the ages—it has always been concerned with language that gets things accomplished. Old rhetorical theories can be adapted to serve modern technical professionals' needs, and new rhetorical theories are developed from experiencing and studying modern communication in these fields.

As in the past, rhetoric is equally concerned with the education of communicators, with analyzing communication texts and situations, and with researching and formulating hypotheses about how communication in these contexts can be better: more efficient, more ethical, and more useful to its audiences and sponsors. These functions of training, criticism, research, and theory work together in a symbiotic fashion.

Technical and Professional Communication has been served by the continuing efforts of rhetorical educators who teach technical and professional writing and speech. Technical communicators themselves, as they encounter rhetorical lore in their professional training and development, become 'rhetors' who apply rhetorical principles every day of their working lives. There are also scholars and consultants who perform the roles of rhetorical critics, researchers, and theorists to technical and professional organizations and universities.

Rhetoricians have always been, and will always be, caught in real ethical and intercultural dilemmas that affect the very shape of our modern life through affecting how knowledge is created, disseminated, and used.

QUESTIONS FOR CRITICAL THOUGHT

1. Smith names her audience explicitly in the opening paragraphs. Is this named audience the same one her essay actually addresses? If yes, how do you know? If not, who is she really speaking to?
2. What is the purpose of Smith's essay?
3. How readable is Smith's essay? If you found it readable, is this assessment a statement about Smith's essay, or about your skill as a reader? On what basis can we make such a judgement?
4. In 'Getting the Story, Telling the Story: The Science of Narrative, the Narrative of Science', Cheryl Forbes argues that all good science writing—and by extension, all good professional writing—in some measure tells a story. Does Smith's essay do so? If so, what is its plot? If not, why not?
5. How does Smith define rhetoric? Compare her definition with those given by Halloran ('Classical Rhetoric for the Engineering Student'), Ryder ('Science and

Rhetoric'), Simons ('Are Scientists Rhetors in Disguise'), Shay ('Aristotle's *Rhetoric* as a Theory of Leadership'), or Urquhart ('Bridging Gaps, Engineering Audiences'). How similar are these characterizations? How might we account for the differences among them?

6. What is a heuristic? Why is such a thing useful? Can you think of any situations in which you have used a heuristic?
7. According to Smith, what's the difference between rhetorical theory and rhetorical practice? Does this distinction hold for other kinds of theory and practice relationships, for example, the relationship between engineering theory and practice? Why is the distinction important? Compare Smith's treatment of this idea with MacLennan's 'Disciplinarity, Identity, and the "Profession" of Rhetoric'.
8. Smith's essay addresses the link between rhetoric as a branch of study and technical communication as a professional specialization. To what extent does her essay develop this theme consistently? Are there any elements in it that do not seem directly related to her principal thesis? If so, what are they? Why has Smith chosen to include them?
9. Can you construct an outline of Smith's main points? What does this tell you about the essay's organization, its arguments, and its coherence?
10. What is the relationship, according to Smith, between rhetorical theory and technical communication?

CHAPTER 15

Classical Rhetoric for the Engineering Student

Stephen M. Halloran

Something called rhetoric has been enjoying a revival in recent years. And, as with so many other academic and intellectual fads, there seems to be somewhat more enthusiasm than real understanding energizing the movement. The confusion is intensified by the fact that in its nearly 2,500 years of existence, rhetoric has been a vast number of different things, many of them only vaguely defined, to a great many different sorts of people. To the Sophists, the educational establishment of ancient Greece, it was the key to success which they contracted to pass on to their pupils for considerable fees. The open, participatory nature of its judicial and political systems made Athens an argumentative society, in much the same sense that contemporary America is a technological society, and men like Georgias and Isocrates, who knew how to win arguments, occupied a position in that society roughly analogous to that of our top academic scientists and engineers. However, to Socrates and Plato, two intellectual renegades whose Academy might be seen as the prototype of today's free universities, rhetoric was trickery, deception, lies.

Rhetoric eventually won the recognition of the antiestablishment Academy when Aristotle composed his *Rhetoric*, which remains to this day the most important single treatise on the subject. Thus the eclipse of the Sophistical tradition by the Peripatetic did not put an end to formal instruction in persuasive discourse, and rhetoric remained at the centre of education down through the Middle Ages and Renaissance, losing its place only in modern times with the rise of science as a new focal point for general education.

Suddenly, at the height of the scientific age, rhetoric, this most ancient and esoteric of the liberal arts, is enjoying a revival. Education is inevitably education for life in a particular society, and it is reasonable to expect that curricular developments sanctioned by the educational establishment should have some relationship to patterns of life in that society. In other words, education ought to be relevant to the needs of the students. The study of rhetoric was clearly relevant to the needs of the students who would later find success of failure in the Roman Senate or the courts of Medieval Europe, but how relevant is it to the needs of modern students, particularly the prospective specialists in our technological society who will seek success in the laboratories of government, industry, and academia? Has classical rhetoric any relationship to patterns of life in modern America?

Some Definitions

To begin with a more basic question, just what is meant by the term rhetoric? Aristotle defined it as 'the faculty of observing in any given case the available means of persuasion' (Aristotle, Book 1, Chapter 2). In classical times, the medium in which these means of persuasion were to be used was oral discourse, but with the invention of moveable type and the transition from oral to print culture, the developing tradition of rhetorical theory was modified somewhat to accommodate the primacy of written discourse. Recently, there has been some effort to further modify the concept of rhetoric, extending its theoretical domain to include media of communication other than spoken or written discourse. There has been at least one attempt to interpret Marshall McLuhan's *Understanding Media* as a contribution to rhetorical theory (Mahony, 1969), and Professor C. West Churchman characterizes the so-called 'systems approach' to problems of management as 'an extension of the old fashioned logic and rhetoric' (Churchman, 1968: 10). It should be added that Professor Churchman is noted not as a rhetorician, but as a systems analyst. It is probably only a matter of time, if indeed it has not already come to pass, before someone publishes an analysis of encounter group therapy as rhetoric.

My present concern, however, is not with the extent of rhetoric as a theoretical discipline for specialists, but rather with the usefulness of rhetoric as a practical discipline for the general student, particularly the general student who will pursue a career in science or engineering. Thus a more traditional, less speculative definition will serve.

In general, rhetoric can be regarded as the study of language, emphasizing its practical rather than its aesthetic qualities; the study of language as an instrument for the achievement of an end by enlisting the cooperation of others. More specifically, it is classical rhetoric, a collection of insights and doctrines about the effective use of language in non-technical discourse accumulated over the centuries in the works of Aristotle, Cicero, Quintillian, St Augustine, Peter Ramus, Henry Peacham, George Campbell, to name just a few.

At first glance, it might seem that what has accumulated is little more than a lexicon of obscure and unpronounceable terms. The principle that one ought not to multiply entities unnecessarily seems not to have carried much weight with rhetoricians: English schoolboys of the sixteenth-century were expected to distinguish nearly two hundred different figures of speech, from *abbaser* to *zeugma*. The tendency persists among many of the contemporary enthusiasts for classical rhetoric to transliterate Greek and Latin terms, frequently at the expense of clarity, and the retain distinctions of questionable value to the general student. For instance, the Aristotelian term *topoi*—literally 'places' or general headings under which arguments can be categorized, such as 'cause and effect', 'statistics', or 'evidence'—is generally rendered in English as 'topics'. This term is bound to prove confusing to the college freshman, who almost certainly learned in high school that a 'topic' is something like 'My Favourite Sport' or 'What is Honesty?'—a subject one has to write about the satisfy the whim of an English teacher.

Unfortunately, one can find literally scores of such confusing or otherwise difficult terms in most of the composition texts based on classical rhetoric. Understatement becomes 'litotes'; inverted word order is 'anastrophe'. The beleaguered student struggling merely to pronounce 'epanalepsis' or 'anadiplosis', is likely to miss the point of the course. For behind that Graeco-Romanesque façade for mystifying terms is a point—the simple yet highly significant insight that the process of writing is an act of communication.

The Classical Theory

Under the influence of an English curriculum almost totally dominated by literature, particularly as viewed by the now middle-aged New Criticism, the process of writing has come to be thought of as self-expression culminating in the creation of a verbal artifact capable of justifying its own existence. By contrast, the rhetorical approach to composition assumes that writing is a deliberate act of communication undertaken for the achievement of a particular aim, and that the verbal artifact exists solely for that aim. This process involves three elements, all of which must be drawn into the document being produced: the writer, or *rhetor*, to use the classical term, who has in mind something he wants to get done; the audience, who must be persuaded to cooperate in the achievement of the rhetor's aim; and a subject or subjects of mutual concern, on which the views or feelings of the audience must be made to coincide with those of the rhetor in order to achieve the desired cooperation.

Classical rhetoric expresses the need for all three of these elements to be drawn into the act of composition by the doctrine of the three modes of appeal. *Ethos* is the classical term for the appeal exerted by the force of the rhetor's character as he 'comes to life' in his writing; *pathos* is the appeal to the feelings and attitudes of the audience; *logos* refers to the appeal generated by the subject-matter under consideration, that about which the rhetor proposes to write.

The process of bringing these three elements together in one verbal artifact is broken down by rhetorical theory into three steps: research leading to the discovery of possible arguments (*inventio*), sorting out and arranging the material in an effective order (*dispositio*), and finding the most appropriate style for presenting these arguments to this audience (*elocutio*). The Greeks and Romans, with their primary concern for oral discourse, added to this list *memory* and *delivery*, completing the five traditional parts of rhetoric. The *topoi* constitute a device to facilitate the process of research or discovery, functioning as a sort of intellectual check-list to be run through in search of arguments. The figures of speech, of which metaphor and simile and such syntactic devices as parallelism and apposition would be the most familiar examples, are simply an attempt to categorize different stylistic alternatives. On the subject of arrangement, classical rhetoricians list five or more sections into which the typical 'speech' can be divided, discuss the function of each, and speculate on such questions as under what circumstances it becomes appropriate to modify the customary order, or to eliminate one or more sections. While it is true that all of this material has too often been presented in an unnecessarily obscure and/or dogmatic manner, the fact remains that classical rhetoric provides a relatively systematic approach at each step in the composition process.

A Systems Approach

Indeed, it should be apparent now why Professor Churchman sees systems engineering as a descendant of classical rhetoric. Like the systems analyst, the rhetorician approaches a task by ascertaining first just what goal is to be achieved, and then demanding of each 'component' in his 'system' that it work effectively with the other components toward the achievement of that goal. Like the systems analyst, he knows that elegance in the isolated component may not contribute to the overall effectiveness of the system; he knows that, under certain circumstances, absolute logical precision might be ineffective, and that, in another situation, a strong emotional appeal might be equally ineffective.

The rhetorician cannot quantify the 'variables' and 'environmental factors' that combine to present a rhetorical challenge, and thus he cannot try out his speech on a computer simulation of the realities he must deal with. He must rely instead on his own sense of those realities—the values, aspirations, and dreams of his audience, whatever authority he can command by virtue of his public image and personal style, and of course the facts of the case at hand. Thus rhetoric might be seen as a point of contact between technology and the liberal arts. It is at once one of the oldest of the liberal arts and the archetype of one of the newest developments of technology. It is a systems approach to problems of human communication, yet an approach that must rely ultimately upon the kind of intuitive sense of oneself, one's fellow man, and the world we share that is the product of a truly liberal education.

I hasten to add that by 'a truly liberal education' I do not mean the sort of musty compendium of literature, classical languages, art history, and medieval philosophy that perhaps still goes by the name at some institutions. I mean rather an education that seeks to produce some sense of man and his world, in addition to whatever vocational training the students may require. And in a world so thoroughly dominated by science and technology as our own, a truly liberal education will necessarily place considerable emphasis on the scientific and engineering disciplines.

Rhetoric for the Engineer

As a systems approach to composition, classical rhetoric would seem to have a certain *a priori* appeal for the engineering student. It will permit him to apply the same practical, systematic cast of mind that led him to choose a career in engineering, to a subject that, because of its apparent lack of system and practicality, he may have found frustrating in high school. The student who groaned at the prospect of having to grind out a 500-word paper on an assigned topic—'air-pollution', let us say—may respond with considerable enthusiasm to the challenge of a rhetorical problem, such as persuading the trustees of the university of establish a centre for environmental studies. Indeed, I have seen it happen; I know of at least one student who pursued his project beyond the completion of the course in which it was assigned, doing further research, writing letters, talking to people in an effort to enlist their cooperation in the achievement of an end he thought important.

But its pedagogical attractiveness is not the most important reason why classical rhetoric ought to be considered as academic fare for the engineering student. Far more significant than the problem of apathy in the composition class is the problem of alienation between the scientific/engineering community and the rest of society.

The ordinary citizen does not understand the complex machinery and techniques of modern science and engineering, and, as a result, he is bewildered by the possibilities for alteration of his physical and social environment created by those machines and techniques. Yet it is that same bewildered citizen who must vote yea or nay, and who must ultimately pick up the check, on projects like the anti-ballistic missile system, interplanetary exploration, and a secondary or tertiary treatment plant for local sewage disposal. The possibilities for disaster—nuclear, ecological, social, etc.—as a result of bad decisions on such projects grow more various and terrifying almost daily. The alternatives for avoiding such a bad decisions are to take power out of the hands of those bewildered citizens and place it in the hands of an informed elite, or to put an end to their bewilderment by establishing free and effective communication between the men who make the machines and the men who pay for them.

One significant and necessary step toward this dialogue would be to improve general education, particularly in the areas of science and engineering—to provide the ordinary citizen with the sort of liberal education I spoke of earlier. But technology advances because men specialize, and specialization means a degree of exclusivity and intellectual isolation. Thus no matter how good a job the schools do of educating the ordinary citizen, he will inevitably be cut off from an understanding of what is happening at the cutting-edge of technological advance.

Unless, that is, there are scientists and engineers who are willing and able to reach across the gap from their side at the same time that our ordinary citizen reaches out from his side. Communication between the scientist/engineer and the ordinary citizen must, if it is to be effective, be a dialogue, a two-way operation, a mutual effort to understand and to be understood, undertaken not by specialists whose narrow fields of interest are mutually exclusive, but by men who share a world with pressing problems, and whose specialized fields of interest may cast light on those problems. And this art of dialogue, of placing objective facts in the context of human needs and aspirations, of merging *logos* with *ethos* and *pathos* in an effort to achieve cooperation is precisely what classical rhetoric is all about.

I do not mean to propose rhetoric as a panacea for the problem of alienation between the scientific community and the rest of society. I do not believe in academic panaceas for social ills. But if we are to deal effectively with our problems within the context of a democratic society, then engineers and scientists must learn to talk not only to each other, but to non-specialists as well.

References

Aristotle. 1954. *Rhetoric and Poetics*. New York: Random House.

Churchman, C. West. 1968. *The Systems Approach*. New York: Delacorte Press.

Mahony, Patrick. 1969. 'McLuhan in the Light of Classical Rhetoric', *College Composition and Communication* 20, 1 (February): 12–17.

QUESTIONS FOR CRITICAL THOUGHT

1. What is the purpose of Halloran's essay? In other words, what is the main point that he hopes to communicate to his readers? Can you locate his thesis?
2. It's likely that you are not a member of Halloran's intended audience. Based on the positioning of his argument and the style in which the article is written, can you identify who the audience might be? What features of the article help to identify the audience?

3. Do you think Halloran's audience includes the English teachers he mentions frequently in the first several paragraphs? Why or why not?
4. Why do you think Halloran is so critical of the English teachers? What purpose does this criticism serve? What is likely to be its effect on his intended readers?
5. Halloran initially defines rhetoric in terms established by Aristotle, and then adds some elaboration of his own. Why does he feel these additions are necessary? How does his definition (found in par. 6) differ from Aristotle's original?
6. What does Halloran mean by a 'systems approach' to writing? Why does Halloran believe that such an approach is especially suitable for engineers?
7. What are the 'three elements' that Halloran says are central to the writing process? Why are these important?
8. Why does Halloran define the term 'liberal education'? How does he modify it from its traditional meaning? Why does he do so?
9. This is a research essay intended for a technical education journal. How does Halloran establish his authority to make the claims he makes?
10. How would you describe the tone of Halloran's essay? Recalling that tone is a reflection of the writer's attitude toward both the audience and the matter under discussion, consider how Halloran treats the various topics he discusses.
11. In what sense is it true that rhetoric is about achieving cooperation?
12. Halloran suggests that there is a problem in the education of technical writers that needs to be changed. In what sense is change needed, and on whose part?

CHAPTER 16

Aristotle's *Rhetoric* as a Handbook of Leadership[1]

Jonathan Shay

Character is a living thing that flourishes or wilts according to the ways that those who hold power use power. Specifically, character has cognitive/cultural content—a person's ideals, ambitions, and affiliations, and the emotional energy that infuses them—what Homer called *thumos*. The leader's own *thumos* is critical to his or her capacity to lead others.

How does a leader get the troops—soldiers, sailors, marines, airmen—to commit themselves to a mission? Aristotle offers a mix of empirical and normative observations in the *Rhetoric* that apply wonderfully to the military situation. For starters, we must understand the context that he thinks his remarks apply to, what it means for a leader to seek trust: It's about dealing with fellow-citizens, where each looks the other in the eye and says, 'you are part of my future, no matter how this turns out.' Some might scoff, and say, 'an infantry company, or a ship, or a squadron is not a deliberative assembly, and decisions are not arrived at by majority vote.' But many of you want a picture of leading without undue reliance on coercion and will see that Aristotle has real food for thought here.

A leader who mentally and in the heart constantly walks away from those he or she is leading and says 'I'm never going to see these jerks again after this assignment is over,' is just faking it from Aristotle's point of view—a sophist for hire, not a true leader, a *rhêtor*.

So having established that the leader and led are part of each other's future, they now have to arrive at a shared, binding commitment to a mission in the face of uncertainty—conflicting, incommensurable goods.

Real military situations requiring real leadership invariably have these two elements. If everything can be done by formula, by the book, what's needed is a supervisor, not a leader. Even in war, many of the things that need to be done preparing for battle can be done by the book (even von Clausewitz acknowledged that). And even in peacetime, many critical decisions cannot be solved by the book, because they involve competing, incommensurable goods and uncertainty. The *Rhetoric* has no Philosopher's Stone that enables you to harmonize conflicting goods or to know what is not known. It provides a descriptive and normative framework for leading one's fellow citizens under these conditions.

Aristotle shows us that leader has three interrelated means of achieving his fellow citizens' trust:

- Appeal to their character (*ethos*),[2]
- Appeal to their reason (*lógos*), and
- Appeal to their emotions (*páthos*).

These three are interrelated, not separate, because the goals of action arise from the troops' ideals, ambitions, and affiliations—their character. Reason concerns the means to reach those goals. And the emotions arise primarily from their cognitive assessments of the real-world improvement or deterioration of their ideals, ambitions, affiliations, and how fast they are changing in the world.

Aristotle has useful comments on the leader's need to build trust through appeal to the troops' character and emotion. He even explains how it is possible to be 'too rational', losing the trust of those you are trying to lead (see Garver's, 'Making Discourse Ethical: Can I Be Too Rational?').

Aristotle goes on to say what the troops are looking for in a leader. What makes the leader trustworthy in their eyes? Aristotle provides another triad. The troops extend trust to someone whose explanations (what he called 'arguments'), training exercises, and decisions provide evidence for

- Professional competence, spirited personal integrity (*aretê*),
- Intelligent good sense, practical wisdom (*phronêsis*), and
- Good will and respect for the troops (*eúnoiâ*).

The centrality of rational explanation ('argument'), rather than coercion or deception, shows the leader's respect for the troops, who are his or her fellow citizens. You can't separate respect from good will. What reasons, examples, and maxims the leader chooses from the infinity available provide evidence for *phronêsis* and *aretê*. The persuasive power that comes when a leader appeals to reason comes more from the degree to which it provides evidence for the leader's respect toward the troops than from the power of reason to compel assent, or having compelled assent, to guide or restrain behaviour.

So as Aristotle famously says in *Rhetoric* I.ii.3, it is the *ethos*, the character of the leader that is most compelling to the troops. I want to connect the old Homeric word *thumós* to what I now want to say about character. This word is most often translated by the single word 'spirit'. In modern times this has become rarified and if you forgive the play on words, spiritualized, so that we lose the sense that is still preserved when we speak of a horse as spirited or an argument as spirited. Professor Rorty, at Brandeis, gave me her best shot at translating the word as 'the energy of spirited honor'. I want you to listen to Aristotle's explanation of *thumós* in *Politics* VII.6.1327b39ff. He says, '*Thumós* is the faculty of our souls which issues in love and friendship. . . . *It is also the source . . . of any power of commanding* and any feeling for freedom.'

The spirited self-respect that Homer called *thumós* becomes particularly critical to leadership in a combat situation. To trust the leader, the troops need to feel that the leader is his or her 'own person', not a slave. In combat, trust goes to the leaders who give critical obedience, rather than blind obedience, to their own bosses (Zwygart, 1993). A leader giving blind obedience to a militarily irrational or illegal order gets the troops killed without purpose ['wasted'] or irretrievably tainted by commission of atrocities.

Notes

1. See Eugene Garver, *Aristotle's Rhetoric: An Art of Character* (Chicago: University of Chicago Press, 1994).
2. Phenomenology of Spirit. Aristotle fans may balk at this as flying in the face of *Rh*. I.ii.3, but it can be justified from the practice Aristotle shows us. It should be evident that I do not dispute the importance of the *leader's* character.

References

Zwygart, U.F. 1993. 'How Much Obedience Does an Officer Need?', *US Army Command and General Staff* College pamphlet.

QUESTIONS FOR CRITICAL THOUGHT

1. Shay treats rhetoric—the ancient art of persuasion—as a method of leadership. On what basis does he make this move? What is the connection between leadership and persuasion?
2. Shay argues that rhetoric provides an excellent foundation for understanding leadership; several other writers in this book, including Smith ('What Connection does Rhetorical Theory have to Technical and Professional Communication'), Halloran ('Classical Rhetoric for the Engineering Student'), Buehler ('Situational Editing'), Campbell ('Ethos: Character and Ethics in Technical Writing'), and Urquhart ('Bridging Gaps, Engineering Audiences'), to name only a few, have argued that it is a sound foundation for the practice and study of technical communication. How can it be both?
3. How does Shay define leadership?

4. The context for Shay's remarks is military. To what extent are his observations on leadership confined to that context? To what extent do they extend to any leadership situation?
5. What is the role of trust in effective leadership? What is its role in persuasion?
6. In his fifth paragraph, Shay distinguishes between leadership and management, and emphasizes that a cookie-cutter, 'by-the-book' approach is insufficient to the demands of leadership. In what respects does he echo Richard Burton's observations about technical writing in 'An Engineer's Rhetorical Journey'?
7. Compare Shay's treatment of logos, ethos, and pathos with Smith's 'What Connection does Rhetorical Theory Have to Technical and Professional Communication?' and with Campbell's 'Ethos: Character and Ethics in Technical Writing'. Where else in this book have you encountered these concepts? Are the treatments similar?
8. What element of communication (or leadership), according to Shay, is most compelling to the audience? Why? In what respects is this element also important to scientific or technical writing?
9. What is 'thumos'? What does this concept have to do with rhetoric, the art of persuasion? What has it to do with leadership?
10. What is the relationship between leadership and technical writing? Why is this essay in the book?
11. How would you describe the style of Shay's essay? How personal is it? How formal? To what extent is its style a function of its context? Explain.
12. Who are Shay's intended readers? Are you included in that group? How do you know?

CHAPTER 17

Are Scientists Rhetors in Disguise? An Analysis of Discursive Processes Within Scientific Communities

Herbert W. Simons

Emboldened by recent attacks on scientific orthodoxies, rhetorically minded critics of science have been vituperative in their characterizations of scientific discourse, particularly the writings of social scientists. An image appearing repeatedly in the 'rhetoric of science' literature has been that of the scientist as a rhetor in disguise—one who falsely pretends to the status of nonrhetor and thereby renders his rhetoric deceptive; one, moreover, who is not above engaging in such allegedly sophistic practices as masking self-serving motives behind a vocabulary of legitimating motives or subtly promoting partisan values in the guise of being informative.

Thus by Andrew Weigert's account, behavioural sociologists are guilty of an 'immoral rhetoric of identity deception' (Weigert, 1970). According to Weigert, journal editors and readers are courted by means of various impression-management techniques, and weaknesses in research or theory are covered over or rationalized away. Grant-givers are wooed by appeals to prejudice and by the framing of social problems in the distinctive jargon of the discipline. And students are indoctrinated by means of exaggerated claims, in survey, theory, and methodology texts, about the discipline's capacities and achievements. The themes of status enhancement and duplicity are sounded once again in Jack Douglas's castigation of social scientists for relying uncritically upon a 'rhetoric of statistics', in Jeanine Czubaroff's illustrations of 'the rhetoric of academic respectability', in Berger and Luckmann's depiction of medical jargon, white coats, and framed diplomas as the rhetorical stock-in-trade of physicians, and in Thomas Szasz's references to psychiatric labels as a justificatory rhetoric of rejection and oppression.[1]

What are we to make of these characterizations of scientists and of scientific discourse? Clearly they stand in marked contrast to the traditional view of science as falling outside the province of rhetoric. Less clear, however, is whether they add up to an inherent case against science in general, or even against the discourse of social scientists. Are the indictments offered by rhetorically minded critics valid in principle or do they apply only to atypical practices by isolated scientists?

One way to gain clarity on the issue of inherency is to cast 'pro' and 'con' positions in the form of a debate, and that is what I shall be doing in this paper. By juxtaposing competing views the debate format should also permit differentiation between easily refutable claims and counterclaims and those which resist clear resolution. In this way it should serve to advance the questions, even if in the process it also makes it more difficult to render clear-cut, unequivocal conclusions.

What would it take to mount an inherent case, one that establishes in principle that scientists are rhetors in disguise? In other words what is the affirmative's burden of proof in this debate?

To begin with, the affirmative must successfully undermine the traditional notion that scientists are in some way different from ordinary rhetors—different by dint of their alleged capacity to provide objective, unambiguous, uncontestable tests of assertions (as opposed to the plausible arguments traditionally associated with rhetorical discourse). To do so it must cut to the heart of the claim to scientific objectivity, the presumed capacity of scientific communities as a whole to correct for the foibles and passions of individual scientists. While critics can undoubtedly point to subjective elements in the discourse of individual scientists, defenders of science might concede the point, as Karl Popper has done, and nevertheless insist that what sets scientific discourse apart from that of ordinary rhetors is the error-correcting *process of exchange* that takes place within scientific communities. Here Popper refers to the normatively sanctioned use of a commonly understood technical language, the insistence by the group as a whole that theoretical statements be framed in such a way that they can be tested by observations and experiments, and, above all, the unrestrained criticism of any and all ideas, including those which emanate from the highest authorities.[2] Once Popper's view of science as a communal enterprise is given credence, it becomes logically inappropriate to conclude that scientific discourse is inherently rhetorical simply by citing evidence of passion, prejudice, or other subjective elements in the discourse of individual scientists. In place of an *act-centred* view of scientific discourse a *process-centred* view is needed, one that shows scientific assertions to be eminently contestable or ambiguous (i.e. rhetorical) even after they have been honed and refined by exchanges within scientific communities.

The notion of rhetoric as a communal process figures prominently with respect to the second burden which the affirmative must carry in this debate, that of showing that scientists are guilty of such allegedly sophistic practices as using a vocabulary of legitimating motives to mask self-serving motives or promoting partisan values in the guise of being informative. Even assuming that all of the charges cited earlier are valid, it is unclear whether they add up to an inherent case. Just as evidence of subjectivity in the discourse of individual scientists does not in itself refute the Popperian notion of scientific objectivity as a communal product, evidence of sophistic practices by individual scientists is an insufficient basis upon which to condemn entire scientific communities.

Beyond evidence of sophistic practices, compelling arguments are needed as to why, in principle, these misdeeds are bound to occur. Barring that, defenders of science might reasonably acknowledge misdeeds while insisting that no serious harm is done so long as scientific communities as a whole maintain appropriate institutionalized commitments to the pursuit of knowledge. To return to charges levelled by critics, some individual scientists may be motivated by greed, power, envy, or the need for recognition, but the norms and practices of any given scientific community might well provide an effective counterforce to these motives. Similarly some individual scientists may exaggerate discoveries and even fabricate them, demean critics and even slander them, ingratiate themselves with institutional superiors and even pander to them, inject personal values into their arguments and even deliberately polemicize: in the long run, however, any given scientific community might well survive these allegedly sophistic practices by dint of an overarching, institutionalized commitment to public testing of testable ideas in a free marketplace of ideas. Unless it can be shown that deceptions, evasions, and distortions *follow* in some way from the very nature of scientific communities, science in its collective, Popperian sense must stand acquitted of the charge of sophistry.

Thus far I have attempted to fix the burden of proof for rhetoricians of science (the affirmative in the debate) in establishing that scientists are inherently rhetors in disguise. In so doing I have stressed the need for a process-centred view of scientific discourse, as opposed to an orientation focused on individual acts. At this point we may let the promised debate begin. In constructing the debate I have attempted to act as an honest spokesman for both sides by giving strong voice to each. Following the debate I offer a 'judge's verdict', consisting of an attempted reconciliation of extreme views as well as a list of questions that continue to nag.

The Debate: Resolved That Scientists are Rhetors in Disguise

THE AFFIRMATIVE

There are two major issues in this debate. First, are scientific communities incapable of providing objective tests of assertions? Second, do patterns of sophistic practices follow in some way from the very nature of scientific communities?

Objective tests of assertions

Since the communal aspect of science figures so prominently in the defense of the notion of scientific objectivity, it is worth emphasizing that the tests made by scientists may be considered unambiguous or uncontestable only if one buys the value and belief premises that scientific communities share in common. That science is not value free has been attested to by many scientists themselves. Alfred Whitehead observed some time ago that 'without judgments of value there would be no science.' And John F.A. Taylor remarked about the scientific commitment to objectivity itself that 'the rule which guarantees the disinterestedness of inquiry is not itself neutral with respect to the matter of disinterestedness' (1938: 229).[3] In his excellent treatise on *A Sociology of Sociology*, Friedrichs suggested that scientists

> shoulder the responsibility of settling upon one of a number of logical systems, none of which is able to validate itself in terms of itself. . . . If a sociologist follows an essentially Aristotelian logic, he will paint social experience in colors quite different from those that would come through given an initial commitment to a dialectical logic-pigments that impinge upon issues of ideology. (1970: 152)

The general point, made by mathematician Kurt Godel, is that 'no single logistic system . . . can tenably claim to embrace only logical truth *and* the whole of logical truth.' It follows, of course, that science's assumptive underpinnings are not themselves scientifically demonstrable. Rather, as Willem F. Zuurdeeg has suggested, science functions 'within a larger framework which is convictional'. The process by which consensus is reached on the so-called transcendent values of science is inherently rhetorical, as Thomas Farrell has argued.[4] And that consensus is also limited. From a perspective outside the scientific, religious revelation or mystical intuition or extrasensory perception may appear to be far truer paths to knowledge than those accepted as a matter of scientific faith.

Even assuming that the foregoing philosophical considerations were somehow irrelevant to the scientific quest, it would still be true that scientists cannot provide objective tests of assertions, at least at the level of paradigm choice. In addition to agreeing on philosophical premises, scientific communities are committed to roughly the same criteria for evaluating theories: problem-solving ability, manageability, and parsimony. But there are inevitably sharp disagreements between adherents to competing paradigms about the relative weights to be assigned to each criterion and about how they should be interpreted in any given case. Moreover, as Thomas Kuhn has argued in his now familiar treatise, *there are no neutral algorithms in terms of which such disagreements can be resolved.*

> When paradigms change, there are usually significant shifts in the criteria determining the legitimacy both of problems and of proposed solutions. . . . To the extent . . . that two scientific schools disagree about what is a problem and what is a solution, they will inevitably talk through each other when debating the relative merits of their respective paradigms. In the partially circular arguments that regularly result, each paradigm will be shown to satisfy more or less the criteria that it dictates for itself and to fall short of those dictated by its opponent. There are other reasons, too, for the incompleteness of logical contact that consistently characterizes paradigm debate. For example . . . , paradigm debates always involve the question: Which problems is it more significant to have solved? Like the issue of competing standards, that question of values can be answered only in terms of criteria that lie outside of normal science altogether, and it is that recourse to external criteria that most obviously makes paradigm debates revolutionary. (Kuhn, 1970: 109–10)

There is a sense, too, in which paradigm debates are about different realities. That is why Kuhn insists,

> Communication across the revolutionary divide is inevitably partial. . . . Equally, it is why, before they can hope to communicate fully, one group or the other must experience the conversion that we have been calling a paradigm shift. Just because it is a transition between incommensurables, the transition between competing paradigms cannot be made a step at a time, forced by logic and neutral experience. (1970: 149–50)

The rhetorical implications of Kuhn's 'revolutionary' thesis have been understood, although hardly appreciated, by Israel Scheffler:

> In place of the notion of Peirce that scientific convergence of belief is to be interpreted as a progressive revelation of reality, we are now to take such convergence as a product of rhetorical persuasion, psychological conversion, the natural elimination of unreconciled dissidents, and the retraining of the young by the victorious faction. Instead of reality's providing a check on scientific belief, reality is now to be seen as a projection of such belief, itself an outcome of non-rational influences. The central idealistic doctrine of the primacy of mind over external reality is thus resuscitated once again, this time in a scientific context. (1967: 73–4)

To round out the argument of this subsection, we would suggest further that scientists who share commitments to a given paradigm still do not have complete and unambiguous rules for testing assertions within the framework of that paradigm. As Koch has argued,

> The Scientific process is, in principle and at all stages *under-determined by rule*. . . . Among the re-analyses of inquiry that are now shaping up there is no point-for-point consensus, but must agree in stressing the absurdity in principle of any notion of *full formalization*, in underlining the gap between any linguistic 'system' of assertions and the unverbalized processes upon which its interpretation and application (not to mention its formulation) are contingent, in acknowledging the dependence of theory construction and use at every phase on sensibility, discrimination, insight, judgment, guess. (quoted in Wann, 1964: 21–2)

A major stumbling block to full formalization is language itself. Kenneth Burke has maintained that any symbolic construction, however it reflects reality, is necessarily 'a selection of reality; and to this extent it must function also as a defection of reality' (1968: 45). Notwithstanding the heroic efforts of logical positivists, the persuasiveness of theoretical claims is still highly dependent on stylistic choices, particularly in the social sciences. Try as they might, for example, psychologists have been unable to avoid the use of central metaphors such as 'drive' and 'attraction' in motivational theories.[5]

And values are reflected in stylistic choices as well. Whether one describes the 'high' produced by LSD as 'mind expanding' or 'mind destroying' matters not; a value judgment is being reflected either way. Nor are such judgments avoided by the use of mechanistic reductions: treating reasons as 'causes' and symbolic actions as 'behaviours'; denying mind, will, spirit, and other mentalisms. Implicit in these reductions is a commitment to what Koch calls 'metaphysical materialism' (1968: 6).

Sophistic practices

Thus far the affirmative has tried to show that scientists misrepresent the process of exchange within scientific communities when they insist that it yields objective tests of assertions. That discourse, rather, is inherently rhetorical: it appeals to subjectively shared premises rather than indubitably true premises; it contains extrafactual, extralogical arguments rather than purely factual, purely logical arguments; and it yields judgments, however credible, rather than certainties.

At this point we will argue that Popper's reference to the 'communal' or 'institutional' aspect of science does not save science from the kinds of sophistic practices cited by Weigert, Douglas, Czubaroff, Berger and Luckmann, and Szasz. Rather, we will maintain that various communities of science permit and partially encourage these misdeeds. An outstanding example is provided by the recent history of social psychology. We think it outstanding, not because we have reason to believe the discipline is unique (at least among the social sciences), but only because social psychologists have been unusually candid of late in detailing its record. Thus McGuire has conceded that the claims of progress during the 50s and 60s were largely illusory. Greenwald has found support for mainstay propositions disappearing from view. Schlenker has acknowledged that current theories have not proven highly useful in applications to real world problems. And Katz has characterized as 'surprisingly small' the number of experimental studies supplying new information to a cumulative body of knowledge.[6]

Of special interest to rhetoricians is the persuasion/attitude change area within social psychology. Here Fishbein and Ajzen (1975) have concluded that despite the hundreds of studies conducted in the area, we are left with virtually no empirical generalizations about message or source effects that are not circular, trivial, or false.[7] They have taken their colleagues to task rather severely for having inconsistently operationalized independent variables while neglecting the need for distinctions among dependent variables; for using single item measures of attitude in hastily prepared questionnaires; for failing to report statistically nonsignificant findings while highlighting significant findings; and for drawing unqualified generalizations in reviews of literature from studies yielding inconsistent findings. To this list may be added the heavy reliance on theatrically staged deception experiments, appreciated more for their 'cuteness' than for their capacity to generate consistent results. Also mentioned by critics, among other concerns, have been the persistent use of semantic differential measures, few of which were factor analyzed; the tendency to justify negative findings from loosely designed studies by means of ad hoc explanations; and the willingness to generalize from restricted samples (usually composed of college students).[8]

With such evidence of shoddy designs, discrepant results, and failures to replicate, why were the severe problems in this area not exposed earlier? Why, instead, did the textbook writers and grant applicants of the 50s and 60s boast of 'ironclad laws', 'relevant theories', and 'formulas for persuasive success'?[9] Why did researchers of the 50s and 60s persist in reporting studies that violated established scientific canons?

A state of pervasive innocence is, of course, one explanation for these actions; self-deception is another. Still another perspective on the matter seems at least partially valid, and that is the kind of rhetorical perspective offered by critics such as Weigert. From that perspective these actions are best described as willful and self-serving forms of selling, displaying, indoctrination, and ingratiation, no different from those actions practiced by ordinary rhetors such as advertisers and politicians. Let none of us profess shock at this interpretation for as Albert Einstein once remarked, 'If an angel of God were to descend and drive from the Temple of Science all those . . . motivated by display and profit, I fear the Temple would be nearly emptied.'[10]

But more to the point of this paper, during the 50s and 60s where were the journal editors, foundation consultants, advisors to book publishers, and other guardians of the social psychology community? The answer is that by their own examples as theorists and researchers, and by their actions as 'gatekeepers' of knowledge, they were helping to lead the charge. Thus, for example, it remains common practice for journals to refuse to publish studies containing statistically nonsignificant findings. And foundations effectively discourage attempts at replication by declaring that they will fund 'original' research. Similarly the tendency to generalize broadly from restricted samples persists with few official objections. As Paul Meehl noted, researchers continued to produce shoddy deception experiments because 'cuteness' and 'cleverness' were rewarded by the discipline (1967: 114).

The general point to be made from this extended example is that there is a range of practices at variance with established scientific canons that scientific communities tolerate and may even encourage. Such violations of idealized norms as plagiarism or outright lying are censured, to be sure, but there appear to be informal norms within scientific communities that give license to less extreme violations of the formal code. Here we will offer a general explanation for this tendency, which, although fragmentary and speculative, seems highly plausible.

First, science in the collective sense requires organization. In addition to being employed by various organizations, scientists organize themselves into professions and coordinate their activities through professional associations: chemical societies, communication associations, and the like.

Literature concerning the sociology of professions suggests that professions are by no means all bad. A profession, by definition, is more than a mere trade or occupation, for example. On the basis of his own review of the literature, Pavalko states that professions are distinguished by their public service character, the abstract and highly specialized expertise they require (and which individual professionals acquire through long and arduous training), their strong sense of subcultural identity, and by their commitment to a calling as well as a well-defined code of ethics (Pavalko, 1971).

Yet there is a counterside to professionalization which, although necessary to the functioning of professional collectivities, can be dysfunctional with respect to the advancement of science. Some inkling of what we are getting at was provided by William Goode when he suggested that 'no occupation becomes a profession without a struggle' (quoted in Friedrichs, 1970: 89). More specifically, every professional collectivity is a permeable social system which influences and is in turn influenced by external systems (e.g., government agencies, foundations, the 'public') as well as subsystems within it (individual members, interest groups, subdisciplines). The status and even the very survival of the professional system are dependent upon its interactions with other systems. For example, it must recruit new members from the external system and compete, in the process, with other collectivities for the best available talent. It must have other resources (time, money, equipment) to conduct its work as well as freedom from external interference. And the professional system must be responsive to the multiple needs of its internal subsystems: for example, individual professionals who have families to feed and careers to advance.

Elsewhere Simons (1970) proposed a theoretical framework for understanding and analyzing the rhetoric of social movements. He argued that as a voluntary collectivity operating within a larger system, every social movement has to fulfill certain *rhetorical requirements*. Moreover these requirements (goal demands, value demands, membership demands, power demands, organizational demands) tend to be incompatible, thus giving rise to *rhetorical problems*. Social movements, therefore, devise *rhetorical strategies* to ameliorate these problems and fulfill their requirements. And, as is commonly the case, these strategies remedy some problems but create others in the process.

It should be apparent that a theoretical extension of the 'requirements–problems–strategies' framework to the rhetoric of scientific professions is being proposed here. Scientific professions are not social movements, to be sure, and their rhetorical needs and remedies will perforce be somewhat different. But they are voluntary collectivities, and they are obligated to do more than produce scientific discoveries. Rhetorically speaking they are required to recruit and indoctrinate new members, justify their claims to special expertise before accrediting agencies, plead for freedom from political regulations or other such pressures, mold and reinforce the sense of collective identity among individual members, and, in general, legitimate the profession and its activities before outsiders and insiders. These requirements are not always compatible with the need for scientific advancements, as when inflated claims about scientific method in textbooks must be unlearned by recruits if they are to produce major scientific achievements. They lead, nevertheless, to sophistic practices that gain informal acceptance within scientific communities. It should be apparent that these practices are part of a pattern, one that can be explained, rather than explained away, by viewing scientific discourse as a collective process.

THE NEGATIVE

Objective tests of assertions

In maintaining that scientists are incapable of providing unambiguous tests of assertions, rhetoricians of science reveal unambiguously their own subjectivist colours. In so doing they further undermine their case. For, as Israel Scheffler argues,

> Objectivity is relevant to all statements which purport to make a claim, to rest on argument, to appeal to evidence. Science . . . is not uniquely subject to the demands of objectivity: rather, it institutionalizes such demands in the most systematic and explicit manner. But to put forth any claim with seriousness is to presuppose commitment to the view that evaluation is possible, and that it favors acceptance: it is to indicate one's readiness to support the claim in fair argument, as being correct or true or proper. For this reason, the particular claim that evaluation is a myth and fair argument a delusion is obviously self-destructive. If it is true, there can be no reason to accept it; in fact, if it is true, its own truth is unintelligible: what can truth mean when no evaluative standard is allowed to separate it from falsehood? (1967: 21)

Scientific communities share philosophical commitments to certain belief and value premises, it is true, but these premises ought not to be regarded as mere prejudices. Unlike religious revelation or mystical intuition (cited as alternative 'paths to knowledge' by the affirmative), the utility of these premises is demonstrated daily by the concrete achievements of those who adhere to them. Paradoxically it is the commitment to such values as the public testing of testable ideas that permits science to be value free in its daily operations. Let religious revelation or mystical intuition yield knowledge of how to build a bridge, then these so-called paths to knowledge will be taken seriously. Until then scientists will remain content with their own philosophical premises.

As for the claim that the switch to a new paradigm is akin to a 'religious conversion'—a matter of 'persuasion' rather than 'proof'—Kuhn grossly overstated his case at some points and backed off from it considerably at others. In his postscript, for example, Kuhn admitted that all talk of 'faith' to the contrary, scientists are 'fundamentally puzzle-solvers'. That is, 'the demonstrated ability to set up and to solve puzzles presented by nature is, in case of value conflict, the dominant criterion for most members of a scientific group'. Although value conflicts remain ('puzzle-solving ability proves equivocal in application'), communication between rivals is difficult, and the new paradigm is resisted at first by adherents to the old, eventually the new paradigm proves decisive to the scientific community as a whole because of its superior ability to solve scientific puzzles (Kuhn, 1970).

The general point, then, is that debates between so-called incommensurables are capable of objective resolution. As Scheffler states:

> Lack of commensurability, in the sense here considered, does not imply lack of comparability. Even works of art may be reasonably discussed, criticized, compared, and evaluated from various points of view; the incommensurability of these works does not establish that such critical discussion must consist of empty rhetoric alone. Comparison is not limited to an effort at what Kuhn calls 'communication

> across the revolutionary divide'; it need not translate, step by step, one paradigm into some other, any more than art criticism need translate one work into another. Having appreciated the differing potentials of competing paradigms, the scientist, like the critic or indeed the historian, may step back and consider the respective bearings of the paradigms with regard to issues he holds relevant. Such consideration is itself not formulated within, nor bound by, the paradigms which constitute its objects. It belongs rather to a second-order reflective and critical level of discourse. This is the level on which paradigm debates take place, and the incommensurability of their objects is no bar to their reasonableness or objectivity. (1967: 82–3)

If rhetoricians were as inclined to read Kuhn's critics as they are to read Kuhn, they would recognize that his thesis is fuzzy, inconsistent, and, in any case, not at all damaging to the basic notions of scientific objectivity and progress. Were they to read Kuhn open-mindedly, they would recognize that he was not characterizing science as subjective or relativistic. And were they more attuned to evidence than conjecture, they would discover, as sociologists of science have done, that there has been far less crisis, alienation, and conflict among scientists at the point of fundamental innovation than Kuhn had intimated.[11]

Let us turn now to the affirmative's remaining objection in this series, the claim that the scientific process is 'underdetermined' by rules, even within the framework of a given paradigm, and that the vagaries of language in particular constitute an inherent impediment to objective resolution of competing assertions.

There is once again an ironic and self-defeating twist in the rhetorician's use of language to prove that all claims which rely on language for support are inconclusive. We will concede, however, that the absence of full formalization of rules presents an obstacle to verification and that problems of language are often acute, especially in the social sciences. But these problems are not inherent. When the locus is shifted from the individual scientist to the scientific community as a whole, the objection fades in significance. While one scientist's formulation 'deflects' from a given reality, another's 'reflects' that reality, and the scientific community, by its method of free criticism, stands ready to provide the necessary tests and correctives to both. For example, it may reject 'mind expanding' and 'mind destroying' as descriptors of LSD effects, insisting on less emotionally charged terms as well as operational definitions.

Sophistic practices

Let us turn, finally, to the question of scientific communities being 'required' to countenance, let alone encourage, sophistic practices incommensurate with scientific achievement.

To begin with, the affirmative has not presented a case against science as a whole. At best the affirmative's arguments can be said to apply to the 'softer' social sciences such as social psychology, for at no point has the affirmative offered evidence of a pattern of sophistic practices within the better established sciences such as physics or chemistry. And it is doubtful whether they could. Having initially conjectured about the existence of positively sanctioned norm violations in the 'hard' sciences, sociologists of science have looked at the evidence and have found, to quote a recent review, that 'the mechanisms of social control in science work on the whole according to the institutional norms in spite of individual deviations' (Ben-David and Sullivan, 1975: 207).

Before buying the affirmative's case as applied to communities of social scientists, we had best look more carefully at what they have called sophistic practices. Surely the affirmative has been ungenerous in characterizing every suspect practice as 'willful', 'self-serving', or a 'misdeed'. Undoubtedly some are errors and no more than that. Others, such as the tendency to exaggerate accomplishments in grant applications, are rather innocuous; foundations routinely expect boastful claims and discount them as part of the grantsmanship 'game'. Still other practices that might initially appear to be unjustifiable could be defended on extrinsic grounds. For example, although the disinclination by journal editors to publish statistically nonsignificant findings may distort the overall research picture, it could be argued that studies with nonsignificant findings tend to be methodologically defective, and that, in any case, it would be financially unfeasible to publish all or most of them.

To be sure, there are deviations from idealized norms in every field, and the affirmative's brief history of the social psychology discipline suggests that during the 50s and 60s, the community as a whole was extremely lax in permitting, sometimes even encouraging, unjustifiable practices. However, by their own admissions of error present-day social psychologists further strengthen our case. Scientific communities are not 'required' to lend support to distortions, exaggerations, and misrepresentations; on the contrary, the current acknowledgments of past mistakes are proof that even in a relatively 'soft' discipline, the community's error-correcting norms eventually prove triumphant. These days social psychologists are going so far as to wonder aloud whether, given the ephemeral nature of their data, they can ever be more than contemporary historians of passing cultural fancies. In place of the bland assurances of previous days, they are setting less ambitious goals for their theories and incisively criticizing the studies and generalizations of the past. Indeed the prevailing talk is of a 'crisis of confidence' in social psychology, as scholars confront squarely the problems of individual differences, situational complexity, and intercultural variation (Gergen, 1973).[12] Surely there is little evidence in all this of a discipline continuing to countenance practices incommensurate with scientific achievement.

Rhetoricians perform a useful service when they call attention to the pressures scientists are under to communicate sophistically and to the deceptions and misrepresentations themselves. But sophistic practices are hardly a 'requirement' of scientific professions or a pattern of behaviour characteristic of scientific professionals. Although some deviations from idealized norms may be countenanced for a time, the error-correcting norms of the scientific community eventually prove triumphant.

The Judge's Verdict

Given a view of scientific discourse as a process of exchange within scientific communities, has it been shown in this debate that scientists are in principle rhetors in disguise? Does the process of exchange yield objective tests of assertions or is that process inherently rhetorical? Do the communities of science provide safeguards against sophistic practices, or do they actively permit and encourage them?

After reviewing the 'pros' and 'cons' in this debate, the scientifically orthodox will undoubtedly be left with the conviction that although the scientific donkey may have been pinned with an unbecoming rhetorical tail, it is still capable of carrying a heavy load. Many rhetoricians will be convinced, to the contrary, that science is rhetorical through and through, and that by its denials and deceptions, the donkey appears as an ass of a different kind.

My point in utilizing a debate format to examine the issues has been that the arguments of both sides have merit to a degree; consequently, I will approach this judge's verdict' in a spirit of reconciliation. With the affirmative I maintain that the donkey is rhetorical through and through. With the negative I conclude that the donkey is still capable of carrying a heavy load.

Because 'rhetoric' tends to be a 'devil' term in our culture—often preceded by 'mere', 'only', 'empty', or worse—scientists understandably recoil from it, insisting instead that their discourse is purely 'objective'. Yet in the classical, nonpejorative sense 'rhetoric' refers to reason-giving activity on judgmental matters about which there can be no formal proof. The classical conception permits and even encourages the eulogistic sense of rhetoric as *good* reason-giving on matters of judgment. In the final analysis that is what defenders of science *mean* by 'scientific objectivity'. Though some positivists still stipulate full formalization of rules as a necessary condition of objectivity, most philosophers of science have begged off from it, as Scheffler did when he equated scientific objectivity with 'reasonableness' in his comparison between scientific judgment-making and art criticism. The negative could not fully refute the claim that science is grounded on scientifically undemonstrable premises: that it has no neutral algorithms for evaluating competing paradigms: and that it is 'underdetermined' by rules within the framework of a given paradigm. But they could rightly insist that the process of free criticism, which Popper alluded to often, yields good reasons in support of assertions that survive the scientific community's screening procedures.

To be sure, practices by some scientists are sophistic, and I believe my 'requirements–problems–strategies' framework may help us to understand why they are employed and why scientific communities sometimes countenance them. However, as the negative has reminded us, we had best use the term 'sophistic' with extreme caution, restricting it to acts that are manifestly willful, self-serving, and in clear violation of established scientific canons. Because some acts are harmful while others are relatively innocuous, we might also attempt to differentiate among them on an 'ethical values scale'. rather than lump them together as the affirmative has in this paper.

The differences between opposing sides in this debate might be further reconciled by distinguishing two senses of the term 'scientific community'. There is, on the one hand, the historical sense of 'scientific community', and as we have seen in the case of social psychology, these communities may be temporarily caught up in the needs of their members or in the needs of the community as a whole to demonstrate its legitimacy. There is also the sense of 'scientific community' as a relatively enduring, institutionalized collectivity and, as history has repeatedly shown, these communities have been capable of surmounting such extreme pressures as papal decrees and totalitarian dictates. In the case of social psychology we have seen that its rhetoric could undergo an abrupt transformation in a brief period of time.

The foregoing also suggests what may be the most appropriate stance of the rhetorician of science in his capacity as critic of discourse. Following Burke some rhetorical critics have been prone to label all science as 'magic', 'ideology', or 'secular theology'. There is merit in a skeptical posture (Nietzsch's 'art of mistrust') but rather than carping at science in general, rhetorical critics would be more persuasive were they to reserve their slings and arrows for more limited and vulnerable targets such as individual practitioners or communities in the time-bound sense. Of special concern should be willful violations of scientific canons, and here critics can stand not as enemies of science, but as defenders of its time-honoured norms.

From a perspective 'friendly' to science, there is also much theoretical work to be done. How in particular do paradigms gain acceptance? By what stratagems and through what channels are theoretical innovations diffused? In the absence of fully formalized rules, what role is played by exemplars and by enthymematic appeals? How is the language of one paradigm translated into the language of another? 'Just because it is asked about techniques of persuasion . . . in a situation in which there can be no proof,' said Kuhn, 'our question is a new one, demanding a sort of study that has not previously been undertaken' (1970: 152). Rhetoricians of science should be in an ideal position to pursue these issues.

Notes

1. Jack D. Douglas, 'The Rhetoric of Science and the Origins of Statistical Social Thought: The Case of Durkheim's *Suicide*', in Edward A. Tiryakian, ed., *The Phenomenon of Sociology*, (New York: Appleton-Century-Crofts, 1969), 44–57; Jeanine Czubaroff, 'Intellectual Respectability: A Rhetorical Problem', *Quarterly Journal of Speech* 59 (1973): 155–64; Peter L. Berger and Thomas Luckmann, *The Social Construction of Reality: A Treatise in the Sociology of Knowledge* (Garden City, NY: Anchor, 1967), 88; Thomas Szasz, *Ideology and Insanity* (Garden City, NY: Anchor, 1969).
2. The traditional distinction between rhetorical discourse and scientific discourse is described well (but not supported) by Ch. Perelman and L. Olbrechts-Tyteca. See *The New Rhetoric: A Treatise on Argumentation*, 1–4; Karl Popper, 'The Sociology of Knowledge', in James E. Curtis and John W. Petros, eds., *The Sociology of Knowledge: A Reader* (New York: Praeger, 1970), 653–4. Said Popper, 'If scientific objectivity were founded upon the individual scientist's impartiality or objectivity, then we should have to say good-bye to it.'
3. See also John F.A. Taylor, 'The Masks of Society: The Grounds for Obligation in the Scientific Enterprise', *Journal of Philosophy* 15 (June 1958): 496.
4. *Encyclopaedia Britannica*, 1961, S.V. 'Logic, History of', (quoted in Robert Friedrichs, *A Sociology of Sociology* [New York: Free Press, 1970], 150); Willem F. Zuurdeeg, *An Analytical Philosophy of Religion* (Nashville: Abingdon, 1958); Thomas B. Farrell, 'Knowledge, Consensus, and Rhetorical Theory', *Quarterly Journal of Speech* 62 (February 1976): 1–14.
5. See K.B. Madsen, *Theories of Motivation* (Cleveland: Howard Allen, 1961).
6. William J. McGuire, 'The Yin and Yang of Progress in Social Psychology: Seven Koan', *Journal of Personality and Social Psychology* 26 (1973): 446–56; Anthony G. Greenwald, 'Consequences of Prejudice Against the Null Hypothesis', *Psychology Bulletin* 82 (1975): 1–20; Barry R. Schlenker, 'Social Psychology and Science: Another Look', *Personality and Social Psychology Bulletin* 2 (Fall 1976): 387; Daniel Katz, 'Some Final Considerations About Experimentation in Social Psychology', in C.G. McClintock, ed., *Experimental Social Psychology* (New York: Holt, Rinehart and Winston, 1972), 557.
7. See especially pp. 11–12, 53–9, 114–17, and 513–19 in Martin Fishbein and Icek Ajzen, *Belief, Attitude, Intention, and Behavior: An Introduction to Theory and Research* (Reading, MA: Addison-Wesley, 1975).
8. See Daniel Katz, 'Social Psychology: Comprehensive and Massive', *Contemporary Psychology* 16 (1971): 277; Gary Cronkhite and Jo Liska, 'A Critique of Factor Analytic Approaches to the Study of Credibility', *Communication Monographs* 43 (June 1976): 91–107; Paul E. Meehl, 'Theory Testing in Psychology and Physics: A Methodological Paradox', *Philosophy of Science* 34 (1967): 114. See also Clyde Hendrick, 'Social Psychology as History and as Science: An Appraisal', *Personality and Social Psychology Bulletin* 2 (Fall 1976): 392–403. Psychologists often justify the use of restricted samples on grounds that they are studying 'basic psychological processes'; common to all persons, and can thus predict from the few to the many. But, as Hendricks has pointed out, the notion of 'basic psychological processes' and others in social psychology is probably chimerical. Moreover, even if such processes could be identified, they would have little predictive value as guides to the control of human behaviour (see pp. 394–96).
9. It should be noted that traces of this kind of exaggeration are even manifested in Herbert W. Simons's *Persuasion: Understanding, practice and Analysis* (Reading, MA: Addison-Wesley, 1976).
10. Weigert suggests that, in general, rhetorical perspectives enable us to understand the discourse of any intellectual elite 'as that of a rhetoric of legitimacy, value, and means for the in-group, and a rhetoric of problem discovery, problem-solving, and functional indispensability for the out-group' (112). See also Friedrichs, p. 141.
11. See, for example, Margaret Masterman, 'The Nature of a Paradigm', in Imre Lakatos and Alan Musgrave, eds, *Criticism and the Growth*

of Knowledge (Cambridge, MA: Harvard University Press, 1970); Dudley Shapere, 'The Structure of Scientific Revolutions', *Philosophical Review* 73 (1964): 383–94. See also Kuhn, pp. 205–7; Joseph Ben-David and Teresa A. Sullivan, 'Sociology of Science', in Alex Inkeles, ed., *Annual Review of Sociology* (Palo Alto, CA: Annual Reviews, 1975), 215.

12. See Kenneth J. Gergen, 'Social Psychology as History', *Journal of Personality and Social Psychology* 26 (1973): 309–20. Studying the past: An excellent example, once again, is Fishbein and Ajzen. Their critique of research on persuasion has led them to the development of a theory of persuasion that abandons any pretense of yielding concrete predictions about source or message effects. See also McGuire: 'The temple bell has tolled and tolled again, disturbing the stream of experimental social psychology research and shaking the confidence of many of us who work in the area'. (446). Also see A.R. Buss, 'The Emerging Field of the Sociology of Psychological Knowledge', *American Psychologist* 30 (1975): 988–1002; Allen C. Elms, 'The Crisis of Confidence in Social Psychology', *American Psychologist*, pp. 967–76.

References

Ben-David, Joseph, and Teresa A. Sullivan. 1975. 'Sociology of Science', in Alex Inkeles, ed., *Annual Review of Sociology*. Palo Alto, CA: Annual Reviews.

Burke, Kenneth. 1968. 'Terministric Screens', *Language as Symbolic Action: Essays on Life, Literature, and Method*. Berkeley: University of California Press.

Friedrichs, Robert W. 1970. *A Sociology of Sociology*. New York: Free Press.

Gergen, Kenneth J. 1973. 'Social Psychology as History', *Journal of Personality and Social Psychology* 26: 309–20.

Koch, Sigmund. 1964. 'Psychology and Emerging Conceptions of Knowledge as Unitary', in T.W. Wann, ed., *Behaviorism and Phenomenology: Contrasting Bases for Modern Psychology*. Chicago: University of Chicago Press.

Kuhn, Thomas S. 1970. *The Structure of Scientific Revolutions*, 2nd ed. Chicago: University of Chicago Press.

Meehl, Paul E. 1967. 'Theory Testing in Psychology and Physics: A Methodological Paradox', *Philosophy of Science* 34: 114.

Pavalko, Ronald M. 1971. *Sociology of Occupation and Professions*. Taska, IL: F.E. Peacock.

Scheffler, Israel. 1967. *Science and Subjectivity*. Indianapolis: Bobbs-Merrill.

Simons, Herbert W. 1970. 'Requirements, Problems and Strategies: A Theory of Persuasion for Social Movements', *Quarterly Journal of Speech* 56: 1–11.

Taylor, John F.A. 1938. *Aim of Education*. New York: Macmillan.

Weigert, Andrew. 1970. 'The Immoral Rhetoric of Scientific Sociology', *American Sociologist* 5 (May 1970): 111–19.

QUESTIONS FOR CRITICAL THOUGHT

1. This is likely the most difficult reading in the entire collection, in part because of its elevated diction and complex style, and in part because the context is unfamiliar to the intended audience for this text. Why, then, would the editor have chosen to include this article? What can 'lay' readers learn from reading something that was intended for an audience different from themselves?
2. What is the purpose of Simons's article? What are its main arguments?

3. Why does Simons structure his article in the form of a debate?
4. In his introduction, Simons outlines some of the charges levelled at scientific writers by their critics and presents the research question he plans to answer: 'Are the indictments offered by rhetorically-minded critics valid in principle or do they apply only to atypical practices by isolated scientists?' Does he satisfactorily answer this question? What is his conclusion?
5. How formal is Simons' article? Why?
6. Is Simons' essay in any way personal? Is it technical writing? Why or why not?
7. Simons' essay is one of several research reports included in this book. Compare his work with that of Lloyd Bitzer ('Functional Communication'), Stephen Halloran ('Classical Rhetoric for the Engineering Student'), Tania Smith ('What Connection does Rhetorical Theory Have to Technical and Professional Communication?'), Cheryl Forbes ('Getting the Story, Telling the Story: The Science of Narrative, the Narrative of Science'), Carolyn Miller ('What's Practical About Technical Writing?'), Bernadette Longo ('Communicating With Non-Technical Audiences'), Jennifer MacLennan ('Disciplinarity, Identiy, and the "Profession" of Rhetoric'), Cezar Ornatowski ('Between Efficiency and Politics'), Paul Dombrowski ('Can Ethics Be Technologized?'), John Lorinc ('Driven to Distraction'), and Stephen Talbott ('The Deceiving Virtues of Technology'). What patterns emerge when you compare these works? Do the essays fall into distinct groups? What differences and what similarities can you discern in the way research is carried out and in the way it is reported in the field of rhetoric and professional writing?

PART IV

Observations on Style and Editing

STYLE has been described as putting 'the right word in the right place'. In other words, it means finding the appropriate fit between the form and the content of any message, between the shape and emphasis of the message and its intended audience, between the context in which the message will be received and the treatment of the issues, between the structure and diction of the message and the credibility of its author.

As the essays in this Part reveal, achieving an appropriate style means more than running a spell-checker. Proofreading and editing take place on many levels, and include not only the correctness of the mechanics of writing, but also the appropriateness of the positioning of your argument, the adherence of your document to the standards expected in the organization where you are writing, the internal consistency of layout and visuals, the accuracy of the information contained in the document, and the appropriateness of the footing you have claimed for yourself as a writer.

These essays approach the subject of style from a variety of perspectives, but each provides some guidance for writing more forcefully, clearly, and effectively.

CHAPTER 18

Effective Writing

George C. Harwell

The Importance of Clear Expression

Written and oral expression is an essential part of technical work. Nearly every job is preceded or followed by a report, and often there are reports both before and after. Many engineers estimate that 75 per cent of their time is given not to doing technical work itself but to communicating ideas about that work to other people. No matter what the particular field is, the medium at some point is words.

It follows that the ability to express himself well is a decisive factor in the career of the professional technologist or scientist. He cannot be happy living in the constant dread of being called on to do what he knows he cannot do—namely, write or speak about his work in such a way that others quickly grasp his meaning. On the other hand, the knowledge that he can do it is a source of confidence, and the knowledge that he has done it brings the deep satisfaction that always follows achievement.

A command of clear expression can also lead to more tangible rewards. First, it makes a good impression on a man's superiors by saving them time when they are considering his work. Second, it shows his technical ability to advantage and results in a higher rating of his professional judgment. And third, these things being so, he is given greater responsibility, which leads to promotion and salary increases.

Apropos of the last point is a study conducted some years ago by one of America's largest engineering companies. Its principal finding was that 100,000 dollars is the minimum difference in the lifetime earnings of two men who have equal technical ability but of whom one can express himself well and the other cannot.

The Elements of Good Writing

Whether the field is technical or not, the basic elements of good writing are always the same. They do not vary according to the subject matter. The phrases 'technical communication' and 'technical writing' do not apply to a special way of writing, but pertain to what the writing is about and have little to do with how the writing is done. Even that type of paper most closely associated with technical and scientific fields—the report—is special only in its general make-up, and it is used today in fields not at all technical.

Understanding now that the principal elements of good writing are the same in the technical field as in other fields, let us see what these elements are. We may consider them to be: accuracy, clarity, simplicity, and readability. Naturally they overlap. For instance, accuracy of expression requires clarity; in turn, clarity depends on simplicity, on structure and language that are easy for the reader to follow.

In addition to overlapping each other, these elements will vary in their relative proportions from paper to paper, depending on either the material, the readers the paper is aimed at, or both. For example, no matter who the expected readers may be, data concerning a rudimentary procedure can be set forth in simpler language than data concerning a complex one. Likewise, no matter what its subject a paper addressed to non-experts should be less technical and therefore less exact than one written for experts, because the need of the non-experts for overall clarity calls for some sacrifice of detail.

Still, in every case the four elements—accuracy, clarity, simplicity, and readability—must be present if the writing is to have its fullest effect. We shall examine these elements more closely.

ACCURACY

For the engineer as well as for the scientist, the first concern is for the accuracy of his material—that is, the correctness and the completeness of his data and the soundness of his evaluation of them. This accuracy will depend on his training and experience, his judgment, and his conscientiousness.

Regardless of how accurate his material may be, however, it becomes worthless if it is communicated inaccurately. Long hours of testing and recording, of computing and evaluating have been wasted if this information is presented so carelessly that the reader misconstrues it and does not act upon it properly. Months of painstaking research and design mean nothing if favourable results are stated so ineffectually that the sponsor of the project fails to extend the financial grant necessary for the engineer to continue his work. In modern industry accuracy in reporting the findings is just as important as the findings themselves. Accordingly, the engineer has completed only half of his job when he has made sure that his data are accurate and his appraisal is sound; it remains for him to write a report that mirrors this accuracy.

CLARITY

To be clear should be our constant aim. From the drawing up of the outline down to the choice of the last word in the final draft of the paper, there is no point where we can safely relax in our efforts to be clear. It is true that an idea can be so simple that almost any way of expressing it will be understood by an attentive reader. But there is the rub: we cannot count on all our readers being really attentive—that is, willing to ponder our meaning if our meaning is not apparent at once. For this reason a fixed maxim of composition is: do not write so that you may be understood, but write so that you *must* be.

Clear organization of the paper as a whole is prerequisite to its being fully grasped. The major topics and the sub-topics under them should have a logical connection that the reader can easily discern. He should always know where he is. A highly complicated subject can be made intelligible by lucid organization, by our carefully planning what we wish to say before attempting to say it.

This clarity of the whole should carry into the parts—into the paragraphs and the individual sentences. Each paragraph should be built around a central idea, called the topic sentence. Usually this sentence is an actual part of the paragraph. When it is not, it must be distinctly implied. Many other sentences, carrying many other ideas, may also be in the paragraph; but if the paragraph is to have unity, these sentences must bear directly on the topic sentence, their ideas must develop the central idea. Moreover, if the paragraph is to be coherent, these sentences must be so arranged that the connection between them is obvious and the flow from one idea to the next is smooth. The basis, then, of clarity within the paragraph is, first, unification of its content around a central thought and, second, coherence between successive ideas.

The building of clear sentences—sentences constructed grammatically and effectively and punctuated correctly—is a subject so broad and important as to deserve separate treatment. Nevertheless, here under the general topic of clarity we should observe that in sentences having more than one idea the writer should take extra pains to achieve clear structure. He should be especially alert for two pitfalls.

One pitfall is neglecting to decide the relative importance of the ideas and thereby failing to indicate to the reader the relative emphasis to be given them. Consider a simple case of two ideas:

a. A new alloy is used in the camshaft.
b. Because of it the camshaft can have a smaller diameter.

There are four possible ways to distribute the emphasis, and it is through sentence structure that the writer indicates which of the four he considers proper under the particular circumstances.

1. If he wants to give the facts equal weight, he should express them in the same kind of construction. Here it would be two independent clauses:

> A new alloy is used in the camshaft; it permits the shaft to have a smaller diameter.

2. If he wishes to retain equal emphasis on the alloy and the smaller diameter but to play down the actual use of the alloy, he should convert the first clause into some type of subordinate construction:

> A new alloy used in the camshaft permits it to have a smaller diameter.

Or he can eliminate the detail about use altogether:

> A new alloy permits the camshaft to have a smaller diameter.

3. If he does not regard the two ideas as of equal significance and wants to stress the first one, he should express it as an independent clause and subordinate the second idea to it—that is, word the second idea as a clause or a phrase dependent on the first one:

> A new alloy, which permits a smaller diameter, is used in the camshaft. (dependent clause)
> A new alloy, permitting a smaller diameter, is used in the camshaft. (phrase)

4. If he wants the stress on the second idea, he reverses the subordination:

> Because a new alloy is used, the diameter of the camshaft can be smaller. (dependent clause)

> Through the use of a new alloy the diameter of the camshaft can be smaller. (phrase)

The other pitfall is failure to arrange the ideas in such a way that the logical connection between them is instantly apparent.

> The machine should be dismantled before it is shipped by a competent mechanic.

This says that the machine is to be shipped by a competent mechanic, but the intended meaning is that it should be dismantled by one. Commas before and after the time clause would help, but it is poor practice to resort to punctuation for unscrambling careless construction. A rearrangement of the ideas is the only remedy:

> Before the machine is shipped, it should be dismantled by a competent mechanic.

> The machine should be dismantled by a competent mechanic before it is shipped.

Another example of a misplaced modifier is:

> A small crack in the bottom of the cylinder, which is a result of overheating, is causing the loss in pressure.

The simplest correction is to use two independent clauses:

> A small crack, caused by overheating, is in the bottom of the cylinder; the crack explains the loss in pressure.

> A small crack in the bottom of the cylinder was caused by overheating; the crack explains the loss in pressure.

> The loss in pressure is due to a small crack in the bottom of the cylinder; the crack was caused by overheating.

The passion to be clear should extend down to the choice of the individual words. The general organization, the paragraph development, and the sentence structure of a paper may all be perfectly sound, and yet its effectiveness can be marred or even destroyed by obscure language. The object of all oral and written expression is to set up in another person's mind the same image that is in one's own—or, to put it another way, to induce the same chemical action in another's brain that is taking place in one's own. Since words and the emphasis given them are the primary means of doing this, success is in direct ratio to the appropriateness of the diction.

This is especially true if the mode of expression is writing, for unlike a speaker, a writer is seldom present to clarify his meaning if it is questioned; he has to rely on his original phrasing. If the language is precise, the image he summons up in the reader's mind will be true and sharp; if the language is hazy, the image will be blurred. In short, whether the image is in focus or out of focus, and to what degree, will be determined by the words chosen to evoke it.

Unfortunately, there is no infallible means by which a writer may be sure that what he has put down will be readily understood by all his readers; and excepting most letters and the simplest reports, he will have more than one reader. They will vary in intelligence, in their power to concentrate, and in their knowledge of the field; and he cannot anticipate the needs of each one separately. Still, he can assure himself of maximum success by adopting a simple procedure: writing the first draft and revising it several days before the final draft is prepared.

The first draft should be written from his own standpoint to make certain he is saying what he meant to say. The first revision should be made primarily from the same viewpoint but with some awareness of a cross section of his readers, especially those readers who know least about the subject. The final revision should be entirely from the viewpoint of these readers; and the more time that elapses between the first draft and the final one, the easier it will be for the author to project himself out of his own thinking and to read his work as others will read it.

The following sentences may be clear to the writer as he puts them on paper, but he will hardly let them stand if he later reads them from someone else's point of view.

> New conduits should be ordered as the old wires are being removed.

This is subject to two interpretations. The first is that the actions are to be simultaneous:

> New conduits should be ordered while the old wires are being removed.

The second indicates cause and effect:

> New conduits should be ordered because the old wires are to be removed.

Another equivocal sentence is:

> Only two days will be needed to fix the generator in the main building.

Does 'to fix' have the meaning here of 'to repair' or 'to install'? Again:

> The boiler cannot be used in its present shape.

Does this mean that the customary cylindrical shape of the boiler must be altered if it is to fit into the space allotted for it, or does 'shape' have its slang meaning of 'condition'? The context would probably indicate the answer, just as it might in the two examples before it. Even so, there is no getting away from the fact that the diction could be, and therefore should be, more precise.

The next sentence illustrates poor choice of structure as well as careless wording:

> The walls must be removed with Lolly columns to support the roof.

Only when the reader gets to the end does he see that the writer is not suggesting a new use for Lolly columns—namely, as a means for tearing down walls. The writer brought the suspicion upon himself by carelessly choosing 'with' to introduce the second idea. This idea, it turns out, is of equal weight with the first one and therefore should be a parallel construction—here, an independent clause:

> The walls must be removed and Lolly columns will be used to support the roof.

Each of the preceding examples of ambiguous diction makes sense as it stands, but merely making sense is not enough. A statement should make the exact sense the writer intends it to make and should be open to no other interpretation. A reader cannot be blamed if he does not take time to grope around until he thinks he has found the correct meaning; the person who should put the most time on a sentence is the writer, not the reader—and he should put enough time on it to ensure that its true meaning not only *may* be understood, but *must* be.

SIMPLICITY

Clearness and simplicity are practically inseparable virtues. That which is clear is usually simple in structure and diction, and that which is simple can nearly always be understood quickly. It is the involved paragraph and the complicated sentence that require more than one reading, and the long, unfamiliar word that causes delay. In view of what has already been said about paragraph and sentence structure and what is to be said about them later, the emphasis here will be on simplicity of language.

The words used to express an idea should be chosen on one basis: to convey that idea as quickly as is practical. With this aim, the words should already be known to the reader or they should be used in such a way that their meaning is obvious. In addition, they should be chosen for conciseness.

Generally speaking, the short word is more forceful than the long one, even when the reader knows both. Yet every day one can observe dozens of instances in which the longer synonym is used. An example prevalent in technical writing is the substitution of the polysyllable 'encounter' for the forthright 'have' and 'find':

> We expect to encounter no trouble.

'Encounter' is a good word. It does mean 'to meet with' or 'to face' and is known to most people, but it is not as direct in this sentence as its equivalents. And in the following one it is downright wrong because its connotation is wholly ignored:

> During the overhaul we encountered several blown gaskets.

One can encounter an abstract like 'trouble' but not that abstract made concrete in 'blown gaskets'. The habit of the word permits its use with a concrete thing only in referring to people: *to encounter Jones on turning the corner, to encounter one's rival.*

In the following sentence the diction is so heavy and strained as to be almost meaningless:

> Inferior gasoline will induce metallic agitation in the combustion chambers.

In language that every layman understands, this says:

> Inferior gasoline will cause the engine to knock.

Natural wording has also been sacrificed in the next sentence:

> We lack the knowledge of why the costs are ascending.

This is a stiff and affected way of saying:

> We do not know why the costs are rising.

The aim of simple diction, to repeat, is to convey a thought as quickly as is practical. This means conciseness as well as selectivity. It means that expression can be impeded by too many words as well as by heavy and unfamiliar ones.

The movement of the following sentences is slowed down by superfluous words; they are just so many pounds of fat hindering the thought:

a. The mechanism can be repaired easily because of the fact that it is constructed very simply.
b. The drum contains a sufficient amount of oil.
c. The frame sprang a leak because of the lack of adequate braces between the sides.

Nowhere is there a downright error; the objection is that each thought is delayed by padding. Compare:

a. The mechanism can be repaired easily because of its simple construction.
b. The drum contains sufficient oil.
c. The frame sprang a leak because the sides lacked adequate braces.

The next example was one of the recommendations in a report describing a possible site for a gasoline terminal. A spur track needed by the terminal would have to cross the property of a brick company, which was willing to grant right of way in exchange for the privilege of using the spur itself. The sentence fairly groans under its burden of fat:

> Use of the track to be constructed should be available to the neighbouring brick company

for any spur desired for its own use at any time requested by it.

The thought is not altered at all if 15 words are deleted:

The proposed track should be available to the neighbouring brick company at any time.

One of the easiest circumlocutions to slip into is the impersonal construction in which 'there' substitutes for a delayed subject;

1. Sometimes there may be leakage at the second joint.
2. There was a tendency for the coil to overheat.
3. There is only one man in the plant who can operate the press, and that is Garrison.
4. There are extra parts that come with the assembly

In every instance none of the meaning is lost and considerable strength is added if the expletive construction is removed:

a. Leakage may occur at the second joint.
b. The coil tended to overheat.
c. Garrison is the only man in the plant who can operate the press.
d. Extra parts come with the assembly.

Of course, the expletive use of 'there' is not always a fault. Indeed, it is to be preferred when it is less clumsy than an alternative construction or when it gives an idea a desired emphasis:

If there is a better means of distributing the load, we should employ it.

One word can be saved by using 'exists', but the saving is hardly worth the less natural reading that results:

If a better means of distributing the load exists, we should employ it.

In the following instance, the expletive gives the first idea an emphasis and the sentence a balance that justify its use:

There are many possibilities, and we intend to study all of them.

To close the discussion of wordiness, the next example seems fit:

Labour is both abundant and inexpensive, making the over-all conditions conducive to constructing the proposed plant in the city of Centreville.

This sentence violates every principle discussed so far: it is not accurate, it is not clear, and it is not simple. It is not accurate because labour, being a single aspect of the subject, cannot make the overall conditions favourable; it can affect only one. If the intention is to say that labour, the major factor in locating the plant, is satisfactory, the wording should be changed. The sentence is not clear because the relative importance of the two ideas is not reflected in the structure; instead, it is reversed. The sentence is not simple because the words are both heavier and more numerous than they need be. As an added criticism, they are unpleasantly alliterative. It is not difficult to eliminate all these faults and get a clean statement:

Because labour, the major consideration, is plentiful and cheap, the plant should be built in Centreville.

READABILITY

If a piece of writing is accurate, clear, and simple, the chances are strong that it is also readable. When these qualities are missing, reading becomes a chore. Conversely, when they are present, it becomes a pleasure because it is easy. The reader's attention is concentrated on what is being said, not distracted by how it is being said. In short, accuracy, clearness, and simplicity in themselves make for readability.

Still, the writer should not stop with them. There are other ways to increase the readableness of a paper. One is to diversify the sentence patterns. A series of sentences all built alike can be monotonous and can cause the reader's attention to flag. On the other hand, variation in the length and in the structure of sentences keeps his attention whetted.

The following paragraphs are the introduction of a report on Nualloy cartridge cases. The writer has made it interesting while still observing the requirements of the introduction of a formal report. The writing is accurate, clear, and simple. Yet the author has not relied on these qualities alone to make it readable. He has gone a step more and freely varied the length and the structure of his sentences. The result is an added vitality.

> Recent studies of Belson .22-caliber rimfire ammunition have shown that brass should no longer be used in rimfire cartridge cases because it is not strong enough to withstand the pressure produced by modern high-velocity loads. When loaded for high velocities brass cases have two major faults. The first is excessive expansion of the cases when fired, causing improper functioning of the weapons. The second fault is occasional bursting of the cases. Shrapnel from bursting cases not only damages the guns, but often it injures shooters and bystanders. These two faults occur most often in semiautomatic pistols and rifles.
>
> An extensive testing and research program has been undertaken by the Belson Arms Company to determine a suitable replacement for brass in rim fire cartridge cases. An acceptable metal must not have the faults of brass, must be easily formed into cartridge cases, and must be reasonably priced; moreover, cartridges made from this metal must have a low rate of misfiring. Since brass is an excellent cartridge case alloy except for the faults previously mentioned, its characteristics are used as standards in judging the alloys which are tested.
>
> Recently the C.K. Bond Company of Baltimore, Maryland, announced the development of Nualloy, an alloy that it recommends for use in cartridge cases. To determine the advisability of using it as a replacement for brass in rimfire cartridge cases, the Belson Arms Company has tested it. This report covers the following matters: the results of these tests, the availability of Nualloy, its manufacture, and its cost. Special attention is given to the retooling necessary to produce Nualloy cartridge cases and to the firing characteristics of these cases.
>
> It is recommended that the Belton Arms Company use Nualloy instead of brass as a rimfire cartridge case metal.

The variety in sentence patterns is more evident when it is reproduced in diagram. The horizontal lines below represent independent clauses; the solid diagonal lines, adjective and adverb clauses; the broken diagonal lines, phrases that are not integral parts of clauses.

A reader's enjoyment of a paper is further quickened if this variety in sentence pattern is carried into the very language. Repetition of words and the use of clichés blunt the effectiveness of a piece of writing regardless of its excellence in other respects. This does not mean that a writer should altogether avoid repeating a word or using a common idiom. If he does so, he is sure to use language that is artificial and inaccurate. What it does mean is that he should avoid needless repetition by using pronouns and close synonyms and that he should convert clichés—stale, overworked phrases which were once vivid—into fresh wording of his own or into plain, literal wording that never loses its force.

Another way, and possibly the best way, to make a paper readable is for the writer to enjoy writing it. Enthusiasm is contagious; so is boredom. For this reason the man engaged on a project should not make the mistake of letting his enthusiasm wane when the time comes to write about it. Rather, he should understand

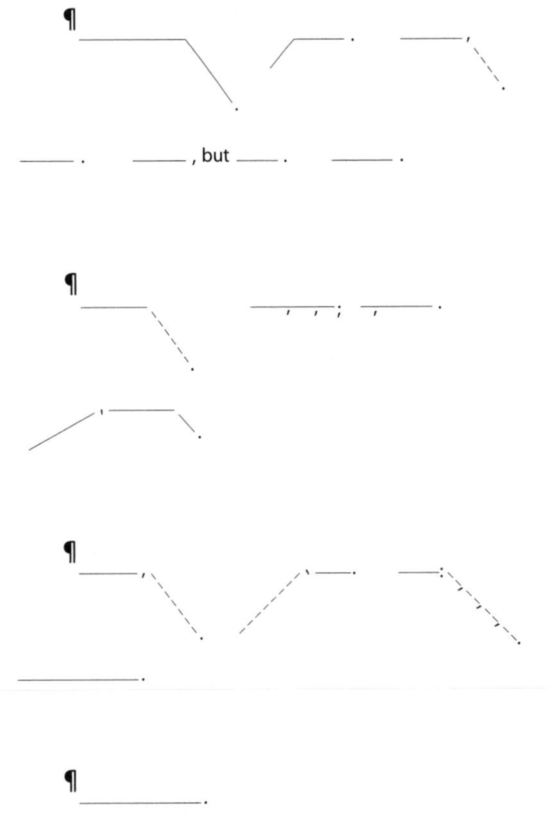

Figure 18.1 'Sentence Pattern in Four Paragraphs'

that through describing his work in a report or a magazine article, he is bringing it to the notice of persons who otherwise might never learn of it. With the composition student, of course, there has probably been no real project; his assigned letter, report, or article must deal with a fictitious situation. Even so, he can generate interest in himself and transmit it to his writing simply by pretending that the assignment is 'real' and that he is going to have 'real' readers.

All too often such adjectives as 'dry', 'stiff', 'flat', and 'dull' are applied to scientific and technical writing—and all too often with justification. But it need not be so. Technical writing allows the same freedom of structure and diction as any other kind of exposition, and the technical man has the same opportunity to exercise versatility—to write with 'style'—as any other non-professional writer has.

Exercises and Assignments

A. A series of phrases appears after each of the following statements. Number the phrases in the order of their accuracy in giving the gist of the statement.

1. In a shipment of 2,000 transistors, 1,806 met specifications.
   ( ) a majority of the transistors
   ( ) many of the transistors
   ( ) most of the transistors
   ( ) about 90 per cent of the transistors
   ( ) slightly more than 90 per cent of the transistors
   ( ) nearly all of the transistors

2. Half of a four-week job has been completed in two weeks.
   ( ) excellent progress
   ( ) satisfactory progress
   ( ) good progress
   ( ) fairly good progress

3. Heat losses were cut from 10 per cent to 2 per cent.
   ( ) reduced considerably
   ( ) reduced appreciably
   ( ) reduced somewhat
   ( ) nearly eliminated

B. Make the following sentences clearer by improving their structure or their diction or both.

1. The first officer following the vice-president is the comptroller.
2. One instance which I recall, an engine change had to be made in sub-zero weather.
3. We are glad you sent us your order and will make every effort to give you our best.
4. Modifications such as multiple carburetion and ignition, high lift cams, superchargers,

and increased compression are made on cars as well as radical body designs being employed.
5. Six evenly-spaced ducts should be used on the two longer walls with four on the other two.
6. An emergency source will be provided by a small Diesel engine driven turbogenerator.
7. The minimum gasket thickness should be such that the filler thickness is at least twice the thickness of the combined layer thickness of the jacket.
8. When a bearing has the capacity for a higher thrust on one side than the other, that side will be marked by a red circle.

C. Reword the following sentences so that the language is simpler and more concise.

1. This adjustment will lead to the elimination of the noise problem.
2. There are two major setups which require a change of punches and three minor setups which call for a change in guide-rule.
3. The inventory shows that we are in need of spare parts.
4. Our lack of sufficient spare parts could cost us a considerable amount of time.
5. We conducted an examination of the machinery located in the powerhouse and are ready to submit a report on it at the present time.
6. The conductors are of inadequate size to efficiently accommodate the quantity of current that is transmitted.
7. We shall make arrangements to proceed with the beginning in the near future.
8. There are certain precautions that must be taken if the results are to be sound and reliable.

D. Revise the sentence pattern and wording of the following paragraphs to make them more varied and readable.

1. The method of eliminating waste is not adequate to handle the present needs. At the present time the shavings are transported from the cabinet shop by a network of suction pipes. The pipes lead to the different machines. The suction in the pipes is supplied by a 2-ft fan. The fan operates at the end of the cabinet shop. The power of this fan is boosted by a larger fan. This second fan operates in the adjoining room.
2. The reactions proceed in the following manner. Lime is heated with coke. The lime decomposes to calcium oxide. The calcium oxide reacts with the coke to give calcium carbide. The calcium carbide is then reacted with water to yield acetylene and calcium hydroxide. The calcium hydroxide is usually considered waste material. The acetylene is then run into a column. There it is mixed with steam and oxygen. Acetylene is hydrolized to acetaldehyde in the presence of mercuric sulfate. The acetaldehyde is oxidized by the oxygen in the mixture to acetic acid.

QUESTIONS FOR CRITICAL THOUGHT

1. What is the purpose of Harwell's essay? How effectively does he fulfill it?
2. For many of the writers in this collection, good writing is contextual—that is, it is determined by its appropriateness to audience and situation. Is that true of Harwell's essay, or does he treat writing quality as independent of situation? What might account for this difference?
3. Like MacLennan ('Why Communication is Important'), Burton ('An Engineer's Rhetorical Journey'), Urquhart ('Bridging Gaps, Engineering Audiences'), and others,

Harwell claims that the ability to communicate effectively is 'a decisive factor' in the career of professional technologists and scientists. What evidence does he provide?

4. Is Harwell's assertion that 'the basic elements of good writing are always the same' challenged by any of the other readings? How does it square with the situational approach advocated by Buehler in 'Situational Editing'? What, exactly, does Harwell mean by his statement?
5. What are the four 'basic elements' of which Harwell speaks?
6. Harwell seems to focus primarily on grammatical elements as the means to good writing. However, he cannot avoid speaking also of audience adaptation. To what extent, for example, is the notion of clarity really a matter of accommodating the audience?
7. In her essay 'Situational Editing', Buehler distinguishes between a rhetorical approach to writing and a grammatical one. Consider the extent to which the four 'basic elements' of good writing are rhetorical, rather than grammatical, issues.
8. Describe Harwell's style. How accessible is it? To what extent is readability a matter of audience adaptation? What can be learned by reading something originally addressed to an audience different from you?
9. Compare Harwell's passage with Glaser's 'Voices to Shun'. What differences do you see? What similarities?
10. How useful are the diagrams of sentence structure that Harwell provides? Do they make his point any clearer?
11. Harwell asserts that 'the best way to make a paper readable is for the writer to enjoy writing it.' Compare this assertion with the advice offered by Peter Elbow in 'Three Tricky Relationships to an Audience'. How does this square with Harwell's advice about his 'four basic elements'?

CHAPTER 19

Clutter

William Zinsser

Fighting clutter is like fighting weeds—the writer is always slightly behind. New varieties sprout overnight, and by noon they are part of American speech. Consider what President Nixon's aide John Dean accomplished in just one day of testimony on television during the Watergate hearings. The next day everyone in America was saying 'at this point in time' instead of 'now'.

Consider all the prepositions that are draped onto verbs that don't need any help. We no longer head committees. We head them up. We don't face problems anymore. We face up to them when we can free up a few minutes. A small detail, you may say—not worth bothering about. It is worth bothering about. Writing improves in direct ratio to the number of things we can keep out of it that shouldn't be there. 'Up' in 'free up' shouldn't be there. Examine every word you put on paper. You'll find a surprising number that don't serve any purpose.

Take the adjective 'personal', as in 'a personal friend of mine', 'his personal feeling', or 'her personal physician'. It's typical of hundreds of words that can be eliminated. The personal friend has come into the language to distinguish him or her from the business friend, thereby debasing both language and friendship. Someone's feeling is that person's personal feeling—that's what 'his' means. As for the personal physician, that's the man or woman summoned to the dressing room of a stricken actress so she won't have to be treated by the impersonal physician assigned to the theatre. Someday I'd like to see that person identified as 'her doctor'. Physicians are physicians, friends are friends. The rest is clutter.

Clutter is the laborious phrase that has pushed out the short word that means the same thing. Even before John Dean, people and businesses had stopped saying 'now'. They were saying 'currently' ('all our operators are currently assisting other customers'), or 'at the present time', or 'presently' (which means 'soon'). Yet the idea can always be expressed by 'now' to mean the immediate moment ('Now I can see him'), or by 'today' to mean the historical present ('Today prices are high'), or simply by the verb 'to be' ('It is raining'). There's no need to say, 'At the present time we are experiencing precipitation.'

'Experiencing' is one of the worst clutterers. Even your dentist will ask if you are experiencing any pain. If he had his own kid in the chair he would say 'Does it hurt?'. He would, in short, be himself. By using a more pompous phrase in his professional role he not only sounds more important; he blunts the painful edge of truth. It's the language of the flight attendant demonstrating the oxygen mask that will drop down if the plane should run our of air. 'In the unlikely possibility that the aircraft should experience such an eventuality', she begins—a phrase so oxygen-depriving in itself that we are prepared for any disaster.

Clutter is the ponderous euphemism that turns a slum into a depressed socioeconomic area, garbage collectors into waste disposal personnel and the town dump into the volume reduction unit. I think of Bill Mauldin's cartoon of two hoboes riding a freight car. One of them says, 'I started as a simple bum, but now I'm hard-core unemployed.' Clutter is political correctness gone amok. I saw an ad for a boys' camp designed to provide 'individual attention for the minimally exceptional'.

Clutter is the official language used by corporations to hide their mistakes. When the Digital Equipment Corporation eliminated 3,000 jobs its statement didn't mention layoffs; those were 'involuntary methodologies'. When an Air Force missile crashed, it 'impacted with the ground prematurely'. When General Motors had a plant shutdown, that was a 'volume-related production schedule adjustment'. Companies that go belly-up have 'a negative cash-flow position'.

Clutter is the language of the Pentagon calling an invasion a 'reinforced protective reaction strike' and justifying its vast budgets on the need for 'counterforce deterrence'. As George Orwell pointed out in 'Politics and the English Language', an essay written in 1946 but often cited during the wars in Cambodia, Vietnam, and Iraq, 'political speech and writing are largely the defense of the indefensible. . . . Thus political language has to consist largely of euphemism, question-begging and sheer cloudy vagueness.' Orwell's warning that clutter is not just a nuisance but a deadly tool has come true in the recent decades of American military adventurism. It was during George W. Bush's presidency that 'civilian casualties' in Iraq became 'collateral damage'.

Verbal camouflage reached new heights during General Alexander Haig's tenure as President Reagan's secretary of state. Before Haig nobody had thought of saying 'at this juncture of maturization' to mean 'now'. He told the American people that terrorism could be fought with 'meaningful sanctionary teeth' and that intermediate nuclear missiles were 'at the vortex of cruciality'. As for any worries that the public might harbour, his message was 'leave it to Al', though what he actually said was: 'We must push this to a lower decibel of public fixation. I don't think there's much of a learning curve to be achieved in this area of content.'

I could go on quoting examples from various fields—every profession has its growing arsenal of jargon to throw dust in the eyes of the populace. But the list would be tedious. The point of raising it now is to serve notice that clutter is the enemy. Beware, then, of the long word that's no better than the short word: 'assistance' (help), 'numerous' (many), 'facilitate' (ease), 'individual' (man or woman), 'remainder' (rest), 'initial' (first), 'implement' (do), 'sufficient' (enough), 'attempt' (try), 'referred to as' (called), and hundreds more. Beware of all the slippery new fad words: paradigm and parameter, prioritize and potentialize. They are all weeds that will smother what you write. Don't dialogue with someone you can talk to. Don't interface with anybody.

Just as insidious are all the word clusters with which we explain how we propose to go about our explaining: 'I might add', 'It should be pointed out', 'It is interesting to note'. If you might add, add it. If it should be pointed out, point it out. If it is interesting to note, make it interesting; are we not all stupefied by what follows when someone says, 'This will interest you'? Don't inflate what needs no inflating: 'with the possible exception of' (except), 'due to the fact that' (because), 'he totally lacked the ability to' (he couldn't), 'until such time as' (until), 'for the purpose of' (for).

Is there any way to recognize clutter at a glance? Here's a device my students at Yale found helpful. I would put brackets around every component in a piece of writing that wasn't doing useful work. Often just one word got bracketed: the unnecessary preposition appended to a verb ('order up'), or the adverb that carries the same meaning as the verb ('smile happily'), or the adjective that states a known fact ('tall skyscraper'). Often my brackets surrounded the little qualifiers that weaken any sentence they inhabit ('a bit', 'sort of'), or phrases like 'in a sense', which don't mean anything. Sometimes my brackets surrounded an entire sentence—the one that essentially repeats what the previous sentence said, or that says something readers don't need to know or can figure out for themselves. Most first drafts can be cut by 50 per cent without losing any information or losing the author's voice.

My reason for bracketing the students' superfluous words, instead of crossing them out, was to avoid violating their sacred prose. I wanted to leave the sentence intact for them to analyze. I was saying. 'I may be wrong, but I think this can be deleted and the meaning won't be affected. But you decide. Read the sentence without the bracketed material and see if it works.' In the early weeks of the term I handed back papers that were festooned with brackets. Entire paragraphs were bracketed. But soon the students learned to put mental brackets around their own clutter, and by the end of the term their papers were almost clean. Today many of those students are professional writers, and they tell me, 'I still see your brackets—they're following me through life.'

You can develop the same eye. Look for the clutter in your writing and prune it ruthlessly. Be grateful for everything you can throw away. Reexamine each sentence you put on paper. Is every word doing new work? Can any thought be expressed with more economy? Is anything pompous or pretentious or faddish? Are you hanging on to something useless just because you think it's beautiful?

Simplify, simplify.

QUESTIONS FOR CRITICAL THOUGHT

1. What is the purpose of Zinsser's essay?
2. This selection is from Zinsser's advice book, *On Writing Well: An Informal Guide to Writing Nonfiction*. Advice-giving is always directive to some degree; how directive is Zinsser? Why or why not?
3. Zinsser opens with an analogy between writing and gardening, between editing and weeding. How vivid is this analogy? How effective is it?
4. According to Zinsser, 'writing improves in direct ratio to the number of things we can keep out of it that shouldn't be there.' Is this true? How do you know? How does Zinsser prove his point?
5. Compare Zinsser's advice to the advice given in the other essays in this chapter, or to the others in the rest of the book. How close is he, for example, to Orwell, Lutz, Casselman, or Plotnik?
6. Why does Zinsser use so many examples?
7. Zinsser equates getting rid of clutter to 'being [your]self' in your writing. What is the relationship between the two?
8. To what extent does Zinsser treat clutter as identical to deliberate obfuscation (the bafflegab spoken of by Casselman in 'Bafflegab and Gobbledygook' and Lutz in 'Doublespeak')? Why?
9. What is the purpose of 'verbal camouflage'? Why does Zinsser declare it 'the enemy'? The enemy of what?
10. Zinsser argues that most first drafts can be cut by 50 per cent; does this surprise you? What guidance does he provide on how to eliminate the clutter from your own writing?
11. At one point, Zinsser seems to equate useful and beautiful. Compare his assertions to Rapoport's account of mathematical language as 'poetic' (see 'The Language of Science') or Snow's views of the 'aesthetic' appeal of science ('The Moral Un-Neutrality of Science'). What is the relationship between utility and aesthetic appeal? To what extent are these values in tension?
12. As the essay progresses, Zinsser's garden metaphor is replaced by a combat metaphor. Why? What, exactly, is being warred upon? Who are the combatants?
13. Does Zinsser mean to equate the deliberate obfuscation of politicians like George W. Bush with the kind of clutter that he finds in his students' writing? If not, why are they juxtaposed? What does such juxtaposition imply?
14. Is Zinsser's article formal? Is it personal? Why or why not? To what extent does Zinsser's purpose control his choices in these areas?

CHAPTER 20

Getting It Together: Strategies for Writing Cohesively

Jennifer M. MacLennan

Sooner or later, nearly everyone who writes on the job learns how important it is to produce reports that get read, briefs that communicate clearly to their intended readers, policies that provoke action and understanding. And sooner or later, too, nearly everyone learns another lesson: how frustratingly difficult it can be to achieve that kind of readability and impact in your own writing.

There are many factors that can make writing ineffective, and just as many possible 'fixes' for what ails you, but one thing is sure: if the reader can't follow your reasoning nothing else you try will rescue your prose from the trash heap.

The goal of this essay is to let you in on a little-known secret about writing well, and in the process teach you what is probably the single most important and useful lesson about writing that you will ever learn, which is: from this moment on, in everything you write, consciously and explicitly link every thought or idea to the thought or idea that preceded it. If you practice applying the strategies you'll learn in this essay, you will instantly double the impact and understandability of your writing, and increase your persuasiveness and your credibility. You can do all these things, in the very next document you write, using tools you will learn in the next couple of pages.

What we are talking about is a feature known as cohesiveness—the way the individual parts of any message are connected. The simple fact is that cohesive messages are more logically coherent, and thus easier to understand. Readers generally experience them as better written, and perceive their authors to be more competent. In short, paying attention to cohesiveness will help ensure that your message is taken seriously.

In part, cohesiveness is a product of something called the 'known-new contract'. This is the understanding that, in any argument, information that is known to the reader will be presented both logically and chronologically *before* information that is new to the reader. That's all there is to it.

Virtually all messages provide *some* new information, but too many fail to show readers why they should care, or how these new ideas connect to what the readers already know. Here's the simple kernel of truth at the heart of the known-new rule: if you want to explain an idea to someone, start with what he already knows and move from there to the part that he doesn't yet understand; if you want to persuade someone, start with things she already cares about and show how your message addresses those concerns. Building in just the appropriate amount of such 'planned redundancy'—a kind of review of what the reader already knows—will help you make your meaning clearer to those you hope to influence.

To test out the pervasiveness of the known-new contract, look at any effective document you have at hand. If it's well written, you'll discover that every sentence follows this pattern, presenting the known information in the subject of the sentence and the new information in the predicate. And every well-written paragraph begins with familiar or established facts before moving on to new data. Finally, individual segments of a well-crafted document (summary, introduction, discussion, and so on) also follow this general principle. If you want your writing to be understood and accepted by your readers, every message you create should *always* be arranged with familiar information first.

Here's how the known-new contract works. Suppose you are introducing your reader to a new concept or idea. The first time you use the term you will need to define it and provide synonyms or examples. The next time you use the word, provide a brief reminder of its meaning to reinforce the initial explanation. Finally, when the new term itself has become familiar enough to be considered known information, you can in turn use it to introduce other new information. Good writers in all fields typically exploit the benefits of the known-new contract to help their readers get the message. Mathematical arguments, for example, begin with what is given—the known information—and proceed to the proof of a theorem—the new information. Writers who ignore the known-new formula typically confuse and alienate their readers, and their messages are ignored as a result.

In addition to paying attention to the known-new contract, you can improve the readability and clarity of your writing by using one or more of the specific strategies described below to link every single sentence in some *explicit* way to the sentence that went before it, and to connect every single paragraph in some *explicit* way to the paragraph that went before. Placing your linking devices close to the beginning of every sentence or paragraph actually helps the audience to pick up the thread of your argument and so stay on track to the end.

Although you may never have noticed them before, you will find at least one, and frequently more than one, of the following devices in *every* sentence of any document that you consider to be well written, from a Harry Potter novel to an engineering report. Interestingly, though these devices are easy to see once you know what to look for, and may strike you as cumbersome when you are learning to use them, the average reader, who simply experiences the writing as especially clear, does not notice them. You can make your own writing just as lucid by mastering the seven explicit strategies explained below. All of the examples shown were taken from this essay.

1. *Repeat a key word from the previous clause or sentence in the new clause or sentence.* Do not overuse the device of repetition, but employ enough 'planned redundancy' to make your message understandable.

    Example: We are concerned here with a feature known as cohesiveness—the way the individual parts of any message are connected. The simple fact is that cohesive messages are more logically coherent and thus easier to understand. Readers generally experience them as better written, and perceive their authors to be more competent. Devices of cohesiveness help to ensure that your message will be taken seriously.

2. *Use a pronoun in the new clause or sentence to refer to a specific noun or noun phrase in the previous clause or sentence.* Avoid using pronouns such as 'it' or 'this' unless they clearly point to a preceding noun; pronouns with no identifiable antecedent can actually undercut the cohesiveness of your text.

    Example: The goal of this essay is to let you in on a little-known secret about writing well, and in the process teach you what is probably the single most important and useful lesson about writing that you will ever learn, which is: from this moment on, in everything you write, consciously and explicitly link every thought or idea to the thought or idea that preceded it.

3. *In the new clause or sentence, use a synonym for a key word or phrase in the previous clause or sentence.* Make sure that the synonym fits the context, and that your reader will connect the two; remember to place the synonym near the beginning of the second clause or sentence.

    Example: Suppose you are introducing your reader to a new concept or idea. The first time you use the term you will need to define it and provide synonyms or examples. The next time you use the word, provide a brief reminder of its meaning to reinforce the initial explanation.

4. *In the new clause or sentence, use an antonym for a key word or phrase in the previous clause or sentence.* The principle of contrast helps your audience to link two thoughts and to follow the thread of your argument. Contrast may be combined with parallelism (device #7) for greater effect.

    Example: Good writers in all fields typically exploit the benefits of the known-new contract to help their readers get the message. . . . Writers who ignore the known-new formula typically confuse and alienate their readers, and their messages are ignored as a result.

    Example: Good writers in all fields typically exploit the benefits of the known-new contract to help their readers get the message. . . . Writers who ignore the known-new formula typically confuse and alienate their readers, and their messages are ignored as a result.

5. *Use a word in the new clause or sentence that is closely associated with a word in the previous clause or sentence*. The link appears similar to the synonym, but differs in that the paired items are words that have become yoked by habit or convention, rather than synonyms in meaning: cats and dogs; salt and pepper; trains, planes, and automobiles; lions and tigers and bears. They may also name two parts of a process: question and answer; problem and solution; speaker and audience.

    Example: Good writers in all fields typically exploit the benefits of the known-new contract to help their readers get the message.

6. *Use a specific connecting word or phrase*. If you have organized your ideas well, cohesiveness (and hence coherence) can be enhanced by adding explicit connectives. Some that you may wish to use include:

since	as well as	in addition to
therefore	however	on the other hand
nautrally	of course	as a matter of fact
also	nevertheless	for instance
for example	once again	furthermore
moreover	thus	first this; second, this
as a result	then	for this reason
because	after	following

There are many more, of course—a complete list would be much too long to include here. Any word that explicitly signals a movement in your thought from cause to effect, problem to solution, first to last, front to back, and so on, is considered a connective.

    Example: The next time you use the word, provide a brief reminder of its meaning to reinforce the initial explanation. Finally, when the new term itself has become familiar enough to be considered known information, you can in turn use it to introduce other new information.

    Example: Although you may never have noticed them before, you will find at least one, and frequently more than one, of the following devices in *every* sentence of the document. Interestingly, though these devices are easy to find once you now what to look for, and may strike you as cumbersome when you are learning to use them, they pass unnoticed by the average reader, who simply experiences the writing as especially clear.

7. *Place ideas in a parallel structure*, which means that phrases, clauses, sentences, paragraphs, or sections of the document should be arranged in the same structural or organizational pattern. Parallel structure can be used in a list, as in

the first example, or it can unify sentences and paragraphs, as shown in the second example.

> *Example*: Sooner or later, nearly everyone who writes on the job learns how important it is to produce reports that get read, briefs that communicate clearly to their intended readers, policies that provoke action and understanding.

> *Example*: If you do this, you will instantly improve your writing and double its impact and its understandability. You will also increase your persuasiveness and your credibility.

In addition to helping you write clearer sentences and paragraphs, parallel structure can also be used to unify sections of a whole document. When you are writing your resume, for example, you create several sections, such as education, volunteer activities, and work experience. Within each of these sections, you organize the information following a similar pattern: for example, your employment might be organized using a date-location-title-duties pattern, while your education follows a similar pattern of date-school-major-courses taken. Used in this way, parallel structure assists your reader in making sense of unfamiliar information.

Starting right now, with the very next document you write, you can make an immediate improvement in the clarity and readability of all your professional writing. Simply build one or more of these devices into every sentence you write, and into every paragraph you construct. Using these explicit writing strategies to create cohesiveness not only helps out our readers, but also makes your own writing task easier because it helps to keep your meaning from drifting or jumping around. The magic of cohesiveness also provides an added bonus: your readers will perceive you as someone who really understands the business, and whose words make a powerful impact.

QUESTIONS FOR CRITICAL THOUGHT

1. For whom was this essay written? How do you know?
2. Compare the style of this essay with MacLennan's other selections ('Why Communication Matters', 'Communicating Ethically', or 'Diciplinarity, Identity, and the "Profession" of Rhetoric'). What differences do you notice? What similarities? Does MacLennan sound the same in all of her selections? Why do you think this might be?
3. This essay gives advice on writing, and is quite directive in its tone. Identify the commands that MacLennan gives. What is their effect on the tone of the essay?
4. How personal is this essay? Is it more or less personal than MacLennan's other selections? What are the elements of purpose, audience, and context that might have shaped MacLennan's rhetorical choices in this essay?
5. What is the known-new contract? Does MacLennan adhere to it in this essay? In her other essays?
6. Is the writing style in this essay formal or informal? Why? What are the features that influence your judgement?

7. How much certainty is there in MacLennan's tone? What are the markers of certainty?
8. What are the seven strategies for creating textual coherence? Why does MacLennan present examples from her own essay to demonstrate the strategies?
9. Do MacLennan's other essays employ the same devices of coherence?
10. Turn to any other selection in the book; to what extent is MacLennan right that all well-written articles exhibit some or all of these devices of coherence?
11. Why does MacLennan state her purpose so explicitly and so forcefully? Look closely at her wording: what promise does she make to the reader? Why?
12. Do you think this essay gives an accurate impression of MacLennan as a teacher? Why or why not? Does your assessment of her as a teacher change when you consider her other essays?
13. Nearly all of the contributors to this book are teachers. To what extent can you make any assessments about their teaching styles based on what they have written? Why do you think this is so?

CHAPTER 21

Voices to Shun: Typical Modes of Bad Writing

Joe Glaser

While no one sets out to write badly, bad writing is as common as belly-buttons. Strangely, though much bad writing is simply the result of carelessness, nearly as much is learned—one form or another of overwriting. Overwriters think they are doing fine even when the results are awful, in some cases especially when the results are awful. They are proud of every sin they commit. Other bad writers just don't listen to themselves—putting readers to sleep with monotonous breath units or paining them with ugly, grating prose. If you recognize yourself in the examples that follow, don't be discouraged. The toughest step toward improving your writing may be seeing that something is wrong in the first place. Once you've made that breakthrough, correcting your style just takes practice and determination.

The Official Style

Actuaries, accountants, bureaucrats, doctors, lawyers, ministers, professors—the learned world churns out mountains of bad writing, and most writers are proud of it. You hear the self-satisfaction in their voices. Fuzzy, inflated 'professional' styles are the number one writing problem in offices, clinics, and universities. Consider this university memo quoted in Richard Lanham's excellent writing guide *Style: An Anti-Textbook*:

> The Task Force is also concerned that it provide the basis for the faculty of this campus to govern itself, rather than being governed by others less understanding of the nature of the University. Consequently, we are asking you to specify the methods you use to evaluate the effectiveness of your instruction. Likewise, we are seeking your views on the function of evaluation and your suggestions for the implementation of evaluation of instruction on a campus-wide basis. This information will greatly contribute to our recommendations regarding the best possible methods for evaluation of instruction which will at the same time be most acceptable to the greatest number of faculty possible, keeping in mind that diverse forms of evaluation will probably be called for in the face of the diverse functions and characteristics of this institution.

The passage displays fairly complicated sentence structure and Latinate diction. While its syllable/word ratio, 1.75/1, is not outrageously high, its rhythms are deplorable. The breath units leave you wheezing. Only the first two are under 30 syllables, while the longest ('This information will greatly contribute to our recommendations regarding the best possible methods for evaluation of instruction which will at the same time be most acceptable to the greatest number of faculty possible') weighs in at 62. Unless you pause after 'Consequently' and 'Likewise,' there are no short breath units.

Another way of seeing what is wrong with the passage is to realize that everything it says could be said better in a fraction of the space:

> The task force wants our faculty to govern itself rather than be governed by others who know less about the nature of the university. How do you evaluate your own teaching? What ideas do you have for making campus-wide evaluation accurate and acceptable to the widest number of faculty? In replying please remember that different forms of evaluation may be needed for different areas of the university.

The second version is 68 words, half the length of the first, yet it includes all the important ideas of the original and is much easier to read.

Putting things briefly when you can and smoothing the way for your readers are hallmarks of good writing, but the original memo doesn't try to do either. The writer would rather strain to sound 'professional' than communicate effectively. Does the original memo sound professional? In the sense that it could not have been written by anyone but an educational bureaucrat, maybe so. But that doesn't mean it's acceptable to anyone but the proud author. It's still a pain to read. No one looking at the second version by itself would think it was inappropriate for the writing situation, and everyone would agree it's more efficient and crisp.[1]

EXERCISE 1

Rewrite the following passages so they are only half as long and easier to read. Change the sentence structure however you like, but include the essential content of the original. How do your revisions change the writer's voice?

1. Taking up a position in near proximity to a radio or similar device may significantly improve reception inasmuch as degraded radio waves dispersed by reflective surfaces in a typical room or enclosure are regularly present in sufficient quantity to noticeably degrade signal quality. A body adjacent to the reception device acts in effect as a filter to absorb randomly dispersed waves so that the signal selection mechanism can better distinguish the strong primary signal and reject the diminished number of weaker reflective signals.
2. Although environmental activists and scholars, who might be termed the environmentally

concerned community, have evidenced a long-standing commitment to ongoing conflict with resource-based industries in the interest of maximizing the quality of the environment, it now appears that natural resources, especially in the Eastern United States, have rebounded sufficiently from serious lows in the past to make resumption of some forms of environmentally taxing economic activity not only desirable but innocuous.

3. Assuming a flat-tax rate of seventeen per cent, substantially below the current maximal rate of approximately forty per cent—coupled with simultaneous cessation of current taxes levied and collected on income derived from dividends, interest, and capital gains—and also concomitant radical enlargement of basic exemptions from taxation that would effectively render many low-income families tax-free, governmental discretionary and other income appears certain to decrease while the proportionate burden of taxation on middle-tier earners and households appears certain to increase.

The memo about teacher evaluation and the examples in the last exercise illustrate, to adopt their own style, a generalized syndrome of dysfunctional verbal obfuscation (all-purpose bumfuzzlement), but much overwriting may more accurately be characterized or classified as essentially discipline-specific. (See how hard it is to stop writing this way once you get started?) Anthropology, computer science, economics, medicine—all have their own leaden tricks of style, which people who master them find irresistible, even artistically satisfying. Many of these are matters of jargon, or the specialized vocabularies of various interests. For instance, students of education rarely say test. They prefer assessment, as in 'educational outcomes assessment'. Psychologists don't change people; they engage in behaviour modification. When you add in doublespeak, pompous language with little meaning, and euphemism, or substituting a mild-sounding word for a more direct one, the possibilities for miscommunication are endless. For years sociologists took the cake as the worst writers among the learned professions. It takes only a sentence or two from Talcott Parsons, a sociologist legendary for his brain bruising style, to see why:

> The mere fact of the presence of certain genes in the gene pool of a species is not a sufficient determinant of their role in the generation of phenotypical organisms. For this to occur, there must be integration of the genetically given patterning with a series of exigencies defined by the nature of the species' life in its environment.

Meaning? Some available genes never modify a species because the changes they trigger aren't suited to the animals' environment.

Lately other disciplines have challenged sociology for the soggy palm. Here is a random sampling from works in several fields on a library new-books shelf.

> **Management and information services**
> In MCDM [Multiple Criteria Decision Making] we always talk about choosing among non-dominated solutions, solutions from which we can improve one objective only by allowing at least one other objective to deteriorate. Even though under certainty with all objectives expressed and measured correctly, we do want nondominated solutions, we may not want this in practice. We may not have all objectives expressed.
> —Stanley Zionts, 'Multiple Criteria Decision Making: The Challenge That Lies Ahead'
>
> **Social work**
> Much can and has been learned from studying individual families who are in at-risk situations. The primary function of this chapter is to explore the individual family situations of at-risk parents and children, and to share

particular insights on how early childhood professionals attempted to engage in empowering relationships with them.
—Kevin J. Swick and Stephen B. Graves, *Empowering At-risk Families During the Early Childhood Years*

Education
Recursive design is also used to make the attribute of performance management an integral part of the curriculum. Instructors begin each lesson with the student progress assessment for that particular learning episode, practicing the Dialogue for Development model as appropriate. The session concludes with a culminating activity in which participants demonstrate all the competencies of that particular lesson.
—Dian K. Castle and Nolan Estes, *High-Performance Learning Communities*

Physical education
A dominant factor in the American way of life is the ability of the average citizen to know about and understand sports, if not as a participant, then as a spectator. Therefore, it is beneficial to learn the rules and strategies of various sports. In addition, knowledge of etiquette, safety, equipment, history, values, techniques, and other factors can enhance the enjoyment of watching or participating in team, dual, or individual activities.
—Dale Mood, Frank F. Musker, and Judith E. Rink, *Sports and Recreational Activities*

English [!]
Given the dominance of logocentrism at the time, is it appropriate to analyze Renaissance works by applying the conclusions of deconstruction, which dictates an ontology opposed to logocentrism? As we have just seen, logocentrism posits an objective reality identified as an unalterable, immaterial realm of absolute being and described as a nonpalpable system of numbers/forms/ideas in the mind of deity. In opposition, poststructuralist thought pushes us beyond skepticism and relativism, arriving at a subjectivism that annihilates all else.
—S.K. Heninger, Jr, *The Subtext of Form in the English Renaissance*

The writers of all these passages are intelligent people with at least a little something to say, but they trip over their tongues. Why? Probably because they feared they wouldn't sound sufficiently impressive if they wrote plainly or because they thought this sort of writing was required for their professional fields. Unfortunately, they may be partly right about professional expectations. But it's important to realize that even in a field largely mired in stuffy writing, you still can be as clear and efficient as possible. For instance, consider this passage from the brochure for a 'Team Building Workshop':

> Teams that function at an exceptionally high performance level get there by each team member being an integral part of the team, which is demonstrated by the active involvement of each team member and a commitment to be the best the team can be.

Would this really sound less professional if you let some of the air out of it?

> All members of exceptionally effective teams are actively involved and committed to the team's success.

Of course, once you clarify the writer's claim to this extent you may also notice, that it may be untrue, and even if it is not it applies just as well to mediocre or even exceptionally ineffective teams that happen to have enthusiastic members.

EXERCISE 2

Rewrite the social work and physical education passages from the last group of examples so they are only half as long and easier to read. Change the sentence structure however you like, but include the essential content of the original How do your revisions change the writer's voice?

EXERCISE 3

Find a passage of jargon and pretentious diction from a memo, professional book, or textbook. Rewrite it so that it is only half as long and easier to read. How do your revisions change the writer's voice?

EXERCISE 4

Rewrite the following passages in the most over-blown pompous style you can invent. Pretend you're a distinguished professor, a CEO, a brain surgeon, or a bishop. Use a thesaurus.

1. We say the cows laid out Boston. Well, there are worse surveyors.
2. The growing good of the world is partly dependent on unhistoric acts; and that things are not so ill with you and me as they might have been, is half owing to the number who lived faithfully a hidden life, and rest in unvisited tombs.

   —George Eliot, *Middlemarch*
3. You must remember this,
   A kiss is still a kiss,
   A sigh is just a sigh;
   The fundamental things don't change,
   As time goes by.

   —Herman Hupfeld, 'As Time Goes By'

The Creative Genius

In recent years a detail-laden descriptive style has found favour with some writing teachers and been promulgated in workshops. More often than not it's a formula for bad writing: Be dramatic! Specify every detail as narrowly as possible! Load up on adjectives and modifiers! String each sentence out to include more insights!

> Across the room sat Aunt Marney, her care-worn, calloused hands crossed in her denim-aproned lap with a resignation born of half a century of stubborn toil on the ungrateful land, her head inclined at a weary, humble angle above a wrinkled breast once fruitful with milk for her abundant, rosy, clamoring offspring, now dry and barren as the untilled acres that surrounded her ramshackle, unpainted shanty.

If you write this way, don't show Aunt Marney what you said about her. She'd take her care-worn, calloused hands and wring your neck. How do you avoid such as style? View modifiers with suspicion, especially doubled and hyphenated modifiers ('care-worn, calloused', 'denim-aproned'). And once a sentence is grammatically complete ('Across the room sat Aunt Marney'), escape from it with decent haste. Don't let it drag on through half a page of additional details.

The creative genius way of writing is another case in which the writer puts style before communication. Details and sense impressions that might be effective in their place become annoying distractions elsewhere:

> Psychologists (my cousin Beth, for example) often note that people in unsettling, unfamiliar situations—their breathing restricted, palms tacky with nervous perspiration—look around almost feverishly in their churning anxiety to see how others are behaving, to match their own actions to those of others in a return to the warm, comforting conformity of the herd.

It's hard to imagine where such a combination of objectivity (Psychologists . . . note) and perspiration would be effective, but it would never do in a psychology class or any professional setting.

EXERCISE 5

Rewrite the Aunt Marney and psychology passages so they are only half as long, less overwrought, and easier to read. Change the sentence structure however you like, but include all the essential content of the original. How do your revisions change the writer's voice?

EXERCISE 6

Rewrite the following passages in the most overheated style you can invent. Pretend you're a creative genius of great sympathy and compassion. Use a thesaurus.

1. In most of mankind gratitude is merely a secret hope for greater favors.

   —La Rochefoucauld, *Maximes*
2. I opened the door, to find a four-foot black snake banging from a pipe. I'm not much frightened of snakes but I religiously believe in their right to privacy.

   —Rita Mae Brown, *Ariadne's Thread*
3. Because you are constantly reshaping the dome of ice cream with your tongue and nibbling at the cone, it follows in logic—and in actual practice, if you are skillful and careful—that the cone will continue to look exactly the same, except for its size, as you eat it down, so that at the very end you will hold between your thumb and forefinger a tiny, idealized replica of an ice-cream cone, a thing perhaps one inch high.

   —L. Rust Hills, 'How to Eat an Ice-Cream Cone'

The Sleepwalker

Sleepwalkers are the opposite end of the spectrum from creative geniuses. Perhaps someone told them to be detached and impersonal; perhaps they never learned to employ a variety of sentence structures and rhythms; perhaps their metabolism is just low. Whatever the reason, everything they write is flat as Houston. They aren't pretentious. They aren't overdramatic. Just dull.

> Remembering dead people we have known reminds us of who we are. They also make us remember that we will die one day as well which is a fact human beings generally forget whenever possible. This has been remarked on by many writers and thinkers like the French author Montaigne.

It's hard to pinpoint the problem with this bloodless passage. Its sentence structure is straightforward. Its diction is fine, if not memorable (syllables/words = 1.5/1). The best indication of what is wrong may be the breath units, which measure 17, 15, 20, and 21 syllables long, creating an unvarying rhythm unsuited to the subject. While death and dying are emotional topics, the written voice just plods along at the same dispirited pace. Look how another writer handles this material:

> It is the dead who tell us who we are, not just as individuals (though that too) but as a species of animals that needs reminding. They tell us constantly that life is a rough place and nobody gets out of it alive. Or as Montaigne put it, 'Live as long as you please, you will strike nothing off the time you will have to spend dead.'
>
> —Christopher Clausen, 'Dialogues with the Dead'

Clausen's treatment gains by its word music. His breath units measure 10, 8, 3, 13, 21, 6, 6, and 14 syllables long, so the voice slows and accelerates, adding interest and variety to the passage. He also creates a nice contrast between Latinate diction ('individuals', 'species', animals') and colloquial words ('life is a rough place', 'nobody'). Instead of just mentioning Montaigne, Clausen gives us his words, which are memorable. Finally there is the matter of emphasis, which usually falls on the last significant word in each sentence of breath unit. The words emphasized in the first version of the passage (ARE, WELL, POSsible, MonTAIGNE) aren't clearly related to each other or to the topic. Clausen's set of emphasized words is much more relevant and expressive (ARE, inDIViduals, THAT, reMINDing, aLIVE, MonTAIGNE, PLEASE, DEAD).

Differences between sleepwalking and wide-awake writers aren't always this dramatic, but they are always significant. Here's a stubbornly dull treatment of a subject that needs all the help it can get, computer programming:

> The computer language C is smaller than several others. It depends on a large runtime library for many operations. The runtime library consists of object files containing machine instructions for functions for a wide variety of services. The functions are divided into a number of groups. Each group depends on a source file containing instructions necessary to use the relevant function. These header files have names that end in the extension *.h*. The standard group of input and output functions is linked to a header file called *stdio.h*.

Not that bad. But dull. The breath units go 14, 17, 33, 14, 26, 13, and 23, a string in which the only real variety is provided by the too-long 33-syllable unit, 'The runtime library consists of object files containing machine instructions for functions for a variety of services.' The grammar is repetitious, and there's no attempt to vary the diction. This passage would bore a cow.

Here's the same content handled by a couple of writers actively trying to keep readers interested in what they are saying:

> One of the reasons C is such a small language is that it defers many operations to a large runtime library. The runtime library is a collection of object files. Each file contains the machine instructions for a function that performs one of a wide variety of services. The functions are divided into groups, such as I/O (Input and Output), memory management, mathematical operations, and suing manipulation. For each group there is a source file, called a *header file*, that contains information you need to use these functions. By convention, the names for header files end with a *.h* extension. For example, the standard group of I/O functions has an associated header file called *stdio.h*.
>
> —Peter A. Darnell and Philip E. Margolils, *C: Software Engineering Approach*

In my reading the breath units in this version run 12, 19, 15, 28, 10, 4, 5, 6, 9, 7, 8, 5, 14, 4, 13, 4, and 24, providing welcome variety, a lot of which arises from a more complicated grammatical scheme. This passage has three subordinate *that* clauses to break up the other one's monotonous succession of simple sentences. Further variety comes from the parallel series 'such as I/O (Input and Output), memory management, mathematical operations, and string manipulation', which also helps the reader understand what is meant by reducing it to specific examples. The slightly unexpected verb 'defers' in the first sentence and the authors' decision to address readers directly as 'you' are two more touches that make the writing of this second version more interesting and accessible. All in all, the second passage invites us into the subject, while the attitude of the first is 'Take it or leave it'. Clausen and Darnell and Margolis take a specific approach to writing: they're thinking of readers, not just content. 'How can I keep the readers on their toes, interested, even moderately entertained?' They know their success as writers depends on meeting this challenge in sentence after sentence.

EXERCISE 7

What differences in rhythm, diction, and emphasis make the second passage in each of the following pairs more lively than the first?

1. Science had a tendency to give naturalistic answers to questions posed by religions.

   > When scientific explanations are accepted some religious issues disappear.Extinguished theologians lie about the cradle of every science as the strangled snakes beside that of Hercules
   >
   > —T.H. Huxley, *Darwiniana*

2. Artists must please themselves first and take it on faith that what pleases them will please someone else as well.

    I personally make music because I want to ask a question, and I want to get an answer. If that question and answer amuse me, then statistically, there are a certain number of other people out there who will be amused by it, and we will all have a good time.

    —Frank Zappa, 'On "Junk Food for the Soul"'

3. People who are outside the mainstream often behave energetically and entertain others, but sometimes they feel alienated and lost on the inside.

    I was much too far out all my life
    And not waving but drowning.

    —Stevie Smith, 'Not Waving but Drowning'

EXERCISE 8

Rewrite the following passages to relieve the tedium. Vary the length of the breath units so that some are at least twice as long as others. Experiment with colloquial and formal diction. Try to make emphasis fall on expressive, relevant words. Change the sentence structure however you like, but include the essential content of the original. How do your revisions change the writer's voice?

1. The SS doctor Josef Mengele came to Auschwitz after being wounded on the Russian Front. At Auschwitz Mengele selected prisoners to be sent to the gas chambers at regular intervals. Among the prisoners he excused from death by asphyxiation were sets of twins he reserved for medical research. Mengele performed a number of medical experiments on these twins to determine what characteristics were inherited. His experiments on the twins included inoculating them with diseases and dissecting them after they were killed.
2. E.H. Carr was a historian at Cambridge University who was one of the first to ask 'What is history?' Carr thought it was not good enough to say that history is the record of what had happened in the past. He pointed out that most of what happened never gets into history books and what does is interpreted by historians in the act of writing the books. So history is more what goes on in the mind of historians studying the past than it is the past itself.
3. Some people find they can organize their thoughts best by talking about them with someone else. Other individuals think best sitting or reclining by themselves in a secluded place. Some people like to walk outdoors and think about what concerns them at the time. Still other people go about their business as if there was nothing on their minds and hope an answer will come while they are thinking about something else.

The Clunker

A tribe related to the sleepwalkers, writers of aggressively clunky prose are deaf not only to rhythm but to everything else, producing sentences that are not just monotonous but downright ugly:

> One function of examination coordination would be the implementation of a revision of the examination schedule to guarantee that examiners scheduled for the same examination were free at the hour the examination was scheduled.

Besides gracelessly repeating whole words (*examination*, *schedule*), this sentence is hobbled by other unintended repetition of sounds (func*tion* of examina*tion* coordina*tion* . . . implementa*tion* of a revision of the examina*tion*). It grates like a rusty hinge, and moves about as briskly, too. Fixing the sentence requires editing out the repetition and breathing a little life into its rhythm:

> Coordinating the examination schedule will mean assigning new times. That way we can guarantee all examiners will be free when needed.

The worst offender when it comes to making ugly sentences is the suffix *-tion* or *-sion*. Because it converts verbs into Latinate nouns, *-tion* causes other stylistic problems as well. But the beginnings of words and internal syllables can also chime against a writer's intentions:

> Next, let's inspect the spectre of special regulation in the private sector.

The best defense against clunky prose is to read your sentences over to yourself silently or aloud, listening carefully to the way they sound. If you have to twist your tongue around the syllables (in*spect* the *spect*er), revise.

Listening to your own prose is not easy at first. Sometimes it helps to tape yourself reading what you've written. Another trick is to analyze sentences in a well-edited magazine like *Harper's* or *Atlantic Monthly*. Forget the meaning and think about sentence structure, diction, and long and short breath units. Or try making good sentences monotonous and ugly. You'll soon grow sensitive to the difference.

EXERCISE 9

What sound problems make the first passage in each of the following pairs read awkwardly compared to the original?

1. The declination of each declivity shall be decreased.

   Every valley shall be exalted.

   —Isaiah 40:4[2]

2. There's a pervading crusading spirit masquerading as an unfading moral superiority and parading its values in a way that is grating and degrading.

   Judge not, that you be not judged.

   —Matthew 7:1

3. Now the reptile had guile and beat the other beasts by a mile in wile.

   Now the serpent was more subtle than any beast of the field.

   —Genesis 3:1

EXERCISE 10

Rewrite the following passages to make them sound better. If necessary, vary the length of the breath units so that some are at least twice as long as others. Eliminate awkward repetition of sounds. Change the sentence structure however you like, but include the essential content of the original. How do your revisions change the writer's voice?

1. Even in the 1950s we had the chemical capacity to make new chemicals and distribute old chemicals like never before. We could put more of the chemical sulfur dioxide that produces sulfuric acid and acid rain in the air than volcanoes do, at a great cost to acid-sensitive plants.
2. You know I was clueless until you clued me in to your conclusions on how concussions can contribute to delusions.
3. While paroxysmal trachycardia, or abnormal accelerated heartbeat, is rarely fatal, fatalities can be expected if heartbeat accelerates to fatal levels.

EXERCISE 11

Rewrite the following passages to make them sound monotonous or ugly. Monotony results from a series of breath units of roughly the same length, usually 15–25 syllables. Ugliness most often means awkward repetition of sounds. Change the sentence structure however you like, but include the essential content of the original. How do your revisions change the writer's voice?

1. I have learned
   To look on nature, not as in the hour
   Of thoughtless youth; but hearing often-times
   The still, sad music of humanity,
   Nor harsh nor grating, though of ample power
   To chasten and subdue.
   —Wordsworth, 'Tintern Abbey'

2. Later that night I loosed my hair from its braids and combed it smooth—not for myself, but so the village girls could play with it in the morning.
   —Annie Dilliard, *Teaching a Stone to Talk*

3. I sympathize with Rodney King. But I will say this: Someone had to be Rodney King. Someone had to force us to begin the next phase of our national morality play. We have always needed martyrs. I am just sorry it was Rodney King.
   —Maya Angelou, 'The Arc of the Moral Universe is Long, but It Bends Toward Justice'

Your Writing

Scan a recent piece of your writing for the traits discussed in this chapter. Is the style overcomplicated? (A syllable/word ratio over 2 or a syllable/breath unit ratio over 25 are danger signs.) Is it overexcited? (Watch for doubled or hyphenated modifiers.) Is it dull? (Too many breath units of the same length are likely culprits.) Is it ugly? (Look for unintentional chiming, 'Philistine philanderer Phil'.) The traits you find represent stylistic tendencies you need to be aware of in order to control them.

POINTS TO REMEMBER

1. Don't put style before substance by trying to sound overly learned or overly creative.
2. Listen to what you've written, if only with your inner ear, to make sure it avoids monotony and places emphasis effectively.
3. Watch out for unintended repetition of words and sounds that can make your sentences ugly.

Notes

1. In professional writing it is important to distinguish between your own writing and others'. By all means make your own style as accessible to readers as you can, given the situation. They'll appreciate it. But if you work with colleagues or bosses who write like the memo writer above (or the examples in the next few pages), critique their styles as you would their children: only when they ask you to and circumspectly even then. Remember that professional overwriting is learned behaviour and people who don't know any better are often inordinately proud of it. You don't want bosses and colleagues angry with you for telling them how to write, even if it seems someone should.
2. The 'Authorized' or 'King James' version of the Bible (1611), more familiar today as the 'Revised Standard Version' of the 1880s, was prepared by some 50 scholars and divines who worked for more than three years in committees to adapt and correct the work of earlier translators, especially William Tyndale, whose incomplete and outlawed translation (1525–31) formed the backbone and set the style for all later efforts. Designed for common readers, the 1611 translation was among the first works of modem English prose and remains arguably the greatest. Like Shakespeare, who first showed what the language was capable of at full stretch, the King James Bible taught us how to structure English prose, and how it should sound—in general, simple, plain, and direct.

QUESTIONS FOR CRITICAL THOUGHT

1. What is the purpose of this selection? For whom does it appear to have been written?
2. Glaser seems to make readability a matter of ratios (of syllables to words) and measurements (of breath groups). How helpful is this construct?
3. Glaser condemns writing that 'strains to sound professional' at the expense of communicating clearly. To what extent is his diagnosis of the problems in technical writing the same as that offered by Buehler ('Situational Editing'), Urquhart ('Bridging Gaps, Engineering Audiences'), Burton ('An Engineer's Rhetorical Journey'), and others?
4. What is 'bumfuzzlement'? Why does Glaser choose to employ such an amusing word? What does it do to his tone?
5. Glaser points to doublespeak, euphemism, and jargon as deadeners of style and understanding. What are these things? Why does Glaser condemn them?
6. Why does Glaser draw examples from so many different professions?
7. What is the purpose of the exercises Glaser provides? What is their effect on the passage?
8. What, exactly, are the voices that a thoughtful reader should shun? Why?
9. How would you describe Glaser's own style? Consider the following examples, and cite some of your own:

    a. 'This passage would bore a cow.'
    b. 'Here's a stubbornly dull treatment of a subject that needs all the help it can get, computer science.'
    c. 'It grates like a rusty hinge, and moves about as quickly too.'

10. One of the great faults of technical writing, according to Glaser, is obfuscation. What is the source of this phenomenon? How many reasons does Glaser offer for its predominance in professional writing?
11. Do the 'points to remember' adequately summarize the article? Why or why not?

CHAPTER 22

Situational Editing: A Rhetorical Approach for the Technical Editor

Mary Fran Buehler

At present, the field of technical editing lacks a comprehensive rhetorical theory: almost all the books and articles on technical communication are directed to the technical writer rather than the technical editor. The information in these publications is often useful to editors, but it cannot completely fill their needs because the rhetorical situation of the editor is different from that of the writer.

In this article, I propose and demonstrate a rhetorical theory of technical editing that is based on a situational approach to teach individual task. The rhetorical, or situational, approach is to be used in conjunction with the grammatical, or programmatic, approach.

The grammatical approach is well understood by good technical editors. Many aspects of the rhetorical approach have also been explored in articles and textbooks, primarily for technical writers; for example, excellent advice has been published on audience analysis (Houp and Pearsall, 1980). Some good papers and articles have been written specifically for the technical editor; one of the best, and one that delineates a situational approach for the editor, is Lola Zook's 'Training the Editor: Skills Are Not Enough' (1976), which has exerted a strong influence on the concepts and theory that I am presenting here.

These, then, are the tools and equipment that have been provided for technical editors—grammatical (and other) rules, instructional material directed to technical writers, and isolated examples of advice and guidance on editorial matters. The editor also relies on pragmatic strategies leaned from the real-life exercise of dealing with authors and manuscripts.

It is my purpose in this article to pull some of these elements together; to differentiate the specific problems for the technical editor from those of technical communicators (especially technical writers) in general; to contrast the rhetorical or situational approach to technical editing with the grammatical approach and to demonstrate why the programmatic approach, in itself, is inadequate; to propose a rhetorical theory for technical editors that will incorporate the pragmatic elements of situational editing now being practiced; and to show how such a theory, which is based on the wider situational implications of rhetoric that have been developed during more than 2,000 years, can be a valuable contribution to the editor's effective communication of technical information.

The Editor and Rhetoric

Why should the technical editor be concerned with rhetoric? The common perception of rhetoric is not favourable. References to rhetoric in our newspapers and other mass media are usually to something like 'mere rhetoric, instead of solid achievement'.

This popular misapprehension is not important and need not concern us unduly. For one thing, the denigration of rhetoric is nothing new: rhetoric has been denounced by some very articulate critics, at least as far back as Plato (trans. 1952), and it has survived and flourished. Besides, when newspaper writers or television broadcasters use the term *rhetoric* in this sense, they are really substituting *rhetoric* as a catchier, or more fashionable, or more persuasive, or more important-sounding term for *language*. What they are really saying is 'mere language, instead of solid achievement'. Why, then, do they use the term *rhetoric* instead of *language*? Probably because it sounds better to them, because they think it will persuade or impress their readers, or for some other reason implied in the specific situation. In short, they are using rhetorical statement in deprecating the term *rhetoric*.

Having discarded some popular ideas about the subject, we return to our original question: Why should the technical editor be concerned about rhetoric? To find an answer, we need to understand what rhetoric—or, at least, from what perspective we will consider it.

A review of the ways in which rhetoric has been defined through the centuries reveals that the concept of rhetoric itself is situational; rhetoric has been defined in the context of the many different social, philosophical, political, and technological environments in which people have found themselves using language to communicate ideas.

Some definitions have emphasized the persuasive aspects for rhetoric: Aristotle—'Rhetoric may be defined as the faculty of observing in any given case the available means of persuasion' (trans. 1954, 1335b: 25). (This in the fourth century BC, when persuasive language could be a necessity for survival in a primarily oral culture, when citizens acted as their own lawyers, and when there was a high degree of democratic participation in government.)

Some definitions have emphasized rhetoric as an oral art, characterized as oratory: Cicero, describing the qualifications of an orator—'he will be an orator, in my opinion worthy of so dignified a title, who, whatever the topic that crops up to be unfolded in discourse, will speak thereon with knowledge, method, charm and retentive memory, combining with these qualifications a certain distinction of bearing' (trans. 1942: 64).

Between the time of these classical Greek and Roman definitions and our own day, rhetoric has been considered to consist primarily of the stylistic devices of language—figures of speech like metaphors and similes (often grouped into the category 'figures and tropes'), or what we in technical communication might be inclined to consider 'fancy language'. In the Middle Ages, rhetoric concerned itself with such evolving kinds of communication as the art of letter writing and the art of preaching sermons, each representing a cultural development of the time.

With the invention of the printing press and the spread of written (as well as oral) and informative (as well as persuasive) communication, definitions of rhetoric have often been broadened. Two well-known definitions by contemporary rhetoricians reflect this trend: Donald C. Bryant's definition, 'I take rhetoric to be the rationale of informative and suasory discourse' (1974: 199), and I.A. Richards' statement that 'Rhetoric . . . should be the study of misunderstanding and its remedies' (1936: 3).

Then what is rhetoric? Obviously, there is no single answer. But the qualities inherent in the many definitions or rhetoric cluster around one concept; rhetoric is the use of language that takes into consideration the several elements of what may be called the rhetorical situation: the speaker or writer, the message to be communicated, the purpose of the message, and the person for whom the message is intended. (Because technical editors deal primarily with written material, I will use the terms 'author' and 'reader' for the communicator and the intended audience.)

Thus, the technical editor should be concerned about rhetoric because rhetoric, being situational, offers a coherent framework in which the editor can rationally approach the different individual situations that make up his or her work. Technical communication is—or should be—a set of situations in which rhetorical choices are made.

Situational and Programmatic Approaches

I have defined what I mean by a rhetorical, or situational, approach as one that considers all (or at least most) of the elements for a communication situation. By contrast, a programmatic approach simply applies a set of rules to all situations. The best example of a programmatic approach in technical communication is the application of 'correct' grammar, spelling, and punctuation, without concern for all the varied elements—which may or may not be flexible—is the application of 'house rules'—that is, the rules of a given publication's organization, journal, or the like, governing various aspects of format, style, and usage.

Certainly a good technical editor must have a thorough mastery of such language skills as grammar, spelling, and punctuation; these skills should be taken for granted. But they are not enough for effective editorial work: it is necessary to know the rules, abut it is not sufficient.

'Follow these rules,' says Grammar, 'and you will always be correct.'

'Consider the individual situation,' says Rhetoric, 'and you may be able to communicate effectively.'

Many of us in technical communication, I believe, have thought that the clear, plain, objective style that we normally strive to achieve is not rhetorical. We may have thought, in fact, that when we pruned away extraneous material from a scientific report or eliminated the author's flights of fancy from a commercial brochure, we were eliminating 'rhetoric' from the message.

But the truth is that the spare, objective style of technical writing is, in itself, a rhetorical choice. This style represents the most appropriate choice, in most situations, for conveying a technical message from its creator to its user, with a minimum of distraction.

The history of our plain, concise style may help us to see just how rhetorical was the choice of that style in the beginning. The development of the objective style can be traced back to the early reports of the Royal Society in England. The Royal Society (correctly known as the Royal Society of London for Improving Natural Knowledge), the oldest scientific society in Great Britain and one of the oldest in Europe, was founded in 1660 to perform scientific and technical experiments, investigate natural history, and explore practical applications of these discoveries, among other aims.

Thomas Sprat, author of the *History of the Royal Society*, first published in 1667, clearly shows us that the language used by the Royal Society—which set the pattern for the scientific and technical language that we use today—was deliberately chosen. Sprat inveighs against the 'fine speaking' of his day:

> Who can behold, without indignation, how many mists and uncertainties, these specious Tropes and Figures have brought on our Knowledge? How many rewards, which are due to more profitable, and difficult Arts, have been still snatch'd away by the easy vanity of fine speaking? For now I am warm'd with this just Anger, I cannot withhold myself, from betraying the shallowness of all these seeming Mysteries; on which, we Writers, and Speakers, look so big. And, in a few words, I dare say: that of all the Studies of men, nothing may be sooner obtain'd, than this vicious abundance of Phrase, this trick of Metaphors, this volubility of Tongue, which makes so great a noise in the World. (Sprat, rptd. 1958: 112)

And so the members of the Royal Society chose a plainer kind of language in which to report the results of their studies:

> They have therefore been most rigorous in putting in execution, the only Remedy, that can be found for this extravagance; and that has been, a constant Resolution, to reject all

> the amplifications, digressions, and swellings of style: to return back to the primitive purely, and shortness, when men deliver'd so many things, almost in an equal number of words. They have exacted from all their members, a close, naked, natural way of speaking: positive expressions; clear senses; a native easiness; bringing all things as near the Mathematical plainness, as they can: and preferring the language of Artizans, Countrymen, and Merchants, before that, of Wils, or Scholars. (Sprat rptd. 1958: 113)

Thus we see that the members of the Royal Society were not really banishing 'rhetoric' from their language—they were making a practical and astute rhetorical choice of the language that best suited their situation.

The plainness of the objective style may lead us to think that good grammar is all we need to know. But although the constraints of this style usually constitute appropriate choices, they should not be applied automatically to all situations, even in technical communication: they should be applied only when the choice makes sense.

A simple example, and one that occurs frequently in technical papers and reports, is the acknowledgment section, in which the author extends appreciation to colleagues for their help, encouragement, and the like. This section normally constitutes a different rhetorical situation from the rest of the report: one does not usually thank one's friends in the concise, curt style of technical exposition. Instead, one may want to say something like 'I should like to express my appreciation to my colleagues for their invaluable assistance . . .'. And what is wrong with that, if it fits the situation?

Unless a journal specifies a style for its acknowledgment sections, the only constraint on an acknowledgment should be its rhetorical effectiveness: if it is so flowery that it makes the author sound silly, then it should be pruned. But it should be pruned rhetorically, not grammatically.

What is meant by rhetorical and grammatical pruning? (I am using the terms pruning as a shorthand word for getting rid of words in written material by means of editorial techniques). The demand for conciseness has been so much a part of our technical communication credo for so long that a discussion may by helpful.

Conciseness may seem so self-evidently desirable that it should be impervious to rhetorical considerations (except for isolated instance, as noted above). Certainly the demand for conciseness lends itself to programmatic applications. 'No unnecessary words' has been a rallying cry, both in technical communication and in general expository writing. The highly regarded and much used *Elements of Style* warns against needless words:

> Vigorous writing is concise. A sentence should contain no unnecessary words, a paragraph no unnecessary sentences: for the same reason that a drawing should have no unnecessary lines and a machine no unnecessary parts. (Strunk and White, 1959: 17)

But unnecessary for what? For programmatic 'correctness' or rhetorical effectiveness? To take only one example: a rhythmical sentence, with a word or two that may not, strictly speaking, be irreplaceable, may read much more comfortably, and hence more meaningfully, than a thudding series of 'short, denotative terms' (or whatever other means to conciseness may be recommended), regardless of the saving in words. Take for example, the following clause:

> When the spacecraft lands on the planet Mars . . .

Obviously, this expression can be pruned. Everyone who will be expected to read it will know that Mars is a planet—the word planet is redundant. But the rhythm of the words is pleasing to the ear. If we prune out the word *planet*, we have a more concise statement:

When the spacecraft lands on Mars . . .

But now we have lost the grace of the rhythm we had before. This may not, of course, be a bad thing: in many applications—for example, in a set of instructions—the short, strong statement would usually be preferable. But a choice of this kind is rhetorical. And that's my point.

In short: the editor should not proceed automatically through all types of writing, pruning away any word that is not absolutely essential to the meaning. Programmatic pruning can destroy theoretical effectiveness. Conciseness is desirable, but it is not all-important.

The Editor's Unique Situation

The editor faces a set of unique rhetorical situations because the editor—unlike the author—is squarely in the middle of each situation. An often-used model for technical communication places the writer/sender essentially on one end of the communication process and the reader/receiver on the other end, with the message (and possibly feedback) flowing between.

When editors are involved in the situation, however—if they are performing rhetorically, rather than programmatically—they are working with the author, on the one hand, and projecting themselves into the character of the reader, on the other hand (see Figure 22.1). Editors attempt to simulate the reader's expected feedback, to test the effectiveness of the message transmission. They also attempt to preserve as much of the author's character, or style, as possible. Some examples of what this can mean are discussed below, in terms of both the reader and the author.

SERVING THE READER

Although audience analysis has been emphasized primarily for writers, rather than editors, some advice on audience analysis has been specifically directed toward editors:

Figure 22.1 A model of the editor's relationship with the author and the potential reader(s). The solid lines represent communication and feedback between the editor and the author; the dashed lines represent the editor's empathetic projection into the position of the readers.

> . . . the editor needs to learn to work, consciously, at many different levels . . . If an electronics manual is to be used by men who are technically proficient but whose reading ability is at the eighth-grade level, that manual is not going to be very useful unless it is prepared with that reading level in mind. The difficulty is that some 12th grade words and some college-level concepts will have to be used in that manual, reading ability or no. So the problem becomes one of finding means for presenting these terms and concepts in a way that will help the technician learn their meaning and use them. Not only the language skills of the editor but the various resources of format and typography and visual aids should be marshaled to make the manual readable, visually helpful, and easy to use. (Zook, 1976: 15)

Here the editor, with a college-level literacy ability, must tailor material for readers with an eight-grade level of literacy. However one solves it, the problem clearly calls for rhetorical skills: no amount of correct grammar and concise writing will ensure that the information is transmitted effectively unless the needs of the audience are correctly assessed and fulfilled.

The example is perhaps an extreme one: the distance between the concepts to be conveyed and the level of the audience's understanding is usually not so great. But the special characteristics of the audience, however subtle, can affect communication in important ways and must always be considered.

SERVING THE AUTHOR

It is generally agreed that the editor should 'preserve the author's style'. That is, a piece of written material should reflect the author's patterns of thought and language so far as possible within the confines of effective communication. The editor, it is generally agreed, should not change an author's language arbitrarily or capriciously, and should not make changes in a manuscript without an identifiable reason.

But, again, what kind of reason? A programmatic reason or a rhetorical reason? For some examples, we can return to classical rhetoric.

In a triumphal ceremony celebrating a military victory, Julius Caesar had a banner carried before him that read (in translation): 'I came, I saw, I conquered' (Suetonius, nd). If a technical editor had been called in to work with Caesar and his designer of celebratory graphics, what might have happened?

Could the editor have said something like following?

> I like the dynamic style of your slogan, but it isn't really correct grammatically. What you have here is an A, B, and C enumeration. Now this kind of enumeration normally takes a conjunction before the final element, and a comma before the conjunction—of course, there are two schools of thought about that: some people have been taught to leave the comma out. But in technical communication we advise that you leave the comma in. Besides, you have some unnecessary words: you don't need to say I three times. Once is enough, after that, people will know who is talking. And so your banner should read 'I came, saw, and conquered.'

Would one of us—as technical editors—have said that? I hope not. I hope we would recognize that Caesar was making the language carry his meaning both in form and in content: he was rhetorically emphasizing the speed and dispatch with which he conducted his campaign. (Suetonius, in fact, tells us that this was the real significance of the statement.) Undoubtedly, the language also reflected Caesar's own personal image—that of a direct, hard-hitting military leader.

I hope that we—as technical editors—would also know that Caesar was using a rhetorical figure, asyndeton, in which an enumeration of elements is listed without conjunctions. The figure, as such, is perfectly 'correct'; depending on the situation, it may or may not be effective.

Or, to take another example: the orator Cicero, in describing the kind of knowledge that a good orator should possess, has Crassus (considered to be Cicero's spokesman) explain, in *De oratore*, that

> . . . the genuine orator must have investigated and heard and read and discussed and handled and debated the whole of the contents of the life of mankind, inasmuch as that is the field of the orator's activity, the subject matter of his study. (Cicero, trans., 1942: 43)

Assuming that this English translation faithfully represents Cicero's Latin, what would we have done, as technical editors, if we had been asked to review Cicero's manuscript? Would we have pointed out the excess of conjunctions, the unnecessary words, and the fact that an A, B, and C enumeration should be properly put together with commas and a final conjunction?

I hope not—because Cicero's expansive and encompassing style reflects perfectly the meaning of his statement: that the orator, too, must have such expansive and encompassing knowledge and experience. Like Caesar, Cicero is making the form as well as the content carry his meaning. The statement also reflects Cicero's own image as a man, and the kind of overwhelming ability that he himself possessed as an orator. The sentence could not possibly carry the same weight of connotations if it were recast in the formula of an A, B, and C enumeration.

I hope, too, that we would recognize this rhetorical figure as *polysyndeton*—that is, an enumeration in which the elements are joined with a series of conjunctions (see Lanham, 1969 for a useful glossary of rhetorical terminology). Again, the figure is perfectly correct—the only question should be: is it effective?

A Rhetorical Theory and Its Implications

On the basis of the limited number of examples included in this brief discussion, some general elements of rhetorical theory for the technical editor may be proposed.

1. Programmatic or rule-following editorial procedures are necessary and basic for acceptable editorial quality.
2. However, the programmatic approach will not provide the editor the scope required for considering each editorial situation individually.
3. A rhetorical approach, which allows the editor to consider all the elements in each rhetorical situation, is required for effective communication. Elements to be considered are the author, the message, the purpose of the message, the intended audience, any applicable house rules, and any other pertinent information. These elements will determine the form (within unavoidable constraints) as well as the content of the message.

From the elements of this theory, one can draw implications for the editor's performance. The skills and abilities necessary to perform programmatic editing, while important, are different form those required for the rhetorical approach.

The programmatic approach requires the following in an editor:

1. Knowledge of the rules involved (grammar, punctuation, house rules),
2. Knowledge of how to apply the rules correctly,
3. Ability to apply the rules consistently throughout an editorial project, and
4. Ability to interpret the rules to the author (for example, to cite a rule for any change that is made in a manuscript).

Obviously, these are very desirable qualities: they are basic skills that every good technical editor must possess. But, as I have demonstrated, these abilities are not enough for effective editing. The rhetorical approach, which builds on the basis of the programmatic approach, requires such additional skills as the following:

1. **Breadth of perspective**—the ability to see the rhetorical situation in its entirely,
2. **Investigative persistence**—the ability to keep asking questions about the rhetorical characteristics of any situation until the necessary answers are received,
3. **Flexibility**—the ability to see each situation as an individual one, and to shift easily from one rhetorical level to another,
4. **Rhetorical knowledge and taste**—the ability to judge whether a given language usage is effective, not merely whether it is 'correct',
5. **Empathy**—the ability to understand and communicate with people (authors and potential readers) of varied educational backgrounds, cultural ties, and literacy levels, and
6. **Self-confidence**—the ability to apply rhetorical considerations, as well as programmatic ones, and to justify any editorial action taken in either approach.

Conclusion

To sum up: The editor's position in technical communication is unique since the editor functions in the centre of the series of rhetorical situations, linking the author and the potential reader, and serving the needs of bother. The editor cannot solve the problems of effective communication by using programmatic techniques alone. The rhetorical approach makes more difficult demands on the editor, poses greater challenges, requires a higher level of effort on a less secure basis—but offers greater rewards in terms of effective communication.

For more information on rhetoric, technical editors—and other technical communicators—are invited to review the special issue of *Technical Communication* that was devoted to the subject (Fourth Quarter, 1978). In addition to the articles in that issue, the various reference lists contain a wealth of sources that could form the beginning of a useful bibliography on rhetoric, representing such rhetoricians as Wayne C. Booth, Kenneth Burke, Chaim Perelman, and many more, including, of course, Aristotle himself, the most basic reference of all.

Note

In developing the editorial concepts presented here, I am greatly indebted to Mrs Lola Zook, who until her retirement, was director of the editorial and production division, Human Resources Research Organization, Alexandria, VA, and subsequently served as an editorial consultant in Washington, DC.

I am also indebted to Professor Walter R. Fisher, Department of Speech Communication, University of Southern California, who encouraged me to investigate the role of rhetoric in technical communication.

References

Aristotle. 1954. *Rhetoric and Poetics*, W. Rhys Roberts, trans. New York, NY: The Modern Library, Random House.

Bryant, D.C. 1974. 'Rhetoric: Its Functions and Its Scope', in W.R. Fisher, ed., *A Tradition in Transition*. Lansing, MI: Michigan State University Press.

Cicero. 1942. *De oratore*. Book I. H. Rackham, trans. Cambridge, MA: Harvard University Press.

Houp, K.W., and T.E. Pearsall. 1980. *Reporting Technical Information*, 4th ed. Los Angeles, CA: Glencoe Publishing Company.

Lanham, R.A. 1969. *A Handlist of Rhetorical Terms*. Berkeley, CA: The University of California Press.

Plato. 1952. *Gorgias*, W.C. Heimbold, trans. Indianapolis, IN: The Bobbs-Merrill Company, Inc.

Richards, I.A. 1936. *The Philosophy of Rhetoric*. New York, NY: Oxford University Press.

Sprat, T. 1958. *History of the Royal Society*. St Louis, MO: Washington University Studies.

Struck, W., Jr, and E.B. White. 1959. *The Elements of Style*. New York, NY: The Macmillian Company.

Suetonius. 1923. *History of Twelve Caesars*, P. Holland, trans. New York: E.P. Dutton & Co.

Zook, L. 1976. 'Training the Editor: Skills are Not Enough', in *Technical Editing: Principles and Practices*. Washington, DC: Society for Technical Communication.

QUESTIONS FOR CRITICAL THOUGHT

1. Why does Buehler regard the absence of rhetorical theory as a 'lack' in the study and practice of technical editing?
2. What is the rhetorical situation of an editor, as Buehler describes it? How does it differ from that of the writer?
3. Buehler argues for a 'situational approach' to technical editing. To what degree is her approach anchored in Lloyd Bitzer's theory of situation (outlined in 'Functional Communication')?

4. Buehler contrasts a rhetorical approach with the 'programmatic' or grammatical approach advocated by others. What do these terms mean? What is the difference?
5. Why does Buehler state the purpose of her article so directly?
6. Buehler provides her reader with a quick overview of the history of rhetoric. Why has she done so? How important is this to her main point?
7. How does Buehler define technical editing?
8. Like many of the other contributors to this book, Buehler contends that 'the spare, objective style of technical writing is, in itself, a rhetorical choice.' What evidence does she provide that this is so?
9. Buehler argues that the traditional emphasis on conciseness can actually make prose less effective. As an antidote to this tendency, she offers attention to rhythm. Compare her treatment of rhythm to Glaser's in 'Voices to Shun'.
10. Why does Buehler use so many classical examples as illustrations?
11. What, according to Buehler, are the things an editor needs to know? Do you need to know them? Why or why not?

CHAPTER 23

Escape From the Grammar Trap

Jean Hollis Weber

Too many editors focus on the details and don't pay enough attention to the bigger picture. Editors can—and should—add even more value through substantive, technical, and usability editing.

Copyediting is important, but the details are only part of what an editor can and should be reviewing. After all, a document can be correctly spelled and punctuated, grammatically correct, use only approved terminology, and follow the style guide perfectly—and still not serve the audience's needs.

This article covers some reasons why editors focus on details and not the bigger picture; describes how much attention technical communicators should pay to formal rules of grammar, punctuation, and usage; and describes how we can distinguish between essential and nonessential rules of grammar, punctuation, and usage.

Why Do Editors Have Such a Narrow Focus?

Some reasons for an editorial focus on details have to do with editors themselves; other reasons arise from the perceptions and priorities of managers and writers.

Many editors are in one of these groups:

- They know how to contribute substantively, but they don't have time—or aren't allowed—to do so.
- They are more comfortable enforcing rules than making critical suggestions and then dealing with writers and others who may not appreciate those suggestions.
- They don't believe they can contribute substantively because they haven't been trained in substantive editing or they aren't sufficiently familiar with the subject matter they are editing.
- They lack the skills to do a good job of copyediting, so they never get the chance to go beyond that stage, even though they might be very good at other types of editing (the skills required are quite different).

Many managers, writers, and other clients believe one or more of the following statements:

- Editors are obsessed with nitpicky details; that's what editing is all about.
- An editor's job is not to substantially revise a writer's work or comment on the technical content or usability of that work.
- Substantive, technical, and usability editing take too long and cost too much.
- Editing is done after the manuscript is written, leaving insufficient time to change anything major that an editor might find.

Distinguish Between Essential, Nonessential, and Fake Rules

How much attention should technical communicators pay to formal rules of grammar, punctuation, and usage? Does incorrect grammar, punctuation, or usage detract from the value and usability of your group's publications? Does your audience care, or even notice, if formal rules are broken?

To answer these questions, we need to examine grammar, punctuation, and usage:

- Grammar is the arrangement, relationships, and functions of words and the ways they are put together to form phrases, clauses, or sentences.
- Punctuation marks are signals that help readers to understand the ideas in a passage and to read more quickly and efficiently.
- Usage is the way in which words and phrases are actually used, spoken, or written in a language community.

Rules of grammar, punctuation, and usage can be essential or nonessential—or even fake! Essential rules are those that are necessary for clear, unambiguous communication. Nonessential rules are those that are not required for clarity and unambiguity. Fake rules may actually be matters of word choice, style, or conventional usage, not rules of grammar; or they may be things many of us were taught were wrong, but which are in fact acceptable variations in usage.

Writers and editors need to pay attention to the essential rules, but can spend less time on nonessential rules—particularly in the face of tight deadlines—and they can ignore the fake rules.

Some rules, such as those about dangling participles and not ending sentences with prepositions, are nonessential because readers can figure out the meaning; but they are still important rules to follow in those cases where following the rule would make the writing easier to understand. For example, split infinitives are acceptable in English ('to boldly go'), but if you replace the adverb (boldly) with a long adverbial phrase, the meaning becomes more difficult to decipher.

I'm sure all technical communicators would like to produce perfect documents, but we rarely have the leisure to do so. Business realities too often require compromises from writers and editors, so we place accuracy and usability ahead of minor issues of grammar and punctuation—as I think we should.

Of course, what's a minor issue to me may be a major issue to you; some audiences may have an unusually high percentage of people who won't trust your facts if they think you're misusing the language; and some of your technical reviewers will focus on the grammar instead of the facts. All of these scenarios provide good reasons to pay attention to grammar rules, or at least not abuse them too blatantly or frequently.

Examples of Essential Grammar and Punctuation Rules

Essential rules are those that are necessary for clear, unambiguous communication. Some examples:

- Use of commas, when errors can cause ambiguity or misunderstanding. For example, these pairs of sentences convey quite different messages:

  Injured and abandoned by their travelling companions, they managed to stagger to a ranger station.

  Injured, and abandoned by their travelling companions, they managed to stagger to a ranger station.

  Tomorrow will be overcast and rainy at times.

  Tomorrow will be overcast, and rainy at times.

- Use of apostrophes in possessives and contractions, but not plurals. Incorrect placement of apostrophes changes meaning (often causing confusion or ambiguity) or is completely wrong. Some examples:
  - Changes meaning: It's (contraction of 'it is' or 'it has') or its (possessive of 'it'); who's (contraction of 'who is') or whose (possessive of 'who'); the manager's decision (one manager made the decision) or the managers' decision (more than one manager made the decision).
  - Just plain wrong: Mens', childrens', its' (all intended to be possessives); video's, photo's (when intended to be plural, not possessive).

- Subject–verb agreement (but see notes on 'data' and 'they', below). When the subject and verb are separated by many other words, this agreement may be difficult to sort out. Often the best solution is to rewrite the sentence: If you can't easily decide whether a verb should be singular or plural, chances are your readers will get lost in the sentence anyway.
- Avoiding dangling modifiers, unclear antecedents, and other constructions that can create ambiguity, even when most readers will eventually figure out what's meant. Some examples:
  - Dangling modifier: Can occur at the beginning or end of the sentence. After reading the original study, the newspaper article is unconvincing. (or) The newspaper article is unconvincing after reading the original study. (The article—the subject of the main clause—did not read the original study.)
  - Squinting modifier: Can relate to a word that comes either before it or after it. Players who seek their coach's advice often can improve their game. (What happens often—seeking of advice or improvement?)

Examples of Nonessential Grammar and Punctuation Rules

Nonessential rules are those that are not required for clarity and unambiguity. Some examples:

- The distinction between 'different from', 'different than', and 'different to.'
  - 'Different from' is traditionally used when the comparison is between two persons or things: My writing style is different from yours.
  - 'Different than' is more acceptably used where the object of comparison is expressed by a full clause: This town is different than it was 20 years ago.
  - 'Different to' is chiefly British; in the US 'to' gets little use and is often considered incorrect even though it is an acceptable variation.
- The use of old forms of English: Use of the subjunctive ('if he were to do something'), the pronoun 'whom' as the objective form of 'who', and several other somewhat old-fashioned (though correct) forms of English.
- Many (but not all) rules about the use of commas, given that many punctuation 'rules' are different in US English and UK English. For example:
  - Comma after introductory word or (short) phrase or clause:

    Having chosen nursing as a career, Susan enrolled in many science courses.

    Having chosen nursing as a career Susan enrolled in many science courses.

    When he was in high school, he was known only as an athlete.

    When he was in high school he was known only as an athlete.

My rule of thumb is: If I stumble after an introductory word, phrase, or clause and have to re-read to make sure I understood the sentence, then a comma is probably required (or the sentence needs rewriting), but if I don't stumble, then the comma is probably optional, even if traditional usage says it is required.

- The distinction between 'which' and 'that' in some clauses. Although technically there is a significant difference, in most (but not all) cases readers will not misinterpret the meaning of the sentence, and conventional usage varies between US English and UK English: UK English uses 'which' in most situations.
- Some apostrophe use. For example, does the use of 'user's guide', 'users' guide', or even 'users guide' or 'user guide' lead to any confusion or ambiguity? I think not. (But do pick one variation and use it consistently.) Yes, there's a difference: 'User's guide' means a manual for one user, whereas 'users' guide' means a manual for multiple users. This is a clear grammatical distinction, but to the reader, it's irrelevant: In both cases, the title clearly communicates that the manual is intended to help them use the product. ('Users guide' is technically incorrect but perfectly clear, and 'user guide' is common usage.)

Examples of Usage Rules

Style and usage rules may be written into a style guide as 'the way we do things here', to improve consistency in a company's publications, but editors and writers need to recognize them as choices, not rules of English grammar.

Another good reason to include some usage rules in your style guide is to clarify what's negotiable in your company and what's not negotiable. Some examples:

- Punctuation order, for example whether commas and periods (full stops) go inside or outside a quotation mark. Conventions vary between US English and UK English.
- Punctuation and capitalization rules for vertical lists. Several styles are in common use; pick one style and use it consistently.

- Whether 'data' is a singular or plural noun. Usage varies; in computing, 'data' is typically collective and singular; in mathematics, 'data' is usually the plural of 'datum.' Choose the conventional usage for your audience.

Examples of Fake Rules

- The rules against using split infinitives or ending a sentence in a preposition. You may have been taught these rules in school, but they are based on some decisions made by a few people a century or two ago and are irrelevant to modern communication.
- The rule against using 'they/them/their' as a singular indefinite pronoun. In fact, the singular 'they' has a long history as being acceptable in English.

Summary

What's the bottom line?

- Realize that copyediting is important, but it is only part of an editor's job.
- Distinguish between grammar, punctuation, and usage rules that are essential for clear, unambiguous communication, and those that are not essential or even irrelevant.
- Recognize that many things we were taught to consider as 'rules' are actually style choices or conventions of usage, and that deviations are not necessarily 'wrong' but rather 'not the way we do it here.'
- Include some grammar and punctuation style choices in the style guide to improve grammar consistency.

Notes

Thanks to Geoff Hart and the students in two editing workshops for their comments on an early draft of this article.

QUESTIONS FOR CRITICAL THOUGHT

1. What problem does Weber address in her essay?
2. What should an editor be doing, according to Weber?
3. Editors address a variety of issues, including grammar and punctuation, sentence structure, style, content, organization, factual accuracy, formatting, adaptation to audience, and adherence to the style guide of a particular workplace or profession. Which does Weber think is most important? Why? Does she agree with others in this book?
4. Weber gives a precise survey of her main arguments. Why?
5. Weber accuses editors of having too narrow a focus. What does this statement mean? Does it tell us more about the editors or about Weber? Why?
6. What are 'fake' rules of grammar? Where do they spring from?
7. How can you tell which grammatical principles are 'essential' and which are 'nonessential'? Does Weber provide any means to help you decide?

8. Has Weber provided sufficient examples to make her point? Why or why not?
9. Weber suggests that editors should place accuracy and usability ahead of 'minor' issues of grammar and punctuation. What does this mean?
10. Does Weber give any reasons why it might be important to pay careful attention to grammatical matters? What is the root of these reasons?
11. How engaging is Weber's essay? How readable is it? For whom was it written? How can you tell?
12. What is the distinction between style and usage guides and the rules of English grammar? How does this difference arise?
13. What is copyediting? How important is it?
14. Do you feel confident that you understand the distinctions Weber is making? What might she have done to make it easier for you to do so?
15. How personal is this essay? As advice, how directive is it?
16. What strategies does Weber use to create common ground with her readers? Are these strategies effective? Why or why not?

CHAPTER 24

Sense and Nonsense about Grammar

Brian Bauld

The most common reason given (and refuted) for the teaching of grammar is that of usage. Even though most of my students write correctly 95 per cent of the time, it is clear that ANY mistakes will stand out when it comes to actual speaking and writing. Like a car with one bad sparkplug, or one worn tire, it is the fault that gets the attention and produces the disaster. Applications and resumes with errors are buried in the dustbin. As well, certainly, there are times when one wishes to speak the language correctly. 'I seen' is popular in this area, but any misuse of the language in certain situations (many) is a way of attaching a red flag to one's lack of education. Of course, one can hear:

> 'I want you to come with Mary and I.'

and

> 'I laid on the beach for an hour.' (no object intended), and

and

> 'He is quite an aggravating person.'

and the like from all quarters and all 'educational' levels, but ubiquity is surely a poor guiding principle. Many critics of grammar instruction will point out that we speak various kinds of English relative to our audiences, but if we cannot keep these languages separated through conscious control of our sentences, we are left in a vulnerable position. 'Yo bro' may be cool, ill advised, corny, dangerous, or just plain stupid, depending on the circumstances. Being able to link language to a wide range of audiences is the mark of an educated person. The inability to do so is a pity. The English classroom gets far too little time to perform its far too broad mandate, but failing to provide our children with the power to control something as self-defining as language seems unconscionable.

There are many mountains to climb in literature-land, but the climbers should be adequately outfitted. If the demands on the students involve only the high-plains walking tours of student-directed learning, peer approval, platitudes, and slogans, then sneakers will do fine. No point in being over-equipped if one will do no climbing. The basic equipment should have been supplied by the middle grades. Sadly, I have students in grade 12 each year receiving grammar instruction for the first time. I have taught student teachers for 20 years and have seen them ill-prepared to explain clearly to a student what is wrong with her language. Only rarely did an education professor make the knowledge of grammar a priority for inclusion in the student teacher's toolbox. Have they all 'eaten of the insane root', to deny the usefulness of grammatical language to the study of English, and even the self? What language do they use to explain the error of a student who writes: 'He is as tall as me'? Without the knowledge of 'subject' or 'clause' the answer must be: 'Do it like this because I say it's right. Over time you might recognize a pattern and not make the mistake again.' This is poor way to find and give an education.

Granted, it is possible for a teacher to over-emphasize grammar and keep it isolated from living examples of language. But, this is just poor teaching, an expression of inexperience, or character, and has nothing to say against the principle of teaching students what goes on when they write, or utter, the words which describe and define themselves. Teaching usage is the sort of thing that can be done as the whole language crowd suggests: as the need arises. Some general principles combined with a good handbook (every student with a handbook? Dream on!) should suffice. Rather, it is the very *language of language* which needs to be taught and ingested as early as possible. As math students should know their multiplication tables without having to look it up, indeed, with the immediate ease provided through memorization and frequent application, so, too, students should know the eight parts of speech, the various kinds of phrases, and the clauses. Know, not know of. Here again the obfuscators weigh in with noise about English grammar being a poor borrowing from Latin, as if it were a matter of nationalism; or, that traditional grammar isn't useful anymore because the cutting-edge linguists don't find it altogether to their liking. My experience is that even a rudimentary knowledge of traditional grammar is empowering. If a linguist wants to throw anomalies at me that defy grammatical description, good for him, but how does that negate the overall, and obvious, usefulness of the system? Is this a form of grammatical Utopianism? If I know that 'as' is normally a conjunction and, as such, will introduce a dependent adverb clause, and if I know what *ellipsis* and *subject-of-a-verb* mean, then I can see for myself why 'He is as tall as I' is the correct form. If I don't wish to use the correct form, that is all well and good, but at least I will know what I am doing. I am pro-choice in that I want students to have the power to choose language that will neither embarrass nor impede them.

Still, this is all on the level of 'proper use' and, for me, this is the least interesting of the reasons to study grammar. The second reason involves aesthetics. If a student wishes to comment on the style of a writer, she will refer to diction, tone, point-of-view, and so on. Much of this might be done without recourse to grammar-talk, but when it comes to syntax, the student without grammar is out of luck. No doubt Faulkner and Hemingway will look and sound different, but to actually speak clearly about those differences will require a knowledge of various syntactical combinations. A student may have the unconscious ability to use parallel structure some of the time, but without the language of phrases and clauses it will remain unconscious. Is reading Keats' 'When I have Fears' made any less impressive when the mind sees and the ear hears the adverb clauses marking the structure and rhythm of this single-sentence sonnet? If we really think that in this we 'murder to dissect', then we might as well give up the whole enterprise. No metaphors or assonance or history of the sonnet or meters—let's just read poems without guidance, and find as little as we like. Literary appreciation begins, as with any study, with the acquisition of certain skills. First attempts may be awkward, but it is only through leaning how to use the tools, that the craftsman or artist succeeds in his trade. Standing in the woods with a bird guide in hand straining to fit the fleeting vision with a picture in the book may seem like a poor way to take a walk in the woods. But it is really just a different way. Also, the effort of this early stage of acquiring knowledge leads to that effortless knowledge which is not unlike Love.

'Learning across the curriculum' should mean seeing that grammar and poetry are as related as dribbling and basketball. Just dribbling for a season might get more than a little tiresome, but frequent practice leads to better basketball. We have a generation of athletes who have been offered poor coaching. They have been told that their scoring is as good as Jordan's, their defense as strong as Pippen's, their three point shooting as good as Larry Bird's. Whom do we think we are fooling?

QUESTIONS FOR CRITICAL THOUGHT

1. Bauld uses a number of analogies in his discussion of grammar. Identify as many as you can. What is their purpose?
2. What is the purpose of Bauld's article? Outline his arguments.
3. What is Bauld's central contention? How does he support his position?
4. Is Bauld writing for his students? If so, how has he adapted his discussion to them? If not, then who is the audience for this article? How can you tell?
5. What does Bauld mean by the 'language of language'? Why is it a necessary component of anyone's education?
6. Bauld argues that 'being able to link language to a wide range of audiences is the mark of an educated person.' To what extent is this true for those educated in the technical professions? You may draw upon other essays in the book to answer this question.
7. How personal is Bauld's article? How formal is it? What standard are you using to make this discernment? To what extent do audience and context influence such choices?

8. For Bauld, 'even a rudimentary knowledge of traditional grammar' is an essential tool for any writer. In fact, according to him, it's 'empowering'. What does he mean?
9. Can you tell from this article what Bauld might be like as a teacher? Would you expect to be able to do so? Why or why not?

PART V

Perspectives on Audience and Context

At first glance we might be inclined to think that the most important element of any professional communication is the content and quality of the message itself. But what we may not immediately recognize is that, while it is fair to say that professional communication is driven by goals and objectives, it is equally true to say that it is both context-dependent and audience-centred, since it always takes place in a particular setting and is always addressed to a specific audience. Unfortunately, too many of us are so focused on the contents of what we are writing that we lose sight entirely of what the audience needs to know and of the context in which they will interpret and understand our messages. If we think of the audience at all, we imagine them as having the same interests and priorities as our own, experiencing the context in the same way we do ourselves. Unfortunately, if we take for granted such a shared perspective, our messages will often fail to accomplish our goals, and may even cause serious and damaging misunderstandings.

To make our communication better, we must learn to engage the audience's attention, motivate their concern, and enable them to make the decisions or take the action that we have deemed an appropriate response to the situation in which we find ourselves. We must become equally adept at understanding the context from their point of view so that our messages will be read and understood in the way we intended them to be. Sadly, many messages fail for exactly this reason, because they miss the audience entirely or because they misconstrue the context. The essays in Part V are intended, in part, as an antidote to this common disease of writing, each in its own way offering suggestions for understanding the rhetorical constraints that are imposed on any writer by the demands of both audience and context.

CHAPTER 25

Making Them an Offer They Can't Refuse: How to Appeal to an Audience

Jeanie Wills

I am going to open this article with a rather startling promise: simply reading it can get you a step closer to improving your grade point average or advancing your career. Does this sound impossible? Too good to be true? Maybe. But it is based on a simple principle, one that you probably already know: the more effectively you connect to your audience, the greater your chances of winning their approval of your work. This approval can come in many forms; it may mean good marks if you're presenting work to an instructor; it might guarantee you will be awarded the contract if you're proposing a design to a client; or it can earn you budget approval for a project that needs funding. Unfortunately, though this principle is something we all instinctively understand, it's much harder to put into practice. The key to winning any audience's approval turns on understanding how to convince them to see, understand, and value the ideas you present. Understanding what the audience needs to hear from you to create this connection is the most difficult task you'll take on in preparing and presenting your ideas for others: difficult—but not impossible, because this essay is going to show you a new way of considering audience.

It will do that by building on what you already know about more traditional ways of thinking about the audience, such as with demographic and psychological analyses. For example, a demographic analysis considers how the audience's age and gender or professional, educational, and social status might affect how they hear your message or how they perceive your credibility. A psychological analysis invites you to examine other factors such as the audience's likely beliefs, values, and attitudes in order to gauge how you might begin to build common ground with them.

However, while demographic and psychological analyses can help you understand some features of your audience, there is more to audience analysis than these techniques alone. In fact, this paper will offer you a new way to consider your audience, a way that will deepen your understanding of what motivates people to learn new concepts, to change their minds, to accept new ways of seeing issues, or to take action. Your new understanding will increase your success in presenting your ideas because it will help you be clear and focused on what you are asking your audience *to be*. That's right—what you are asking them *to be*—not just what you want them to think or to believe or to do.

The ability to understand your audience this way is critical no matter how your message is delivered, but because the process is most visible in a public presentation, we will take a speech as our model. Each time you have a speaking opportunity, you have the opportunity to change your audience. Every time you speak, you offer them a new role to play. In other words, each speaking opportunity lets you show people a new way to *be* in their world. Once you understand this one concept, you'll begin to see why and how good speakers succeed in moving their audiences to new ways of knowing, new ways of evaluating, and new ways of seeing.

Most likely, you've never thought about speeches and presentations as events that change an audience, but each speech an audience hears changes them in some way. Even attending a university lecture can change an audience, sometimes in profound ways. Consider that lectures are designed to present facts, information, or ideas. But if those facts are new to their hearers, if the ideas are provocative, if the audiences have been asked to think about things in a way they had not previously considered, then the students who leave that lecture are subtly different people from those who went in an hour earlier. Lectures can change us because they typically explain topics, describe them in innovative ways, or demonstrate perspectives that we may never have considered before, and when our understanding has altered, we change along with it.

This kind of subtle change happens frequently, but our words can have an even more profound effect if we consciously offer the audience a clearly defined role to play. When the role is one the audience knows how to accept, then they can *become* people who are more knowledgeable, who have more intellectual depth, and who have a greater understanding of a subject than they did before. Accepting such a role has a ripple effect. For instance, audiences who have been moved to take on a new role may be more likely to accept other roles that are similar in nature. In other words, the audience learns that they can learn, and they take that knowledge not only into other classes, but also into their professional lives.

University classes are meant to change us and to develop our understanding, but this same effect can happen in other places as well. For example, even listening to someone's thoughtful interpretation of a TV show is an event that can change us. When someone interprets a show in a way we hadn't considered before, she may change how we evaluate and judge this show and thus, how we know or understand our entertainment. In other words, we are offered a new role to play as thoughtful consumers of popular culture. For example, a researcher at the University of Saskatchewan argues that the Canadian 'mocumentary', *The Trailer Park Boys*, actually embodies Biblical values (Polkinghorne, 2006). This surprising interpretation may make us see the program in a new way, perhaps even changing how we interpret and value it. By changing the way we evaluate and judge the show, the speaker is, in fact, offering us a new way *to be* in the world: more able to see past the surface of a message, more open to novel approaches, more aware of subtext in popular entertainment.

Of course, differing speech purposes offer differing roles to audiences. If we are persuaded to attend a 'how to' seminar at the local home building store, the presenters extend to us the invitation to play the role of home renovators, and hopefully, the promise of playing this role successfully. In a very fundamental way, demonstration speeches and presentations change us through their teaching by inviting us to alter our perspectives as we move from a position of incompetence to one of competence. Of course, when we sign up for the 'how to' seminar, we signal our eagerness to accept the role of do-it-yourselfer, but the presenters must still show us how we can successfully play the role, or we won't consider the demonstration or the demonstrators to have been credible.

The role a presenter offers to an audience is not always as clear and practical as the one offered by a demonstration speech. Nevertheless, all presentations offer the opportunity for change. A motivational talk, for example, can change how we feel about a particular issue and impel us to behave differently. On the other hand, proposal speeches or speeches that urge the audience to take a particular and specific action are ones that change us from bystanders into active participants. On their incentive, we become people who have a voice, and who have power to act.

All speeches and presentations are opportunities for a speaker to offer an audience new roles to play or new ways to be in the world. As a speaker, your goal is to have the audience, 'see as I see, know as I know, and value as I value' (Brock and Scott, 1980: 19)—in other words, to play the role of participant in a changed view of the world. However, to do this successfully, you must make sure that the role you offer is one the audience can and will accept, and one they are capable of fulfilling. If you 'miscast' them, the audience will reject the role and your presentation will fail in its purpose, which is to get the audience to see a topic from your point of view, to understand it as you do, and to attach the same significance to it as you do.

Thinking about a presentation as an invitation to the audience to become something more than they already are is a novel idea for most speakers. It's potentially empowering, but it also presents some challenges. Before you can offer the audience a role to play, before you can become the agent for change that the speaking situation encourages you to be, you need to be able to understand where your audience stands *before* they hear your talk. In other words, the way you position your topic for your audience will determine whether you can gain their attention and acceptance by bringing them to a 'psychological meeting place' (Wilson and Arnold, 1964: 69). If you can do this, they will be able to hear your arguments and be moved by them.

But remember, before you can change the audience, you first have to understand how they see themselves now, and how their current role compares with the new role you are inviting them to play. Before you can move them, you and they need to share the same mental space and understand the topic in similar ways. This is where traditional types of demographic and psychological analyses can help you to get some insight into their probable concerns and interests. A breadth of experience helps too. Even when you can't know their thoughts, values, beliefs, and desires with absolute certainty, you can make thoughtful judgements about what their attitudes are likely to be, and this is the first step. Once you know what their attitudes are and what psychological territory they occupy, you have a better chance of motivating them to change, to move into the new territory that you offer them, and to accept a new role.

To ensure the audience is successful in adapting to their new role and in making the journey to the new territory you are inviting them to, you need to be clear about what it is you are asking of them. I reiterate this point because it's crucial to have clarity yourself. Ask yourself how the journey that you are guiding them through will result in them *becoming* something else, and ask if that something else, as you have presented it, will be desirable to them. If you're asking your audience to become something new and different and to move from their familiar territory to a new one without demonstrating how the change will benefit them, they may decline your invitation and reject the role. And, if you ask the audience to explore new territory, but don't show them how to get there, they'll refuse the journey.

To illustrate what I've said about roles and territories, let me share some speech examples with you. In the following three examples, I'll describe the importance of clearly envisioning the role you are asking the audience to play and the territory you are inviting them to move to. These examples come from real speeches given by students in a professional communication course. As part of the course requirements, the students are asked to give a proposal speech in which they ask or urge the audience to take a very specific action. The students must discover ways to make the action immediate and urgent because the more immediately the action can be taken, the more vivid and moving the speech will be.

One novice speaker proposed that the 17 young men and three young women in his audience of professional students commit to breastfeed their children or any future children they *might* have. Of course, some problems with the topic are immediately obvious: to date, males of the species do not lactate, making it impossible for 17 of the 20 audience members to assume the role he offered them. But the topic failed in other ways too, primarily because the speaker failed to prepare a psychological meeting ground with the audience. He didn't position his topic appropriately, partially because he never once mentioned his own investment in the topic, and partially because the nature of the topic put an uncomfortable focus on the three women in the class. He failed to take into account the current role of these *non-parents*, and to consider the viability of the new role he was offering—that of a breastfeeding mom.

The speech did offer a survey of health benefits to the baby, but it also spent what seemed to be an inordinate amount of time talking about physical benefits for the new mom. The males in his audience were completely excluded, while the female students were so discomfited that they couldn't *hear* the speaker offering them a viable role to play. Without providing any link to where they currently stood, the speaker tried to invite them to play a role in which they would be urgently concerned with infants they hadn't even thought about yet. What they heard instead was him singling them out from the rest of their classmates based on their physiology. Not only were they not prepared to accept to the role he offered, they rejected his offer to accompany him imaginatively into this future territory because they found his topic and, consequently, the speaker himself, offensive.

The failure wasn't in the audience's inability to project themselves into the future; the failure instead was in the speaker, because he did not offer them the means, motive, or opportunity to see themselves as he invited them to do. In fact, what many of the audience members saw was a colleague who exploited a public speaking situation to try to legitimize what they considered to be an inappropriate topic. Given the speaking context, there was probably no way for this student to succeed with this topic and this particular audience. However, had he considered *where they were* and compared it to *where they needed to be* to benefit from the speech, had he considered *what they were* and compared it to what he wanted them *to be*, he may have discovered a way to present the topic so that, at the very least, the audience's reaction wasn't hostile.

Another novice speaker in a professional communication classroom for engineering students presented a speech that also failed because he couldn't clearly see what role he was asking his audience to play. This speaker was concerned about the dangers that caffeine poses to a coffee drinking audience. He decided that the best way to get the audience's attention and spur them to action—to stop drinking coffee—was by encouraging them to 'Drink beer instead of coffee in the morning.'

Unfortunately, the speaker was asking his audience to play roles that could have had a seriously negative impact on their futures. He didn't stop to consider how inappropriate it is to ask aspiring professionals to start their day inebriated, an action that could compromise their relationships with their professors and with their classmates. Had they followed his advice, they would have established an unprofessional and potentially dangerous habit with consequences that might have ranged from academic suspension to impaired charges. Even worse, he was implicitly inviting them to develop a cavalier attitude toward their university education, their professional relationships, their future careers, and their own safety.

The students who heard this speech completely rejected the role the speaker offered them. They did so for two reasons. First, he made no attempt to create a psychological meeting place with his fellow students who take their professional identity very seriously. He could even have repositioned the speech by simply asking them to stop drinking coffee in the morning. He could have made the action dramatic and vivid by saying 'Drinking beer in the morning would not be as harmful to your health as drinking coffee is.' Second, he was disrespectful of their professional aspirations, their pride in being students in a professional college, and all the privilege that accrues to them because of their roles here.

Finally, because he treated the speech as a kind of joke, many in his audience felt that he had wasted their valuable class time. Because this speaker failed not only to create common ground, but also to offer a role the audience could see themselves playing, he destroyed his chances of success. They resolutely refused to move from where they were to where he wanted them to go, and with good reason. He did leave a lasting impression on his audience, in that it was one of the few speeches that actually aroused their hostility!

On the other hand, when a speaker builds common ground, when he demonstrates that he and the audience share psychological space, when he offers them new ground to move to, shows them how to get there, and presents them with a role they simply can't refuse, that speaker will succeed with that audience. For example, one concerned student in the College of Engineering told the audience they could fight economic and environmental exploitation, help coffee producers, and drink great coffee all with one simple action: buying Fair Trade coffee.

The speaker first built common ground by referring to the engineering code of ethics that will govern by all the students who wish to work as professional engineers after they graduate. He reminded his audience that the code states, 'All members and holders of temporary licenses shall recognize this code as a set of enduring principles guiding their conduct and way of life and shall conduct themselves in an honorable and ethical manner, upholding the values of truth, honesty and trustworthiness, and shall safeguard human life and welfare and the environment' ('Professional Conduct', nd). By referring to this code of ethics, the speaker connected his topic to the aspirations of the audience whose dream of becoming professional engineers included a willingness to embrace the professional code. The speaker reminded them that they all would be expected to 'hold paramount the safety, health and welfare of the public and the protection of the environment and promote health and safety'. He presented arguments that demonstrated the economic injustice suffered by coffee producers of the world and the environmental hazards imposed by economic pressures. The speaker's insight into his audience's desire to embrace the role of the professional engineer became a way of motivating them to take the action he asked them to take and to accept the role he offered them.

The student speaker positioned his audience first as people connected to his topic by the ethical code of their chosen profession. He offered them the opportunity to act on their ethical concern for others, to take advantage of their position of privilege and power as university students and future engineers, and to become part of a 'global community'. He envisioned them as people who had both the power and the desire to change the world. As he saw them so they saw themselves.

In addition to presenting the audience with the motive to take on a new role as buyers of Fair Trade coffee and upholders of fairness and justice, the speaker also presented them with the means and the opportunity to act, making it easy for the audience to accept a new role for the benefit both of themselves and others. He gave them the means of identifying the brand by showing them the 'Fair Trade' logo. He provided the opportunity to take the action by noting that this brand was on all fairly traded coffee sold in grocery stores, and he named the locations on campus that brewed and sold fair trade coffee.

This speaker's presentation succeeded because he offered the audience a role they could embrace. As future engineers, they believed they had a duty and obligation to uphold the code of ethics of their profession. The speaker motivated them to stop participating in the exploitation of coffee producers and to acknowledge that with their position of privilege comes responsibility. In essence, he offered to meet them in student territory, offered them a heroic role to play, and then showed them the territory or scene where they could play that role. They eagerly accepted.

When next you have the opportunity to address people, remember that you also have the opportunity to change who they are and how they act. No matter what you're presenting or where you're speaking, you have this power. But, to make full use of this power, you must think clearly about what your audience would have *to be* in order to accept the message and the role such acceptance entails. As speakers, we invite people not just to believe something, but *to be* something (Black, 1993). When you invite the audience to play a role, to be something more than what they are now, visualize clearly in your own mind what it is you are asking them to become. When you learn to recognize the roles you are offering to an audience, you'll learn to make them offers they won't refuse.

References

Black, Edwin. 1993. 'The Second Persona', in Thomas W. Benson, ed., *Landmark Essays in Rhetorical Criticism*, pp. 161–72. Davis, CA: Hermagoras Press.

Brock, Bernard, and Robert Scott. 1980. *Methods of Rhetorical Criticism*. Detroit, MI: Wayne State University Press.

Polkinghorne, Silas. 2006. 'Seeing the Holy Trinity on TV', *On Campus News* 8 September. Available at http://www.usask.ca/communications/ocn/06-sept-08/6.php (accessed 30 September 2006).

'Professional Conduct'. nd. Available at http://www.apegs.sk.ca/Default.aspx?DN=722 (accessed 2 October 2006).

Wilson, John, and Carroll C. Arnold. 1964. *Public Speaking as a Liberal Art*. Boston, MA: Allyn and Bacon.

QUESTIONS FOR CRITICAL THOUGHT

1. What is the purpose of this essay? Who is its intended audience? How do you know?
2. What does Wills promise her audience? How does she engage them?
3. What does Wills mean by 'role'? Does she mean play-acting, or is she using the term in some other sense?
4. What novel idea does Wills offer?
5. What do you need to know to change an audience, according to Wills? Where can you obtain this information?
6. Wills uses two predominant metaphors in this essay—role and territory. Why? How do these metaphors function? What are they for? What is their relationship to each other?
7. Why does Wills include so much detail about the student speeches?
8. What do speeches have to do with writing? Is the role of audience the same in both situations? Why or why not?

9. How audible is Wills's voice? What features of style make it possible for you to 'hear' a voice?
10. Is this essay formal or informal? What are the markers of formality in writing style?
11. Wills is giving advice. How directive is it? Compare her essay with MacLennan's 'Getting It Together: Strategies for Writing Cohesively'. In what ways do the two differ? Why?
12. Compare Wills's advice with that given by Elbow in 'Three Tricky Relationships to an Audience'. Is their advice consistent? Why or why not?

CHAPTER 26

Bridging Gaps, Engineering Audiences: Understanding the Communicative Situation

Burton L. Urquhart

David is an engineer with an innovative design, and he's very excited about the prospect of manufacturing his product for sale. Unfortunately, David doesn't have the money he needs to get his product into the market, so to make this possible he has called in a few favours. An acquaintance in the banking business has managed to arrange a meeting for David with a group of likely investors.

David is determined to impress the investors so they will back the production of his design. As he prepares his pitch for them, he builds a portable model of the prototype and makes copies of the scientific studies documenting his design's soundness. He makes speaking notes that explain the design's operation, the need that it fills, and its proven scientific validity. He practises until he is confident of his message and his delivery.

On the morning of his presentation, David is feeling very positive, not only about the potential of his design, but also about the presentation he has prepared. When he arrives at the meeting and begins to speak, however, he runs into a few unanticipated surprises. The investors prove to be a much more challenging audience than he had expected. They ignore his attempts to elaborate on scientific details, and instead want to know about market surveys, advertising pitches, profitability, and product positioning. These are questions that David hasn't thought about and can't begin to answer; in fact, this is the kind of information he had expected to get from *them*. He tries to point them to his research studies, but without success.

Despite his initial confidence in his design and the supporting material he has provided, in the end David's pitch is rejected by the investors, and he leaves the meeting empty-handed. Frustrated by his failure and bitter at the setback, he concludes that his audience is just too short-sighted or too stupid to understand his product.

There is no doubt that David didn't foresee how the meeting with the investors would go. He had prepared for questions about the technical aspects of his design, and he had plenty of valid scientific research on his side. Where David failed, however, was in understanding how to position his message so his audience would hear and respond with enthusiasm—and a cheque. Unfortunately for David, even the best technical and scientific support couldn't solve his central problem, which has little to do with science and everything to do with human nature and motivation.

David's failure lay in his inability to persuade those whose help he desperately needs to get his product to market. The success of David's design depends only partly on its technical soundness; in David's case—as in the case of many engineers who wish to get their work approved and funded—success depends just as much on the skills of persuasion. David's failure provides a case study of what happens when a would-be persuader misreads the audience and overlooks their needs, expectations, or values. In such a case, the technical sophistication of the design is insufficient to secure the investors' confidence and their commitment. What is needed is the skill of persuasion—mastery of the ancient art known as rhetoric.

Rhetoric is the formal study of how our practical messages exert their influence. It teaches us not only to understand what's going on in the messages of others, but also how to craft our own messages in more effective ways. David's investors were interested in the possibilities of his design; if they hadn't been, they would not likely have agreed to a meeting. But they are also likely to be somewhat skeptical and in need of assurances; after all, if the business goes belly-up, they stand to lose their investment. David's audience is willing to be persuaded, but they must hear a message that lays their concerns to rest.

The task of persuading an audience has much in common with the design process itself. When an engineer creates an engineering design, he is using his technical abilities to change the world as it is into the world as he envisions it. He develops a physical intervention to bridge the gap between the design problem and its solution. Similarly, when that same engineer pitches a design solution to clients, management, or the public, he is engaged in an equally challenging bridging activity, and one that is just as important to his success. He must develop a discursive intervention to bridge the equally challenging divide between people's understanding and their willingness to cooperate in making the design a reality. Unfortunately, while most of an engineer's training focuses on the technical skills needed to bridge the gap between idea and reality, relatively little attention is paid to the essential communication skills needed to bridge the communication gap.

Bridges in understanding, like other kinds of bridges, are not built by one person alone. Any engineer who hopes to transform ideas into reality needs more than technical savvy. Such a person also needs the cooperation of other people: investors, contractors, technical support staff, labourers, and management. But these people need to be encouraged to participate fully, need to be engaged with the project, and need to be committed to its success.

Any engineer who hopes to be successful must also bring together a whole range of competing priorities. The successful engineer must find a way to conjoin the goals of management, marketing, and technical staff, unite the needs of the clients with a suitable and feasible design, and negotiate the sometimes incompatible pressures of safety and economy, functionality and appearance, deadlines and workability. And that same engineer must find a way to get the clients to see the problem—and the solution, the resulting design—as the engineer sees it.

Unfortunately, while every engineer recognizes how challenging the design process is, many fail to understand that their most significant challenges can arise from bridging the communication gap that divides understanding from action. Such *communication* challenges are just as much a part of the engineer's job as designing is. If David had realized this, he might be much closer to marketing his design. Instead, he is left without investors, and is still trying to figure out what went wrong.

David understands that he failed to secure the financial support he needs because he was unable to get the investors to see his design as he saw it, to know its usefulness and benefits as he knew them, or finally to share in his vision, to believe in his solution as he so passionately did. However, he hasn't yet gotten beyond blaming them for their short-sightedness and stupidity. To him, the merits of his design are obvious, and he can't understand why they weren't equally obvious to the investors. What David hasn't grasped is that there is a gap between his priorities as a technical specialist and the priorities and values of his audience of finance and commerce specialists, whose focus is business viability rather than technical research and innovation. Or rather, he has failed to understand what such a gap in priorities *means* for him as a client who is attempting to secure needed funding. In order to bridge this gap and create shared understanding, David needs better insight into the context in which he is communicating. A clear understanding of the situation he is entering will help him reduce the distance between his own priorities and those of the investors.

In order to determine how wide these gaps in understanding are David needs to be able to put himself in the shoes of the investors. Like any speaker, he needs to learn how to accommodate his audience's perspective and priorities. He needs to design his message with the same goal that shaped his technical design: to bridge the divide between the world as it is and the world as he envisions it.

What caused the investors to reject David's project was not a failure of imagination or insight on their part, but a failure on David's part to create a shared understanding. Unless David is able to recognize and correct this shortcoming, he is doomed to fail with the next audience of investors, and the next. To be successful, he needs to learn to master his communication challenge—not only because it's important to his audience, but also because it is vital to his own professional interests.

However, shared understanding isn't achieved automatically. Figuring out how to create shared understandings with those whose technical expertise is different from our own is difficult for everyone. Specialists in any field often have a difficult time comprehending how others will view similar situations. For David, as for all of us, the challenge lies in being able to put aside his specialized understanding and see the problem from a completely different vantage point, in a different context, and with priorities that may differ from his own. If he is to be successful, David needs to talk to his investors in terms that they understand, and to position his invention as an answer to the questions the investors might bring, rather than explaining it in terms that an engineer finds interesting or satisfying.

It's obvious that David did not think this carefully about the communication situation, about his audience, or about his purpose, before meeting with the investors. If he had done so, the outcome of his meeting could have been more positive. David might have secured the investors' good will, their understanding, and perhaps even some of their investment dollars. He failed to do so this time, but he need not face failure every time. An analysis of what happened can help him to build a more effective communication strategy, so that he can bridge not only the technical gap between the design problem and his solution, but also the gaps between the understanding, motivation, values, and purposes of the investors and himself.

What can David do differently when he next has a chance to address a group of investors, and where should he begin? For an answer, let's take a closer look at David's experience with his first investors' meeting. David went in knowing that he had a winning design. Extensive testing had demonstrated its effectiveness, and a successful prototype had validated the test results. Since the project's infancy, David and his team had had the support of national academic and government research funding. Clearly his project is based on solid scientific groundwork.

Even more important, David was confident that his design would fill a demonstrated need, thus solving an already existing problem. Before the meeting, he was certain that this point would secure the investors' interest and funding. Not only is his design effective, but it is also sophisticated enough to remain a standard in the industry for years to come. David believed that he was offering the investors an incredible opportunity to invest in a new product that is both needed *and* scientifically sound.

Consequently, he was astounded when the investors seemed to doubt the viability of his proposal without even seeming to consider the scientific arguments. In his shock, he challenged the soundness of the investors' judgement, telling them that they would be 'missing out on an opportunity of a lifetime'. He could not understand why they seemed not to trust his expertise. He became impatient with what seemed to him irrelevant questions, insisting that the investors were missing the point; after all, he argued, the design is really quite simple to understand and to use. Rather than steering the investors' attention back to the functionality of his prototype, however, this strategy backfired, and at this point, the investors informed David that they had heard enough. They were simply not interested in pursuing his project any further.

The flat-out rejection surprised David so much that he found himself unable to resist a parting jab; he told the investors that he pitied them for their misguided judgement, called their decision 'short-sighted', and criticized them for being 'unable to think big enough'. He would, he declared, look forward to proving them wrong while making another, more savvy investor very rich.

What David didn't see is that, after he had left the room, the investors expressed their own surprise and irritation at how utterly he had failed to comprehend the purpose of the meeting: he was supposed to demonstrate why they should invest their money in his project, not provide its scientific pedigree. That much they had been willing to accept on the basis of his expertise; it was his complete failure to outline a business plan, combined with his arrogant tone, that caused them to dismiss him and his project. From their perspective, they had come to the meeting with good will, only to have David completely waste their time.

What happened to David is what happens to many technical specialists who must present their ideas to non-technical audiences. Despite a solid design supported by rigorous engineering science and testing, David was unable to obtain the investors' confidence and their financing. From a scientific point of view, David's idea is a sure-fire winner, but his inability to convince business and management representatives of its appeal means his product will never get beyond the prototype stage.

If he is to be successful the next time he faces an audience of investors, David needs to understand how to identify the things he needs to say so that his audience will see what he sees and embrace his solution. In order to engage his audience and secure their confidence, David needs a much clearer definition of their shared purpose. He needs to understand and respect the investors' priorities, and he needs to throw off his narrow engineering focus so that he can better provide the information his audience needs.

Fortunately for David, there is a method that can help him to avoid repeating the same mistakes the next time he faces his investors. He can use a series of questions developed from communication theorist Lloyd Bitzer's (1968) model of the rhetorical situation to help him understand how to position his message so that the investors can share his enthusiasm and confidence. Bitzer advances a model that focuses attention on the purpose of any communication, the needs and attitudes of the audience, and the potential gaps in understanding. It provides a means for David to better understand the situation facing him.

The first question David needs to consider is, what is the purpose of the meeting? David mistakenly believed that the purpose of his meeting with the investors was to demonstrate the scientific validity and quality of his design. He assumed, again wrongly, that the obvious need for such a product would be enough to convince the investors to buy into a potentially risky venture. His mistake, not surprisingly, was to view the problem as an engineer would view it rather than trying to see it from the perspective of his audience.

Interestingly, the investors may be just as committed as David is to solving social or technological issues, but whatever their personal feelings might be, in this context they have other priorities. They want to know that the device or solution is marketable. They need to be convinced that there is an audience for the product, and that it can be profitably produced or deployed. Even though it will sound strange to an engineer like David, in a context of such priorities, proving scientific validity is of secondary concern. David's real purpose in this meeting—the one he overlooked entirely—should have been to demonstrate his design's potential for marketability and profitability. In such a context, its scientific validity and its usefulness are supporting details rather than the focus of the meeting.

The second question that should shape David's message is: What does the audience need to hear? Because David failed to recognize his audience's priorities and thus misunderstood the purpose of the meeting, he consequently failed to provide answers to the investors' most pressing questions: Why should I invest money in this product? What return can I expect, and over how long a term? David's audience for this meeting consisted of people who were being asked to risk some of their assets on an untried product. They needed some reasonable assurance that they would not lose their investment on this venture. Public need and scientific validity may be compelling arguments for securing a government research grant, but these factors do little to help investors make a decision, because they do not demonstrate that the product is feasible from a production and marketing point of view.

At this point, the investors' confidence depends more on the quality of David's business plan and marketing research than on the quality of his scientific results. This is the information the investors need and are interested in. In his preparation, David should have considered what the audience needed to hear rather than what he wanted to say, and what would be most likely to convince them to take action.

The final question that David must consider is: What are the obstacles he faces in making his case? It may surprise him to learn that the greatest obstacle he faces in this situation is his own point of view—his engineering assumptions and values. It is his own perspective that creates the largest gap between himself and the investors. As an example of how this might be so, consider how each party views the idea of 'viability'.

For David, viability means that the design works, that the device will do the job it was created to do, that it is a sound application of engineering principles. From this perspective, David's design meets, even exceeds, the requirement of viability, and David is justifiably pleased. For investors, however, 'viability' is something else entirely. From their perspective, it has nothing to do with science and everything to do with business. The research that they need to hear is not scientific data, but financial and marketing data, since it is this information on which the investors will largely base their decision. When it comes to viability, profit margins and rates of investment return mean more to them than elaborate scientific testing procedures and results. They will also want to determine how much effort and thought have gone into David's business plan.

The investors will, in part, judge David's credibility on these considerations. They need to know if the product can be produced at a reasonable cost to ensure a large enough profit margin. It is likely that their questions will sound like these: How much does it cost to make this? How much can you sell it for? Can the materials be easily sourced? Is there a ready-made market? What segment of the population will be targeted in marketing? What about the competition? Have you located a production facility for this product? Does this facility have the capacity for expansion? David's scientific research is important only insofar as it gives the investors confidence in the product itself, but it is the business and investment potential of the product that will ultimately convince them of the soundness of the venture. Understanding what the relevant research is for this situation was a major obstacle for David in his first investors' meeting.

If he is to successfully persuade a future group of investors, David must also explain the competitive market or any solution for market share retention in this area. His failure to provide such information most likely caused the investors to question his judgement—not as a scientist, but as a business partner.

The more David considers the investors' point of view, the more he closes the gap between the investors and himself, and the greater the understanding he will have of the obstacles that could jeopardize his persuasive communication with them. David's future success will depend on his ability to translate his design into a profitable venture for those investors in the hour he is in that boardroom.

By carefully analyzing his situation using these three questions as a guide, David can develop a fuller understanding of the communication situation that he is entering. This situational understanding will help David to achieve the perspective he needs to effectively bridge the gap between him and the investors. If he takes the time to do his analysis carefully, he can learn to see outside of his disciplinary limits and understand how to meet the investors in their own territory.

Adapting arguments to meet an audience's expectations and concerns is a daunting task, especially if the audience doesn't share our technical expertise. However, we can learn to approach our communication situations thoughtfully and avoid the same mistakes that David made. Like David, we need to analyze our communication situation by asking these same three questions: What is the purpose of my communication? What does my audience need to know to enable them to make a decision? And what are the potential obstacles to our achieving mutual understanding?

This process of understanding your audience and strategizing a persuasive message is a crucial bridging activity for all professionals, even if they never have to face a panel of investors. The strategies David has learned, and the questions he now knows to ask, are equally useful for understanding any audience. David's experience teaches us that it is not enough that we see how a technical problem and the solution may be brought together; what is equally important is creating shared understanding, eliciting cooperation, and meeting the audience's needs as we invite them to 'see as we see, know as we know, and value as we value' (Brock and Scott, 1980: 19).

While we may never find ourselves in exactly the same position as David was with his audience of investors, we can all benefit from his experience. Every professional at some point must face an audience who are not familiar with or knowledgeable about the details being presented. We 'sell' ourselves at job interviews, we work in teams with people from other backgrounds, and we communicate ideas to clients and management. In all these situations, we need to be able to bridge the gaps between our own assumptions and needs and the attitudes and concerns of our audience. Knowing what they want, how they are likely to view the situation, and what obstacles in understanding might exist, our audience analysis efforts will help us develop more effective persuasive messages. We will, like David, stand a better chance of success if we take the time to think through our situation by focusing on how it appears to those whose approval and support are necessary to make our goals a reality.

References

Bitzer, Lloyd. 1968. 'The Rhetorical Situation', *Philosophy and Rhetoric* 1: 1–14.

Brock, Bernard L., and Robert L. Scott, eds. 1980. *Methods of Rhetorical Criticism*. Detroit: Wayne State University Press.

QUESTIONS FOR CRITICAL THOUGHT

1. Why does Urquhart open with the story of David's pitch to his investors? What is the role of this anecdote/case study in the development of Urquhart's argument?
2. In what respects could this paper be considered an answer to the question posed by Tania Smith in 'What Connection does Rhetorical Theory Have to Technical and Professional Communication'?
3. What is Urquhart's thesis? Where is it first stated?
4. How does Urquhart define rhetoric? Does his definition differ in any significant respect from the other definitions in this book?
5. Why exactly do the investors reject David's design?
6. What is 'viability'? What does this example teach us about the nature of language use, about the assumptions that colour our communication, and about the role of values in the way we determine meaning? What are the implications of these insights for our own communication?
7. In what respects does Urquhart's paper serve to confirm the assertions made by Ryder ('Science and Rhetoric'), Smith ('What Connection does Rhetorical Theory Have to Technical and Professional Communication?'), and Burton (An Engineer's Rhetorical Journey')?
8. How effective is the 'bridge' metaphor? How does Urquhart use it? What does it mean?
9. Urquhart's article points to a common failing among specialists in all fields. What is that failing?
10. Who is the intended audience for this article? How do you know? What elements of the article itself point to the identity of the audience?
11. Urquhart identifies a problem common to professional communication; to what extent does he offer a solution to the problem?
12. Compare the style of Urquhart's essay with that of Smith ('What Connection does Rhetorical Theory have to Technical and Professional Communication?'), Halloran ('Classical Rhetoric for the Engineering Student'), or Burton ('An Engineer's Rhetorical Journey'). What differences do you see?

13. What is credibility, according to Urquhart? On what does David's credibility depend? How does he lose his credibility with the investors?
14. Urquhart proposes three questions that can help a writer or speaker to frame the message appropriately. What are they? Do they constitute a heuristic, in the sense that Smith uses the term?
15. Why does Urquhart spend so much time discussing the solution to David's plight, rather than directly advising us about how *we* might improve our communication?

CHAPTER 27

Communicating with Non-Technical Audiences: How Much Do They Know?

Bernadette Longo

Professional communicators all ask themselves this question in some form sooner or later: How much background knowledge does my reader have about the subject of this text I'm writing? This question is of paramount importance if technical writers want to communicate effectively with non-technical readers.

Whether the writer knows the reader personally or must invoke an imagined person, the issue of communicating with this reader comes down to accommodating his or her needs through words on paper. Using this model, the audience analysis process can be seen as two steps: determining who the reader is, then stating technical information in a form that accommodates the reader's background knowledge.

The First Step: Determining Who the Reader Is

In determining who the reader is, the writer may in some rare instances know the reader personally, in which case it will be a relatively easy task to determine how much this person knows about the subject at hand. In most cases, though, the writer must invoke some imagined reader to address and this reader will be a composite of the rhetorical situation in which the writer and reader find themselves. The demographic details of this reader will not be as important as the community standards which he or she embodies.

In the background of questions about audience is a rhetorical consideration of the larger context for writing, which is primarily a consideration of the cultural context of the communities in which the writer and reader operate. These communities define communication conventions which writers need to understand in order to produce what will be accepted as authoritative by members of those communities. Patricia Bizzell (1982) explained how conventions evolve within discourse communities: 'Groups of society members can become accustomed to modifying each other's reasoning and language use in certain ways. Eventually, these familiar ways achieve the status of conventions that bind the group in a discourse community'

(214). Joseph Bocchi (1988) expanded on the process through which writers learn about their community:

> Each time writers make inferences about the conventions of their community, each time writers interact with other members of their community, they are reading and defining role boundaries as well as gaining information about how individuals act individually within these boundaries and about how these individuals perceive their own roles. Each time writers communicate to others, they are composing and testing and redefining their knowledge by applying it to a specific rhetorical situation. (5)

Writers use information about their social context in all stages of their writing, from deciding on issues for discussion and how to organize that discussion to conceptualizing the audience and rhetorical situation.

Because an understanding of the rhetorical context of community conventions is so crucial to producing a successful text, it is not sufficient to come up with a list of audience attributes unless you also understand what to do with them (Atlas, 1979; Warren, 1993). An audience is made up of readers who are more than their individual attributes; they are members of overlapping community cultures at work, at home, in the community at large. Their individual attributes are shaped and influenced by these cultures. A writer needs to understand the influences of cultures on intended readers in order to hazard a prediction about how a text will be accepted by those readers.

Asking Questions to Help You Understand Your Reader

The writer can use questions like these to imagine the typical reader of a particular text: How does this reader's community value my subject? What writing conventions operate in this reader's community? How much background knowledge is a member of this community likely to have?

HOW DOES THIS READER'S COMMUNITY VALUE MY SUBJECT?

Taking these questions one at a time, the first deals with how the intended reader's community values the subject about which you will be writing. As Young, Becker, and Pike (1970) show, this consideration of values is important because it is one way the writer can use shared information to build bridges between herself and the reader:

> The motive for communication arises from an awareness of difference and a desire to eliminate it or at least to modify it. But there can be no interaction between writer and reader . . . unless they hold certain things in common, such as shared experiences, shared knowledge, shared beliefs, values, and attitudes, shared language. Things that are completely separated from each other cannot interact. (172)

Sources for exploring a community's values about your subject can be articles in journals and newspapers, interviews with community members, and your interpretations of graphic material. When analyzing this material, here is an important consideration: Each time something is chosen to be included in one of these media, something else is excluded. How members of your target community make these choices is indicative of the values they place on the chosen and unchosen elements.

In seeking to learn more about community values regarding your subject, Young, Becker, and Pike suggest considering whether your reader's values are hierarchically structured and whether some values are more important than others. Then choose to highlight those shared values that have priority over some unshared values and make several connections regarding shared values to solidify the positive interaction (174).

WHAT WRITING CONVENTIONS OPERATE IN THIS READER'S COMMUNITY?

The second question explores the writing conventions that operate in your reader's community, which help to determine what your reader will expect from the text you are writing. This question is important because, as Clark and Haviland (1977) explained, the reader's decision whether or not to participate in a cooperative contract with the writer and the outcome of this participation are governed by communication conventions that help us understand what has been written. Martin Nystrand (1986) saw the text's role this way:

> [W]riters and readers interact not only with the text but also with each other by way of the text. Writers gauge their intentions in terms of the expectations of their readers, and readers measure their understanding in terms of the writer's intentions. Communication between writers and readers requires that the text they share configure and mediate these respective interests and expectations. This requirement means that the skilled writer's choices and options at any point in the composing process are determined not just by what the writer wants to say but also by what the text has to do and, in turn, by what the reader may reasonably expect the text to do. (ix)

Again, looking at text produced by respected members of your target community will help you to understand what types of writing your intended reader normally considers authoritative. By looking at such things as the structure, tone, style, and arguments of these texts, you will develop some idea of how you can adapt your text to your reader's standards.

HOW MUCH BACKGROUND KNOWLEDGE IS MY READER LIKELY TO HAVE?

The third question examines the level of background knowledge a member of the target community is likely to have. This is an especially important question when writing to non-technical audiences, since your reader needs to have enough background information to understand your technical subject matter. Young, Becker, and Pike (1970) argue that this background knowledge can be a bridge of shared information over which a writer and reader can meet to communicate. They further advise that when building these bridges of shared information, the writer must be sure to attach the unfamiliar information to an image that is familiar to the reader (173). In this way, the reader can internalize the new information, transforming it into shared information on which to build further bridges.

In writing to non-technical audiences, you are safe to assume that your reader has very little background knowledge about your technical subject. But if you have researched the previous two questions, you will probably have some idea concerning how much your intended reader knows about your topic and what community values are associated with it. You may find that your reader's community has some background knowledge about your subject, but you want to modify this information and/or the values associated with your subject. This is done through writing strategies that emphasize the information you share with your intended reader.

THE SECOND STEP: WRITING TEXT THAT ACCOMMODATES YOUR READER'S NEEDS

Once these audience analysis questions have been explored, the writer can then turn to the task of composing text that meets a non-technical reader's needs based on two considerations: finding an effective appeal for this non-technical reader and building bridges of shared information.

In choosing an appeal, the writer can look to standard topics based on the genre of the intended text. For scientific texts, Jeanne Fahnestock (1986) has identified two 'basic appeals' used when addressing popular audiences: wonder and application. The 'wonder' appeal, which Fahnestock likens to the deontological argument of classical rhetoric, 'attempts to praise or excoriate something by attaching to it a category that has a recognized value for an audience' (279). In using this appeal, you might choose to highlight a shared value inherent in your subject and explain the significance of this to make it explicit for your reader. The 'application' appeal, likened to the teleological argument, holds that 'something has value because it leads to further benefits' (279). In using this appeal, you might focus on a benefit that has value for your reader's community and make its significance explicit. When addressing non-technical readers, it is important to make the significance of your subject explicit because your audience probably does not have sufficient background knowledge to necessarily recognize its inherent significance.

While the wonder and application appeals are common to scientific texts written for popular audiences, a review of Aristotle's *Rhetoric* yields additional standard topics which can be employed by technical writers. (See also Corbett, 1971 and D'Angelo, 1986, for further discussion of topics from classical rhetoric.)

When the appeal is determined, the writer can build bridges of shared information with the reader's background knowledge by gradually including new information with the given (shared) information in such a way as to help the reader understand the new ideas (Young, Becker, and Pike, 1970; Clark and Haviland, 1977; Weissberg, 1984). Joseph Williams explains the given/new strategy like this:

> Put at the beginning of a sentence ideas that you have a/ready stated, referred to, implied; ideas that you can safely assume your reader is familiar with, will readily recognize. . . . Put at the end of your sentence the newest, most surprising, the most significant information, information you want to stress, perhaps the information that you will expand on in the following sentence. (40)

This given/new approach works to build bridges with your non-technical reader when you begin by addressing the values or background knowledge of your reader's community. As you add new information and explain it for your reader, it becomes shared information. You can then continue building incrementally on this additional shared information to move your reader's understanding of your technical topic from a little knowledge to enough knowledge to understand your subject matter.

By analyzing the intended reader's community culture, then choosing an interesting appeal and building on the reader's background knowledge, a technical writer can effectively present even complex information to non-technical readers.

References

Aristotle. 1991. *On Rhetoric*, George A. Kennedy, trans. New York: Oxford University Press.

Atlas, Marshall. 1979. 'Addressing An Audience: A Study of Expert-Novice Differences in Writing', *Technical Report No. 3*. Washington, DC: ERIC.

Bizzell, Patricia. 1982. 'Cognition, Convention, and Certainty: What We Need to Know About Writing', *Pre/Text* 3, 3: 213–43.

Bocchi, Joseph. 1988. 'The Collective Concept of Audience in Nonacademic Settings.' 39th Annual Convention of College Composition and Communication. St. Louis, MO, 17–19 March.

Clark, Herbert H., and Susan E. Haviland. 1977. 'Comprehension and the Given-New Contract', in Roy O. Freedle, ed., *Discourse Production and Comprehension*. Norwood, NJ: Ablex.

Corbett, Edward P.I. 1971. *Classical Rhetoric for the Modern Student*, 2nd ed. New York: Oxford University Press

D' Angelo, Frank J. 1986. 'Topoi and Form in

Composition', in Donald A. McQuade, ed., *The Territory of Language*. Carbondale, IL: Southern Illinois University Press.

Fahnestock, Jeanne. 1986. 'Accommodating Science: The Rhetorical Life of Scientific Facts', *Written Communication* 3, 3: 275–96.

Nystrand, Martin. 1986. *The Structure of Written Communication: Studies in Reciprocity between Writers and Readers*. New York: Harcourt Brace Jovanovich.

Warren, Thomas L. 1993. 'Three Approaches to Reader Analysis', *Technical Communication* 40, 1: 81–8.

Weissberg, Robert C. 1984. 'Given and New: Paragraph Development Models from Scientific English', *TESOL Quarterly* 18, 3: 485–99.

Williams, Joseph M. 1989. *Style: Ten Lessons in Clarity and Grace*, 3rd ed. New York: Harper Collins.

Young, Richard E., Alton L. Becker, and Kenneth L. Pike. 1970. *Rhetoric: Discovery and Change*. San Diego, CA: Harcourt Brace Jovanovich.

QUESTIONS FOR CRITICAL THOUGHT

1. According to Longo, the question of audience is 'of paramount importance' for technical writers. Why?
2. The question of audience comes up repeatedly in the contributions to this book; what does this frequency tell you about its importance?
3. What does Longo mean by 'determining who the reader is'? Do you actually have to know your reader personally to write effectively for that person?
4. What is the relationship between the audience and the rhetorical situation?
5. For Longo, the important feature of audience is community. What does she mean by community? What is its role in the identification of audience?
6. Longo treats some of the same issues discussed by Elbow in 'Three Tricky Relationships to an Audience'. Read his essay and consider what differences emerge in the ways each discusses audience. What are these differences, and why do you think the two authors have used such different approaches to the same topic? Are there any similarities in their observations?
7. Longo provides a series of questions designed to help the writer identify the reader, then offers advice on 'what to do' with the information developed from the questions. How appropriate is this approach in fulfilling her purpose?
8. Forbes argues in 'Getting the Story, Telling the Story: The Science of Narrative, the Narrative of Science' that good technical writing has a narrative flavour, that it tells a story. Is this true of Longo's passage? Is it 'good' writing? Why or why not?
9. How personal is Longo's article? How formal is it? Does it display an audible voice?
10. Longo's essay is all about audience. How effectively has she adapted to her own audience? Who do you think they are?

CHAPTER 28

Three Tricky Relationships to an Audience

Peter Elbow

Perhaps you've noticed there are two distinct *kinds* of difficulty in writing. One kind feels as though you are straining to lift a heavy load of bricks onto your shoulder or struggling to carry something unwieldy across a stream with only slippery stepping-stones to walk on or trying to thread a needle whose eye is almost invisibly small. Taxing or scary or frustrating but also clean, hard work.

But there's that other kind. You are trying to fight your way out from under a huge deflated silk balloon—layers and layers of light gauzy material which you can bat away, but they always just flop back again and no movement or exertion gets you any closer to the open air. Or you are lost in a dense fog with no sense of direction—or rather just enough sense of direction to realize you are going in circles. Or you are sinking slowly into swamp mud and every effort to crawl or swim gets you in deeper. Or you are trying to saw through a thick plank and the harder you try the tighter your saw gets stuck in the cut. It's this kind of difficulty that makes you feel helpless and angry and finally stops you.

I have learned that when my writing feels difficult in the first way, it is a sign that I am indeed wrestling with the difficulties of writing itself: figuring out my thoughts, working out the logic, finding language for what is just barely emergent in my mind, or finding the right approach for a difficult audience. But when I experience that second kind of difficulty, it means I haven't yet managed to get my teeth into the writing task itself. There is some mix-up. Often it is because I am going about my work in a self-defeating way—perhaps trying to edit my words carefully while I'm only just writing out my earliest tentative thinking. But often it's a mix-up about audience. This feeling of working at cross-purposes to my goal—this continual racing of the motor while the gears refuse to engage—often comes from being afraid of the audience or confused about who it is or mistaken about what I am trying to do to it.

There are certain relationships to an audience that are inherently tricky because at the same time that they make it hard to write well, they also keep you from realizing what is causing the difficulty. Here are the three that I have noticed as I watch myself and others struggle: writing when you are trying to persuade readers, when readers are compelling you to write, and when your writing is entirely uninvited.

Persuasion

'Can't you see how wrong you are?'

There is nothing tricky about those occasions when you can use what could be called straightforward persuasion. You can jump right in and give good information, argue with reason, and season the whole thing with good manners. For example:

- Your committee (company, neighbourhood, school) has to choose between three plans. You have been appointed to study and recommend one. Your report will go to people in an audience who have not made up their minds yet. Indeed they are really asking you to help them make up their minds. They don't want tricky tactics or emotional appeals, they just want the best information and arguments. Your task is similar if you want to persuade them but are not yourself on the committee.

- You are writing a letter to the newspaper to persuade readers to vote for a certain candidate or measure, but you are trying primarily to sway the undecided readers, not the enemy. (Some studies show that more people read letters to the editor than any other section of most newspapers.)
- You are writing a job application or applying for a scholarship. You know the reader has to give someone the job and is trying to find the best candidate and so will read your qualifications with interest. It's important in such pieces of writing not to be bashful, roundabout, defensive, or coy in telling your strengths. In a kind of neutral, disinterested, and succinct way, you must frankly brag. (If the reader has a huge stack of applications to read; he will probably make a lot of 60-second eliminations in order to cut the number of applications down to manageable proportions before reading them carefully. Therefore, you must summarize your best material in your opening paragraph or cover sheet—don't include there anything questionable that could be used to eliminate you.)

What makes these occasions for straightforward persuasion is that your readers are open to your words either because they have not made up their minds or because you have some kind of authority on the topic or because they need to make a decision and are therefore open to new information and arguments in the most reasonable and human way.

Before going on to tricky persuasion I suggest this one simple but deep strategy for straightforward persuasion. Try hard to find good arguments for your position, but then try even harder to find arguments to refute yours. Then figure out how to answer those refutations. That is, the doubting game or the dialectical process turns out to be a powerful way to generate good persuasion. The strength of your argument depends more than anything else on your willingness to be a smart lawyer for the opposition. The only problem with this strategy is that you sometimes discover your original position is wrong. But that's useful information, too.

What concerns me is this chapter, however, are tricky audience situations and, in this case, I am thinking about the many times when you are trying to persuade someone in a straightforward way but actually you are wasting your time.

- You are writing a letter to the newspaper to persuade readers about the certain bill or candidate or situation, but this time your position is a minority one. Perhaps you want to argue for an end to all armaments—or income taxes or welfare. Or perhaps you are writing about a polarized issue like some of the recent bottle-deposit bills and your are not satisfied just to write to the relatively few middle of the roaders with open minds. If your bill is to win, you've got to *change* some of your opponents' minds.
- You are trying to dissuade someone from dropping out of college or hitchhiking around the country or divorcing you. Or trying to persuade your reader to accept your decision to do one of those things.
- You are writing an article or pamphlet or leaflet to persuade workers at a nuclear plant that nuclear power is a bad thing; or to persuade intellectual undergraduate women that abortions should be illegal.

What makes these attempts at persuasion tricky is that you are addressing your words to people who have a stake in what you are trying to refute. You are caught in a bind. The more you try to persuade them, the more their stake in their view causes them to dig in their heels. For you to win they must define themselves as losers. You can't argue without making your readers into your enemy, and enemies can't be persuaded—only beaten. But you can't beat people with words—or at least not if they don't consent to be beaten—because of that brute fact about reading: words only work if they are inflated with human breath and it's the reader who has to do the blowing. Why should the enemy pedal if you are steering where he doesn't want to go? 'Let me come up to your tower and show you that you are stupid for opposing deposits on bottles,' but your reader has to haul you up in the hand-crank elevator. Why should he? 'Let me show you movies to prove you are a murderer for condoning abortion,' only the reader has to crank the generator to make electricity for your movies.

So what can you do? Trick them? Say 'I have a wonderful trip I want to show you, you'll love it,' and get them to pedal while you steer and then suddenly take a turn down the path they hate? Keep your destination secret? 'Have you ever thought about the fact that all men are mortal? Odd, isn't it? And perhaps you haven't ever looked at it this way before, but, you know, Socrates is a man. HA HA! GOTCHA! Socrates is mortal!'

If your readers have a stake in what you are arguing against, you cannot take straightforward persuasion as your goal. You must resist your impulse to change their beliefs. You have to set your sights much lower. The best you can hope for—and it is hoping for a great deal—is to get your readers just to understand your point of view even while not changing theirs in the slightest. If you can get readers actually to entertain or experience your position for just a moment, you have done a wonder, and your best chance of getting them to do so is not by asking them to believe or adopt your point of view at all.

In short, stop trying to persuade the enemy and settle for planting a seed. If you think about the way people actually do change their beliefs—which is rarely—it is usually a gradual process and depends on a seed lying dormant for awhile. Something has to get them to a position where they might say, 'Imagine that. He actually believes that stuff and he's not crazy. I never could imagine a sane person thinking the country could get along without an army. I always thought it was some kind of emotional hang-up—something odd said by people who have a thing about uniforms or guns or something. I didn't realize that there really were coherent arguments. Of course they are all wrong, deeply misguided arguments, but now I can see why they appeal. It's interesting to know what it's like for a person to actually see things that way.'

If you can get a reader to take your point of view for just that one conditional moment—to inflate your words with his breath—then future events will occasionally remind him of the experience. Contrary views are inherently intriguing. And if your position has any merit, your reader will begin—very gradually of course—to notice things that actually support it. For the first time, for example, he will begin to notice specific incidents when armies or armaments increase danger to his country rather than decrease it. A seed is the best you can hope for.

So how do you plant a seed? You do it by getting the person actually to see through your eyes. There are many ways of doing this, but I think they all depend on one essential inner act by you: seeing through his eyes. And it's not enough just to do it as an act of shrewd strategic analysis: 'Let's see what actually passes for thinking in the minds of those rednecks.' For them to experience your point of view even for a moment, they must let down their guard. You can't get them to do so unless you let down yours, too: actually experience *their* point of view for the inside, not just analyze it. Though persuading can employ the doubting game, planting a seed calls for the believing game.

What does this mean in practice? If you relinquish your effort to make readers change their beliefs and settle instead for trying to get them merely to entertain yours for a moment, and if you start with an honest attempt to see things through their eyes, you will find a whole range of specific ways to write your letter, article, or report—depending on your skills and temperament. You can trust your instincts once you understand your goal: somehow to persuade readers to work *with* you rather than against you in the job of breathing life into your words. For example, if I were writing a short article or leaflet to readers with a stake in what I'm trying to refute, I wouldn't say, 'Here's why you should believe nuclear power is bad.' How can I get them to invest themselves in words which translate 'Here's why you've been bad or stupid'? I would take an approach which said, 'Here are the reasons and experiences that have made me believe nuclear power is bad. Please try to understand them for a moment.'

There are various ways to try to get readers to work with you. Your best choice depends upon your temperament and the circumstances. But if you are trying to change deeply held beliefs, autobiography, biography, and fiction turn out to be among the most effective types of writing. After all, changing a belief requires having an experience, not just getting some information or logic, and it's not surprising if imaginative and experiential writing sometimes prove more effective than argument.[1]

It's no accident that people so often use arguments on the enemy that only work on allies. Most of the things that feel like good arguments only work on people who agree with you or are at least open-minded. It's all too easy, as you are writing along in your room, to start hammering home arguments which prove resoundingly that that the enemy is *wrong*! These feel like good arguments because of a mix-up about audience. We have let ourselves forget the real audience and started to write a speech about the evils of nuclear power that is just perfect for people who already believe nuclear power is evil. It would bring down the house at an anti-nuclear rally. But unfortunately it will make no headway at all on someone who doesn't already agree.

So what works on opponents? There is no simple answer. You need feedback to find out. Very few people get accurate honest feedback from an opponent as to how their arguments are working—feedback that says, 'Here's what it felt like being your opponent and reading your words. Here are the places where you actually made a dent on me, made me listen, made me actually consider your words seriously, and here are the places where you just made me dig in my heels all the harder against you.' The only occasion when we are likely to get sincere, thoughtful feedback from an opponent is when we write something for a teacher who happens to disagree. But teachers usually don't give you 'here's-what-it-felt-to-be-your-enemy' feedback. Usually they try to extricate themselves from combat and give you more theoretical feedback on exactly those techniques of persuasion that won't work here because they only work on disinterested readers with no stake in the issue.

What you need then more than anything else is feedback from opponents. It's not easy to get, but it's possible. Find a friend who is an opponent on your issue and coax him to give you honest feedback. To get a helpful opponent you may have to ask a favour of a friend's friend. And if you can't make a friendly contact with someone who disagrees strongly with you on the issue you are writing about (shouldn't that be cause for concern?) you can practice on other topics where you and your friends actually disagree.

SUMMARY AND ADVICE

- For any persuasive writing, take time to think carefully about your relationship to your audience and what you are asking of it. Can you really hope to make those people *agree with you* or should you settle more realistically for just trying to get them to *listen to you*? Have they made up their minds yet? If so, how much stake do they have in the view you want them to abandon? Do they have any special reasons to listen to you? Is there some authority you have which they will accept? Is there some new decision or action they must perform that might make them willing to consider new information and arguments? In short, are you trying to persuade or to plant a seed?

- How much do you have at stake in the issue? If you are arguing for one of your important beliefs, you will probably have an almost irrepressible urge to make readers agree with you—an urge that may destroy any chance of success.
- Get accurate feedback—especially from the enemy. Find readers who will tell you honestly what their position was before they read your piece, what happened to them as they read, and what changes, if any, were finally produced in their views. It's often discouraging feedback because words seldom produce change of position, but if you are trying to persuade, perhaps the most useful thing you can learn is how seldom it is possible.
- There's one more strategy that does wonders whether you are trying to get someone to agree or just to listen: be right. If you're right you can sometimes succeed even though your writing has serious weaknesses. Reality helps you make your case. (It's not foolproof, of course, since sometimes being right makes you so insufferable that people are willing to stay wrong just for a chance of disagree with you.) It sometimes helps you to define you task of persuasion as part of a larger task of finding out the truth.
- Whether you are trying to persuade an open-minded reader to agree with what you are saying or trying to get an enemy reader simply to experience what you are saying, there is one essential thing you must learn: How to enter wholeheartedly into the skin of your readers and see or argue as they would.

Compulsory Writing

'I think I'll just hold this gun to your head till you finish.'

Much of the writing we do is compulsory. It starts in school and continues on the job. Writing an important thank-you letter as an adult can feel just as compulsory as when your mother sat you down and forced you to write a letter to Grandma for a birthday present you didn't like. If you write at all as an adult, it's probably because you have learned to be stoical and resigned about compulsory writing. 'I wish I didn't have to write this thing this weekend. I'd like to be outdoors. Still, that's the way it goes, this is always happening to me.' But as you work on the writing, you have a particularly hard time. You take all weekend and don't finish till late Sunday night. And all the while you tend to say to yourself, 'I'm so *bad* at writing. I wish I had *skill* in writing.'

It is hard for you to see that you ruined your whole weekend needlessly. You could have gotten the job done in half the time, in fact you could have gotten it done at work before the weekend even started. You think your weekend was ruined by your difficulty in writing but what ruined it was your difficulty in dealing with compulsory tasks. You were so busy complaining about how bad a writer you are, you didn't remember the times when writing went much better. You may not have had many good writing experiences—but then you may not often write without a gun at your head.

Or perhaps you aren't so stoical. You get so furious that you fume and stamp your feet and bang your fist all weekend. And yet you may not realize how much that impedes your writing. That blankness in your mind when you try to think of ideas, that difficulty you have in just letting yourself write down sentences at all, that pressure in your head when you try to organize what little you have to say: you tend to experience these as lack of intelligence or lack of skill in writing when really they come from your inability to deal with compulsory tasks.

I don't mean to imply that this analysis makes things easy. Solving the problem for your reactions to compulsory tasks is probably harder than learning how to write well. But at least there is hope of progress if you can tell which one is holding you back—if you can feel the difference between trying to saw through that plank with your own imperfect saw and trying to saw through that plank when your own efforts are binding the saw. If you persist in thinking your only problem is a writing problem, you block progress on both fronts.

If you have to do a piece of compulsory writing it helps to face the central issue squarely: are you going to consent or refuse? To consent is not necessarily to cave in. You don't have to like the task or the taskmaster, you don't have to grovel, but if you want the writing to go well, you have to invest yourself in the job whole-heartedly. If this is hard for you, it is probably because it *feels* like grovelling or caving in. You may not be able to put your full strength into the job—to consent—unless you feel you feel you *could* refuse. And this is a matter of power. It feels as though 'they' have all the power. It is true that they have authority and therefore they probably have sanctions. They can fire you or flunk you. Or hate you. But the final power is yours. You are in charge of whether you consent or refuse. What feels compulsory is not compulsory. Even people 'compelled' with actual guns have sometimes insisted on their power to refuse. I am thinking of the successful nonviolent resistance by Norwegian school teachers during World War II.[2]

Does it help, you may well ask, to portray your harassed supervisor or your bumbling teacher as a sadistic TV Nazi pointing a gun at your head, when what you are trying to learn is to consent (when appropriate) to compulsory task? But if you can feel, underneath your alleged difficulty in writing, your older feelings left over from the many times 'they' twisted your ear or somehow compelled you to give in, you will have much better luck in stepping beyond those past feelings and getting this present job done quickly. (Those TV movies with Nazis wouldn't have such appeal if they weren't really about the universal childhood experience of being helpless before superior power.)

But you may not believe in your power to refuse unless you really use it—openly and with full responsibility (instead of fooling yourself into being sick or having an emergency or 'trying as hard as you can' and somehow not succeeding). Perhaps refusing is not the ideal solution, but it's better than that familiar worst-of-both-worlds compromise: you don't get the fun of saying No or the satisfaction of doing the job quickly with investment. All you get is a ruined weekend and a sense of powerlessness.

SUMMARY AND ADVICE

- Figure out whether the writing is compulsory. Is someone else really demanding it? If not, it's not compulsory. If so, it's not still compulsory: you can refuse.
- Are you sure the price of refusal is too high? Will you really be fired? Are you sure you want that job? Will they hate you for life? Are you sure you care? It is easy to assume the world will come to an end if you say no.
- If you finally decide to consent—if you decide it's not worth whatever the price is just to get out of doing this piece of writing—then consent! Do the job wholeheartedly without fighting it. You don't have to love the job to invest your best efforts in getting it done quickly and getting some pleasure from it.
- If angry resentful feelings hold back your writing, stop, recognize those feelings for what they are, scream them out or write them down for ten minutes, and then get back to your job. Insist on your power to write efficiently.
- But don't forget the advantages of compulsory writing. Sometimes you learn things because

people 'make' you. Children seem to be aided in learning self-control by internalizing the control exercised over them by others. When you sign up for a writing course, what you may well be doing is simply paying someone to make you write every week. You realize you cannot yet get yourself to write every week, but you are willing to pretend the teacher can make you do it. There's nothing wrong with putting that make-believe gun into his hand if it will help you learn faster. But, remember, it's make-believe.

Uninvited Writing

'Pssst! Hey Mack. You wanna buy my novel?'

What a relief, then, to write, not because someone is demanding it, but because you want to. Even if it is a tricky letter, even if it is a piece of persuasion that will be hard for you because you lack the professional training you need, or even if it is a novel you know will keep you in the woods for years; still it gives enormous satisfaction to feel that you have made the decision to expend your time and effort this way. You know you will have frustrations, but you want to write this thing and so you find it easy, comparatively speaking, to put up with them. The main psychological fact about uninvited writing is that you naturally invest yourself in the writing task.

Or do you? For if uninvited writing always goes so well, how come everyone doesn't do it? Part of the problem may be that most people are introduced to writing in school where it is compulsory. 'Who would even write if they didn't have to?' But, in addition, uninvited writing has a built in difficulty of its own. It takes arrogance, *chutzpah*, *hubris*. 'Uninvited writing' is just another way of saying 'no audience'. You have to walk up to strangers on the street and tap them on the shoulder and say, 'Excuse me, would you please stop what you are doing and listen to me for a few hours? I have something I'd awfully much like to tell you.' You know the reply you will get.

Why engage in uninvited writing if you have to put up with that? And so most of us don't. Which would be fine except for one small fact: we *do* have things we want to tell people even if they haven't invited us to do so. But there is another fact. We are all capable of stopping people on the street and fixing them with our eye and getting them to listen and making them glad they did. We are, that is, capable of writing things which make readers want to read and glad they did. We just have to *do* it, and probably put up with a lot of rude refusals for a while. But we can insist on being heard.

Insisting on being heard. I remember the particular moment when I saw clearly how essential that feeling is for all writing, but especially for uninvited writing. I hadn't yet, I think, published anything—and no one had asked me to write this piece I was struggling with, but I was trying to say some things in it that were very important to me about teaching and learning. I had already managed to get down on paper in one form or another a lot of what I wanted to say. (In other words, my fear of tapping strangers on the shoulder wasn't so overwhelming that I pretended I had *nothing* to tell the world.) But the writing was going terribly. The whole thing was a mess, and no matter how hard I tried I couldn't seem to get things clear. And then finally things went better. I stopped to reflect on what had happened, and I wrote a note to myself (shortened and cleaned up here):

> 6/11/71. I'm correcting a near-to-final draft. Finally I'm making it much clearer and better. I'm rearranging sentences and points so they finally work. I had it all screwed up—my interpretation all mixed in with my information in an ineffective way—and my information unclear. Then a series of rearrangings made things fall into place with a click.

So what made this possible? It can't be any new knowledge about logic or sentence or rhetoric. I was already trying as hard as I could to use all of that knowledge I had. I was struggling over and over again—writing and rewriting, arranging and rearranging—and it was still mud. It didn't work. All my best knowledge didn't help.

But finally I can see what did help. It was the feeling 'Damn it, I've got to be done with this thing and I know goddamn well most people won't really hear it and thus they won't accept what I'm saying—it will all roll off their backs—even if they read it, which they probably won't do because it is such a mess—but if they do they will think it's just a fuzzy harebrained scheme of Elbow's. I'm tired of that. I'm not willing for that to go on any longer.' In short, what made the difference was a *decision* I made about my stance toward the reader. That inner act of readjusting my transaction with readers *caused* the words and ideas finally to come out in a different and better order.

It was like my readjustment to my lecture audience where I got mad at students saying they couldn't hear me and I moved from behind the podium to the front of the stage. A combination of frustration and anger made me finally insist on being heard and this made me suddenly able to do something with language I hadn't been able to do till then.

The essential question for writing, then, is this: how long are you willing to be unheard?

It would be impossible to avoid all compulsory writing and sad to run away from all uninvited writing. But having a gun at your head and having to go out and tap strangers on the shoulder are not your only ways of relating to the audience. Readers can invite you: call you up and say 'Will you come out to dinner with me? I'll pay if you will tell me about your trip.' Or 'It's on me if you'll tell me your thinking about the project you did last year. I have one now just like it.' What better way to make you enjoy communication and to bring out the best thinking. An audience that invited your words but doesn't demand them acts like suction.

Ten years ago I had only a vague sense that I might write a book. It was sort of a fantasy that I didn't take seriously. But when a publisher's representative knocked on my door to show me books for the courses I was teaching and asked at the end whether I had any writing projects in mind, and when he said that his editor might like to talk to me about my idea, and when after some negotiation the editor was willing to offer me a contract, suddenly I started to take the idea seriously. Because someone was willing to publish me, I started to have more ideas and, more important, I started to write them down like mad.

If you want to see the vivid effect of an inviting audience, think back to occasions when people wanted your thinking or advice about something you'd never thought about. At first you had nothing to say but the fact of their asking probably put things in your head.

Writing's greatest reward, for most of us anyway, is the sense of reaching an audience. Ideally, the audience should love what we write, but in the last analysis, it's enough if we can feel them reading. The fisherman falls in love with fishing because of that unpredictable wiggle, that moving pressure on his hand, even if the fish gets away. At least you felt them tasting your bait, at least you made contact with someone on the other end of the line. This experience makes you want to pick up the pencil and try again. *This* time you'll hook them. But it is you who are hooked.

The usual way to get yourself invited to write something is by doing well under the two previous conditions: writing something uninvited or compulsory that's good enough to make them call you up and ask for something else. (A good reason to learn to deal with uninvited and compulsory writing.) It seems unfair. The rich get richer. The best racers get the best starting place. You don't get the delightful encouragement of an invitation till you have already had a success.

But you don't have to wait for the invitation. Without having to muster all the courage it takes to stop strangers on the street, you can nevertheless find friends or make acquaintances who will want to read your words. In effect, publish: find an inviting audience, even if you have to copy out your writing in two copies or ditto it or pay for xeroxing; even if you have to start with friends who read it partly because they like you and care about what's in you head. Invite them over to read or listen, even if part of the incentive is a nice dinner or good refreshments. And you can find others who will want to read your writing because they want someone to read theirs. However you get it, a willing audience does wonders. It causes you suddenly to write more easily, to think of more, and get more satisfaction from writing.

Many people sabotage their hunger for an audience by sending off their stuff to highly competitive magazines or publishers who will almost certainly reject it. Too many rejection slips can make you so discouraged that you give up. Don't attempt large unknown audiences till you have made full use of a small known group of willing readers: connected with it, gotten pleasure from it, gotten feedback and learned to improve your writing on the basis of it. Only then are you in a good position to decide what to send off into the unknown and how much rejection you are will to put up with.

People also sometimes sabotage their instinct for finding a real audience by feeling they need to get permission from an expert before giving their writing to the real audience. If experts are the real audience for your writing, by all means give it first to them. But if, for example, you are writing up some important insights you've learned about how to be a better parent, you are likely to have the impulse to give your writing first to a psychologist or therapist or university professor in the field. You feel you need an expert to check out your words before they go to the real audience of parents. It's natural impulse. I've certainly acted on it numerous times. We seek someone with authority to tell us if we are right or to give us suggestions. Most of all, we seek a midwife to usher our child into the world.

But watch out. Checking your writing with an expert often turns out to kill the whole project. First you have to find the right expert. That can be a problem. Then the expert may not respond. Experts are busy. Even if they respond positively, their response may actually stymie you: 'This is very interesting. I think you should read Smith and Jones, oh yes, and Abernathy'—just three people to him but a year's reading or more for you—and if you do start reading, you are liable to conclude, 'Oh dear, I have nothing really new to say,' or 'Oh dear, there's so much I don't know about this field, I can't write till I master it.' And your project withers and dies even though you have already written a piece with lots of good insights—a piece that might in fact be more useful to real parents than Jones or Smith or Abernathy if only you get a little feedback from parents and do a little revising.

And, of course, the expert may discourage you in a much more straightforward manner. Once I sent off an essay about learning that I was excited about to an expert I thought would see the genius in it and give me some good suggestions. I got a reply which said nothing more than 'I wish people wouldn't use the word "concept" unless they really understood what it meant.' But how could it be otherwise? The authority is tired of reading about child rearing. He's read too much already. He is not a willing audience for your words. At best he reads out of duty or as a favour. He will simply notice the differences between what you have written and what he believes to be the best writing in the world about the topic.

I paint a bleak picture. Of course it *can* work out well. The expert might give you just the encouragement you need—along with a few suggestions which are just right for helping you revise and give your writing to the audience. But I'm deeply suspicious of the impulse that makes so many people feel they must get clearance from readers for whom the words are not intended before giving them to readers for whom they are intended. Experts are experts because they know a lot, but the one thing they cannot tell you is what it is like to read your words as a non-expert—for example as a curious or baffled parent who has read very little about child-rearing.

'But what if my thinking is false,' you may say, 'and my advice about child care is wrong?' But if you were riding on a bus or talking to friends you would tell them what you have to say about child-rearing if they were curious to know. So why do you need permission now from an expert to do the same thing? To engage in the essential audience transaction in writing—directing words to people who are interested in what you are saying? Speaking would be a curious business if we felt we had to get permission from listeners who are not likely to want to hear our words before directing them to people who are likely to want to hear them.

You are in a good position to go to experts *after* you have road-tested your words—after you have seen what works in practice and what doesn't and then done some revising. At this point you will have a crucially different relationship to experts than if you sent it to them first. You won't be saying, 'Please sir, may I have permission to let this thing out into the world?' (as though your writing were a new drug that might turn out to be thalidomide). You will feel more like a colleague saying, 'Look, I've got something interesting here, something that works. I wonder if you would be willing to tell me where you agree and where you don't.'

SUMMARY AND ADVICE

- Don't wait for an invitation. You probably have writing you want to give the world, even if the world hasn't gotten around to asking you yet. Write it and give it to the world uninvited. Insist on being heard.
- But work things out so you also get invitations. Find a willing audience of real people who are interested in what you are writing about and who will actually enjoy reading it. If you start by sending your writing to magazines or publishers who are unlikely to take it or by trying to get experts to stop what they are doing and take you in hand, you are likely to snuff out your instinct to be heard.
- After you are getting the help and nourishment that comes from having a real audience, then make use of experts and try to expand your audience by wider publication.
- Look for writing situations that are half-way between invited and uninvited. For example, write letters to newspapers and magazines. They didn't specifically ask *you* for *your* thinking—they won't necessarily publish your letter—but they did ask for people like you and thinking like yours.

Notes

1. There is an interesting literary problem opening up: how can you write what could be called propaganda, but is honest and doesn't make the reader feel manipulated—in short, good literature? It no longer seems as self-evident to me as it once did that good literature and propaganda must be contradictory categories.
2. See 'Nonviolent Resistance and the nazis: The Case of Norway' in *The Quiet Battle*, Mulford Sibley, editor (Boston, 1963). Also the second section of Part III of *Conflict Regulation* by Paul Wehr (Boulder, Colorado, 1979).

QUESTIONS FOR CRITICAL THOUGHT

1. What is the effect of the analogies in the opening paragraph of Elbow's essay? Why does he use them?
2. What are the three relationships to an audience that Elbow describes? To what degree are they really descriptions of three rhetorical situations? What does this tell us about the relationship between audience and context?
3. Elbow uses the first person pronoun and speaks directly to the reader as 'you'. What is the effect of this strategy on the readability of the passage?
4. What do you make of Elbow's assertion that 'words only work if they are inflated with human breath'? Explain what he means.
5. Why does Elbow suggest that fiction is a very powerful means of persuasion? If he is right, does this add anything new to Cheryl Forbes's view (in 'Getting the Story, Telling the Story: The Science of Narrative, the Narrative of Science') that good science writing should be narrative?
6. Elbow argues that direct persuasion isn't always possible, particularly when the audience has a vested interest in the opposite view. At such times, he argues, the best you can hope for as a writer is to 'plant a seed' of understanding, to get the audience to see through your eyes. In order to do this, you must be able to see through theirs. To what extent is this an apt analysis of the case study described by Urquhart in 'Bridging Gaps, Engineering Audiences'?
7. According to Elbow, appeals to logic work only on disinterested readers who have no stake in the situation. How similar is this to Dombrowski's contention (in 'Can Ethics Be Technologized?') that facts by themselves cannot persuade?
8. In what sense, according to Elbow, can your own investment in your topic be an impediment to effective persuasion? Does he contradict himself when he later advises that, to write well you have to invest yourself wholeheartedly in what you're doing?
9. Why does Elbow take persuasion as the exemplar for his discussion of a writer's relationships to an audience?
10. Consider Elbow's advice about compulsory writing; to what extent is it applicable more broadly to life?
11. Does Elbow's essay provide any inkling about the function and popularity of blogging?

12. Why does Elbow advise against getting expert advice? Does he include expert advice on writing? If so, is he violating his own rule of thumb by having written the book from which this selection comes?
13. How readable is Elbow's passage? What qualities make it so?

CHAPTER 29

What's Practical About Technical Writing?

Carolyn R. Miller

Courses and programs in technical writing are both praised and damned for being 'practical'. Other writing courses are practical, to be sure: in general, practical rhetoric emphasizes that discourse is a means for pursuing a goal. Thus, freshman composition aims to help students be more effective as students, technical writing aims to help them be more effective as engineers or accountants or systems analysts, and the writing instruction that accompanies many literature courses aims to help them to be more effective as reader-critics. But since technical writing is singled out for being practical, it is worth considering what makes it so.

The Meaning of 'Practical'

Most immediately, the practical seems to be concerned with getting things done, with efficient and effective action. Furthermore, efficiency and effectiveness seem more important for some types of action than for others; that is, some actions themselves have practical aims (rather than aesthetic or ritual ones), actions concerned with the material necessities of making a living or managing a household. One can thus *be* practical (or impractical) *about* practical action. *Being* practical suggests a certain attitude or mode of learning, an efficiency (or goal directedness) that relies on rules proved through use rather than on theory, history, experience, or general appreciation. Practical rhetoric therefore seems to concern the instrumental aspect of discourse—its potential for getting things done—and at the same time to invite a how to, or handbook, method of instruction. Technical writing partakes of both these dimensions of practical rhetoric.

The rhetoric of the early Greeks also involved both dimensions. They emphasized that rhetoric was an art (or techne). This meant (to Aristotle, at least) that rhetoric was conceptualized and teachable (not a knack, as Plato had feared) but neither certain nor absolute (not a science, as Plato had hoped). Greek rhetoric thus initiated both a handbook tradition of instruction and a counterposed theoretical appreciation for the multiplicity of relations between means and ends.

Richard Bernstein has suggested that there are both 'low' and 'high' senses of 'practical', two senses that parallel the handbook and theoretical traditions of rhetoric. It is the low sense, Bernstein says, that calls to mind 'some mundane and bread-and-butter activity or character. The practical man is one who is not concerned with theory (even anti-theoretical or anti-intellectual), who knows how to get along in the rough and tumble of the world' (1971: x). The high sense, which derives from the Aristotelian concept of praxis and underlies modern philosophical pragmatism, concerns human conduct in those activities that maintain the life of the community. One of the many reasons for the discrepancy between these two senses of the practical highlights the dilemma of technical writing, which is usually called practical in the low sense (by both its friends and its enemies, incidentally). This reason has to do with the social structure of the Greek city-state, which permitted the free citizen to be concerned with the good of the polis without being much concerned with bread-and-butter activities. The reason, of course, is the institution of slavery. Manual labour and most commercial activity were performed by noncitizens—slaves, foreigners, women. These activities were 'preconditions' to the fulfillment of human potential in self-government, according to Nicholas Lobkowicz: 'One would almost be tempted to say that the Greeks considered all "prepolitical" activities prehuman and that only in the political life were they able to see a way of life which transcended the animal realm' (1967: 22). Technical writing, the rhetoric of 'the world of work', of commerce and production, is thus associated with what were low forms of practice from the beginning. In a world in which it is more dishonourable to own slaves than it is to work for a living, we might question whether this association should prevail

A Conceptual Contradiction

Before trying to suggest what it might mean to apply the higher sense of practical to technical writing, I want to indicate some difficulties in accepting the low sense uncritically, as many technical writing teachers have. These difficulties are revealed by a contradiction within the self-justifying discourse of technical writing pedagogy: the attempt to hold both that nonacademic rhetorical practices are inadequate (and therefore need improvement through instruction) and that they serve as authoritative models (and therefore define goals for instruction). We seem, that is, uncertain about where to locate norms, about whether the definition of 'good writing' is to be derived from academic knowledge or from nonacademic practices. Most teachers will recognize the contradiction in the familiar dilemma of having to admit to students the discrepancy between practices that are supposed to be effective and those that are actually preferred and accepted.

The first side of the contradiction is the familiar justification for teaching technical writing. We teach it because when students graduate and begin writing on the job, they do not do very well. In the technical writing textbook I use, the first chapter, 'Why Study Technical Communication?' documents the 'inadequate communication skills of many technical professionals' (Olsen and Huckin, 1983: 7). For example, it quotes a survey about recently graduated civil engineers showing that writing and speaking are the areas of competence most important to civil-engineering practice but that about two-thirds of recent graduates are judged 'inferior' in these areas; results for mechanical and electrical engineers are similar. Complaints about technical writing from senior officials in science and industry include 'foggy language', failures of emphasis and coherence, illogical reasoning, poor organization—a familiar litany. Most technical writing textbooks begin with the same rationale, that nonacademic rhetorical practices are wanting. The justification for academic instruction is that academics know something that can help improve professional practices.

The second side of the contradiction derives from the research that interested faculty members have begun to do on rhetorical practices in business, industry, and science. This research is justified not only by the academic assumption that knowledge is a good thing but also (and often primarily) by the belief that knowledge of nonacademic practices is necessary to define goals for teaching practical rhetoric. As Paul Anderson puts it, 'We [educators] must first understand the profession, then design our curricula accordingly. Only if we understand intimately the job we intend to prepare our students to perform can we create effective professional programs' (Anderson, 1984b: 161).

One of the favourite research projects is the survey, which can show what kinds of work-related writing the population surveyed does, how important it seems to be, what its common problems are, and what qualities and features are valued. In reviewing selected surveys, Elizabeth Tebeaux notes discrepancies between instructional assumptions and industrial practices and concludes that 'several curricular changes are clearly mandated' in order to 'meet the communication needs of writers in industry' (1985: 422). Anderson reviewed 50 surveys, because they can provide 'teachers with important insights they can use as they design courses in business, technical and other forms of career-related writing' (Anderson, 1985: 4). Many surveys, such as those by Marcus Green and Timothy Nolan and by Bill Coggin, have been proffered as authoritative sources of information about what a curriculum should accomplish for its graduates. Ethnographic research has also been justified in instructional terms: according to Stephen Doheny-Farina, for example, 'By learning more about nonacademic contexts for writing, we are learning more about the kinds of rhetorical demands faced by many of our college graduates,' and this knowledge 'can inform the teaching of writing' (1986: 159).

Major national grants have gone to researchers engaged in work justified in these same ways, a clue to the institutionalization of this line of reasoning, as well as to its extension from technical writing to composition in general. The Fund for Improvement of Post-Secondary Education (FIPSE) sponsored a project on writing-program evaluation at the University of Texas; the project produced a report saying that 'before any college writing program can be judged effective or ineffective, we must know first if what it teaches has value to its graduates in later life. Like any educational program, the overall effectiveness of writing programs must be judged according to the needs of the population they serve' (Faigley, et al., 1981: 1–2). Another FIPSE grant went to Wayne State for a university-industry collaborative effort on research and curriculum development in professional writing. The researchers present cooperation between academics and practitioners as the way to 'ensure that students are prepared for the diverse communication tasks outside the university' (Couture, et al., 1985: 392–3). FIPSE has also sponsored research on collaborate writing in the workplace by Lisa Ede and Andrea Lunsford, who cite as a major problem 'the dichotomy between current models and methods of teaching writing . . . and the actual writing situations students will face upon graduation'; this dichotomy results, in part, from 'our lack of detailed understanding about on-the-job writing' (Ede and Lunsford, 1985: 69). The National Institute of Education earlier sponsored work by Lee Odell and Dixie Goswami on writing in nonacademic settings; their study also suggests that our ability to teach writing will be 'enhanced' by more complete understanding of how people come to write successfully on the job (Odell and Goswami, 1984: 257).

Practice as Descriptive or Prescriptive

In its eagerness to be useful—to students and their future employers—technical writing has sought a basis in practice, a basis that is problematic. I do not mean to suggest that academics should keep themselves ignorant of nonacademic practices; indeed, much of the research I cited above has been extremely illuminating. But technical writing teachers and curriculum planners should take seriously the problem of how to think about practice. The problem leads one to the complex relation between description and prescription. Odell warns against mistaking one for the other: 'we must be careful not to confuse *what is* with *what ought* to *be* . . . We have scarcely begun to understand how organizational context relates to writing, and we have almost no information about which aspects of that relationship are helpful to writers and which are harmful' (1985: 278). Anderson also warns us about this mistake: in presenting a model of the technical writing profession for use in designating curricula, he cautions that the model 'represents an ideal. It is built around the *best* practices of the profession, not around common practice—or malpractice' (Anderson, 1985: 165). He gives as examples usability resting (not common but good) and readability formulas (common but bad). Neither Odell nor Anderson, however, gives us much help in understanding what is helpful and what is harmful, what is good practice and what is malpractice. Even David Dobrin's discussion of the contradictions involved in teaching to the standards of employers, although it recommends both curricular and corporate reform, relies finally on accepting practices of the workplace on their own terms; teachers should 'make people at work better able to deal with others' (1985: 159).

At this point, it is worth recalling an earlier (unfounded) study of writing in nonacademic settings, 'Writing, Out in the World,' a chapter of Richard Ohmann's *English in America* (1976). Ohmann avoids the contradiction of taking practice as both imperfect and authoritative by positing a wider perspective from which to make such judgments; he requires, as Odell and Anderson and Dobrin do not, a basis for evaluating a practice other than that of the practice itself. The nonacademic writing Ohmann examined is that of futurists and forecasters, of foreign-policy analysts, and of the government officials who wrote the memorandums we call 'The Pentagon Papers'. Ohmann sought to establish, not that academic writing is different from writing in the workplace, but that they are dangerously similar; he concludes that academic instruction in writing 'has helped, willy nilly, to teach the rhetoric of the bureaucrats and technicians' (1976: 205). He claims that the

> writing of the powerful and influential shares some characteristics with the required writing of their college-age sons and daughters; that these characteristics are fairly important to the style of thinking and planning that guides the most powerful country in the world; and that this style has some systematically dangerous features when it operates not in the classroom bur on the stages of history. (1976: 173)

A similar and more direct charge has been made recently by Susan Wells, who claims that the ideology of technical writing explicitly assents to its instrumental subordination to capital; the aim of the discipline as a whole is to become a more responsive tool' (1986: 247). Being useful is not necessarily good, according to these Marxist critics, but little in the discourse of technical writing allows for this conclusion or explores its consequences. Because the Marxist critique features practical activity as a central concept, it raises questions that are particularly germane to technical writing, questions about whose interests a practice serves and how we decide whose interests should be served.

Practice and Higher Education

The uneasy relation between nonacademic practice and academic instruction has been part of academic discussions about technical writing from their beginnings in the late nineteenth century, as Robert Connor's historical work has shown. Connors documents recurrent debates over whether practical or humanistic goals should prevail in technical writing courses (or, as they were commonly called, 'engineering English'), whether, that is, such study should prepare technical students for work or for leisure. Moreover, these debates reflect a larger debate in American higher education, about the appropriate relation between vocational preparation and cultural awareness. In mid-nineteenth century, this debate transformed the American college curriculum, according to the educational historian Frederick Rudolph, who points specifically to the Morrill

Act of 1862 and the founding of Cornell in 1866. The first president of Cornell, Andrew White, 'confronted all the choices that had been troubling college authorities: practical or classical studies, old professions or new vocations, pure or applied science, training for culture and character or for jobs' (1982: 117). White opted for pluralism, for providing many courses of study in preparation, for many kinds of lives: 'the Cornell curriculum . . . multiplied truth into truths, a limited few professions into an endless number of new self-respecting ways of moving into the middle class' (1982: 119). In a similar vein, Laurence Veysey's study of the emergence of the American university in the nineteenth century traces the development of 'utility' as a basis for education. During this period, according to Veysey, 'America was a scene of vocational ambition,' both in terms of individual aspirations and in terms of the desire for public service. At the same time, the notion of public service broadened to include practical and technical occupations, not just the gentlemanly occupations for which earlier education had been preparatory. 'Vocational training,' says Veysey, 'directly affected the undergraduate curriculum of the new university' (1963: 66).

Other commentators have emphasized that the relation between instruction and practice is part of a more general condition, the subsistence of higher education in a socioeconomic matrix. Clark Kerr, in *The Uses of the University*, says that 'the life of the universities for a thousand years has been tied into the recognized professions in the surrounding society, and the universities will continue to respond as new professions arise' (1966: 111). (This view, of course, implies that the classical curriculum served as preparation not for leisure but for the upper-class vocations of law, politics, and the ministry.) John Kenneth Galbraith has noted that 'it is the vanity of educators that they shape the educational system to their preferred image. They may not be without influence, but the decisive force is the economic system' (1971: 236). More specifically, in his critique of nonacademic writing, Ohmann comments that

> the constraints upon English from the rest of the university and especially from outside it are strong . . . . [T]he writers of the textbooks and the planners of courses . . . can hardly ignore what passes for intellectual currency in that part of the world where vital decisions are made or what kind of composition succeeds in the terms of that part of the world. (1976: 206)

Current enthusiasm for 'industry–university collaboration' in applied research and development is perhaps the most recent manifestation of this general and necessary relation. But there is also a repertoire of accepted mechanisms for channelling the relation-internships, advisory councils, certification of graduates, and procedures for justifying and accrediting programs. These mechanisms are used in educational programs for the established professions, like law, medicine, engineering, and teaching, as well as in several areas of practical rhetoric with relatively long curricular histories, like journalism and public relations. For the most part, the channels these mechanisms create are one-way: influence flows primarily from nonacademic practices to the academy. The gradient is reflected, in the language at the industry-university interface, which includes, on the one hand, 'demand', 'need', 'value', and, on the other, 'response', 'service', and 'utility'. My own university, a land-grant institution, provides a case in point. Its 'Mission Statement' declares that the university 'has responsibility for the academic, research, and public service programs in areas of primary importance to the State's economy'. University policies concerning proposals for new degree programs require statements concerning the proposed program's relation to the institutional mission, to student demand, and to 'manpower' needs in the state.

Teachers of technical writing have advocated applying the mechanisms of nonacademic influence to their new programs, using the same kinds of language. Internship programs should be adopted in technical communication programs, according to a recent review of literature, because they encourage students to relate their study of theory to practice, permit faculty members to 'keep in touch with' current practices, and enable employers 'to influence college programs' (Gloe, 1983: 18–19). Advisory councils are advocated because they 'integrate the endeavors of the two worlds [academic and business-industrial] directly and in a[n] . . . effective manner' (Brockmann, 1982: 137). (Certification has been discussed within the Society for Technical Communication, but there is insufficient consensus in the profession to arrive at standards [1980: 6]; accreditation is now being investigated by the society [1986].)

Such language echoes the discourse of other professional programs, programs that have provided precedents for technical communication.

Library science
It is widely believed and reported that a chasm of mutual ignorance and indifference separates librarians and library educators from one another . . . . All sectors of practice regularly and strongly express a desire for more influence over the content and character of professional education. (Clough and Galvin, 1984: 2)

Public relations
Practitioners and educators must act in concert to guide public relations in the direction of professionalism. (Commission on Graduate Studies in Public Relations, 1985: 5)

Information science
Lack of communication between the employers of information professionals and the institutions that educate and train them is one reason that educational institutions are not meeting needs and demands of the changing environment and new technologies. (Griffiths, 1985: abstract)

Business
MBA curricula must be reevaluated and, perhaps, restructured if they are to meet business expectations, and—from the point of view of business—if they are to better prepare students for the real world in which they will build their careers. (Jenkins and Reizenstein, 1984: 24)

Journalism
What training and preparation do radio and television journalists consider important for a career in their field? Answers . . . should contain valuable insights for the broadcast journalism educator. (Fisher, 1978: 140)

Training and Development
Training activities involve a wide variety of skills, abilities, knowledge, and information . . . . An interdisciplinary approach to T&D preparation is important, given the range of competencies required. (Reed, 1984: 11)

This discourse is infected by the assumptions that what is common practice is useful and what is useful is good. The good that is sought is the good of an existing industry or profession with existing structures and functions. For the most part, these are tied to private interests, and to the extent that educational programs are based on existing nonacademic practices, they perpetuate and strengthen those private interests—they do indeed make their faculties and their students 'more responsive tools'. As the minutes of one meeting of the advisory council co the School of Engineering at my university indicate, regular contact between the university and industry 'makes students more valuable to industry'.

Praxis and Techne

My discussion so far has relied on a set of related oppositions that pervade the discourse of higher education:

theory versus practice
academy versus industry
ivory tower versus marketplace
idle speculation versus vocationalism
inquiry versus action
gentleman-scholar versus technician-dupe
contemplation versus application
general versus particular
knowing-that versus knowing-how
science versus knack

In this form the oppositions are probably unresolvable, and the best we can hope for is Anderson's notion that they should form a 'creative tension' (1984a: 6).

Another approach is to suspect the worst: that a dichotomy so widespread must be (at least partly) false. And in fact, Aristotle's characterization of rhetoric as an art, rather than a science or a knack, cuts through these oppositions with a middle term—techne. As he defines it in the *Nicomachean Ethics*, 'a productive state that is truly reasoned' (1955: VI, iv), techne requires both particular and general knowledge, both knowing-how and knowing-that; techne is both applicable and conceptualized. Donald Schön's recent critique of professional education relies on the same middle term: it is 'art', he says, that professionals display in practice, and it is art that unifies theory and application in a process he calls 'reflection-in-action' (1983). Aristotle's *techne rhetorike*, or treatise or rhetorical art, joins theory and practice by deriving knowing how from knowing that, prescription from description. Although positivist philosophy claims that this derivation is fallacious ('you can't get "ought" from "is"'), one of the major insights of Marx, according to Bernstein, is to deny the positivist fallacy. Marx (as well as Aristotle) is able to derive from description of existing social practices the shape of human need and potential—which provide the basis for prescription.

But to understand Aristotle's *Rhetoric* only as a techne is to miss what Aristotle himself has to say about practice. Understood as techne, Aristotle's treatise would fall within the handbook tradition, as a set of instructions that helps one produce texts. Such a treatise would concern productive knowledge, or *episteme poietike*, one of three kinds of knowledge in Aristotle's system: theoretical (concerned with knowing for its own sake), practical (concerned with doing), and productive (concerned with making). According to George Kennedy, Aristotle does not make the connection between rhetoric and productive knowledge (as he does for poetics) but treats rhetoric as theoretical knowledge concerned with 'discovering' the available means of persuasion (1980: 63).

The remaining alternative—that Aristotelian rhetoric is practical, rather than theoretical or productive—has been argued by Richard McKeon (1952), and its implications have been explored by Eugene Garver (1985). To see rhetoric as practical, in Aristotle's system, is to emphasize action over knowledge or production; rhetoric becomes a form of conduct, like the related practical realms of ethics and politics, which are constant background presences in the *Rhetoric*. Aristotle distinguishes carefully in the *Nicomachean Ethics* between production and practice, poiesis and praxis: as distinct from 'science', or theoretical knowledge, both concern the variable, or that which can be other than it is; but they differ in that production 'aims at an end other than itself', the product, and practice aims at its own performance, at 'doing well'. The reasoning appropriate to production takes the form of techne, art or technique, and the reasoning appropriate to performance, or conduct, takes the form of *phronesis*, prudence; for Aristotle there can be no art, or technical knowledge, of conduct. Prudence is the reasoning that makes one 'capable of action in the sphere of human goods' (1955: 6, v). Like techne, prudential reasoning is situated to undermine the oppositions that plague discussions of professional education, for it necessarily concerns both universals and particulars: it applies knowledge of human goods to particular circumstances (1955: 6, vii; Garver, 1985: 645). Unlike techne, however, which is concerned with the useful (that is, with the quality of a product given a set of expectations for it), prudence is concerned with the good (that is, with the quality of the expectations themselves).

Aristotle's concept of praxis has also informed some recent thinking about human action. As the central concept in Marx, praxis highlights the way in which the human person 'is the result of his [or her] own work' (Bernstein, 1971: 39; see also Lobkowic, 1967: 418–20). Human belief structures and social relations are understood to be used in practical relations between human beings and objects. Schön's account of professional practice emphasizes the 'knowing inherent in intelligent action' (1983: 50). Moreover, practices, as Alasdair MacIntyre has insisted, create not only knowledge but their own goods, and because practices are necessarily social, these goods require 'subordinating ourselves within the practice in our relationship to other practitioners' (1984: 191). The insights for the academic are that practice creates both knowledge and value and that the value created comprehends the good of the community in which the practice has a history.

Understanding practical rhetoric as a matter of *conduct* rather than of production, as a matter of arguing in a prudent way toward the good of the community rather than of constructing texts, should provide some new perspectives for teachers of technical writing and developers of courses and programs in technical communication. For example, it provides a reasonable basis for the necessary combination of academic and nonacademic contributions to curriculum. If praxis creates knowledge, academics should indeed know about nonacademic practices. But the academy does not have to be just a receptacle for practices and knowledge created elsewhere. The academy itself is also a set of practices, including those of observation, conceptualization, and instruction—practices that create their own kind of knowledge. Such knowledge allows the academy to provide a standpoint for inquiry into and criticism of nonacademic practices. We ought not, in other words, simply design our courses and curricula to replicate existing practices, taking them for granted and seeking to make them more efficient on their own terms, making our students 'more valuable to industry'; we ought instead to question those practices and encourage our students to do so too. Wells's 'pedagogy for technical writing' suggests that we should aim 'to work within the structures of technical discourse so that students can negotiate their demands but also be aware of the limited but real possibility of moving beyond them' (1986: 264). My own earlier sketch of a new pedagogy similarly suggested the need to promote both competence and critical awareness of the implications of competence (Miller, 1979: 617). I might now supplement critical awareness with prudential judgment, the ability (and willingness) to take socially responsible action, including symbolic action.

An understanding of practical rhetoric as conduct provides what a techne cannot: a locus for questioning, for criticism, for distinguishing good practice from bad. That locus is not the individual or any particular set of private interests but the human community that is created through conduct; this community is the basis for practice in Bernstein's 'high' sense. While the good that praxis in this higher sense creates may include the interests of individuals and industry, it is larger and more complex; the relevant community is not the working group or the corporation but the larger community within which the corporation sells its products, pays taxes, hires employees, lobbies, issues stock, files lawsuits, and is itself held accountable to the law.

Through praxis we make ourselves and each other in interaction: Aristotle emphasizes the political dimension of this interaction, Marx the economic. But whether our everyday activities are primarily those of governing a community or those of making a living, they have both political and economic dimensions. If technical writing is the rhetoric of 'the world of work', it is the rhetoric of contemporary praxis. In teaching such rhetoric, then, we acquire a measure of responsibility for political and economic conduct.

References

Anderson, Paul V. 1984a. '"Introduction" to Special Issue on Education', *Technical Communication* 31: 4–8.

———. 1984b. 'What Technical and Scientific Communicators Do: A Comprehensive Model for Developing Academic Programs', *IEEE Transactions on Professional Communication* PC-27 (September): 161–7.

———. 1985. 'What Survey Research Tells about Writing at Work', in Lee Odell and Dixie Goswami, eds, *Writing in Nonacademic Settings*, pp. 3–83. New York: Guilford.

Aristotle. 1932. *Rhetoric*, Lane Cooper, trans. Englewood Cliffs, NJ: Prentice-Hall.

———. 1955. *Nichomachean Ethics*, J.A.K. Thompson, trans. New York: Penguin.

Bernstein, Richard J. 1971. *Praxis and Action*. Philadelphia, PA: University of Pennsylvania Press.

Brockmann, R. John. 1982. 'Advisory Boards in Technical Communication Programs and Classes', *Technical Writing Teacher* 9 (Spring): 137–46.

Clough, M. Evalyn and Thomas J. Galvin. 1984. 'Educating Special Librarians: Toward a Meaningful Practitioner-Educator Dialogue', *Special Libraries* 75 (January): 1–8.

Coggin, Bill. 1980. 'Better Educational Programs for Students of Technical Communication', *Technical Communication* 27: 13–17.

Commission on Graduate Studies in Public Relations. 1985. *Advancing Public Relations Education: Recommended Curriculum for Graduate Public Relations Education*. New York: Foundation for Public Relations Research and Education, inc.

Connors, Robert H. 1982. 'The Rise of Technical Writing Instruction in America', *Journal of Technical Writing and Communication* 12: 329–52.

Couture, Barbara, Jone Rymer Goldstein, Elizabeth L. Malone, Barbara Nelson, and Sharon Quiroz. 1985. 'Building a Professional Writing Program through a University/Industry Collaborative', in Lee Odell and Dixie Goswami, eds, *Writing in Nonacademic Settings*, pp. 391–426. New York: Guilford.

Dobrin, David N. 1985. 'What's the Purpose of Teaching Technical Communication', *Technical Writing Teacher* 12: 146–60.

Doheny-Farina, Stephen. 1986. 'Writing in an Emerging Organization', *Written Communication* 3, 2: 158–85.

Ede, Lisa, and Andrea Lunsford. 1985. 'Research into Collaborative Writing', *Technical Communication* 32, 4: 69–70.

Faigley, Lester, Thomas P. Miller, Paul R. Meyer, and Stephen P. Witte. 1981. *Writing after College: A Stratified Survey of the Writing of College-Trained People*. Technical Report No. 1. FIPSE Grant No. G008005896.

Fisher, Harold A. 1978. 'Broadcast Journalists' Perceptions of Appropriate Career Preparation', *Journalism Quarterly* 55: 1940–4.

Galbraith, John Kenneth. 1971. *The New Industrial State*, 2nd ed. New York: New American Library.

Garver Eugene. 1985. 'Teaching Writing and Teaching Virtue', *Journal of Business Communication* 22: 51–73.

Gloe, Esther M. 1983. 'Setting up Internships in Technical Writing', *Journal of Technical Writing and Communication* 13: 7–27.

Green, Marcus and Timothy D. Nolan. 1984. 'A Systematic Analysis of the Technical Communicator's Job: A Guide for Educators', *Technical Communication* 31: 9–12.

Griffiths, J.M. 1985. 'Competency Requirements for Library and Information Science Professionals', *Information Science Abstracts* 20: 85–2017.

Jenkins, Roger L. and Richard C. Reizenstein. 1984. 'Insights into the MBA: Its Contents, Output and Relevance', *Selections* (Graduate Management Admissions Council) 1 (Spring): 19–24.

Kennedy, George A. 1980. *Classical Rhetoric and Its Christian and Secular Tradition from Ancient to Modern Times*. Chapel Hill, NC: University of North Carolina Press.

Kerr, Clark. 1966. *The Uses of the University*. New York: Harper.

Lobkowicz, Nicholas. 1967. *Theory and Practice: History of a Concept from Aristotle to Marx*. Notre Dame, IN: University of Notre Dame Press.

MacIntyre, Alasdair. 1984. *After Virtue*, 2nd ed. Notre Dame, IN: University of Notre Dame Press.

McKeon, Richard. 1952. 'Aristotle's Conception of Language and the Arts of Language', in R.S. Crane, ed., *Critics and Criticism*, pp. 176–231. Chicago: University of Chicago Press.

Miller, Carolyn R. 1979. 'A Humanistic Rationale for Technical Writing', *College English* 40: 610–17.

Odell, Lee. 1985. 'Beyond the Text: Relations between Writing and Social Context', in Lee Odell and Dixie Goswami, eds, *Writing in Nonacademic Settings*, pp. 249–80. New York: Guilford.

Odell, Lee, and Dixie Goswami. 1984. 'Writing in a Nonacademic Setting', in Richard Beach and Lillian S. Bridwell, eds, *New Directions in Composition*, pp. 233–58. New York: Guilford.

Ohmann, Richard. 1976. *English in America*. New York: Oxford University Press.

Olsen, Leslie A., and Thomas N. Huckin. 1983. *Principles of Communication for Science and Technology*. New York: McGraw-Hill.

Reed, Jeffrey G. 1984. 'Preparing the Training and Development Specialist: Skills and Knowledge Essential for Practice', *Journal of Business Education* 60: 8–13.

Rudolph, Frederick. 1977. *Curriculum: A History of the American Undergraduate Course of Study Since 1636*. San Francisco: Jossey-Bass.

Schön, Donald. 1983. *The Reflective Practitioner: How Professionals Think in Action*. New York: Basic.

Society for Technical Communication. 1986. *Strategic Plan, 1986–1990*. Washington: Society for Technical Communication.

———, Ad Hoc Committee on Certification. 1980. 'Certification of Technical Communicators', *Technical Communication* 27, 1: 4–6, 15.

Tebeaux, Elizabeth. 1985. 'Redesigning Professional Writing Courses to Meet the Communication Needs of Writers in Business and Industry', *College Composition and Communication* 36: 419–28.

Veysey, Laurence R. 1963. *The Emergence of the American University*. Chicago: University of Chicago Press.

Wells, Susan. 1986. 'Jurgen Habermas, Communicative Competence, and the Teaching of Technical Discourse', in Cary Nelson, ed., *Theory in the Classroom*, pp. 245–69. Champaign, IL: University of Illinois Press.

QUESTIONS FOR CRITICAL THOUGHT

1. What does it mean to call technical writing 'practical'? Why does Miller spend time defining such an ordinary word?
2. Miller refers to the 'self-justifying discourse of technical writing pedagogy'. Do any of the selections that you have read in this book fit this charge? Why or why not?
3. What is the most common rationale offered by the technical writing texts to justify the teaching of writing? Why does Miller object to it? To what extent does this book use the same rationale?
4. Why is it difficult to figure out what makes for good practice in technical writing?
5. In what sense does the teaching of writing entrench 'dangerous features' in the writing of professionals? Does Miller have a solution?
6. What are the oppositions that have informed Miller's discussion? Why does she list them explicitly?

7. What is the distinction between practical goals (that is, vocational preparedness) and humanistic ones (that is, cultural awareness) in the teaching of professional writing? Does Miller allow for any middle ground between these two oppositions?
8. What is the meaning of *techne*? *Phronesis*? *Praxis*? Why does Miller reach for such unfamiliar words to name her middle ground?
9. Like several other contributors to this volume, Miller offers Aristotelian rhetoric as a solution to the dilemmas facing those who teach and theorize about technical writing. Why does she see rhetoric as a means of resolving the dichotomy she has identified?
10. How readable is Miller's essay? What are the features that make it so (or not so)?

CHAPTER 30

These Minutes Took 22 Hours: The Rhetorical Situation of the Meeting Minute-Taker

David Ingham

With the obvious exception of the ubiquitous Inter-office Memo, the minutes of meetings arguably represent the most widespread form of professional and business writing, both in the private and the public sector: virtually all serious organizations (and some not-terribly-serious ones, for that matter) have meetings, if only Annual General Meetings, and except for the most casual of these, minutes are kept. Indeed, I've belonged to half a dozen organizations where the 'Minutes' files take up at least a linear foot of shelf or filing cabinet space. You will look in vain, however, at Communication textbooks (including [ahem] those by the editor of the present collection) for more than a casual mention of meeting minutes.[1] If so many get written, then, why has so little attention been paid to them?

For one thing, it is doubtful that any other form of writing has evoked such uninspired, even insipid prose, limping along primarily by means of the passive impersonal ('the point was made', 'it was decided', and so forth). For another, minutes aren't seen as particularly useful—as necessary, perhaps, and occasionally handy for checking exactly what motions were passed and when, but beyond that, little or nothing. Moreover, writing them is almost invariably seen as a chore (or even a form of punishment—one organization [which shall remain nameless] established a policy that the last member to arrive at the meeting was saddled with the task of keeping the minutes).

Why, then, have I gone on for two paragraphs about them—and had the effrontery to imply that they're worth spending an entire article over? Quite simply, they represent one of the most complex rhetorical situations imaginable.

The two fundamental aspects of any writing situation, surely, are the twin questions of *audience* and *purpose*. For whom are minutes written? For everyone who was at the meeting, obviously, but also for members who weren't at the meeting, and possibly others as well. Ah, yes, 'possibly others'. In fact, the potential audience is *everyone who could* ***conceivably*** *read them*. Future members (especially recording secretaries) of the organization—almost *everyone* higher up in the organization, including the President—the news media—lawyers looking for something on which to base a lawsuit—nasty government officials ('officious' by definition)—even forensic auditors (that is, accountants who are looking for evidence of financial wrongdoing to see if criminal charges can be laid)—all of these constitute the potential audience.

You think I'm joking? I know of one group—a sub-group of a much larger organization—whose members were hugely frustrated at the actions of the directors of that larger group. An overzealous minute-taker wrote that 'several members speculated whether [a particular action by the larger group] was due primarily to ignorance, or to malice.' Sure enough, those minutes were read by a Director of the larger group, who promptly threatened to sue for libel. (Note to potential litigants: the strongest possible defence to an action of libel is to prove the truth of the assertion; only slightly less strong is to prove that the writer had reasonable cause to believe in the truth of the assertion—in this particular case, the minute-taker, citing this jurisprudence, rather impudently said, 'Which do you want me to prove?' Cooler heads prevailed, the minutes were revised, and apologies issued for mistakenly circulating an inaccurate draft.)

If the audience can be huge and varied—not to mention the potential risks, as well—then the potential purposes of minutes are even more so. Presumably, minutes are intended primarily to be a permanent, accurate record of what transpired at a meeting. Even here, though, they can't help but fall short of the mark: in practical terms, they need to sum up the gist of discussions—some of which can be quite heated—and cannot help but take some sort of stance. And even quoting accurately doesn't always work: selective quotation can make a speaker look bad—has indeed been used deliberately for that purpose.[2] Ideally, minutes provide (as a former colleague of mine put it) 'letters to ourselves'—they form a sort of institutional memory bank.

This means that the minute-taker must consider the reaction of *everyone* who might read them. Are the minutes not only accurate, but also fair, collegial, and respectful? Would anyone be offended to read this? Do these minutes demonstrate that the members present were actively adhering to the mandated goals and objectives of the organization—not merely doing so, but *being seen* to be doing so?

This sort of 'positioning' (OK, in some cases it will slide into posturing) leads to another rhetorical locus: ethos. For the purposes of this discussion, this term underscores the fact that meeting minutes construct in the mind of the reader(s) an impression not only of the minute-taker, but also of every member present and even (less directly) of all members of the organization and of the organization itself.

As an aside, let me note that the best way to ensure circulation of minutes beyond a quick skim by the narrow circle of members who were present is to use a large running header: 'Confidential' in 16-point type or so. Furthermore, such a strategy can be used effectively for some indirect forms of communication, where direct communication might not work. For example, the 'stupidity or malice' remark noted above, while not entirely successful, did get the attention of someone in a position of power and communicated the smaller group's frustration and resentment, where several more direct forms of complaint had been ignored or brushed off. Being *overheard*, rather than heard directly, subtly changes the rhetorical situation; for a skilled minute-taker, this strategy can be used to considerable advantage. And as a last point in this digression, employing such a strategy frequently involves protecting members of the group: to say 'everyone at the meeting expressed anger [or frustration or resentment or whatever]' or to attach inflammatory opinions to specific members, can leave them open to reprisals. The strategy of saying 'several members present expressed' or even 'considerable frustration was expressed', while it lacks rhetorical incisiveness, can be in fact a conscious exercising of prudence.

I trust that the foregoing has offered convincing evidence that the humble minute-taker plays a key role in any organization, and that complex rhetorical considerations lie at the very heart of meeting minutes. At least two questions remain, however: can a verbatim transcription of all that was said at a meeting guarantee objectivity, and do minutes always have to be prudent, careful, and therefore dull to read (and write)?

The answer to the first—perhaps surprisingly—is 'no'. Besides being incredibly labour-intensive, such transcriptions are either edited, which represents a form of falsification, or they are utterly faithful to what was actually said, invariably making the speakers look inarticulate. Most of us don't speak in paragraphs, or even sentences, and our discourse is, um, full of, er, 'word-whiskers'—breaks in thought—full of repetitions—which in general do not take anything like the form of written discourse. The 'answer' is for the transcriber to make editorial decisions—to 'clean up' the speaker's inarticulateness. But exactly how far should this go? Isn't the audience *entitled* to see that a particular speaker was unusually inarticulate? How much editing is 'the right amount'? And doesn't a more-or-less verbatim account risk not catching a speaker's emphasis and intention—something a 'regular' minute-taker could capture? The best-known 'verbatim' transcript is the 'minutes' of the meetings of the Canadian House of Commons (*Hansard*). Even here, anyone who has watched a session and then read *Hansard* will be struck by the lack of congruity: where *Hansard* says, 'SOME HON. MEMBERS: Oh, oh,' the reality is that this moment was likely one of virtual chaos, with insults, cat-calls, and taunts shouted across the aisle. (One giveaway is that the next record is generally of the Speaker saying 'Order, please.')

As for the second question, the more formal the proceedings of an organization, the more formal (and probably tame) the minutes will (and should) be. Moreover, with so many rhetorical balls to keep in the air, the minute-taker can be excused for (pardon the mixed metaphor) keeping her or his head down. But life would be boring indeed if there weren't some opportunity to have fun with meeting minutes—and who wouldn't smile if the list of members present at a meeting of (say) a college English department included 'HRH Queen Elizabeth II' or 'Gov. Aaaaahnold Schwartzenegger'? One of the most effective ways of lightening minutes, though, is irony (subject to all the usual cavils about its use), from hyperbole to litotes—and it can be both one of the safest and most dangerous devices, since the degree of irony is rarely clear. But among colleagues and friends (and with no lawyers likely involved), how much better than the usual dry prose would it be to read something like the following [names omitted to protect the guilty and innocent alike]:

> Registration Committee: Prof.— , speaking for the Chairperson of the committee who, as secretary (for once in his life) was too preoccupied to utter a word, reminded members that registration had occurred, and would again. This fact being beyond the ability even of the Department to debate, the meeting went on to loftier concerns without moving or approving anything (to the visible chagrin of Profs. — and —).
>
> Other Business: [. . .] After this orgy of pleasant and praiseworthy sentiments, members cheerfully agreed at 3:00 pm to adjourn for the summer, leaving the impressed secretary the happy task of giving written articulation to the wise and thoughtful utterances that had filled Arts 213 that afternoon.
>
> Submitted,
> T. J. Matheson (Secretary *pro tem, ad hoc*, and *de facto*)

Long live the minute-taker, 'humble' no more!

Notes

1. For example, Guffey et al. discusses minutes only in a very brief section called 'Following up on Meetings', and offers one quotation, cited indirectly (1999: 517). The longest discussion of minutes I could find was in *Bourinot's Rules of Order*, though here most of the material is procedural, and the advice rather general: the minute-taker should aim for 'completeness, clarity and succinctness' (Stanford, 1995: 59).
2. One organization, in response to a complaint by a member who was in fact quoted accurately but wasn't happy about it, 'solved' the problem by passing a motion that minutes would in future contain only the wording of motions and whether carried or defeated, and not a record of discussions, leading to the absurdity of minutes that said only 'The meeting was called to order at 11:00 am' and 'The meeting was adjourned at 12:30 pm.'

References

Guffey, Mary Ellen, et al. 1999. *Business Communication: Process and Product*, 2nd ed. Toronto: Nelson.

Matheson, Terence J. 1990. 'Minutes' of the meeting of the Department of English, University of Saskatchewan, 11 April.

Stanford, Geoffrey. 1995. *Bourinot's Rules of Order*, 4th ed. (revised). Toronto: McClelland and Stewart.

QUESTIONS FOR CRITICAL THOUGHT

1. One of the qualities of good writing is liveliness or audibility—a quality that makes us believe we are listening to an actual human voice. Ingham's writing style has this quality in abundance. Can you find any specific devices or strategies he uses that create this effect?
2. In the corporate world a secretary or administrative assistant frequently takes minutes, but there are organizations and occasions in which any member of the group may be asked to take minutes. Is Ingham's passage intended as advice to those of us who need help with this occasional task? If not, what is it intended to do?
3. How would you characterize Ingham's tone and stance? Recall that tone refers to the attitude that the writing displays toward its audience, while stance reflects its attitude toward the subject. Is Ingham respectful of his audience and his subject matter? Why or why not?
4. Is Ingham correct that minute-taking represents 'one of the most complex rhetorical situations imaginable'? (Consider other standard forms of professional writing, such as reports, proposals, user manuals, or log books.) If he is incorrect about the relative complexity of minute-taking, why might he have made this assertion? If he is

correct that this activity is carried out in the most complex of rhetorical situations, why have other contributors not mentioned minutes at all?

5. Ingham notes that the minute-taker must consider the reactions of 'everyone who might read them'. Is this uniquely true of minutes? What, then, must the writer do about it?
6. Is editing really 'a form of falsification' as Ingham charges? How do you think Mary Ann Buehler ('Situational Editing') might respond to this assertion? What about William Zinsser ('Clutter')? If it is indeed so, could the same be said of writing? Why or why not?
7. Ingham writes about 'a sub-group of a much larger organization' for which the minutes of meetings became contentious. He observes that minutes might then be used to convey indirectly to the larger organization the feelings and views that the sub-group might be unable to convey directly. Does this view of organizational writing surprise you?
8. Do you think that Ingham has himself been a minute-taker? What makes you think so?
9. Why does Ingham use so many asides—comments in parentheses or set off by dashes—to break up his sentences? What is their rhetorical effect? Take another look at their content; what do they do to Ingham's tone? To his credibility?
10. Ingham speaks of strategies that might make the prose of minutes livelier and more readable. Among the devices he recommends is irony. What is irony? What are its dangers, especially when the audience is multiple? Why do Elbow ('Three Tricky Relationships to an Audience'), Buehler ('Situational Editing'), Campbell ('Ethos: Character and Ethics in Technical Writing'), and others who are concerned with the readability of writing not feature it as a strategy? Why does Ingham do so, then?
11. How personal is Ingham's style? How formal is it? Can you describe the voice that you are hearing in this essay?

PART VI

Language

LANGUAGE, that uniquely human ability to think abstractly and to communicate in symbols, is central to most of our definitions of what it means to be a human being; it is also part of what makes each individual unique. Language is also what provides most of our knowledge of the world we live in; although we often think otherwise, relatively little of what we know comes through our own direct observation or experience. Instead, we obtain most of our understanding of the world through the words of others. Even what we directly observe for ourselves is structured and organized for us by our language.

We use language to serve a variety of purposes, both practical and symbolic. It is one of the ways in which we accomplish the tasks necessary to living, and it is *the* way in which we obtain the cooperation of others in achieving our goals. Language is what allows human beings to organize into societies and cultures. Through its power of abstraction, language frees us from the here-and-now world, enabling us to learn the lessons of the past and dream of possibilities for the future. It is what has enabled advances in human understanding and technology.

However, because it is so powerful in shaping our understanding of the world, language is sometimes the source of struggles over who can say what, to whom, and under what circumstances. In fact, many of our contentious social issues are disputes over the meaning of words: remember the public arguments in the early 2000s over the meaning of the word 'marriage'? Americans argue over what is meant by the phrase 'the right to bear arms' in their constitution. Canadians struggle over what 'universal access' to health care really means. Legal battles, particularly at the supreme court level, often depend on the interpretation of a word in law. The struggle over the word 'engineer', detailed in the selection by John Speed, is one such instance.

The way we use language is also an indicator of how we reason and how we hope to influence the thinking of others. Several of our selections reveal the ways in which language can be used deliberately to deceive, and argue that sloppy language use is both a cause and an effect of sloppy thinking. Others demonstrate the process of argument by definition, one of the most common approaches to learned disputation. The goal of this section of the book, then, is to provide a taste of some of the debates over language, and some of the possibilities of its study, and to stimulate your thinking on this fascinating subject.

CHAPTER 31

The Language of Science: Its Simplicity, Beauty and Humour[1]

Anatol Rapoport

The most advanced branches of science, namely, the physical, have developed their own special languages, such as the mathematics of theoretical physics and the molecular structure symbolism of chemistry. Even in these highly formalized sciences there is a residue of 'ordinary' language; and, of course, many sciences are still written entirely in the general purpose idiom. My remarks will be largely concerned with this vernacular component of the language of science.

There are two widely held notions about the language of science: first, that the vocabulary of this language consists largely of words difficult to spell and to pronounce; second, that it is an extremely prosaic language, a language of factual reports, one that excludes both sentiment and imagery. These two notions are held by most non-scientists—both by those who hold the language of science in disdain and those who are awed by it. The anti-intellectuals, when they ridicule scientific vocabulary, often accuse the scientist of deliberately obscuring his language, either because of snobbism or exhibitionism. They say that the long, difficult expressions are often no more than needlessly complicated ways of saying simple things, such as can be frequently demonstrated in translations from Federalese and Legalese.

In rebuttal, the defenders of scientific terminology point out that the scientist must make distinctions that are of little consequence to the layman. For example, when the layman says 'grouse', the ornithologist must say *Lagopus scoticus* or *Lyrurus tetrix* or *Bonasa umbellus*, because all of these grouse do not even belong to the same genus, and the distinctions are important to the ornithologist. Since there are hundreds of thousands of known species of animals and plants, the vocabulary of the naturalist must be immense. To avoid further confusion, Latin has been adopted as the universal language of biological taxonomy. The necessity for the long and difficult vocabulary is thus demonstrated in this instance. The argument, however, is not immune to a counter-rebuttal, which is not without justification. Many examples of academic discourse can be cited where precision of meaning or fineness of distinction are not at all in evidence but an obscure vocabulary is very much so. However, we shall leave this argument at this point and turn to the other widespread notion about the language of science, namely, that it is dry and matter of fact.

Paul H. Oeser in his charming article entitled 'The Lion and the Lamb: An Essay on Science and Poetry' (1955: 89–96) offers this dramatic example of contrast between prose and poetry. The first is from Henry S. Canby's *Definitions: Second Series*:

> Tail nearly as long as to slightly longer than wing, more or less rounded, flat (not vaulted), the retrices relatively broad, with broadly rounded tips. Tarsus less than one-fourth to about one-third as long as wing, the acrotarsium with a single row of large transverse scutella, the planta tarsi usually with a single row of small scutella along outer side and smaller, irregular scutella on inner side; lateral toe reaching to or slightly beyond . . .

Contrasted with this is a stanza by Lew Sarett on the common loon (*Gavia immer*):

With mournful wail from dusk to dawn
He gibbered at the taunting stars—
A hermit-soul gone raving mad,
And beating at his bars.

Again we recall the familiar attacks upon and the defenses of the language of science. The attackers point to scientific descriptions as desiccative, as something that acts like a destroyer of spiritual meanings. I suppose these people find a similarity between the language of science and that perversion of the commercial mentality, which is aware of the price of everything and of the value of nothing.

And, of course, there are the standard defenses. We must separate knowledge from sentiment. Recall that compassion may comfort the sick, but it will never cure them as prosaic medical knowledge will. In politics, sentiment leads to demagogy, while dispassionate analysis encourages statesmanship, etc.

Here, too, both sides have a point. Unquestionably to gain and communicate knowledge a language of detachment is indispensable. But often a language which excludes affect also excludes insight, an important ingredient of all knowledge that matters.

My concern will not be with the issue of whether the often outlandish vocabulary and the graceless style of scientific writing is a good or a bad thing. Instead I will make a case against the assumptions made both by the foes and by many of the friends of the scientific attitude, namely, that the language of science necessarily depends on a large and esoteric vocabulary and that this language is necessarily devoid of poetry. I will argue that just in those areas of science where the richest and the most profound insights abound, the reverse is true: (1) vocabularies are small and consist of short, commonly used words; (2) the elements of poetry, such as symmetry of expression, figures of speech, and rich imagery abound.

I am referring to mathematics and physics, which certainly deserve to be viewed as the epitome of scientific achievement, not only in themselves but for their increasing influence upon the methodology of all natural science and increasingly even upon social science. The character of scientific language varies widely among the disciplines, but it is certainly defensible to take the language of mathematics and physics as important manifestations of the language of science.

The key terms in the vocabulary of modern higher algebra are words like *group*, *ring*, *field*, *ideal*, *trace*, *norm*, *normal*, *simple*. The vocabulary of analysis (another of the main branches of mathematics) depends heavily on words like *limit*, *function*, *converge*, *continuous*, *regular*, *analytic*, *pole*. In topology we have different kinds of *sets*, for example, *open*, *closed*, and *perfect*; different kinds of *spaces*, for example, *compact*, *ordered*, and *separable*. A term of crucial importance in topology is *neighbourhood*. Fundamental words in physics are *force*, *mass*, *work*, *power*, *action*, *energy*, *field*, *charge*, *current*, *potential*, *flux*, *heat*, *pressure*, *temperature*, *nucleus*, *particle*, *orbit*, *spin*.

Certainly none of these words appear to have been designed to impress or intimidate anyone. However, here is the rub. It is impossible to explain the meaning of a single one of the words in the mathematical list and of most of the words in the physics list to anyone who has not had literally years of certain kinds of experience.

As an example, there is no practical way even to begin to explain to a non-mathematician the meaning of the homely word *neighbourhood* as it is used in topology. The usual way to explain the meaning of a word is 'backwards', that is, in terms of other words believed to be more directly related to experience. If these are still not understood one continues the regression until words linked to experience are reached. But if such a process were started with *neighbourhood* in its topological sense, the chain of reduction would have to go on and on. Before commonly understood terms were reached, the explanation would become so long that its beginning would be forgotten.

There remains only the 'forward' type of explanation. That is, one starts with commonly understood mathematical terms and compounds their meanings into more complicated notions to which one then gives names. Then one compounds these notions into still more complex ones, etc., until the term to be defined is reached. This chain will, of course, be just as long as the 'backward' chain, but one can hope that in the forward process, the intervening concepts will be digested or, to change the metaphor, will be used as successive springboards. But this is nothing but training. At the end of the process, the trainee with his internalized compounded concepts will have become a mathematician. Therefore one must conclude that the meaning of terms like *neighbourhood* can be imparted only to mature mathematicians. Exactly the same situation exists with respect to the word *simple* as it is used in the theory of linear algebras, and with most of the other 'easy' words that make up the vocabulary of higher mathematics.

Thus it appears that in important areas of science, far from using complicated words to denote simple notions (as many believe) the language of science uses very simple words to denote exceedingly complex notions. Moreover, the compounding of notions makes large vocabularies unnecessary. The latter point is easy to illustrate in the language of physics. The logic of the language of physics is the logic of mathematical operation. In this respect it is quite unlike the logic to which our Aristotelian heritage has accustomed us, namely, the logic of taxonomy. In the latter, newly defined terms stand largely for intersections of classes, and classes are defined by syndromes of properties. Thus in classical logic horses are defined as quadrupeds with certain equine properties, and human beings as bipeds with certain anthropoid properties, etc.

Most of the terms in physics are defined not by classification but by mathematical operation. *Mass* is *force* divided by *acceleration*; *power* is *work* divided by *time*; *action* is *energy* multiplied by *time*, etc. To some one who thinks only in terms of classificatory logic these definitions must seem strange. True, we tend to become used to them, exposed as many of us are to elementary engineering and to mass production (who has not heard of 'man-hours'?). But just because we are becoming used to this sort of thinking we often fail to realize what a profound difference in semantics the operational language of exact science has brought about.

Think for a moment about *acceleration*. A car salesman tells you that the car he is selling will attain a speed of 60 miles per hour from dead start in 11 seconds. Its acceleration during this period is therefore 88 feet per second per 11 seconds, or eight feet per second per second. Instead of using the awkward expression 'per second per second', the physicist says 'per second squared.' In my school days I remember a bold and honest boy in the physics class who asked the teacher just what he meant by a 'square second'. The teacher's sarcastic reply indicated that he did not know.

It is, of course, foolish to try to evoke an 'image' of a square second. But that is just the point. The literal-minded are inevitably stymied, because they insist on thinking in terms of visual images and cannot get rid of a feeling that what cannot be visually imagined is not quite real. I am not referring to the dullards and to those who sacrifice imagination on the altar of so-called hard-headed realism. I am referring to the best minds in classical Greece, the first honest-to-goodness mathematicians in the Western World. They *could* think mathematically, but almost exclusively in terms of visualized geometry. To them a 'square' *was* a square, not the result of a mathematical operation. Therefore the cube of a quantity might still have made sense to them, because it could be represented as a real cube. But a fourth power would have appeared as unreal to the Greek geometer as the fourth dimension appears to the layman of today. Neither is translatable into a visual image.

Here, then, is the first tremendous conceptual innovation which we owe to the language of science. Science has freed the intellect from dependence on concrete visualizable conceptualizations—largely through introducing mathematical operation as a generator of concepts. It is this semantic device which makes a comparatively small vocabulary sufficient for physics. Indeed the technical glossary of the physicist is quite small, and the majority of terms in it are borrowed from common usage. The vast complexity of the language of physics stems from the richness of mathematical manipulation, which makes the coinage of new and complicated terms largely unnecessary. One might say that truly scientific terms are to common language words as the letters of the alphabet are to ideograms. The letters are simpler, more abstract, and endowed with far greater potential for combinations.

There are certainly other areas of thought where abstract unvisualizable concepts abound, for example, depth psychology, philosophy, and theology. These areas, however, are by no means characterized by simple vocabularies, as physics is. I believe this is because the discipline of strictly operational definition, the compounding of simple concepts into complex ones by clearly prescribed rules, is lacking in these areas. Consequently, meanings and distinctions are not clear and a great proliferation of terminology, even of private jargons, clutters the language of those areas. The esoteric languages of the speculative disciplines are, in a way, the very antithesis of the characteristic language of science.

I shall argue against the prevalent notion, namely, that the language of science is prosaic. Examples supporting this view, such as excerpts from handbooks and catalogues are well known. Examples to the contrary are hardly known at all. This is not surprising. If science has a language of its own, its poetry is also its own. This poetry speaks to the initiates, and, like most poetry and indigenous humour, it does not lend itself well to translation. Nevertheless I will try, at the risk of sacrilege, to translate some instances of scientific poetry and of scientific humour.

Poetry, I suppose, is a mixture of music, imagery, and metaphor. I will begin by giving an example of a scientific metaphor. First consider the notion of a *spectrum* as it is used in optics. Light, as is generally known, is pictured in physics as electromagnetic vibrations in a certain range of frequency. Monochromatic light consists of a single frequency, describable by a simple sine wave. Combinations of sine waves of different frequencies and amplitudes characterize the actual light of our experience. A beam of such light, passed through a prism, will decompose into the constituent monochromatic parts, i.e., into different pure colours. These colours (i.e., wave lengths) with their associated amplitudes make up the spectrum of the beam. Mathematically speaking, the spectrum is a table in which each of the frequencies of which a beam of light is composed is associated with its amplitude. To the eye, a spectrum appears as a gradation of colours in space. In particular, a rainbow is a spectrum. Here it would seem, the contrast between the prose of scientific language and poetry is particularly crass, for the rainbow is usually held to be a poetic object, and its reduction to a table of numbers is conventionally viewed as a debasement of esthetic value. But please follow me further.

It is shown mathematically that any periodic vibration is analyzable into constituent frequencies with associated amplitudes. Indeed, the frequencies of a strictly periodic process must all be whole number multiples of a single fundamental frequency. A musical tone is just such a process. Every musical tone, therefore, has a *spectrum*, perceived by the ear as the 'quality' of the tone, that which makes the difference between the tone of the oboe and that of a flute, or the difference between the violin tone of a beginner and that of a master. In every musical tone the frequencies are all simple multiples of a single fundamental frequency. The spectra of such tones (called discrete spectra) are therefore particularly simple. A noise, however, is not a strictly periodic process and so cannot be broken up into a series of evenly spaced frequencies. The spectrum of a noise is a continuous one, in which all the frequencies, not just the multiples of a fundamental, may be represented.

Now the spectrum of white light is also a continuous one, specifically one in which *all the frequencies have the same amplitude*. In view of all these analogies, one can justify the metaphor 'white noise', i.e., *a noise whose frequency spectrum contains all the frequencies with equal amplitudes*.

But time-dependent processes are not confined to light and sound. Electric current or the motion of an object can also be expressed as mathematical functions. If the mathematical functions are periodic, the associated spectra are discrete; otherwise they are continuous. In particular, *any* process whose frequency spectrum has the same mathematical form as the spectrum of white light can be christened 'white noise'—a term used in modern communication engineering, a truly interdisciplinary metaphor, whose parentage stems from optics and acoustics respectively, but whose general meaning now implies a precise *mathematical* definition of 'chaos'. It is 'chaos' because it can be shown that a white noise process is one where it is most difficult to guess on the basis of what has happened what is going to happen next!

Now you may well ask why this example is offered as an illustration of poetry in the language of science. To answer this question, let us recall that the effectiveness of a metaphor is gauged by two things: by the degree of superficial diversity and the degree of inherent similarity of the things identified. If a metaphor is extremely far-fetched but at the same time calls attention to a profound relatedness of two phenomena, it is a powerful metaphor. A poetic metaphor is supposed to impart insight—the insight of recognition. And one of the principal functions of poetry is to impart such insights, which poetry does largely through inspired uses of metaphor. I submit that the language of science does just this.

Science, we see, has its own peculiar method of metaphor construction. Like the metaphors of poetry, those of science stem from insights. But these insights, unlike those of poetry, are rooted not in personal or cultural perceptions of relatedness but in the actual structure of the world to the extent that this structure reflects itself in the mathematical description of reality.

One more example of this sort is instructive, because the metaphor it gives rise to is so far-fetched as to seem humorous. Several years ago workers concerned with the statistical characteristics of verbal output discovered striking regularities in the relationship between the frequency with which words occur in large samples of verbal output and the number of different words associated with each frequency. Thus the words which are used with the greatest frequency like 'the', 'of', and 'to' have the fewest representatives, while words which occur least frequently have the most representatives. To put it another way, of the words that occur frequently, there are few; of the words that occur rarely, there are many.

Bénoit Mandelbrot showed that these statistical regularities in the frequency of words are just what one would expect to find if one assumed that the users of language used it in such a way as to convey the most 'information' at a given fixed average 'cost' per word. Here 'information' is used in the technical sense of the communication engineer, and 'cost' is analogous to the energy expended in encoding, producing, or decoding a word. It was also shown by workers concerned with the mathematical theory of information (particularly by Claude E. Shannon and Norbert Wiener) that the communication engineer's measure of information is mathematically analogous to what is known as entropy in thermodynamics. The statistics of verbal output approaching its equilibrium (in the evolution of language) then becomes entirely analogous to the statistical behaviour of a physical system, say a gas, approaching its equilibrium. There too, the greatest possible entropy (analogous to information) is attained under the restraint of constant total energy (or average energy per molecule), i.e., the temperature of the gas. The mathematical isomorphism of the two mathematical theories being complete, it becomes possible to identify a parameter in the formula describing the frequency distribution of words in a given sample with the parameter describing the temperature of a gas which has reached equilibrium. Indeed, Mandelbrot makes such a metaphorical identification. He speaks of the 'temperature' of a language sample. It turns out that one of the 'hottest' examples of English prose is Joyce's *Ulysses*, while the language of children of schizophrenics turns out to be characteristically 'cool'!

I think these examples match some of the weirdest identifications to be found in poetry, and I suppose they serve a somewhat similar purpose. They are forms of play, but in the case of the mathematical metaphor, they are strictly disciplined play.

Nor does the language of science lack ordinary metaphors, whose function is simply to add to the vividness of discourse. In mathematics one finds an *osculating* curve, one which has very close-fitting contact with another. (*To osculate* means, of course, *to kiss*.) One finds in mathematics also *pathological* functions, i.e., relations among variables with bizarre paradoxical properties. The language of American electronic technology is full of slang with its *feedbacks*, *dashpots*, *choppers*, and *pip-trappers*. The engineer has also introduced the *black box*, the hidden inner arrangement of a system, which must be inferred by observing the relations between the inputs and the outputs. The psychologist has enthusiastically embraced this term, for it fits his own methodological problem. He observes the stimuli and the responses of an organism; what happens between them is shut in a *black box*, the organism's nervous system.

Threshold is a common term in neurophysiology. It has become so common that we forget its metaphorical origin. The physicist speaks of a *degenerate gas*, a metaphor borrowed from the mathematical metaphor *degenerate curve*, a curve characterized by a special value of a parameter which reduces it to a more primitive curve. The word *primitive* is, of course, also a metaphor, related to a hierarchy of complexity in mathematical expression. The biologist speaks of *mutation pressure* and of *selection pressure*, which are not pressures at all in the literal physical sense but which produce effects analogous to those of pressure, and of an *adaptation landscape*, over whose *hills* and *valleys* a population can be said to *wander* as its genetic *makeup* undergoes evolutionary changes. *Pay-off*, whose origin is possibly rooted in underworld jargon, is today a key term in the highly sophisticated mathematical theory of games.

As we pass to psychiatry, the metaphors of its technical discourse become, as is well known, so profuse that one wonders whether this language has not already crossed the boundary between science and poetry.

It is most difficult to speak of those esthetic qualities of the language of science which are most directly related to direct perception of beauty, difficult in the same sense as it is for a European to explain to a non-European wherein lies the haunting charm of a Schubert song. Some of the grandest generalizations of theoretical physics appear remarkably symmetrical when expressed in mathematical equations. For example, Hamilton's magnificent summary of the entire scope of Newtonian mechanics looks like this:

$$\frac{\delta H}{\delta q_j} = p_j \qquad \frac{\delta H}{\delta p_j} = -\dot{q}_j$$

As with the 'easy' words of advanced mathematics, it is useless to explain the meaning of the symbols. Just to understand H requires thorough familiarity with advanced theoretical physics. Therefore nothing is gained by identifying the p_j with the momentum of the jth particle of a system, the q_j with a generalized spatial coordinate, and the dot as a symbol of differentiation with respect of time. The words in the definition are equally meaningless to one not at home in the language of physics. However, an empathetic attitude may lead one at least to a faint understanding of the sort of feeling that a scientist might have in contemplating the almost perfectly symmetrical arrangement of the symbols on the page ('almost' because of the minus sign, which appears in the second equation only), while he ponders the fact that the knowledge accumulated over 24 centuries is concentrated in those symbols. Perhaps the Chinese philosopher, contemplating the exquisite calligraphic expression of an ancient aphorism on the printed page experiences a similar emotion.

In summary, I have tried to emphasize not the well-known aspects of the language of science, namely, its precision, its algorithms of deduction, its ideal of objectivity, but rather the characteristics not so widely appreciated: its pithiness (rather than verboseness of which it is often accused), and its esthetic qualities.

The semanticists often say that our language does our thinking for us. This is true, of course, of the language of science. Our language is also a source of erotic activity—I take erotic here in its generalized sense of playful creativeness. The expressive richness of dialects, slangs, and jargons attests to this. The language of science is by no means lacking in such elements. It even gives birth to its own special jokes. To appreciate these one needs a sense of humour rooted in an intimacy with the special situations and states of mind which occur in scientific activity.

I will now tell three jokes and will violate the ethics of humour by explaining them. The explanations will kill the jokes, but this cannot be helped. Vivisection is sometimes a necessary evil.

The first is an example of the vilest form of humour, the pun. One can only say in its defense that this one is a quadruple pun. On releasing the animals from the ark, Noah bid them to go forth and multiply. Suddenly two little snakes spoke up, 'But we can't multiply—we are adders.' Thereupon Noah constructed a table from roughhewn lumber and said, 'Here is a log table. Now you adders can multiply.'

The puns on 'adders' and 'multiply' are obvious. However there are two more puns, barbs aimed at the viscera of the mathematician: 'log table', i.e., a logarithmic table is a device which reduces multiplication to addition.

The next joke, in the form of a wisecrack definition, is more sophisticated mathematically. 'A topologist is a guy who doesn't know the difference between a coffee cup and a doughnut.' The definition hits at the very essence of topology, in which all configurations are considered equivalent if they can be deformed into each other without 'tearing'. A little reflection reveals that coffee cups and doughnuts belong to the same topological genus (the 'torus'), which distinguishes them from baseballs (a lower genus) and lidless teapots (a higher one).

The last joke concerns the mental patient who was convinced that he was dead. The psychiatrist, intent on trying the rational approach, got the patient to admit that dead men don't bleed and then pricked the patient's finger and pointed triumphantly to the emerging drop of blood. The patient's response was, 'I was wrong, doctor. Dead men apparently do bleed.' This joke, of course, needs no explanation. But it may be pertinent to point out that its humour taps a most profound question of scientific philosophy—the metaphysical underpinnings of the language of science: where is the border between knowledge and faith?

Notes

1. Address delivered at a symposium, 'The Structure of Science', at the Wistar Institute, Philadelphia, 17 April 1959.

References

Oeser, Paul H. 1955. 'The Lion and the Lamb: An Essay on Science and Poetry', *American Scientist* 63: 89–96.

QUESTIONS FOR CRITICAL THOUGHT

1. Why does Rapoport refer to mathematics as the 'language' of theoretical physics? Is mathematics a 'language'? In what sense? If it is, what is a language?
2. Rapoport frames his discussion around 'two widely held notions' about scientific language. What are these notions?
3. What is Rapoport's purpose? What is the structure of his argument? (In other words, into what sections is the argument divided, and how are they arranged?)
4. What sort of evidence does Rapoport provide to support his argument?
5. For what kind of audience does this address appear to have been written? How do you know? What view of the world would the audience have to entertain for this article to make sense to them?
6. To what extent is Rapoport's article really a definition?
7. This selection was originally an oral address. Does it retain any stylistic features appropriate to an oral presentation? Why or why not?
8. According to Rapoport, what is the challenge to the understanding that faces the reader of scientific language?
9. Rapoport spends most of his discussion on the claim that scientific language is 'dry and matter of fact'. To rebut this view, he turns to the language of mathematics, which he argues employs many of the elements of poetic expression. What are these elements? How does Rapoport demonstrate his point?
10. How personal is Rapoport's essay? How formal is it?
11. Does Rapoport employ the 'objective narrative' style common to most technical writing? Why or why not?

CHAPTER 32

Digitariat

Bill Casselman

The word digitariat is new to me but is at least nine years old. I don't know who coined it and have no proof that it is Canadian. The earliest clearly dated citation of digitariat that I have found is from *Wired* magazine online, issue 4.11, November 1996, in an article by Jason Sheftell:

> As the digital revolution shifts into warp speed, some cyber soothsayers worry that average Americans will be left tinkering in the garage while the wealthy zip around in virtual spaceships. Technological apartheid, they say, could further widen socioeconomic inequities, creating a 'digitariat' of unwired have-nots.

Digitariat was formed by analogy with proletariat. Pro-le-ta-rius is the immediate nounal origin in Latin. Proletarius arose as a label according to a division of the Roman people by Servius Tullius. A *proletarius* was a citizen of the lowest class, who served the state not with his property, but only with his children (Latin, *proles*), a *proletary*. As an adjective it meant low or common quite early, for the Roman comic dramatist Plautus refers to *sermo proletarius* 'vulgar gossip.' The basic Latin root of the word is proles 'offspring'.

Karl Marx used the word proletariat as a synonym for 'working class', men and women without wealth who sold their labour to the rich owning classes. You may review definitions of the term proletariat at this well-written website: http://www.answers.com/topic/proletariat.

One reason the word may not endure too long is its contradictory definitions. Some print and electronic commentators use digitariat to refer to the elite entrepreneurs and computer mavens who operate and control the digital world, a complete *volte-face* from the term's original meaning.

Here is a citation of that usage. In a November 2001 article entitled 'Towards a People's Alternative to "Intellectual Property Rights"', Pio Verzola, Jr, a Philippine writer on technology, used the word in this way:

> On the other hand, the same technological revolution has in fact enhanced the social character of production and the role of the working class in leading the way out of the capitalist crisis. Contrary to the pipedream that a new elite 'digitariat' has pulled the rug out from under the proletariat, droves of white-collar, computer-literate workers are in fact joining the category of industrial workers. If at all, the proletariat is absorbing unto itself the intellectual qualities of the so-called 'digitariat'.

Such a topsy-turvy tumble in the meaning of a word is not new in the history of English. Consider the adjective 'nice', which began its English life as a word meaning 'stupid'. But when a new word has opposing definitions its chance of survival is diminished.

QUESTIONS FOR CRITICAL THOUGHT

1. What is the purpose of this short passage?
2. Contrast Casselman's style in this excerpt with his style in 'Babblegab and Gobbledygook'. What differences are apparent? Why do you think the style has changed?
3. What is the meaning of the word 'digitariat'? Where did it come from?
4. Why might the coinage 'digitariat' not survive for long?
5. It has been estimated that the English language may contain as many as two million words, with approximately 1.4 million of these being technical in nature. What does this short passage teach us about the development of such language?
6. What is the value of including such a selection in a reader for technical writers?

CHAPTER 33

Politics and the English Language

George Orwell

Most people who bother with the matter at all would admit that the English language is in a bad way, but it is generally assumed that we cannot by conscious action do anything about it. Our civilization is decadent and our language—so the argument runs—must inevitably share in the general collapse. It follows that any struggle against the abuse of language is a sentimental archaism, like preferring candles to electric light or hansom cabs to aeroplanes. Underneath this lies the half-conscious belief that language is a natural growth and not an instrument which we shape for our own purposes.

Now, it is clear that the decline of a language must ultimately have political and economic causes: it is not due simply to the bad influence of this or that individual writer. But an effect can become a cause, reinforcing the original cause and producing the same effect in an intensified form, and so on indefinitely. A man may take to drink because he feels himself to be a failure, and then fail all the more completely because he drinks. It is rather the same thing that is happening to the English language. It becomes ugly and inaccurate because our thoughts are foolish, but the slovenliness of our language makes it easier for us to have foolish thoughts. The point is that the process is reversible. Modern English, especially written English, is full of bad habits which spread by imitation and which can be avoided if one is willing to take the necessary trouble. If one gets rid of these habits one can think more clearly, and to think clearly is a necessary first step toward political regeneration: so that the fight against bad English is not frivolous and is not the exclusive concern of professional writers. I will come back to this presently, and I hope that by that time the meaning of what I have said here will have become clearer. Meanwhile, here are five specimens of the English language as it is now habitually written.

These five passages have not been picked out because they are especially bad—I could have quoted far worse if I had chosen—but because they illustrate various of the mental vices from which we now suffer. They are a little below the average, but are fairly representative examples. I number them so that I can refer back to them when necessary:

1. I am not, indeed, sure whether it is not true to say that the Milton who once seemed not unlike a seventeenth-century Shelley had not become, out of an experience ever more bitter in each year, more alien [sic] to the founder of that Jesuit sect which nothing could induce him to tolerate.

   —Professor Harold Laski (essay in *Freedom of Expression*)

2. Above all, we cannot play ducks and drakes with a native battery of idioms which prescribes egregious collocations of vocables as the basic 'put up with' for 'tolerate' or 'put at a loss' for 'bewilder'.

   —Professor Lancelot Hogben (*Interglossia*)

3. On the one side we have the free personality: by definition it is not neurotic, for it has neither conflict nor dream. Its desires, such as they are, are transparent, for they are just what institutional approval keeps in the forefront of consciousness; another

institutional pattern would alter their number and intensity; there is little in them that is natural, irreducible, or culturally dangerous. But on the other side, the social bond itself is nothing but the mutual reflection of these self-secure integrities. Recall the definition of love. Is not this the very picture of a small academic? Where is there a place in this hall of mirrors for either personality or fraternity?

—Essay on psychology in *Politics* (New York)

4. All the 'best people' from the gentlemen's clubs, and all the frantic fascist captains, united in common hatred of Socialism and bestial horror at the rising tide of the mass revolutionary movement, have turned to acts of provocation, to foul incendiarism, to medieval legends of poisoned wells, to legalize their own destruction of proletarian organizations, and rouse the agitated petty-bourgeoise to chauvinistic fervor on behalf of the fight against the revolutionary way out of the crisis.

—Communist pamphlet

5. If a new spirit is to be infused into this old country, there is one thorny and contentious reform which must be tackled, and that is the humanization and galvanization of the BBC. Timidity here will bespeak canker and atrophy of the soul. The heart of Britain may be sound and of strong beat, for instance, but the British lion's roar at present is like that of Bottom in Shakespeare's *A Midsummer Night's Dream*—as gentle as any sucking dove. A virile new Britain cannot continue indefinitely to be traduced in the eyes or rather ears, of the world by the effete languors of Langham Place, brazenly masquerading as 'standard English'. When the Voice of Britain is heard at nine o'clock, better far and infinitely less ludicrous to hear aitches honestly dropped than the present priggish, inflated, inhibited, school-ma'amish arch braying of blameless bashful mewing maidens!

—Letter in *Tribune*

Each of these passages has faults of its own, but, quite apart from avoidable ugliness, two qualities are common to all of them. The first is staleness of imagery; the other is lack of precision. The writer either has a meaning and cannot express it, or he inadvertently says something else, or he is almost indifferent as to whether his words mean anything or not. This mixture of vagueness and sheer incompetence is the most marked characteristic of modern English prose, and especially of any kind of political writing. As soon as certain topics are raised, the concrete melts into the abstract and no one seems able to think of turns of speech that are not hackneyed: prose consists less and less of *words* chosen for the sake of their meaning, and more and more of *phrases* tacked together like the sections of a prefabricated henhouse. I list below, with notes and examples, various of the tricks by means of which the work of prose construction is habitually dodged.

Dying Metaphors

A newly invented metaphor assists thought by evoking a visual image, while on the other hand a metaphor which is technically 'dead' (e.g., *iron resolution*) has in effect reverted to being an ordinary word and can generally be used without loss of vividness. But in between these two classes there is a huge dump of worn-out metaphors which have lost all evocative power and are merely used because they save people the trouble of inventing phrases for themselves. Examples are: 'ring the changes on', 'take up the cudgel for', 'toe the line', 'ride roughshod over', 'stand shoulder to shoulder with', 'play into the hands of', 'no axe to grind', 'grist to the mill', 'fishing in troubled waters', 'on the order of the day', 'Achilles' heel', 'swan song', 'hotbed'. Many of these are used without knowledge of their meaning (what is a 'rift', for instance?), and incompatible metaphors are frequently mixed, a sure sign that the writer is not interested in what he is saying. Some metaphors now current have been twisted out of their original meaning without those who use them even being aware of the fact. For example, 'toe the line' is sometimes written as 'tow the line'. Another example is 'the hammer and the anvil', now always used with the implication that the anvil gets the worst of it. In real life it is always the anvil that breaks the hammer, never the other way about: a writer who stopped to think what he was saying would avoid perverting the original phrase.

Operators or Verbal False Limbs

These save the trouble of picking out appropriate verbs and nouns, and at the same time pad each sentence with extra syllables which give it an appearance of symmetry. Characteristic phrases are 'render inoperative', 'militate against', 'make contact with', 'be subjected to', 'give rise to', 'give grounds for', 'have the effect of', 'play a leading part (role) in', 'make itself felt', 'take effect', 'exhibit a tendency to', 'serve the purpose of', etc., etc. The keynote is the elimination of simple verbs. Instead of being a single word, such as break, stop, spoil, mend, kill, a verb becomes a phrase, made up of a noun or adjective tacked on to some general-purpose verb such as prove, serve, form, play, render. In addition, the passive voice is wherever possible used in preference to the active, and noun constructions are used instead of gerunds (by examination of instead of by examining). The range of verbs is further cut down by means of the *-ize* and *de-* formations, and the banal statements are given an appearance of profundity by means of the *not un-* formation. Simple conjunctions and prepositions are replaced by such phrases as 'with respect to', 'having regard to', 'the fact that', 'by dint of', 'in view of', 'in the interests of', 'on the hypothesis that'; and the ends of sentences are saved by anticlimax by such resounding commonplaces as 'greatly to be desired', 'cannot be left out of account', 'a development to be expected in the near future', 'deserving of serious consideration', 'brought to a satisfactory conclusion', and so on and so forth.

Pretentious Diction

Words like 'phenomenon', 'element', 'individual' (as noun), 'objective', 'categorical', 'effective', 'virtual', 'basic', 'primary', 'promote', 'constitute', 'exhibit', 'exploit', 'utilize', 'eliminate', 'liquidate', are used to dress up a simple statement and give an air of scientific impartiality to biased judgements. Adjectives like 'epoch-making', 'epic', 'historic', 'unforgettable', 'triumphant', 'age-old', 'inevitable', 'inexorable', 'veritable', are used to dignify the sordid process of international politics, while writing that aims at glorifying war usually takes on an archaic colour, its characteristic words being: 'realm', 'throne', 'chariot', 'mailed fist', 'trident', 'sword', 'shield', 'buckler', 'banner', 'jackboot', 'clarion'. Foreign words and expressions such as *cul de sac*, *ancien régime*, *deus ex machina*, *mutatis mutandis*, *status quo*, *gleichschaltung*, *weltanschauung*, are used to give an air of culture and elegance. Except for the useful abbreviations i.e., e.g., and etc., there is no real need for any of the hundreds of foreign phrases now current in the English language.

Bad writers, and especially scientific, political, and sociological writers, are nearly always haunted by the notion that Latin or Greek words are grander than Saxon ones, and unnecessary words like 'expedite', 'ameliorate', 'predict', 'extraneous', 'deracinated', 'clandestine', 'subaqueous', and hundreds of others constantly gain ground from their Anglo-Saxon numbers. The jargon peculiar to Marxist writing ('hyena', 'hangman', 'cannibal', 'petty bourgeois', 'these gentry', 'lackey', 'flunkey', 'mad dog', 'White Guard', etc.) consists largely of words translated from Russian, German, or French; but the normal way of coining a new word is to use Latin or Greek root with the appropriate affix and, where necessary, the *-ize* formation. It is often easier to make up words of this kind (deregionalize, impermissible, extramarital, non-fragmentary and so forth) than to think up the English words that will cover one's meaning. The result, in general, is an increase in slovenliness and vagueness.

Meaningless Words

In certain kinds of writing, particularly in art criticism and literary criticism, it is normal to come across long passages which are almost completely lacking in meaning. Words like 'romantic', 'plastic', 'values', 'human', 'dead', 'sentimental', 'natural', 'vitality', as used in art criticism, are strictly meaningless, in the sense that they not only do not point to any discoverable object, but are hardly ever expected to do so by the reader. When one critic writes, 'The outstanding feature of Mr X's work is its living quality,' while another writes, 'The immediately striking thing about Mr X's work is its peculiar deadness,' the reader accepts this as a simple difference opinion. If words like black and white were involved, instead of the jargon words dead and living, he would see at once that language was being used in an improper way.

Many political words are similarly abused. The word Fascism has now no meaning except in so far as it signifies 'something not desirable'. The words 'democracy', 'socialism', 'freedom', 'patriotic', 'realistic', 'justice' have each of them several different meanings which cannot be reconciled with one another. In the case of a word like democracy, not only is there no agreed definition, but the attempt to make one is resisted from all sides. It is almost universally felt that when we call a country democratic we are praising it: consequently the defenders of every kind of regime claim that it is a democracy, and fear that they might have to stop using that word if it were tied down to any one meaning. Words of this kind are often used in a consciously dishonest way. That is, the person who uses them has his own private definition, but allows his hearer to think he means something quite different. Statements like Marshal Petain was a true patriot, The Soviet press is the freest in the world, The Catholic Church is opposed to persecution, are almost always made with intent to deceive. Other words used in variable meanings, in most cases more or less dishonestly, are: 'class', 'totalitarian', 'science', 'progressive', 'reactionary', 'bourgeois', 'equality'.

Now that I have made this catalogue of swindles and perversions, let me give another example of the kind of writing that they lead to. This time it must of its nature be an imaginary one. I am going to translate a passage of good English into modern English of the worst sort. Here is a well-known verse from Ecclesiastes:

> I returned and saw under the sun, that the race is not to the swift, nor the battle to the strong, neither yet bread to the wise, nor yet riches to men of understanding, nor yet favour to men of skill; but time and chance happeneth to them all.

Here it is in modern English:

> Objective considerations of contemporary phenomena compel the conclusion that success or failure in competitive activities exhibits no tendency to be commensurate with innate capacity, but that a considerable element of the unpredictable must invariably be taken into account.

This is a parody, but not a very gross one. Exhibit (3) above, for instance, contains several patches of the same kind of English. It will be seen that I have not made a full translation. The beginning and ending of the sentence follow the original meaning fairly closely, but in the middle the concrete illustrations—'race', 'battle', 'bread'—dissolve into the vague phrases 'success or failure in competitive activities'. This had to be so, because no modern writer of the kind I am discussing—no one capable of using phrases like 'objective considerations of contemporary phenomena'—would ever tabulate his thoughts in that precise and detailed way. The whole tendency of modern prose is away from concreteness.

Now analyze these two sentences a little more closely. The first contains forty-nine words but only sixty syllables, and all its words are those of everyday life. The second contains thirty-eight words of ninety syllables: eighteen of those words are from Latin roots, and one from Greek. The first sentence contains six vivid images, and only one phrase ('time and chance') that could be called vague. The second contains not a single fresh, arresting phrase, and in spite of its ninety syllables it gives only a shortened version of the meaning contained in the first. Yet without a doubt it is the second kind of sentence that is gaining ground in modern English. I do not want to exaggerate. This kind of writing is not yet universal, and outcrops of simplicity will occur here and there in the worst-written page. Still, if you or I were told to write a few lines on the uncertainty of human fortunes, we should probably come much nearer to my imaginary sentence than to the one from *Ecclesiastes*.

As I have tried to show, modern writing at its worst does not consist of picking out words for the sake of their meaning and inventing images in order to make the meaning clearer. It consists in gumming together long strips of words which have already been set in order by someone else, and making the results presentable by sheer humbug.

The attraction of this way of writing is that it is easy. It is easier—even quicker, once you have the habit—to say, in my opinion it is not an unjustifiable assumption that than to say I think. If you use ready-made phrases, you not only don't have to hunt about for the words; you also don't have to bother with the rhythms of your sentences since these phrases are generally so arranged as to be more or less euphonious. When you are composing in a hurry—when you are dictating to a stenographer, for instance, or making a public speech—it is natural to fall into a pretentious, Latinized style. Tags like a consideration which we should do well to bear in mind or a conclusion to which all of us would readily assent will save many a sentence from coming down with a bump.

By using stale metaphors, similes, and idioms, you save much mental effort, at the cost of leaving your meaning vague, not only for your reader but for yourself. This is the significance of mixed metaphors. The sole aim of a metaphor is to call up a visual image. When these images clash—as in The Fascist octopus has sung its swan song, the jackboot is thrown into the melting pot—it can be taken as certain that the writer is not seeing a mental image of the objects he is naming; in other words he is not really thinking.

Look again at the examples I gave at the beginning of this essay. Professor Laski (1) uses five negatives in fifty-three words. One of these is superfluous, making nonsense of the whole passage, and in addition there is the slip—alien for akin—making further nonsense, and several avoidable pieces of clumsiness which increase the general vagueness. Professor Hogben (2) plays ducks and drakes with a battery which is able to write prescriptions, and, while disapproving of the everyday phrase put up with, is unwilling to look egregious up in the dictionary and see what it means; (3), if one takes an uncharitable attitude towards it, is simply meaningless: probably one could work out its intended meaning by reading the whole of the article in which it occurs. In (4), the writer knows more or less what he wants to say, but an accumulation of stale phrases chokes him like tea leaves blocking a sink. In (5), words and meaning have almost parted company.

People who write in this manner usually have a general emotional meaning—they dislike one thing and want to express solidarity with another—but they are not interested in the detail of what they are saying. A scrupulous writer, in every sentence that he writes, will ask himself at least four questions, thus: What am I trying to say?

What words will express it? What image or idiom will make it clearer? Is this image fresh enough to have an effect? And he will probably ask himself two more: Could I put it more shortly? Have I said anything that is avoidably ugly?

But you are not obliged to go to all this trouble. You can shirk it by simply throwing your mind open and letting the ready-made phrases come crowding in. They will construct your sentences for you—even think your thoughts for you, to a certain extent—and at need they will perform the important service of partially concealing your meaning even from yourself. It is at this point that the special connection between politics and the debasement of language becomes clear.

In our time it is broadly true that political writing is bad writing.

Where it is not true, it will generally be found that the writer is some kind of rebel, expressing his private opinions and not a 'party line'. Orthodoxy, of whatever colour, seems to demand a lifeless, imitative style.

The political dialects to be found in pamphlets, leading articles, manifestoes, White papers, and the speeches of undersecretaries do, of course, vary from party to party, but they are all alike in that one almost never finds in them a fresh, vivid, homemade turn of speech. When one watches some tired hack on the platform mechanically repeating the familiar phrases—'bestial', 'atrocities', 'iron heel', 'bloodstained tyranny', 'free peoples of the world', 'stand shoulder to shoulder'—one often has a curious feeling that one is not watching a live human being but some kind of dummy: a feeling which suddenly becomes stronger at moments when the light catches the speaker's spectacles and turns them into blank discs which seem to have no eyes behind them. And this is not altogether fanciful. A speaker who uses that kind of phraseology has gone some distance toward turning himself into a machine. The appropriate noises are coming out of his larynx, but his brain is not involved, as it would be if he were choosing his words for himself. If the speech he is making is one that he is accustomed to make over and over again, he may be almost unconscious of what he is saying, as one is when one utters the responses in church. And this reduced state of consciousness, if not indispensable, is at any rate favourable to political conformity.

In our time, political speech and writing are largely the defense of the indefensible. Things like the continuance of British rule in India, the Russian purges and deportations, the dropping of the atom bombs on Japan, can indeed be defended, but only by arguments which are too brutal for most people to face, and which do not square with the professed aims of the political parties.

Thus political language has to consist largely of euphemism, question-begging, and sheer cloudy vagueness. Defenseless villages are bombarded from the air, the inhabitants driven out into the countryside, the cattle machine-gunned, the huts set on fire with incendiary bullets: this is called *pacification*. Millions of peasants are robbed of their farms and sent trudging along the roads with no more than they can carry: this is called transfer of population or rectification of frontiers. People are imprisoned for years without trial, or shot in the back of the neck or sent to die of scurvy in Arctic lumber camps: this is called elimination of unreliable elements.

Such phraseology is needed if one wants to name things without calling up mental pictures of them. Consider for instance some comfortable English professor defending Russian totalitarianism. He cannot say outright, 'I believe in killing off your opponents when you can get good results by doing so.' Probably, therefore, he will say something like this:

> While freely conceding that the Soviet regime exhibits certain features which the humanitarian may be inclined to deplore, we must, I think, agree that a certain curtailment of the right to political opposition is an unavoidable concomitant of transitional periods, and that the rigors which the Russian people have been

> called upon to undergo have been amply justified in the sphere of concrete achievement.

The inflated style itself is a kind of euphemism. A mass of Latin words falls upon the facts like soft snow, blurring the outline and covering up all the details. The great enemy of clear language is insincerity. When there is a gap between one's real and one's declared aims, one turns as it were instinctively to long words and exhausted idioms, like a cuttlefish spurting out ink. In our age there is no such thing as 'keeping out of politics'. All issues are political issues, and politics itself is a mass of lies, evasions, folly, hatred, and schizophrenia. When the general atmosphere is bad, language must suffer. I should expect to find—this is a guess which I have not sufficient knowledge to verify—that the German, Russian, and Italian languages have all deteriorated in the last ten or fifteen years, as a result of dictatorship.

But if thought corrupts language, language can also corrupt thought. A bad usage can spread by tradition and imitation even among people who should and do know better. The debased language that I have been discussing is in some ways very convenient. Phrases like a not unjustifiable assumption, leaves much to be desired, would serve no good purpose, a consideration which we should do well to bear in mind, are a continuous temptation, a packet of aspirins always at one's elbow. Look back through this essay, and for certain you will find that I have again and again committed the very faults I am protesting against.

By this morning's post I have received a pamphlet dealing with conditions in Germany. The author tells me that he 'felt impelled' to write it. I open it at random, and here is almost the first sentence I see: '[The Allies] have an opportunity not only of achieving a radical transformation of Germany's social and political structure in such a way as to avoid a nationalistic reaction in Germany itself, but at the same time of laying the foundations of a co-operative and unified Europe.' You see, he 'feels impelled' to write—feels, presumably, that he has something new to say—and yet his words, like cavalry horses answering the bugle, group themselves automatically into the familiar dreary pattern. This invasion of one's mind by ready-made phrases ('lay the foundations', 'achieve a radical transformation') can only be prevented if one is constantly on guard against them, and every such phrase anaesthetizes a portion of one's brain.

I said earlier that the decadence of our language is probably curable. Those who deny this would argue, if they produced an argument at all, that language merely reflects existing social conditions, and that we cannot influence its development by any direct tinkering with words and constructions. So far as the general tone or spirit of a language goes, this may be true, but it is not true in detail. Silly words and expressions have often disappeared, not through any evolutionary process but owing to the conscious action of a minority.

Two recent examples were 'explore every avenue' and 'leave no stone unturned', which were killed by the jeers of a few journalists. There is a long list of flyblown metaphors which could similarly be got rid of if enough people would interest themselves in the job; and it should also be possible to laugh the *not un-* formation out of existence, to reduce the amount of Latin and Greek in the average sentence, to drive out foreign phrases and strayed scientific words, and, in general, to make pretentiousness unfashionable. But all these are minor points. The defense of the English language implies more than this, and perhaps it is best to start by saying what it does not imply.

To begin with it has nothing to do with archaism, with the salvaging of obsolete words and turns of speech, or with the setting up of a 'standard English' which must never be departed from. On the contrary, it is especially concerned with the scrapping of every word or idiom which has outworn its usefulness. It has nothing to do with correct grammar and syntax, which are of no importance so long as one makes one's meaning clear, or with the avoidance of Americanisms, or with having what is called a 'good prose style.' On the other hand, it is not concerned with fake simplicity and the attempt to make written English colloquial. Nor does it even imply in every case preferring the Saxon word to the Latin one, though it does imply using the fewest and shortest words that will cover one's meaning.

What is above all needed is to let the meaning choose the word, and not the other way around. In prose, the worst thing one can do with words is surrender to them. When you think of a concrete object, you think wordlessly, and then, if you want to describe the thing you have been visualizing you probably hunt about until you find the exact words that seem to fit it. When you think of something abstract you are more inclined to use words from the start, and unless you make a conscious effort to prevent it, the existing dialect will come rushing in and do the job for you, at the expense of blurring or even changing your meaning. Probably it is better to put off using words as long as possible and get one's meaning as clear as one can through pictures and sensations. Afterward one can choose—not simply *accept*—the phrases that will best cover the meaning, and then switch round and decide what impressions one's words are likely to make on another person.

This last effort of the mind cuts out all stale or mixed images, all prefabricated phrases, needless repetitions, and humbug and vagueness generally. But one can often be in doubt about the effect of a word or a phrase, and one needs rules that one can rely on when instinct fails. I think the following rules will cover most cases:

1. Never use a metaphor, simile, or other figure of speech which you are used to seeing in print.
2. Never us a long word where a short one will do.
3. If it is possible to cut a word out, always cut it out.
4. Never use the passive where you can use the active.
5. Never use a foreign phrase, a scientific word, or a jargon word if you can think of an everyday English equivalent.
6. Break any of these rules sooner than say anything outright barbarous.

These rules sound elementary, and so they are, but they demand a deep change of attitude in anyone who has grown used to writing in the style now fashionable. One could keep all of them and still write bad English, but one could not write the kind of stuff that I quoted in those five specimens at the beginning of this article.

I have not here been considering the literary use of language, but merely language as an instrument for expressing and not for concealing or preventing thought. Stuart Chase and others have come near to claiming that all abstract words are meaningless, and have used this as a pretext for advocating a kind of political quietism. Since you don't know what Fascism is, how can you struggle against Fascism? One need not swallow such absurdities as this, but one ought to recognize that the present political chaos is connected with the decay of language, and that one can probably bring about some improvement by starting at the verbal end. If you simplify your English, you are freed from the worst follies of orthodoxy. You cannot speak any of the necessary dialects, and when you make a stupid remark its stupidity will be obvious, even to yourself.

Political language—and with variations this is true of all political parties, from Conservatives to Anarchists—is designed to make lies sound truthful and murder respectable, and to give an appearance of solidity to pure wind. One cannot change this all in a moment, but one can at least change one's own habits, and from time to time one can even, if one jeers loudly enough, send some worn-out and useless phrase—some 'jackboot', 'Achilles' heel', 'hotbed', 'melting pot', 'acid test', 'veritable inferno', or other lump of verbal refuse—into the dustbin, where it belongs.

QUESTIONS FOR CRITICAL THOUGHT

1. Orwell identifies four common faults that make much modern writing unintelligible. How well do his categories correspond with the audience relationships identified by Elbow in 'Three Tricky Relationships to an Audience'?
2. Orwell asserts that much of the trouble with contemporary writing arises from simply 'gumming together long strips of words which have already been set in order by someone else' without regard to 'picking out words for the sake of their meaning and inventing images in order to make the meaning clearer'. To what extent is this a problem of audience adaptation, which so many of the other contributors to this book have identified (including Glaser ['Voices to Shun'], Elbow ['Three Tricky Relationships to an Audience'], Buehler ['Situational Editing'], Urquhart ['Bridging Gaps, Engineering Audiences'], Burton ['An Engineer's Rhetorical Journey'], and others). Why does Orwell never explicitly address the issue of audience adaptation?
3. Orwell provides a four-question guide to better writing. What four questions does he advise that we should ask about our own writing? Where do these questions focus the writer's attention?
4. What does Orwell mean when he says that 'political speech and writing are largely the defence of the indefensible'? Can you think of any contemporary examples that confirm this statement?
5. Orwell warns against the over-use of clichés and ready-made phrases largely because they can have the effect of 'concealing your meaning even from yourself'. Take another look at Casselman's discussion of bafflegab and Glaser's notion of 'bumfuzzlement' (in 'Voices to Shun'). Why do so many rhetorical and linguistic theorists warn of the dangers of this kind of language?
6. Orwell argues that a writer who uses language carelessly and thoughtlessly 'has gone some distance toward turning himself into a machine', in part because his use of such bafflegab allows him to be 'almost unconscious of what he is saying'. Compare Orwell's argument on this point with Talbott's definitions of 'machine' and 'device' in 'The Deceiving Virtues of Technology'.
7. 'Inflated style is itself a kind of euphemism,' says Orwell. What is a euphemism? How does inflated style perform this function? Why is euphemism dangerous?
8. Is Orwell correct in his belief that sloppy language produces sloppy thought? Why?

9. Orwell offers advice in the form of six points to writers who wish to improve their style. What are the six points? If we consider Buehler's distinction (in 'Situational Editing') between a 'rhetorical' and 'grammatical' approach to writing, which are these?
10. Is Orwell's advice rhetorical in the sense the other contributors have used the term? Why or why not?
11. Does Orwell adhere to his own principles?
12. How personal is Orwell's style? How formal is it? Is it appropriately adapted to purpose and context? How do you know?
13. Orwell charges that political language 'is designed to make lies sound truthful and murder respectable, and to give the appearance of solidity to pure wind'. Consider Ornatowski's account (in 'Between Efficiency and Politics') of the engineer Stephen, whose challenge was to make his point clear while avoiding political fallout. Does Orwell's admonition apply to Ornatowski's advice? If so, what are we to make of it?

CHAPTER 34

The World of Doublespeak

William Lutz

There are no potholes in the streets of Tucson, Arizona, just 'pavement deficiencies'. The Reagan Administration didn't propose any new taxes, just 'revenue enhancement' through new 'user's fees'. Those aren't bums on the street, just 'non-goal oriented members of society'. There are no more poor people, just 'fiscal underachievers'. There was no robbery of an automatic teller machine, just an 'unauthorized withdrawal'. The patient didn't die because of medical malpractice, it was just a 'diagnostic misadventure of a high magnitude'. The US Army doesn't kill the enemy anymore, it just 'services the target'. And the doublespeak goes on.

Doublespeak is language that pretends to communicate but really doesn't. It is language that makes the bad seem good, the negative appear positive, the unpleasant appear attractive or at least tolerable. Doublespeak is language that avoids or shifts responsibility, language that is at variance with its real or purported meaning. It is language that conceals or prevents thought; rather than extending thought, doublespeak limits it.

Doublespeak is not a matter of subjects and verbs agreeing; it is a matter of words and facts agreeing. Basic to doublespeak is incongruity, the incongruity between what is said or left unsaid, and what really is. It is the incongruity between the word and the referent, between seem and be, between the essential function of language—communication—and what doublespeak does—mislead, distort, deceive, inflate, circumvent, obfuscate.

How to Spot Doublespeak

How can you spot doublespeak? Most of the time you will recognize doublespeak when you see or hear it. But, if you have any doubts, you can identify doublespeak just by answering these questions: Who is saying what to whom, under what conditions and circumstances, with what intent, and with what results? Answering these questions will usually help you identify as doublespeak language that appears to be legitimate or that at first glance doesn't even appear to be doublespeak.

FIRST KIND OF DOUBLESPEAK

There are at least four kinds of doublespeak. The first is the euphemism, an inoffensive or positive word or phrase used to avoid a harsh, unpleasant, or distasteful reality. But a euphemism can also be a tactful word or phrase that avoids directly mentioning a painful reality, or it can be an expression used out of concern for the feelings of someone else, or to avoid directly discussing a topic subject to a social or cultural taboo.

When you use a euphemism because of your sensitivity for someone's feelings or out of concern for a recognized social or cultural taboo, it is not doublespeak. For example, you express your condolences that someone has 'passed away' because you do not want to say to a grieving person, 'I'm sorry your father is dead.' When you use the euphemism 'passed away', no one is misled. Moreover, the euphemism functions here not just to protect the feelings of another person, but to communicate also your concern for that person's feelings during a period of mourning. When you excuse yourself to go to the 'restroom', or you mention that someone is 'sleeping with' or 'involved with' someone else, you do not mislead anyone about your meaning, but you do respect the social taboos about discussing bodily functions and sex in direct terms. You also indicate your sensitivity to the feelings of your audience, which is usually considered a mark of courtesy and good manners.

However when a euphemism is used to mislead or deceive, it becomes doublespeak. For example, in 1984 the US State Department announced that it would no longer use the word 'killing' in its annual report on the status of human rights in countries around the world. Instead, it would use the phrase 'unlawful or arbitrary deprivation of life', which the department claimed was more accurate. Its real purpose for using this phrase was simply to avoid discussing the embarrassing situation of government-sanctioned killings in countries that are supported by the United States and have been certified by the United States as respecting the human rights of their citizens. This use of a euphemism constitutes doublespeak, since it is designed to mislead, to cover up the unpleasant. Its real intent is at variance with its apparent intent. It is language designed to alter our perception of reality.

The Pentagon, too, avoids discussing unpleasant realities when it refers to bombs and artillery shells that fall on civilian targets as 'incontinent ordnance'. And in 1977 the Pentagon tried to slip funding for the neutron bomb unnoticed into an appropriations bill by calling it a 'radiation enhancement device'.

SECOND KIND OF DOUBLESPEAK

A second kind of doublespeak is jargon, the specialized language of a trade, profession, or similar group, such as that used by doctors, lawyers, engineers, educators, or car mechanics. Jargon can serve an important and useful function. Within a group, jargon functions as a kind of verbal shorthand that allows members of the group to communicate with each other clearly, efficiently, and quickly. Indeed, it is a mark of membership in the group to be able to use and understand the group's jargon.

But jargon, like the euphemism, can also be doublespeak. It can be—and often is—pretentious, obscure, and esoteric terminology used to give an air of profundity, authority, and prestige to speakers and their subject matter. Jargon as doublespeak often makes the simple appear complex, the ordinary profound, the obvious insightful. In this sense it is used not to express but impress. With such doublespeak, the act of smelling something becomes 'organoleptic analysis', glass becomes 'fused silicate', a crack in a metal support beam becomes a 'discontinuity', conservative economic policies become 'distributionally conservative notions'.

Lawyers, for example, speak of an 'involuntary conversion' of property when discussing the loss or destruction of property through theft, accident, or condemnation. If your house burns down or if your car is stolen, you have suffered an involuntary conversion of your property. When used by lawyers in a legal situation, such jargon is a legitimate use of language, since lawyers can be expected to understand the term.

However, when a member of a specialized group uses its jargon to communicate with a person outside the group, and uses it knowing that the nonmember does not understand such language, then there is doublespeak. For example, on 9 May 1978, a National Airlines 727 airplane crashed while attempting to land at the Pensacola, Florida airport. Three of the fifty-two passengers aboard the airplane were killed. As a result of the crash, National made an after-tax insurance benefit of $1.7 million, or an extra 18¢ a share dividend for its stockholders. Now National Airlines had two problems: It did not want to talk about one of its airplanes crashing, and it had to account for the $1.7 million when it issued its annual report to its stockholders. National solved the problem by inserting a footnote in its annual report which explained that the $1.7 million income was due to 'the involuntary conversion of a 727'. National thus acknowledged the crash of its airplane and the subsequent profit it made from the crash, without once mentioning the accident or the deaths. However, because airline officials knew that most stock-holders in the company, and indeed most of the general public, were not familiar with legal jargon, the use of such jargon constituted doublespeak.

THIRD KIND OF DOUBLESPEAK

A third kind of doublespeak is gobbledygook or bureaucratese. Basically, such doublespeak is simply a matter of piling on words, of overwhelming the audience with words, the bigger the words and the longer the sentences the better. Alan Greenspan, then chair of President Nixon's Council of Economic Advisors, was quoted in *The Philadelphia Inquirer* in 1974 as having testified before a Senate committee that 'It is a tricky problem to find the particular calibration in timing that would be appropriate to stem the acceleration in risk premiums created by falling incomes without prematurely aborting the decline in the inflation-generated risk premiums.' Nor has Mr Greenspan's language changed since then. Speaking to the meeting of the Economic Club of New York in 1988, Mr Greenspan, now Federal Reserve chair, said, 'I guess I should warn you, if I turn out to be particularly clear, you've probably misunderstood what I've said.' Mr Greenspan's doublespeak doesn't seem to have held back his career.

Sometimes gobbledygook may sound impressive, but when the quote is later examined in print it doesn't even make sense. During the 1988 presidential campaign, vice-presidential candidate Senator Dan Quayle explained the need for a strategic-defense initiative by saying, 'Why wouldn't an enhanced deterrent, a more stable peace, a better prospect to denying the ones who enter conflict in the first place to have a reduction of offensive systems and an introduction to defensive capability? I believe this is the route the country will eventually go.'

The investigation into the Challenger disaster in 1986 revealed the doublespeak of gobbledygook and bureaucratese used by too many involved in the shuttle program. When Jesse Moore, NASA's associate administrator, was asked if the performance of the shuttle program had improved with each launch or if it had remained the same, he answered, 'I think our performance in terms of the liftoff performance and in terms of the orbital performance, we knew more about the envelope we were operating under, and we have been pretty accurately staying in that. And so I would say the performance has not by design drastically improved. I think we have been able to characterize the performance more as a function of our launch experience as opposed to it improving as a function of time.' While this language may appear to be jargon, a close look will reveal that it is really just gobbledygook laced with jargon. But you really have to wonder if Mr Moore had any idea what he was saying.

FOURTH KIND OF DOUBLESPEAK

The fourth kind of doublespeak is inflated language that is designed to make the ordinary seem extraordinary; to make everyday things seem impressive; to give an air of importance to people, situations, or things that would not normally be considered important; to make the simple seem complex. Often this kind of doublespeak isn't hard to spot, and it is usually pretty funny. While car mechanics may be called 'automotive internists', elevator operators members of the 'vertical transportation corps', used cars 'pre-owned' or 'experienced cars', and black- and-white television sets described as having 'non-multicolour capability', you really aren't misled all that much by such language.

However, you may have trouble figuring out that, when Chrysler 'initiates a career alternative enhancement program', it is really laying off five thousand workers; or that 'negative patient care outcome' means the patient died; or that 'rapid oxidation' means a fire in a nuclear power plant.

The doublespeak of inflated language can have serious consequences. In Pentagon doublespeak, 'pre-emptive counterattack' means that American forces attacked first; 'engaged the enemy on all sides' means American troops were ambushed; 'backloading of augmentation personnel' means a retreat by American troops. In the doublespeak of the military, the 1983 invasion of Grenada was conducted not by the US Army, Navy, Air Force and Marines, but by the 'Caribbean Peace Keeping Forces'. But then, according to the Pentagon, it wasn't an invasion, it was a 'predawn vertical insertion'. . . .

The Dangers of Doublespeak

These examples of doublespeak should make it clear that doublespeak is not the product of carelessness or sloppy thinking. Indeed, most doublespeak is the product of clear thinking and is carefully designed and constructed to appear to communicate when in fact it doesn't. It is language designed not to lead but mislead. It is language designed to distort reality and corrupt thought. In the world created by doublespeak, if it's not a tax increase, but rather 'revenue enhancement' or 'tax base broadening', how can you complain about higher taxes? If it's not acid rain, but rather 'poorly buffered precipitation', how can you worry about all those dead trees? If that isn't the Mafia in Atlantic City, but just 'members of a career-offender cartel', why worry about the influence of organized crime in the city? If Supreme Court Justice William Rehnquist wasn't addicted to the pain-killing drug his doctor prescribed, but instead it was just that the drug had 'established an interrelationship with the body, such that if the drug is removed precipitously, there is a reaction', you needn't question that his decisions might have been influenced by his drug addiction. If it's not a Titan II nuclear-armed intercontinental ballistic missile with a warhead 630 times more powerful than the atomic bomb dropped on Hiroshima, but instead, according to Air Force Colonel Frank Horton, it's just a 'very large, potentially disruptive reentry system', why be concerned about the threat of nuclear destruction? Why worry about the neutron bomb escalating the arms race if it's just a 'radiation enhancement weapon'? If it's not an invasion, but a 'rescue mission' or a 'predawn vertical insertion', you won't need to think about any violations of US or international law.

Doublespeak has become so common in everyday living that many people fail to notice it. Even worse, when they do notice doublespeak being used on them, they don't react, they don't protest. Do you protest when you are asked to check your packages at the desk 'for your convenience', when it's not for your convenience at all but for someone else's? You see advertisements for 'genuine imitation leather', 'virgin vinyl', or 'real counterfeit diamonds', but do you question the language or the supposed quality of the product? Do you question politicians who don't speak of slums or ghettos but of the 'inner city' or 'substandard housing' where the 'disadvantaged' live and thus avoid talking about the poor who have to live in filthy, poorly heated, ramshackle apartments or houses? Aren't you amazed that patients don't die in the hospital anymore, it's just 'negative patient-care outcome'?

Doublespeak such as that noted earlier that defines cab drivers as 'urban transportation specialists', elevator operators as members of the 'vertical transportation corps', and automobile mechanics as 'automotive internists' can be considered humourous and relatively harmless. However, when a fire in a nuclear reactor building is called 'rapid oxidation', an explosion in a nuclear power plant is called an 'energetic disassembly', the illegal over-throw of a legitimate government is termed 'destabilizing a government', and lies are seen as 'inoperative statements', we are hearing doublespeak that attempts to avoid responsibility and make the bad seem good, the negative appear positive, something unpleasant appear attractive; and which seems to communicate but doesn't. It is language designed to alter our perception of reality and corrupt our thinking. Such language does not provide us with the tools we need to develop, advance, and preserve our culture and our civilization. Such language breeds suspicion, cynicism, distrust, and, ultimately, hostility.

Doublespeak is insidious because it can infect and eventually destroy the function of language, which is communication between people and social groups. This corruption of the function of language can have serious and far-reaching consequences. We live in a country that depends upon an informed electorate to make decisions in selecting candidates for office and deciding issues of public policy. The use of doublespeak can become so pervasive that it becomes the coin of the political realm, with speakers and listeners convinced that they really understand such language. After a while we may really believe that politicians don't lie but only 'misspeak', that illegal acts are merely 'inappropriate actions', that fraud and criminal conspiracy are just 'miscertification'. President Jimmy Carter in April of 1980 could call the aborted raid to free the American hostages in Teheran an 'incomplete success' and really believe that he had made a statement that clearly communicated with the American public. So, too, could President Ronald Reagan say in 1985 that 'ultimately our security and our hopes for success at the arms reduction talks hinge on the determination that we show here to continue our program to rebuild and refortify our defenses' and really believe that greatly increasing the amount of money spent building new weapons would lead to a reduction in the number of weapons in the world. If we really believe that we understand such language and that such language communicates and promotes clear thought, then the world of *1984*, with its control of reality through language, is upon us. . . .

The Doublespeak of Graphs

Just as polls seem to present concrete, specific evidence, so do graphs and charts present information visually in a way that appears unambiguous and dramatically clear. But, just as polls leave a lot of necessary information out, so can graphs and charts, resulting in doublespeak. You have to ask a lot of questions if you really want to understand a graph or chart.

In 1981 President Reagan went on television to argue that citizens would be paying a lot more in taxes under a Democratic bill than under his bill. To prove his point, he used a chart that appeared to show a dramatic and very big difference between

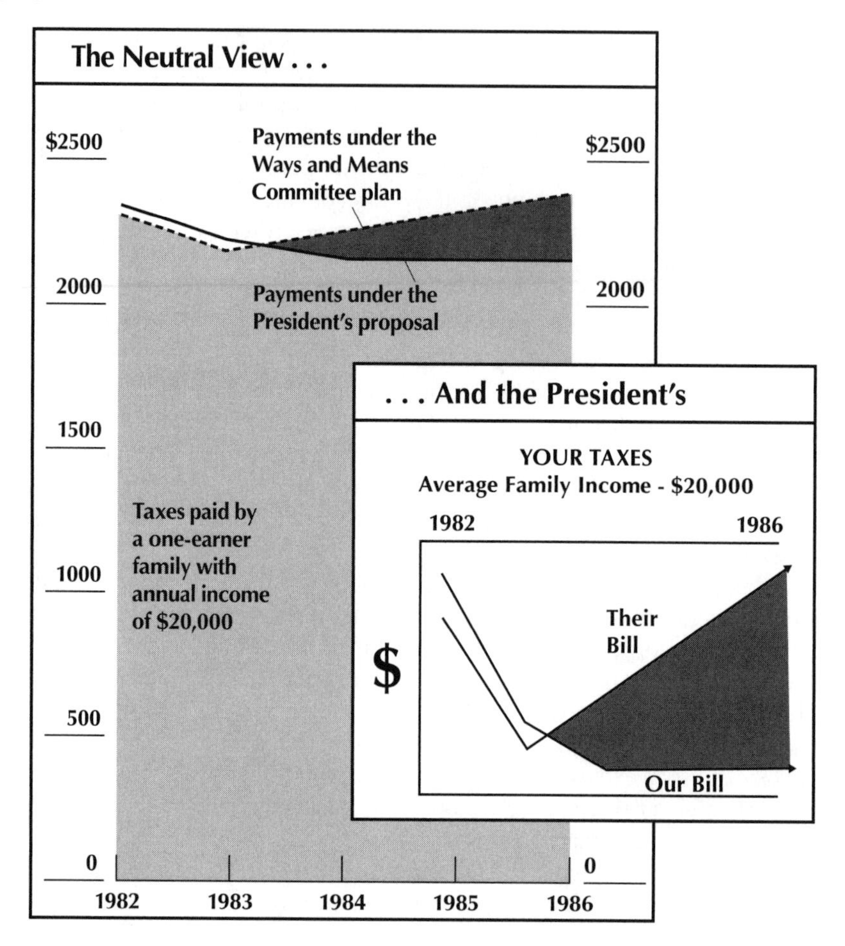

Figure 34.1 President Reagan's misleading and biased chart, compared with a neutral presentation regarding the same tax proposals.

the results of each bill (see Figure 34.1). But the president's chart was doublespeak, because it was deliberately designed to be misleading. Pointing to his chart, President Reagan said, 'This red space between the two lines is the tax money that will remain in your pockets if our bill passes, and it's the amount that will leave your pockets if their bill is passed. On the one hand, you see a genuine and lasting commitment to the future of working Americans. On the other, just another empty promise.' That was a pretty dramatic statement, considering that the maximum difference between the two bills, after five years, would have been $217.

The president's chart showed a deceptively dramatic difference because his chart had no figures on the dollar scale and no numbers for years except 1982 and 1986. The difference in tax payments was exaggerated in the president's chart by 'squashing' or tightening the time scale as much as possible, while stretching the dollar scale, starting with an oddly unrounded $2,150 and winding up at $2,400. Thus, the chart had no perspective. Using the proper method for constructing a chart

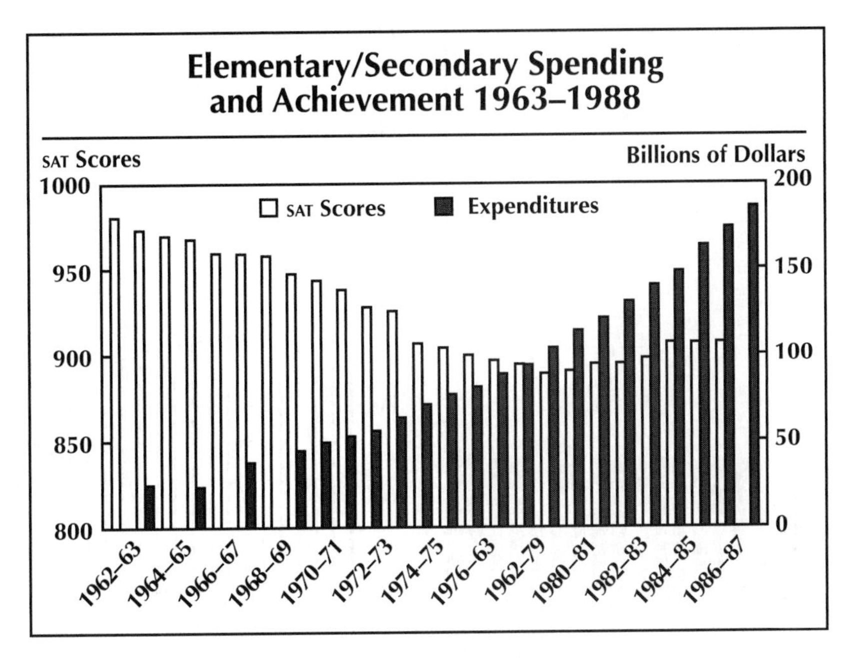

Figure 34.2 Misleading graph from the Department of Education, showing school spending relative to SAT scores.

would have meant starting at $0 and going up to the first round number after the highest point in the chart, as done in the 'neutral view' in Figure 34.1. Using that method, the $217 seems rather small in a total tax bill of $2,385.

What happened to the numbers on the president's chart? 'The chart we sent over to the White House had all the numbers on it,' said Marlin Fitzwater, then a press officer in the Treasury Department. Senior White House spokesperson David Gergen said, 'We took them off. We were trying to get a point across, not the absolute numbers.' So much for honesty.

In 1988 the Department of Education issued a graph that seemed to prove that there was a direct connection between the rise in elementary and secondary school spending and the decline in scores on the Scholastic Aptitude Test (see Figure 34.2). The Reagan Administration had been arguing that spending more money doesn't improve education and may even make it worse. But the chart was doublespeak. First, it used current dollars rather than constant dollars, adjusted for inflation. Because each year it takes more money to buy the same things, charts are supposed to adjust for that increase so the measure of dollars remains constant over the years illustrated in the chart. If the Department of Education had figured in inflation over the years on the chart, it would have shown that the amount of constant dollars spent on education had increased modestly from 1970 to 1986, as Figure 34.3 shows.

Second, scores on the Scholastic Aptitude Test go from 400 to 1,600, yet the graph used by the Education Department (Figure 34.2) used a score range of only 800 to 1,000. By limiting the range of scores on its graph, the department showed what appeared to be a severe decline in scores. A properly prepared graph, shown in Figure 34.4, shows much more gradual decline.

The Department of Education's presentation is good example of diagrammatic doublespeak.

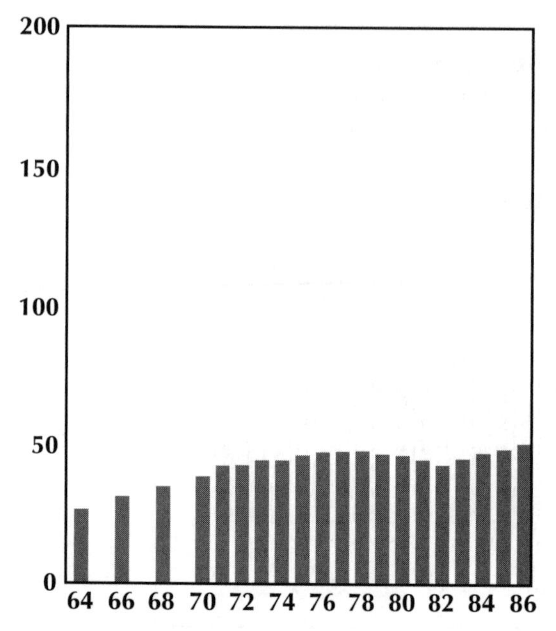

Figure 34.3 Elementary/secondary education spending in constant dollars (billions).

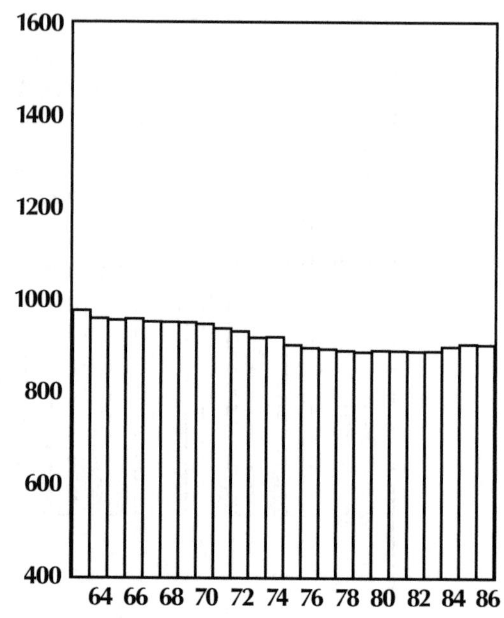

Figure 34.4 SAT scores, 1963–1986.

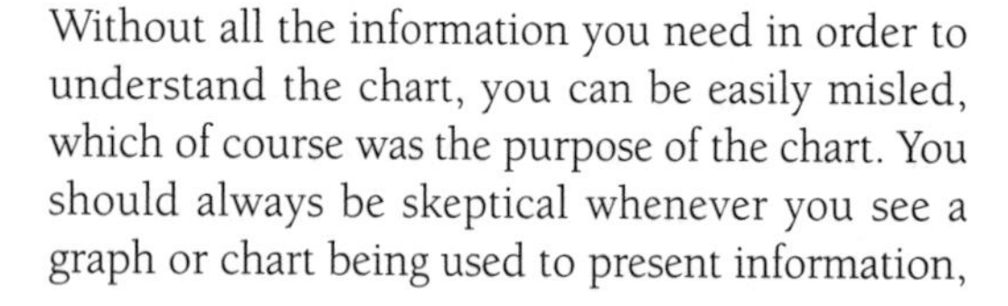

Without all the information you need in order to understand the chart, you can be easily misled, which of course was the purpose of the chart. You should always be skeptical whenever you see a graph or chart being used to present information, because these things are nothing more than the visual presentation of statistical information. And as for statistics, remember what Benjamin Disraeli is supposed to have said: 'There are three kinds of lies—lies, damn lies, and statistics.'

QUESTIONS FOR CRITICAL THOUGHT

1. What is doublespeak?
2. What are the four kinds of doublespeak described by Lutz? Are graphs a fifth kind?
3. Lutz implies that doublespeak is not communication. Why not?
4. Compare Lutz's treatment of doublespeak with Casselman's 'Bafflegab and Gobbledygook' and with Orwell's 'Politics and the English Language'. Do you see any differences?
5. Describe Lutz's tone. Does he regard his audience primarily as consumers of doublespeak or as producers of it? How do you know?
6. How can we identify doublespeak in the messages we receive? Does Lutz have any advice on how we might avoid writing doublespeak ourselves?
7. How, exactly, do graphs function as doublespeak?

8. Is doublespeak determined by vocabulary or by intention, or both? How can we determine intention in a written message?
9. How accessible is Lutz's style? What features make it so?
10. Is Lutz's passage personal? How formal is it? Why did Lutz make the stylistic choices he did? To what extent do audience and context influence these choices?

CHAPTER 35

Bafflegab and Gobbledygook: How Canadians Use English to Rant, to Lie, to Cheat, to Cover up Truth, and to Peddle Bafflegab

Bill Casselman

What Is Gobbledygook?

Gobbledygook is language used to deceive, not to communicate. It is language spoken or written to evade responsibility, not to answer a question but to defuse it by coating the answer in a syrop of verbal glop. What follows comes from some of the worst cases of gobbledygook ever uttered or written by a Canadian.

For example, a current major phoney-baloney word is bycatch. Green Peace and The Suzuki Foundation have both expressed concern over this fishermen's cover-up word. The Pacific Seafood Processors Association told the *Vancouver Sun* (7 December 1988) about their 'bycatch'. It looks and sounds like a fisherman's honest and sturdy term. In fact, it is a cover-up word that refers to thousands of seabirds, seals, dolphins, whales, and sea lions that die strangled in ocean driftnets set to catch surface-swimming squid and other edible 'sea-mass'. Whenever you see the word used with approval, you know some kind of creepy lying is going on, because honest people don't use the word. Gobbledygook is often that simple.

Acid rain in Canada becomes 'atmospheric deposition of anthropogenetically-derived acidic substances'. Such language means to suffocate stark fact in a muffling blanket of mumbo-jumbo. A medical services coordinator for a fire department in Canada reported that emergency personnel had found a victim 'in a non-viable condition. He had no pulse and was not breathing.' Well now, Mr Coordinator. Can I call you Lem? That'd be what? Dang near ready fer a white nightie and little wings? Leastways, tuggin' at the Grim Reaper's hem, eh? I mean, you get a citizen up there in the higher realms of non-viability, you're saying: the buzzards are circling, right? I think what you're suggesting, Lem, is: this dude was rapidly approaching non-dudedom. Bone-yardwise, ready to be planted. Would I be close? Lem, seriously now and all kidding towards one side, could this guy be dead?

Gobbledygook can be euphemism for the lowest motive. A Canadian funeral parlour is now a 'Bereavement Resource Centre'. As someone driving by their tasteful, glow-in-the-dark plastic sign wondered, 'Gee, do they still take the dead?'

A CBC TV interviewer asks a surgeon about the source of a transplanted kidney. 'From a cadaveric donor', says the pussyfooting sawbones. If he had not used euphemistic gobbledygook, think how influential this doctor (who had just benefited along with his patient from the process) could have been, by reminding people to sign the donor card on the back of their driver's licence and by stating that spare parts of the newly dead can keep some people alive. But no, he chose not to offend viewers and fudged the source with the rare adjective 'cadaveric'. Such misdirected gentility makes some want to jump up and yell, 'You mean you cut it out of a stiff, eh, Doc?'

A bureaucrat from the British Columbia Ministry of Education comes to a school with bad news for its teachers, but never once uses the now context-sensitive word 'teacher'. Instead, throughout his speech, this evader speaks of 'on-site facilitators of pupil learning'. Plenty of this spin-doctor-spiel is actually bad English and semi-literate, like the verbal garbage in the last sentence.

Origin of the Word Gobbledygook

No surprise is it to learn that this term sprang to the lips of a man who had to listen to politicians blabbing. During World War II, Congressman Maury Maverick of Texas made the word up one day in Washington, DC, after listening to more verbal bamboozlement than he could abide. In May of 1944 Maverick told the *New York Times* magazine: 'Perhaps I was thinking of the old bearded turkey gobbler back in Texas who was always gobbledygobbling and strutting with ludicrous pomposity. At the end of this gobble there was a sort of gook.' The new word was so echoic and fitting that it passed immediately into popular speech.

The congressman has an interesting last name. Could it be the origin of the word 'maverick' to name anyone unorthodox and not part of a group? Yes, the congressman's grandfather was Samuel E. Maverick (1803–70) who was a Texas rancher and state politician who refused for certain practical reasons to ever brand his stock, vast herds of longhorn cattle. He then playfully claimed that all unbranded range stock might belong to him, following an agricultural precedent of pioneer America that unbranded animals on the open range, not rustled and unclaimed, belonged to whoever first branded them. Texas ranchers took to calling any unbranded cattle that wandered from a herd 'mavericks'. Then use of the term in Texas politics followed, and a maverick became any politician who would not follow his party's policy line.

Gobbledygook on the Job

In August 1987 the Health Sciences Centre in Winnipeg placed a Help Wanted ad for a 'Co-ordinator, Occurence [sic] Screening, Quality Assurance Department'. Uh-huh. Not too assuring a start. The word 'occurrence' is misspelled, not once but four times more in the explanation that followed. Help is indeed wanted. But in the next paragraph are the real meat and potatoes of the advertisement. Should I wish to write 'meat and potatoes' in gobbledygook, I could say: the veridical carnal comestibles and solanaceous tuberosities.

The ad reads: 'Occurence [sic] screening is an objective, criteria-based review of medical records conducted concurrently and retrospectively to identify and flag, confirm, analyze, trend, and report instances of suboptimal care attributable to health care disciplines. Under the general direction of the Director, Quality Assurance, the incumbent will co-ordinate the development, implementation and maintenance of multidisciplinary and integrated systems of occurence [sic] screening; will assess and review adverse patient occurence [sic] data; will assist with the identification of existing and/or new resources required to conduct occurence [sic] screening.'

Now I am a mere medical layman in a state of mystification, but it appears all they want is a clerk to tote up procedural mistakes. But is that the job offered? I will never know, due to the ad's muzzy verbiage. By the way, would death be an adverse patient occurrence? I do want to say something positive. It is very assuring to be 'under the general direction of the Director', as long as he's not directing their spelling.[1]

BUREAUCRATIC BAFFLEGAB, GOVERNMENT-STYLE

Must gobbledygook always consist of many-syllabled senselessness? No, often it appears in language plain and bare. In 1996, Prime Minister Jean Chrétien's Liberal government brightened hopes of school-leavers with a 'First Jobs' program. Canadian corporations were to help young people get into the work force by hiring them for one year. 'First Jobs' emerged from a brain-storming session with federal Treasury Board chairman Art Eggleton and the Boston Consulting Group which then canvassed corporations for their ideas. Many of these same corporations continue in the process of massive layoffs designed to increase profit at the expense of Canadian workers. During the years 2002 to 2005 Ontario alone lost 60,000 jobs in manufacturing. The young people in Chrétien's 'First Jobs' scheme—really temporary help who can be let go after one year—were going to be paid near-minimum wages.

What 'First Jobs' amounted to was the Liberal government, elected on a promise to create jobs, giving sanction and perhaps financial assistance to corporations to fire older, full-time employees working for higher salaries and to replace them with minimum-wage internships by temporary workers. 'First Jobs' was a weasel term. A more accurate title would have been 'Short, Meaningless Jobs Designed by the Liberals to Woo Corporate Canada'.

An Ottawa civil servant once said that it had taken years to learn to write in a style such that his government department could not be held accountable for any error.

In 1995, a knife-wielding intruder slipped past those guarding our Prime Minister's residence and penetrated the mansion as far as the Chrétiens' bedroom. Although the trespasser was caught before physical harm came to the Prime Minister and his wife, the RCMP sentinels were lax. This sloppy guarding called forth a paper blizzard of departmental studies. Here is part of one public RCMP document, *Security Breach—24 Sussex Drive. Final Report, 1995-11-17*. One section of this report entitled 'Shift Scheduling' begins with basic information: 'The members attached to the Prime Minister's/Governor-General's Uniformed Security Detachment work on 12-hour shift rotations consisting of 2 days, 2 nights and 4 days off. . . . A member rotates from post to post with commensurate breaks depending on resource numbers.'

Then comes a review of whether or not a 12-hour shift is too long for an individual RCMP guard. Here the reader asks himself how attentive anyone is after doing anything for twelve hours straight. The report continues: 'Schedules impact morale and health, as well as operational efficiency. Often the greatest levels of efficiency improvement will result in unacceptable attenuation of "quality of life" considerations. Hence a schedule that is acceptable to both the workers and the managers will likely consist of a negotiated trade-off and will often be less than maximally efficient. The demarcation of reasonable efficiency ranges that are also consistent with psychological, social, and physical health is important. Quality of life [sic] considerations . . . translate into such issues as minimizing the number of consecutive duty tours in nights, maximizing the number of complete or partial weekends off during the year, and maximizing the predictability of working tours.'

In plain English: Do cops get careless, tired, and sick after too many consecutive night shifts? Yes. Just like all the rest of us other human beings who have evolved as day-active mammals designed to sleep at night. But the report's author cannot bring himself to utter such clarity. It might sound like the RCMP was whining. Awwwww.

Clear statements about too few guards getting dopey from too long, poorly designed shifts would not be stern, police-like, and mature. But that is the excuse offered in this report for the bad police work.

In order not to make the RCMP sound like a bunch of complaining sucks then, the report puffs itself up with deceptive language: semi-sociological jargon mixed with a stuffy bluster of pseudo-psychological bureaucratese. A gobbet of gobbledygook like 'unacceptable attenuation of "quality of life" considerations' might read on the surface as brisk and concerned, until one pauses to ask what it means. In the context of the paragraph quoted above, it means: put guards on long shifts and they'll get clumsy, inattentive, maybe injured, or shot by intruders.

Scientific reports in correct English don't make formal mistakes like 'unacceptable attenuation of "quality of life" considerations'. When a noun phrase is used as an adjective before another noun, one does not surround it with quotation marks; one hyphenates the phrase so that it appears like this: quality-of-life considerations. The very act of putting 'quality of life' in quotations is also a macho dismissal of the phrase, as if to say: I'm putting this in quotes because I'm a big, brave policeman and quality of life is only for weaklings. Yes, it's a tiny point. But little mistake after little mistake sticks in the alert reader's mind, until very early in perusal of the report, the reader sniffs the acrid reek of prose sweat, the stench of English sentences straining to be what they are not.

This prose wants to appear scientific. But it is only a cheap imitation of the general sound of scientific prose, the best of which communicates ideas, reports facts, and clearly labels suppositions and hypotheses. Such copy also loves itself. There is a self-congratulatory smugness in the repetition of clichés borrowed from business reports and scientific abstracts, words like 'commensurate', 'impact' as a verb, 'maximally efficient', 'predictability'. This is not English clear and simple and eager to tell you something. This is a ready-to-burst colostomy bag of words. It's full of crap, and the person who wrote it knows that. But the report had to be written. Orders descended from Ottawa politicians, from the Prime Minister's Office, from RCMP HQ as faces reddened and leaders became vulnerable to bad press about the break-in. Yes, after the break-in and after this report, some changes occurred in the way the RCMP guards our Prime Minister. But better prose, more clearly expressed, in an earlier report, might have prevented the frightening duration of the trespass.

Small Print Alert!

Be suspicious of small print. Use a magnifying glass always. Copy is not set in teeny type to save space, but to discourage people from reading the gobbledygooked message. Here is part of an application for a joint checking account once used by a Canadian chartered bank. By signing it, you agreed:

> that the Bank is hereby authorized to credit the said account with all monies paid to the Bank (i) at the branch of account or (ii) at any branch other than the branch of account for the credit of any one or more of us, the proceeds of any orders or promises for the payment of money, of bonds, debentures, coupons or other securities, signed to be drawn by or payable to or the property of, or received by the said Bank (i) at the branch of account or (ii) at any branch other than the branch of account for the credit of us or any one or more of us, and to endorse any of such instruments on behalf of us of any one or more of them.

Whew! Any reader of that would wave the olive branch and sue for semantic peace. That is the worst English I've ever seen in small print. Would you entrust financial responsibility to an organization that approved such incomprehensible bank-speak? I would not. It could be rewritten in clear English using lists, clarifying punctuation, and nixing the not-legally-necessary repetitions. This is old-fashioned, pompous legalese, imitating the clotted English of Victorian contracts, and even as such it is an amateur's botched attempt at contractual precision.

Legitimate Uses of Jargon

The making of new words is a sign that a language is vital. But never be under the illusion that because a new label has been plastered on something, because it has been dubbed with a scrap of pseudo-scientific jargon, we will somehow know it more deeply.

On the other hand, all sciences need a private vocabulary. Some jargon is necessary. In one sense, all professional terminology is jargon, including that of medicine, law, education, business, sports, theology, etc., because the unique word-hoards of these fields contain terms not familiar to the general population. Linguists estimate that one-half of the vocabulary of all major world languages consists of scientific and technical terms. In Modern English, with more items of vocabulary than any other language that has existed, technical vocabulary makes up closer to 80 per cent of all words. Jargon, of course, is not gobbledygook. The legitimate use of private vocabularies in science is meant to make complex statements precise and clearer than ordinary language would permit. Gobbledygook wants information made imprecise and unclear, in order to befuddle the reader sneakily.

Notes

1. This example and several others in this chapter are drawn from Rick Coe's excellent and unfortunately discontinued 'Doublespeak Update' which appeared twice a year (1987–90) in the academic journal *English Quarterly*, published by the Canadian Council of Teachers of English when Coe was a professor of English at Simon Fraser University in Burnaby, BC.

QUESTIONS FOR CRITICAL THOUGHT

1. What is bafflegab? Is it distinct from gobbledygook, or are the two synonyms?
2. Casselman advances his argument primarily by example. How effective is this method?
3. What is Casselman's purpose?
4. Who is Casselman's intended audience? How do you know?
5. To what extent is Casselman's essay primarily one of definition?
6. Why does Casselman employ such strong phrases as 'spindoctor spiel', 'syrop of verbal glop', 'verbal garbage', 'phoney baloney', and the like? What is their effect?
7. Why does Casselman provide a history of the word 'gobbledygook'?
8. How would you characterize Casselman's tone? How is it established? Point to some specific characteristics of the passage in formulating your answer.
9. What is a weasel term?

10. Casselman characterizes one of his examples as 'a ready-to-burst colostomy bag of words. It's full of crap.' What is the function of such directness and such colourful language?

11. What is the purpose of gobbledygook and bafflegab? Why do such phenomena exist? What, if any, is their relationship to the objective narrative style of technical writing described by so many of the contributors to this book?

CHAPTER 36

Gasping for Words

Arthur Plotnik

Something moves you to express your thoughts. The subject is love. Or beauty. Mortality. Some poignant experience.

You hesitate—and well you should. Launching ideas as messages is not exactly blowing kisses from a train. You are putting thoughts into words, which is more like flapping the tongue to escape gravity. We work our tongues endlessly, but lift-off is so rare it's a miracle we don't keel over like some NASA dud.

Yet we go on flapping rather than fall silent or simply moo at one another. We struggle with words because they separate us from the lowing beasts and tell the world who we are, what we want, and why.

No one will dispute the need for verbal expression, because no one will sit still to listen. The need is assumed, but it is never more clearly illustrated than when Americans visit foreign lands of funny-speaking people. Even with a stock of their funny phrases we find it difficult to express our individuality. We barely distinguish ourselves from the wash jerking on the clotheslines. We feel like babies, unable to express the nuances of pleasure and discontent. And babies hate that feeling.

In foreign travel I often find myself, oh, about fifty thousand words short of being interesting to anyone but the local pickpockets. I remember one moody trip when, travelling alone, I dined night after night talking to my cheeses and such mistakenly ordered dishes as pickled cow's face with hairy nostrils. One evening, a sensitive-looking young couple gestured for me to join them. We exchanged basic phrases, but what I wanted to express was an overflow of feeling, something like this:

> My dear companionable saviors—For the last three weeks a shadow of melancholy has obscured my perceptions, dimming the beauty of your countryside and the conviviality of its inhabitants. Solitude, when no longer self-imposed, soon deepens into isolation and near madness. Now, however, as your kind concern and sensitivity restore my spirit, all that I have perceived unscrolls and engulfs me in its majesty. I exult in your land and its people.

What came out was the equivalent of 'Me like here. Food good. Everything very good. You go America?'

They'd sat a three-year-old at their table.

It is so often like that, even at home in our native language. We ache for the radiance of expressiveness—of vivid expression. We grope for words to light up the cosmos or the written page or the face across the table. But the harder we try, the more we seem to darken the waters, like a squid in its ink.

This is not a book to scold us squid. We are all in the dark. Only for the most routine messages do our habitual means of expression suffice. We have at hand the words and gestures to convey that it's raining outside, that we feel hung over, that we wish the @#! television turned down. We can describe a head cold ('miserable'), an appliance sale ('unbelievable'), or last night's extra-inning ball game ('great') to our satisfaction. These are generic events, events that perfectly fit a type, and for all anyone cares, we can package them in our most generic language.

Reacting to life's nongeneric stimuli, however, we find ourselves grossly unprepared. So many thoughts and passions stampede inside us, as mad for release as the bulls of Pamplona. But open the gates and see how our puny words scatter, overwhelmed and impotent. We gasp for the telling language and we choke. We rant. We go mute. We claim linguistic fifth amendments: 'Words fail me.' 'It defies description.' We buy sympathy cards the size of tombstones just to say, 'WORDS CANNOT EXPRESS. . . . '

Let No Gallo Write My Epitaph

Unable to express the nuances of experience, we fall back on generic language in ways that mock our humanity. When wine producer Julio Gallo died in 1993, his brother and sixty-year business partner Ernest offered this statement, according to *The New York Times*:

> Julio was a great brother, a great partner, and a great human being. His passing is a great personal loss to me and both of our families.

Can words no longer express human sentiment? Have our lives grown so subtle and complex as to outrun our 600,000-word heritage? Or has the American experience become such that a few flabby expressions and thirty million handguns say it all?

Perhaps Ernest Gallo was just exhausted, as are most people.

Yet even in supercharged moments we succumb to generic drag. Recently three hyperenergetic teens from good schools appeared as guest critics on a television show. Reacting to a remake of the film *Of Mice and Men*, they fairly quivered with insights and jostled for a turn to express them. Yet the expression 'Great!' was the sole adjective in their arsenal. Not one of the bright-eyed three could escape the tug of that word.

Each day, even as specialized vocabularies grow, fewer words seem to serve for larger gulps of experience. Every parent knows the tendency of kids to describe the world according to the most fashionable dyad of the day: neat/dumb, excellent/gross, dude/dork, whatever. Millions of adults are already fixated in this stage. At a textbook publishing house in Chicago, for example, one programmer categorizes all phenomena as 'cool' or 'bummer'. Such bipolarity—friend/foe, food/nonfood—smacks of organisms that lick the environment rather than describe it.

But, bummer! Who wants to reverse the miracle of human expression? The more complex our experience, the more we long to unravel it in words. The more generic our lives, the more we yearn to express our individuality. When it comes to language, however, our floundering prompts the saying '*tanto nadar para morir en la orilla*'—so much swimming only to perish at the shore. Twelve, sixteen, twenty years of schooling, inundated by verbiage every day, and we can't speak our hearts.

Generic Patterns

What's the answer? Vocabulary building? That is a partial answer, if one's vocabulary is freshened and not merely encumbered. Hauling around words like *xeric*, *succedaneum*, or *quaquaversal* can be counterproductive. Word tonnage isn't the point. With no more than eighteen thousand different words, Shakespeare's writings have stimulated the Western world for four centuries; the average American commands some twenty thousand words and about four minutes of attention.

We have lively words, but we are stuck in patterns developed in an age of standardization and mass media. As we parrot standard models that reach for the lowest common denominator, we express ourselves in phrases used by everyone for everything. Often, to compensate for the banality, we pour on pop phrases and jargon like Tabasco over cornflakes. Language to describe distinctive moments or thoughts is parked somewhere in our heads, but the synapses to carry it forward are choked with babble.

'Having an incredible time in the Himalayas,' reads a postcard from a young traveller. 'The scenery is so excellent. We stayed at some fantastically neat monasteries. The Buddhist-type thing is awesomely mind-boggling and bizarre and could totally change my whole lifestyle. I'm like, where am I?'

A postcard needn't aspire to revelation, but back home its writer will deliver longer versions of these adventures in language no more revealing, even though this traveller has experienced an epiphany of sorts and yearns to describe it. In all areas of expression we find ourselves speaking and writing postcards, stepping back to frame everything in stock generalities and panoramic vistas. And the response we can expect is the one we give to postcards: flip, flop, and into the trash.

Earlier Eloquence

One hears much lamenting about the decline of average language skills in the United States. Sometimes the lamenting itself is lamentable. Comparing writing and speaking skills out of context is slippery business, yet who doesn't marvel at the expressiveness of soldiers, travellers, frontier women, and so many other nonliterary Americans of earlier centuries? Were they better educated? Not necessarily. Were they exposed to superior models? Not always, though perhaps fewer corrupting forces pounded away at them. Were they driven by incentives now meaningless? Perhaps.

Jacob Shallus (1750–96), born in Philadelphia to German speaking parents, was an ordinary early American I happened to study for a biography I was writing. A clerk and copyist for the Pennsylvania Assembly, he struggled against poverty yet aspired to the lofty style of the documents he inscribed on parchment (including the Constitution of the United States). In 1788, even as he grovelled for a loan, he grovelled eloquently. Imagine Visa receiving this note:

> Sir
> Being arrested for the sum of Twelve pounds some odd shillings and likely to be under the necessity of going to confinement, I have taken the freedom to address you for my relief as well as that of a wife and eight children who must otherwise feel the effects of my situation—Good Sir I can restore it to you as soon as the Assembly meet, therefore in treat you will estend your friendship to a person altho' little known to you but who will exert himself upon every occasion for your Service. Pray do not deny me and contemplate that you are lending a small pittance for a few days only and thereby destroy the violent intentions of an unrelenting Creditor.
> I am,
> Sir With great esteem
> Your very hble serv
> JShallus

Not masterly prose or contemporary spelling, but how many civil servants today can toss off these elegant constructions? Before democratization softened class distinctions, a pretty way with words affirmed one's status, helped separate one from the lower orders. Conversely, low-born citizens might mask or even transcend their origins by imitating the diction of the elevated.

Today, however, we speak as blandly or as jargony as we please; we speak the language of our subcultures, and, hey, whoever don't like it can you-know-what. Recently I saw a T-shirt that proclaimed, IF YOU DON'T LIKE MY ATITTUDE, STOP TALKING TO ME! Status rarely drives us to acquire a more 'cultured' or more generally expressive idiom. English teachers may point us toward 'prescriptive' grammar, that is, the 'standard' patterns used by the society's leading white literati, but substandard grammar won't bar us from Martha's Vineyard, a skybox at the Super Bowl, or even the boardroom. Americans judge status by many factors other than expression. Too often one hears, 'How much does that SOB make to be opening his mouth?' and not always, 'His stunning way with words assures us of his authority.'

Still, the longing for expressiveness remains, if not to achieve status then to experience the joy and relief of getting what's inside outside. Unexpressed, the particular richness of a life is lost in the mall of generic memory. We barely know what an experience was until we shape it into words that somehow distinguish it. Sensing this, most of us have resolved to sharpen our powers of expression, just as we've resolved to firm up those hips and thighs; but a thousand obstacles slow us down. We tend toward inertia; we need prodding.

Cyrano de Plotnik

Prodding, of course, can be gentle or otherwise. The most classic example of 'otherwise' occurs in Edmond Rostand's drama *Cyrano de Bergerac*, when the swashbuckling Cyrano tutors an enemy aristocrat in expressiveness.

Having disrupted a theater performance, Cyrano offers 'pointed' appreciation to anyone in the audience who will describe, as a subject of comedy, his famous, prodigious nose.

The Viscount de Valvert steps forward. 'Your nose is big, very large,' he sputters.

Working the crowd, Cyrano responds (in rhymed French): 'That's it? Rather brief, young man. Oh! God! There are so many ways you might have expressed it!'

Cyrano then offers the lesson, challenging himself to invent a pithy insult of his nose in each of nineteen categories, including pedantic, eloquent, dramatic, admiring, lyric, rustic, military, and practical. For example:

> Dramatic: *'Tis the Red Sea when it Bleeds!*
> Admiring: *The very sign a perfume shop needs!*

And so on, in a panoply of self-insult that Cyrano now turns on Valvert—'These, Sir Idiot, are things you might have said, had you some tinge of letters, or wit, within your head.' With that, the swords are drawn. Even as they duel, Cyrano describes the swordplay in an extemporaneous ballad, speeding to the immortal line:

> Then, as I end the refrain . . . thrust home!

The viscount falls, the wiser in matters of expression—but unfortunately, run through.

Now, that's prodding. But there is also the gentler sort. Enter this essay and its author, your very hble serv., APlotnik. Who is APlotnik? On the one hand, I am a career observer of expression, awash in credentials as an author, editor, publisher, and speaker. On the other hand, I remain a struggler after expressiveness, a stumbler, someone who still bleeds over every sentence, pounds tables and slams doors at the frustration of trying to express the most elemental thought. I am as guilty of small talk as any hairdresser and not unknown to say

'Great!' and 'Really!' in the process. Thus there will be no superior tone, no linguistic snobbery, here. In fact, some passages—maybe this one—will be monuments to the peril of opening one's yap.

It should come easier to me. As a younger man I lived and worked in New York City, where snappy expressiveness is the *sine qua non* of getting attention, where even failure to articulate a sandwich order can mean shame and lost opportunity. But when the formidable delicatessen clerks cried 'Next!' I was often the hesitator, the one who lost his turn as he drawled, 'Let me think. . . . '

I'm still thinking. Not only about the power of expression to secure an onion bagel lightly toasted Nova lox no butter with a schmear (of cream cheese)—but beyond! . . .

Not to take anything away from water rifles and computer games, but words—sacred, silly, or profane—still provide some of the best diversion that life has to offer.

QUESTIONS FOR CRITICAL THOUGHT

1. What is the most striking feature of Plotnik's style? Provide some examples from the essay.
2. Plotnik isn't talking directly about technical writing, but about verbal expression in general. What value is there for a technical writer in reading his work?
3. Plotnik claims that 'we parrot standard models [of expression] that reach for the lowest common denominator, we express ourselves in phrases used by everyone for everything.' To what extent is his criticism of contemporary writing similar to Orwell's ('Politics and the English Language'). In what ways is it different?
4. Plotnik describes a 'longing for expression' that he argues is central to human experience. Is he right? In what sense is it true that 'we barely know what an experience was until we shape it into words.'
5. Are words really a 'diversion' to Plotnik, or are they a matter of greater seriousness? How do you know?
6. For whom is Plotnik writing? How can you tell? Is his writing adequately adapted to his audience? How does he engage them and maintain their interest?
7. How personal is Plotnik's article? What is its purpose? Is the level of personal engagement appropriate to that purpose? Why or why not?
8. How audible is Plotnik's voice? What are the elements that make it so?
9. What elements of Plotnik's style would not be suitable to a formal technical report? Why?

CHAPTER 37

What Do You Mean I Can't Call Myself a Software Engineer?

John R. Speed

Your job title is *software engineer*. You've had the title for years. It's on your business card. However, somewhere along the line, you might have encountered someone who said that you couldn't call yourself a software engineer. They said it was against the law. Although you managed to ignore them at the time, you might have secretly wondered if they were right.

At the Texas Board of Professional Engineers, we regularly receive questions arising from such encounters. The callers come from every background—from high-tech to no-tech. Their education and experience levels vary from those without high school diplomas to PhDs, and from basic entry-level jobs to the highest technical ranks. Is it true that they can't call themselves engineers? The question seems straightforward, but the answer is complicated.

Two fundamental legal realities set the stage for the rest of this discussion. Only licensed professional engineers

- have the legal authority to use an engineering title without restriction, and
- can offer engineering services.

Because the term software engineering is so common, the Texas Board of Professional Engineers felt that existing restrictions on the use of engineering could grow into an undue business hardship on software engineering practitioners. In June 1998, the Board established software engineering as a recognized engineering discipline and established licensing criteria specifically suited to software engineers (Speed, 1998). The intent was to streamline the licensing process for a growing new field of engineering.

To get a better understanding of the situation, let's take a crash course on licensing, title use, and engineering practice.

What Do Engineering Licensing Laws Regulate?

In most US states and Canadian provinces and territories, laws prohibit the unlicensed 'practice of engineering'. Texas law is typical, so we'll use it as an example. (See the two boxes for information on licensing and certification information in British Columbia and Ontario, and in the UK.)

Regulation of engineering practice starts with a legal definition. The definition of the 'practice of engineering' generally looks something like this:

> The 'practice of engineering' means any service or creative work, the adequate performance of which requires engineering education in the application of special knowledge of the mathematical, physical, or engineering sciences to such service or work. (*Texas Engineering Practice Act*, rev. 1999)

Despite its circular nature, the definition has several important features.

- Engineering practice is defined in terms of an engineering education
- Engineering practice can be recognized by

Certification in the United Kingdom

by Trevor Burridge

Engineering accreditation in the UK is governed by the Engineering Council, which acts as an umbrella organization for all the individual engineering professional bodies, including the Institute of Electrical Engineers, the British Computer Society, and the Institute of Mechanical Engineers. Each institute has a number of professional or corporate membership grades (member of fellow). For those that cannot qualify for professional membership, there are other grades, such as associate member or companion.

To qualify for professional membership, applicants must demonstrate a sufficient combination of education (usually good grades in an accredited bachelor's program), training (usually two years of approved training), and relevant experience in a position of responsibility. Substantial experience can often be offered in lieu of formal training and, in some circumstances, even education. Applicants must be sponsored and proposed by several existing professional members and must undergo a technical interview by a panel of experts.

Corporate members of each professional engineering organization are eligible to be entered on the Engineering Council Register and can only then use the designator CEng (Chartered Engineer) after their names. Engineers who fail to meet accepted professional standards or fail to abide by the code of conduct and ethics can face disciplinary action, which might result in them being taken off the register. This form of certification is the same for all branches of engineering in the UK.

The IEE and BCS are both full members of the Engineering Council. For almost a decade, both these bodies have been accepting and awarding corporate membership and Engineering Council registration to software engineers. For more information about IEE and BCS membership and entry requirements, see http://www.iee.org and http://www.bcs.org.uk. The Engineering Council website is http://www.engc.org.uk.

A chartered software engineer therefore has similar entry requirements and standing as a chartered civil or mechanical engineer. However, from my experience, only a small minority in the software development field in the UK bothers to apply for any form of membership, let alone submit to the entry procedures for professional membership. I believe the main reason for the low uptake is that employers do not see it as mandatory or even especially desirable. Job advertisements focus almost exclusively on the technical skills with the three-year half-life in C++, Java, NT, UML, Orbix, Oracle, and so on. (See http://www.jobserve.com for typical UK software jobs. Search for 'software engineer' and see what you come up with.) This is even true with more 'enlightened' employers who are genuinely attempting process improvement initiatives.

I suspect part of the problem is the chicken-and-egg situation. Employers think that there are not many chartered software engineers, so they see no point mentioning this as desirable in job advertisements. Applicants see that chartered status isn't important to get a job, and it gives no extra benefits or pay, so why bother with the hassle of becoming chartered?

The UK has had this version of full engineering certification for software engineers for a decade now, but it is still not a factor that most practicing software developers consider important.

Trevor Burridge *is a freelance consultant. Contact him at trevor@parallel-consultants.com.*

its application of mathematical, physical, or engineering sciences to a problem.
- Any problem that cannot be adequately solved without applying an engineering education to it is, by definition, an engineering problem.

US states create licensing boards to establish a system to license qualified engineers and regulate their practice (*Engineering Licensure Laws*, 1999). By law, those rules must concentrate on the protection of public health, safety, and welfare.

Some software engineering activities clearly fall under the legal definition of the engineering practice. For example, the design, testing, and implementation of embedded and real-time systems require a detailed understanding of the engineered electrical or mechanical components. Similar activities performed on software systems for mechanical devices, electrical devices, and power systems are clearly engineering services.

Other activities are harder to classify as engineering or nonengineering, primarily because of the vague nature of the phrase 'public health, safety, and welfare'. Although almost everyone understands the need to protect public health and safety, many gloss over the term *welfare*. After all, it's easy to justify regulations that are designed to keep water clean or prevent explosions. But what does welfare mean in this context? It describes a state of general well-being or prosperity. I suggest substituting words and phrases such as money, resources, or business interests in place of welfare. Instead of protecting public health, safety, and welfare, the regulations are protecting the public health, safety, and business interests of citizens. Because of the many ways in which software affects welfare and the term's encompassing meaning, you can see why most enforcement cases the Texas Board investigates involve engineering's impact on welfare.

British Columbia and Ontario Licensing

by Keri Schreiner

In recognition of the often-unconventional path to software development expertise, the Professional Engineers of Ontario organization has announced new criteria for licensing software practitioners as professional engineers.

Traditionally, an applicant for licensure must have experience and education in the same area. However, the death of accredited software engineering programs makes it difficult for software practitioners to meet the criterion. To address this issue, PEO has until now assessed applicants for software engineering licensure on a case-by-case basis.

PEO President Patrick Quinn said that their new criteria defines the core knowledge required for PEng licensing and provides a way to consistently assess practitioner's qualifications. 'This is an important change and a first step in introducing professional regulation to the software industry,' Quinn said, adding that the Y2K problem has highlighted the need for such regulation and professional accountability.

Provided they meet general criteria, candidates without a degree from an accredited engineering program or equivalent can now be licensed by PEO if they pass written exams or proved relevant work experience beyond the four years required under general licensing criteria. Other general criteria include knowledge of control theory, mathematical foundations, digital systems and computer architecture, and software design and programming fundamentals. Candidates must also show knowledge in three of seven specialization areas and successfully complete the Professional Practice Examination on engineering law and ethics.

PEO's announcement follows an earlier one by the Association of Professional Engineers and Geoscientists of British Columbia, which said in June that it would begin licensing software engineers. The APEGBC's qualifications are a bachelor's degree or higher in electrical or computer engineering, engineering science, physics, or the computer-software field and a minimum of four years of software engineering experience.

Like PEO, the APEGBC cited two primary factors in its decision to extend licensure to software practitioners: their important contributions to the engineering field and the need for professional regulation in the software industry. Stuart Bourhill, chair of the APEGBC's Computer and Software Engineering Task Force, said they have been working closely on licensure issues with the Canadian Council of Professional Engineers, the IEEE, and several Canadian universities. 'We believe that only through close cooperation with these other groups and professional societies will we be able to adequately address the needs of the software engineering professional,' he said.

The APEGBC also announced the formation of a new division for computer and software engineering that will provide a forum for practitioners to discuss professional guidelines, development strategies, and education opportunities. More information on both announcements is available at the PEO website (http://www.peo.on.ca) and the APEGBC site (http://www.apeg.bc.ca).

Kari Schreiner *is a contributing editor for IEEE Software. Contact her at keri@grooveline.com.*

Restrictions on Engineering Titles

Licensing laws are designed to maintain a level of integrity in the engineering profession, restricting use of the term *engineer* to people who have demonstrated a minimum competency (*Texas Engineering Practice Act*, rev. 1999).

Engineer title restrictions are most strict in situations where someone is attempting to offer consumers consulting or contract services. They are also strict in situations where engineering is practiced in the public sector. Licensing boards have designed title restrictions in these circumstances to protect consumers against fraudulent claims. The laws operate on the fundamental premise that a US state government should provide some level of consumer protection against fraud, and they are aimed at those who falsely claim engineering expertise.

If you are reading this article, you are probably most concerned with a few specific titles such as software engineer, systems engineer, or network engineer. Licensing boards are concerned with *any* use of the title engineer. As I mentioned earlier, US state licensing boards allow unlimited use of engineering titles only if you are licensed as a professional engineer in that state (*Engineering Licensure Laws*, 1999).

A person can also break laws related to the use of the term engineering in two ways. One way is to offer or perform engineering services without obtaining the required qualifications. The other is to incorrectly label an activity engineering. In Texas, we see the law broken in both of these ways.

Unqualified Software Engineering

It is illegal to practice software engineering in Texas without a license (or an appropriate exemption, which I'll discuss later). As I mentioned earlier, development of software for engineered systems—including embedded systems, real-time systems, mechanical devices, electrical devices, and power systems—is software engineering. A person who performs software development work in these areas without a professional engineering license or exemption is breaking the law. Many practitioners try to skirt the law by calling their work 'technical service' or 'technical consulting' instead of engineering'. Regardless of what they call it, offering engineering work without a license is illegal.

Calling Software Development Software Engineering

Many kinds of software development affect the public welfare but are not considered software engineering under Texas law. Work on financial systems or business systems might affect the public welfare, but it doesn't require the application of physical sciences, and can be performed without an engineering education. Unless these activities also include the design of hardware interfaces, power systems, or other engineered components, they do not appear to meet the legal definition of engineering practice.

When someone reports illegal software engineering practice, the Board's first task is to determine whether the practice is really engineering. Quite often, our evaluation determines that the activity in question is not. It might be more accurately described as 'code writing', 'product support', or other tasks that the board doesn't regulate. If the practitioner is calling these activities 'code writing' or 'product support', there are no legal problems. However, if the practitioner is calling the activity software engineering, he or she is violating the law. Such use of an incorrect term leads consumers to believe that they are receiving services in full conformance with the legal protections provided by engineering law. Therefore, anyone use such terms illegally should reasonably expect legal problems.

Exemptions

Legal use of an engineer title depends on two variables: how you use the title and the employment conditions under which you practice engineering (*Texas Engineering Practice Act*, rev. 1999).

Almost every US state law has some allowance for the legal use of engineering titles by non-professional engineers (*Engineering Licensure Laws*, 1999), or *exemptions*. The *industrial exemption* is the most commonly claimed. It is available to a well-defined class of engineers, which happens to be the largest class of engineers in the nation. Engineers can claim an industrial exemption if they meet the following conditions:

- They practice engineering only for their full-time employers
- Their practice is limited to work on their employer's facilities or on products that their employer manufactures.
- They do not use an engineering title outside their company.
- They do not claim that they are qualified to offer engineering services to another party.

Let's imagine that you meet the requirements for an industrial exemption. How do you avoid title-use problems? Most of the illegal title-use cases in Texas stem from situations where practitioners misunderstand the industrial exemption. The best way to avoid problems is to practice 'title abstinence'. Never refer to yourself as an engineer outside your company. Never moonlight engineering work for another company or project. Never offer engineering work on a contractual basis.

Most US states have other title exemptions, which are available to individuals such as full-time federal employees, railroad engineers, and engineers working for utility companies. Graduates of recognized engineering programs are often allowed to use a term such as *graduate engineers* (*Texas Engineering Practice Act*, rev. 1999). All states have programs to recognize those who pass the first national examination, the Fundamentals of Engineering exam; they can register in their state to use terms such as *engineering intern* or *engineer-in-training* (*Engineering Licensure Laws*, 1999).

As with titles, there are industrial, federal, and other exemptions for engineering practice, and the circumstances are identical. If you perform your engineering work only for your full-time employer's property or on their manufactured products, and you do not violate title restrictions, you are eligible for an exemption.

How Software Engineers Are Licensed in Texas

For almost four decades, the Texas Board has licensed software engineers under other discipline headings, such as electrical, industrial, or control-system engineering. The reason for licensing under these other disciplines is that software engineering activities were generally performed in support of one of these more 'traditional' engineering disciplines (Speed, 1998).

Once a license has been issued, the discipline designation is almost totally irrelevant. In most states, individuals are licensed as *professional engineers*, not as software engineers, chemical engineers, electrical engineers, and so on (*Texas Annotated Code*, 1999). In the same way that statutes require attorneys or medical doctors to practice only in their areas of competence, professional engineers are required to limit their practice to their areas of expertise. Obviously, as engineers develop new skills, their area of expertise might change.

At a meeting in June 1998, the Texas Board enacted rules that recognized software engineering as a distinct engineering discipline (*Texas Annotated Code*, 1999), which lets individuals with acceptable educational credentials and software engineering experience apply for a license.

Because no nationally recognized software engineering exam exists, only highly experienced software engineers are eligible (Speed, 1998). The licensing scenario outlined below is the only route available to a Texas software engineering license until software engineering examples are developed.

- Possession of an engineering, a computer science, or other high-level math or science degree evaluated by the Board as adequate
- At least 16 years of creditable experience performing engineering work (only 12 years are required for individuals holding a degree approved by the Engineering Accreditation Commission of the Accreditation Board for Engineering and Technology).
- References from at least nine people, five of whom must be licensed engineers
- Required educational and other credentials.

Anticipating need, the Texas Board has also initiated the development of exams for licensing software engineering practitioners with less experience. It intends to seek national use of any exams through the National Council of Examiners for Engineering and Surveying. Once the exam is available, less experienced individuals can apply for a Texas PE license by passing it. Eight years of the experience requirements may be subtracted for those who pass an NCEES exam.

Should you pursue a PE license? Everyone has personal career goals. Are you doing work that meets the legal definition of engineering practice? Do you plan to work in that arena during your career? If so, you should contact your state licensing board for details about how to get started.

Licensing is one of the logical developments of any maturing profession. Software engineering is far more mature than many practitioners realize. As you ask yourself whether you should pursue a professional engineering license, look around your office. In 10 years, what will distinguish you from the people around you? What will set you apart as a professional? For most of this century, the answer has been the professional engineering license.

Licensing boards around the nation will address the need to license software engineers as exams are developed. This process will take several years, but we are moving ever close. Decisions time is approaching. How will you respond?

References

Practice and Procedure, Bylaws and Definitions, Title 22. 1999. *Texas Annotated Code*, Chapter 131. Available at http://www.main.org/peboard/law.pdf.

Speed, J.R. 1998. 'Software Engineering: An Analysis of the Proposed Actions by the Texas Board of Professional Engineers'. Texas Board of Professional Engineers, Austin, TX.

'Summary of Engineering Licensure Laws (Part 11B, United States)'. 1999. *Engineering Licensure Laws*. National Society of Professional Engineers, Alexandria, VA.

The Texas Engineering Practice Act, Article 3271a. 1999. *Texas Civil Statutes*, 1937 (Revised in 1999). Available at http://www.main.org/peboard/law.pdf.

QUESTIONS FOR CRITICAL THOUGHT

1. What is the purpose of this article? For whom was it originally written?
2. What is the effect of the title? Does it adequately represent the article's content?
3. Look up the words 'engineer' and 'engineering' in a reliable dictionary. How many meanings are listed? How many of them are governed by the regulations Speed discusses?
4. What is a 'legal' definition? Who has the authority to regulate such a thing?
5. To what extent is it possible for any individual or group to control the meaning and usage of words? To what extent are organizations of professional engineers interested in doing so?
6. Why does Speed spend so much time 'unpacking' the legal definition?
7. Speed suggests that regulation of the title 'engineer' is for the protection and confidence of the public. To what extent is the process also a matter of turf?
8. Why are some activities harder to classify as 'engineering'?
9. Speed argues that 'licensing is one of the logical developments of any maturing profession.' Does he explain why this is so? Can you make any inferences about why?
10. Does the article satisfy the expectations evoked by its title? Why or why not?
11. Are there any lessons here about the nature of language? If so, what are they?
12. Speed's article is largely a definition. Is the act of defining a term primarily informative or is it persuasive? Is there any such thing as 'purely' informative? Why or why not?
13. Many of the issues that receive contentious public debate—same-sex marriage, euthanasia, and universal health care, for example—come down to squabbles over how words are defined and used. To what extent is definition a matter of power? To what extent is it a matter of influence?

CHAPTER 38

Disciplinarity, Identity, and the 'Profession' of Rhetoric

Jennifer M. MacLennan

After six years as the only rhetorician in the English department at a small undergraduate school, I moved to the College of Engineering at a comprehensive university, where I am the first occupant of an endowed Chair in communication. This move has been accompanied by a few surprises, in particular, the ready acceptance of my interest in practical application as well as in theoretical investigation. This reception has been a welcome contrast to my experiences in the literature department, where a consideration of the rhetorical was typically viewed as 'a commercial blemish on the otherwise spotless face of art' (Booth, 1983: 90), and where an interest in teaching professional communication was considered little better than a form of intellectual prostitution. The juxtaposition of two such very different perspectives has, for me, raised questions about the very nature of rhetoric by foregrounding some surprising parallels between the reflective practice of engineering and the same features of the rhetorical process. This paper explores the tantalizing possibility of a new way of understanding rhetoric; in particular, it proposes that rhetoric, unlike other humanities and social science disciplines, can fruitfully be understood to be a profession—like engineering, like medicine, like law.

A persistent thread of discussion in our scholarly journals has, in varying degrees over the years, explicitly treated the field of communication as a profession, and in some cases overtly distinguished it from other academic pursuits.[1] As well, of course, the relationship of communication to the professions in general has also been the subject of frequent discussion,[2] and fields closely allied with ours, such as broadcasting and journalism, offer an education that typically combines theory with hands-on training. What I am suggesting, however, is more than adoption of a label; it is a change of perspective. I hope to show that rhetoric, as both a theoretical study and a practical art, shares much in structure and approach with the other learned professions, more perhaps than with the other arts and science disciplines with which it is usually housed. I expect that raising this possibility will be unwelcome to some who feel that professionalizing the discipline will somehow undermine its scholarly integrity, but my own experience suggests it is worth considering the extent to which such a paradigm shift would be of value to rhetoricians and to other practitioners of speech communication.[3] In order to illuminate the implications of such a change in self-definition, I would like to examine what the concept of 'profession' involves.

The central issue in understanding the nature of professional activity is recognizing its integration of theory with practice. This issue is also of continuing and significant concern to rhetoricians, who, as Barnett Baskerville points out, must 'keep at least one foot on the ground' as they contend with the practical implications of their communicative choices (1971: 122). Whatever else may be meant by the term, those who have written about the nature of professions agree that, as Donald Schon points out, 'professional practice is an instrumental activity' (1983: 23) directed toward identifiable goals.[4] According to Wilbert Moore, the hallmark of the profession is the application of 'very general principles, *standardized* knowledge, to concrete problems' (Moore, 1976: 56). Nathan Glazer further describes what he calls the 'major' professions—law, medicine, business, and engineering—as being 'disciplined by an unambiguous end, . . . [and] grounded in systematic, fundamental knowledge' (Glazer, 1974: 363). Schon explains: 'although all occupations are concerned . . . with the instrumental adjustment of means to ends, only the professions practice rigorously technical problem solving based on specialized knowledge' (1983: 22). In other words, the professions are distinguished not solely by their practical aspect but by their mastery of theory as a means of conditioning and illuminating practice. I propose to demonstrate that rhetoric, too, employs theory in just such a manner.

For many in the academy, the most visible differences between the professions and other academic disciplines is that, unlike students in arts and science programs, those in professional schools are being prepared to enter a specific, identifiable professional practice when they graduate. This clear focus introduces an element of 'training' that is typically viewed with suspicion and even disdain by academics in the arts and sciences, because it is perceived to place practice above theory. A second significant difference between the professions and other university disciplines is the existence of legislative bodies responsible for the accreditation or certification of professional programs in post-secondary institutions. Such entities, typically professional associations, also set standards of professional conduct, license their practitioners, and perform disciplinary action when necessary.[5] These two factors—an identifiable career path for graduates and an external accreditation process—are the features that, at least to outside observers, most clearly distinguish professional programs from those offered in faculties of arts and science; they are also, typically, the features that cause other academics to reject the concept of professionalism as incompatible with academic freedom and the ends of 'pure' scholarship.

However, they are not the issues that preoccupy professionals themselves when they consider the nature of professionalism. Indeed, they treat issues of career training and accreditation as secondary to the relationship between theory and application that is seen as the main feature of professional education and practice. It is in this regard that the parallel between rhetoric and the professions is most evident, and it is for this reason that the value of professionalizing rhetoric cannot be so easily dismissed. As practitioners and scholars whose discipline inevitably involves application, we might do well to consider what it might mean to rhetoric to be conceived as a profession, not in the rather loose sense in which we currently use the term, but in the same rigorous sense in which those in the professions themselves use it.

If we take the instrumental imperative as central to professionalism, there seems good reason, at least initially, for suggesting that rhetoric, too, can lay claim to professional status as 'a *practical*, instrumental art' (Baskerville, 1971: 123). More than 2,000 years before Nathan Glazer described the nature of professional education, another professional characterized his art as having 'distinct ends in view' and insisted on the pervasiveness and utility of his subject: 'It is clear, then, that rhetoric is . . . universal; it is clear, also, that it is useful' (Aristotle, 1954).

Subsequent theorists, however much they have expanded the scope and boundaries of rhetoric, have tended to retain this emphasis on instrumental utility or practical value.

Throughout the twentieth and twenty-first centuries, rhetoric has been repeatedly described, as it is by Edward P.J. Corbett, as an 'instrument for communication', and its criticism as 'more interested in a [discourse] for what it *does* than for what it is' (Corbett, 1971: xxii). In fact, Herbert Wichelns (1925) inaugurated the twentieth-century practice of rhetorical criticism with this very argument. Robert Brock and Richard Scott

(1980) similarly emphasize that rhetoric, as one of 'the practical arts', can best be understood as an 'instrument to aid [our] living'. Finally, Kenneth Burke, the most influential theorist of the twentieth century, shares this view, describing rhetoric as discourse 'with a definite audience in mind, and for a definite purpose. It [is] literature for *use*.'(1969: 4)

Perhaps the most celebrated definition of rhetoric as an instrument can be found in Lloyd Bitzer's 'The Rhetorical Situation'. Bitzer situates rhetoric within a 'complex of persons, events, objects, and relations', and therefore argues that it cannot be understood without reference to its audience and context:

> a work of rhetoric is pragmatic; it comes into existence for the sake of something beyond itself; it functions ultimately to produce action or change in the world; it performs some task. In short, rhetoric is a mode of altering reality, not by the direct application of energy to objects, but by the creation of discourse which changes thought and action. The rhetor alters reality by bringing into existence a discourse of such a character that the audience, in thought and action, is so engaged that it becomes mediator of change. (1968: 3–4)[6]

In addition to this instrumental emphasis, Nathan Glazer explains that the ends of professional activity are 'unambiguous'—not because they avoid indeterminacy, but because they represent specific and identifiable human aims. Rhetoric, the principal means by which we exert influence over others, is equally pragmatic and goal-oriented. From its beginnings, it has been viewed in precisely this way, as primarily concerned with 'the right method and means of succeeding in the object we set before us' (Aristotle, 1954: 1355b)—that is, the moving of an audience to belief or action. Of course, as a blend of theory and practice, the goal of rhetoric is 'not simply to succeed in persuading' in a given instance, but to understand how it is that people *are* persuaded, and to be able to apply that understanding to practical purposes, to 'discover the means of coming as near such success as the circumstances of each particular case allow' (1954: 1355b). Like all professionals, rhetoricians need more than a grasp of theory; they also need trained judgement that allows them to apply that theory in finding solutions to problems in the real world of human practice.

Clearly it would be difficult to deny this pragmatic aspect of rhetoric; it would be equally difficult to deny its complex and extensive body of theoretical knowledge. It is in this combination of elements that the similarity between professional practice and rhetorical practice is most immediately evident. However, as important as is this relationship between theory and practice, the question of whether the art can be understood as a profession in its fullest sense involves more than just the instrumental application of a systematic body of knowledge. At this point in the discussion, it might be helpful to understand more about the nature of professional knowledge and activity as they are conceived by their practitioners. According to Edgar Schein in *Professional Education*, professional knowledge can be distinguished by three features:

> 1. An *underlying discipline* or *basic [theory]* component upon which the practice rests or from which it is developed.
> 2. An *applied* . . . . or '*engineering*' component from which many of the day-to-day diagnostic procedures and problem-solutions are derived.
> 3. A *skills* and *attitudinal* component that concerns the performance of services to the client, using the underlying basic and applied knowledge. (1992: 39)

The first of Schein's components of professional knowledge involves a body of rigorous and systematic theory that is abstracted from and independent of individual cases. The prototypes of professional expertise in this sense are the 'learned professions' of law and medicine, as well as scientific professions such as engineering. What is important here is not the stability of such a body of theory, since—as Thomas Kuhn has established—even scientific knowledge undergoes paradigm shifts that challenge received theories and allow advancements in understanding as more subtle information becomes available (Kuhn, 1970). Instead, what matters is the rigour and systematicity of underlying principles and their value as foundations for practice. It is clear that each of the learned professions identified by Schein has an underlying discipline or body of knowledge that conditions practice; it seems equally clear that the same is true of rhetoric, and has been since Aristotle wrote his treatise on the art.

Looked at in light of the first principle in Schein's scheme, Aristotle's *Rhetoric* certainly seems to describe a professional activity. He emphasizes that his treatment will consider 'the systematic principles of rhetoric itself' (1355b). Throughout his discussion of the art or *techne* of rhetoric, Aristotle demonstrates that 'the subject can plainly be handled systematically . . . and everyone will at once agree that such an inquiry is the function of an art' (1954: 1354a).[7] Later, Aristotle reminds us again of the universal (or basic, in the sense employed by Schein) nature of the principles he articulates when he insists that 'the same systematic principles apply to political as to forensic oratory,' the two main categories of argumentative rhetoric treated by him in the *Rhetoric* (1354b). The abundance of subsequent discussions of rhetoric, from Ramus to George Campbell to Kenneth Burke, have all offered methodical, systematic treatments of the function and scope of the art.[8]

In structuring his rhetorical system, Aristotle establishes a foundation for the discipline in more than the content he describes. His taxonomy provides elements still useful today,[9] and his scientific, methodical approach has provided a model for subsequent treatments of rhetoric, including some rather surprising applications.[10] First, Aristotle defines the focus of his treatise: 'rhetorical study, in its strict sense, is concerned with the modes of persuasion' (1954: 1354b), and then carefully outlines its component parts, the modes of appeal: 'the first kind depends on the personal character of the speaker; the second on putting the audience into a certain frame of mind; the third on the proof, or apparent proof, provided by the speech itself' (1954: 1356a). He then discusses three requirements of an effective rhetor, which are, of course, coherent with the modes of appeal already presented:

> (1) to reason logically, (2) to understand human character and goodness in their various forms, and (3) to understand the emotions—that is, to name them and describe them, to know their causes and the way in which they are excited. (1954: 1356a)

Finally, Aristotle outlines the three divisions of public communication available in his time—'(1) political, (2) forensic, and (3) the ceremonial oratory of display'—and points out

> three distinct ends, one for each of its three kinds. The political orator aims at establishing the expediency or the harmfulness of a proposed course of action . . . Parties in a law case aim at establishing the justice or injustice of some action . . . . Those who praise or attack a man aim at proving him worthy of honour or the reverse. (1954: 1358b)

As well, although the province of rhetoric has been greatly expanded since Aristotle, particularly during the last century (partly as a result of advancements in communication technology), the basic definition of rhetoric established by his *Rhetoric*, and the fundamental tools of the rhetor (ethos, logos, and pathos) remain influential after nearly 2,500 years, even among those whose goals are mostly immediate, practical, and concrete.[11]

Aristotle's rhetorical principles have even been adapted to the needs of the most pragmatic of pursuits, modern sales (Cosnett, 1996). Subsequent treatments of rhetoric, including those that have broadened the scope of the discipline, have attempted to establish similarly rigorous theoretical frameworks for the understanding (if not the creation) of rhetorical discourse, among them Kenneth Burke's dramatistic pentad,[12] Walter Fisher's narrative paradigm,[13] Northrop Frye's treatment of the subject in his *Anatomy of Criticism* and elsewhere,[14] Ernest Bormann's fantasy theme analysis,[15] and approaches that have grown out of feminism, Marxism, deconstruction, and postmodernism.[16]

Rhetoric, like most other academic disciplines, can easily be shown to meet the first of Edgar Schein's criteria, that of possessing an underlying discipline or body of theoretical knowledge. However, it is the second of Schein's components of professional knowledge that reveals significant differences between rhetoric and most other arts and science disciplines. This second category of professional knowledge involves 'an *applied* . . . or "*engineering*" component' upon which a profession's 'diagnostic procedures and problem-solutions' are based. That rhetoric offers an 'applied' component hardly needs demonstration here. Both its theoretical products and its objects of study (rhetorical 'texts', whether classically defined or more broadly considered)[17] concern the dynamic process of 'adjusting ideas to people, people to ideas' (Bryant, 1963: 402). This process involves both a diagnostic element that analyses audience and context, and a problem-solution component of rhetorical invention that together allow an experienced practitioner to evaluate the demands of the communication situation and produce appropriate, effective discourse. For this reason, theories involving the dynamics of rhetorical response cannot be studied exclusive of pragmatic application, since even when we are at our most theoretically arcane, we are ultimately concerned with the means, motives, and methods of shaping action through symbolic appeals. For rhetoricians as for other professionals, the application and the theory are interdependent, and 'the speaker before an audience, the pamphleteer, any[one] who seeks to influence others with words, stands at the centre of the real world' (Baskerville, 1971: 121).

That Aristotle is very aware of the application of his rhetoric as well as its theoretical aspect is evident in his advice on the 'means and methods' of achieving our rhetorical goals, and on his castigation of other rhetoric teachers for ignoring significant parts of the art while indulging in irrelevancies (1954: 1354a). However, despite the much-lamented rigidity of neo-Aristotelian applications of his theory, Aristotle's rhetoric is no cookie-cutter recipe.[18] Instead, it describes a discipline designed to develop a dynamic quality of judgement that will allow us to 'discover the means of coming as near such success as the circumstances of each particular case allow' (1954: 1355b). It is in the nature of this application that rhetoric, like the other 'learned professions'—including engineering, architecture, and medicine—is an *art*. All such professions, according to Donald Schon, involve a reflective process that, while it relies on previous experience as a guide, does not blindly apply pre-existing solutions in a Procrustean fashion. Interestingly, recent debates about the efficacy of service learning and its value in the field of communication as 'the junction where theory and practice meet' (Katula and Threnhauser, 1999: 248)[19] sound remarkably akin to Schon's discussion of professional education. Says Schon: '[All professionals] engage in a process whose underlying structure is the same: a reflective conversation with a unique and uncertain situation. . . . The art of these practitioners . . . seems to me to be, in considerable measure, a kind of reflection-in-action' (1983: 130). Accordingly, the professional practitioner

> approaches [each] problem as a unique case. He does not act as though he had no relevant prior experiences; on the contrary. But he

> attends to the peculiarities of the situation at hand, . . . seeks to discover the particular features of his problematic situation, and from their gradual discovery, designs an intervention. (129)

Like the rhetorician, whose study of individual cases—'individuations', to borrow Burke's term (1968: 143)—is meant to sharpen her judgement and assessment skills, the reflective professional in other fields learns to assess current problems by drawing on experience with previous interventions in much the same way that Chaim Perelman's rhetor constructs standards of the 'universal audience' based on broad experience with particular audiences (Perelman and Olbrechts-Tyteca, 1971: 34–5). The rhetorician, as much as the physician, the lawyer, or the engineer, employs a similar process of 'reflection-in-action' as she assesses the demands of the rhetorical situation and responds appropriately to them. The key to understanding the reflective approach to problem solving is recognizing its quality of 'artistic performance', which Schon describes as a process of responding to the complexity of individual problems

> in what seems like a simple, spontaneous way. [The practitioner's] artistry is evident in his selective management of large amounts of information, his ability to spin out long lines of invention and inference, and his capacity to hold several ways of looking at things at once without disrupting the flow of inquiry. (1983: 130)

Schon's description echoes Aristotle's characterization of the effective rhetor, who learns from experience to infer general principles and refine her approach to new problems.

Despite the explosion of theory in recent years, the importance of such experience to the mastery of the art of rhetoric remains evident in the fact that training in rhetoric continues to employ pragmatic application along with theoretical and critical analysis. Courses in public speaking and in practical criticism are the mainstay of rhetoric programs everywhere, and other forms of experiential learning are also highly prized by communication scholars and teachers.[20] Most rhetorical theorists recognize that, because it is 'harnessed to the world' (Dillon, 1986: 12), rhetoric has an inescapable quality of application that must be taken into account if the rhetorical interaction is to be understood as a communication process. Barnett Baskerville, for example, expresses concern at what happens to rhetoric in situations where 'prescription has outrun performance', and insists that theorists and critics should understand 'something of the difficulties of the art—and preferably as the result of actual experience' (1971: 117). Full understanding of rhetoric relies on direct hands-on experience and not only on theoretical musings, which cannot provide 'any clear conception of the immense difficulties involved in adjusting people to ideas and ideas to people' (Baskerville, 1971: 21).

Aristotle's definition of the art of rhetoric captures this blend of experience and theoretical understanding in its emphasis on 'the faculty of discerning, in any given case, the available means of persuasion' (1954: 1354b). Such discernment is not a matter of simply following a formula to produce a desired result, but one of applying (and developing) an informed judgement and sensitivity to the specific circumstances of individual cases. It is through this process of universalizing or abstracting basic principles that rhetoric can be understood to involve 'the power of observing the means of persuasion on almost any subject presented to us' (1954: 1355b). The combination of audience analysis and rhetorical invention draws upon fundamental principles of persuasion derived from study and experience. These elements constitute the same process of 'reflection-in-action' described by Donald Schon, and thus it seems clear that rhetorical activity involves the same interplay of experience and judgement that allows all professionals to design appropriate interventions 'in any given case'. Simply put, rhetorical activity relies on the application of 'standardized knowledge to concrete problems' (Moore, 1976: 50) that sets professional practice apart from other forms of 'instrumental adjustment of means to ends' (Schon, 1983: 56).

The final component of professional knowledge outlined by Schein is a skills and attitudinal component that explicitly concerns actual practice by considering the performance of service to the professional's clients, or to society in general. Does rhetoric, like its counterpart professions of medicine, engineering, or business, display such social value? Or, to put this question another way, are communication professionals thought to contribute to their society through the practice of their discipline? There are many current indications that this question may be answered affirmatively. For example, with his recent prediction that, as a discipline, 'we will increasingly view service to society as a critical feature of our academic and scholarly mission, and that we will develop a sense of social responsibility as a discipline-centred objective,' James Chesebro seems both to recognize a trend and to call for some movement that will 'provide a more profound sense of the discipline of communication as a societal force . . . [and] take the discipline . . . beyond its traditional domain within the academy' (Chesebro, 1996: 12). A similar awareness of a dimension of social responsibility is suggested by the theme of the 2000 meeting of the National Communication Association: 'Communication: The Engaged Discipline' recognizes societal pressure to 'justify the huge investment being made in our research and teaching institutions' and the need to respond with '"engaged campuses" that create partnerships with local, state, national, and international communities' (NCA, 1999). The discussion goes on to note that 'as a discipline, we more than most have maintained a central commitment to engaging our basic research and teaching in ways that improve social and communication practices at individual and societal levels' (NCA, 1999: 325). This view of communication is reiterated on the home page of the discipline's largest professional association, the National Communication Association,[21] and in such publications as Applegate and Morreale's 'Creating Engaged Disciplines' (2001). As well, the recent increase in attention to experiential learning as a valuable tool for teaching and learning about communication also suggests further the extent to which scholars already recognize a significant social role for our discipline, and the prevalence of consulting practice among communication faculty reveals a recognition of service to clients and society in general that extends beyond the educational utility of service learning. It also points to yet another element that communication practitioners already share with their colleagues in the professions, since consulting practice, rare in most humanities departments, is widely embraced and even expected in many professional faculties.

The practical or 'engineering' component of rhetoric has been evident throughout its history; indeed, this element of practical consideration is inevitable, since all rhetorical appeals are by necessity adapted to an actual audience who is to be influenced and persuaded. As well, whether their focus has been application, criticism, theory, or philosophy of rhetoric, treatises in the subject have all involved a consideration of how and why audiences can be moved to pleasure, belief, and action by the symbolic efforts of rhetors. Many have even offered practical advice as to how to approach specific audiences. This audience-centredness means that rhetoric can never be entirely independent of its pragmatic origins, and that—like other professions—it must continue to respect the constraints imposed by its context. True to form, Aristotle emphasizes that, 'of the three elements in speech-making—speaker, subject, and person addressed—it is the last one, the hearer, that determines the speech's end and object' (1358a–b).

A similar emphasis on audience permeates nearly all discussions of rhetoric, whether scholarly or popular. The onus on the speaker to know her particular audience, to be familiar with audience response in general, and to understand how the elements of appeal will shape the audience's response remains a central feature not only of rhetorical practice but also of rhetorical theory—whether these elements are considered to be ethos, logos, and pathos; will, imagination, and passions; identification and division; or reciprocity, social proof, commitment, and scarcity (Cialdini, 1993). For the rhetorician, whether speaker, critic, or theorist, the focus is always in the end an audience, be they lay people or experts, who must be convinced through a mix of strategies appropriate to their expectations, experience, values, and beliefs (Cialdini, 1993).[22] This means that, as an essential component of the art of rhetoric, the rhetor must be adept at audience analysis, and in essence must, as Vance Packard (1964b) suggests, know the audience better than they know themselves. The challenge to the professional in the field of rhetoric is best explained by Kenneth Burke's admonition that 'what is in a man's best interests may not be what he is interested in' (Burke, 1969). The rhetor who would persuade effectively must be able to combine experience with careful observation and judgement to bring the two together.

That the study of rhetoric involves a significant and inescapable element of practical application along with an extensive body of theory and a recognition of social responsibility is an inevitable conclusion. However, at the end of our consideration of the three components of professional knowledge and their suitability as a paradigm for understanding the study and practice of rhetoric, we are left with a question about the implications of viewing rhetoric as a professional as well as scholarly pursuit. Can there be value in such a designation?

There are undoubtedly those who fear that 'professionalizing' rhetoric may diminish its perceived scholarly legitimacy (as critics of service learning have similarly feared).[23] It is certainly true that there exists within the academy in general a tendency to privilege 'pure' scholarly activity over 'applied' branches of knowledge, as our colleagues in the field of rhetoric and composition have discovered. This tendency exists even within the professions, as Donald Schon explains: 'the concept of "application" leads to a view of professional knowledge as a hierarchy in which "general principles" occupy the highest level and "concrete problem solving" the lowest' (1983: 24). However, despite what James Chesebro refers to as such 'intellectual snobbery' (1996: 12), there is also reason to expect that reconfiguring rhetoric as a true profession could actually raise our status in the academy, just as the prototypes of learned professions—law, medicine, engineering, and commerce—have enjoyed increased prestige within the university in terms of both scholarly specialization and social value. Rhetoric might be similarly enhanced, particularly at a time when the taxpaying public is demanding greater 'accountability' from universities. Acquiring professional status would also help to end the territorial dispute with literary studies that has characterized so much of our history over the past century and even before.[24]

Indeed, professionalizing our activity could enhance both our scholarship and our status by providing formal recognition for what we do. For instance, in my professional college, in addition to the standard evaluative categories of 'teaching, research, and service', faculty can be assessed according to a fourth, equally important component (College of Engineering, nd). 'Professional practice' is a category that would allow for the assessment of valuable work performed by many rhetoricians as a legitimate part of their scholarly and pedagogical enterprise—work that often remains invisible and unrewarded under the current tripartite assessment tool. Today's 'demand that higher education and the disciplines within it reconsider [their] role' (NCA, 1999: 325) was actually foreseen nearly 40 years ago by Vance Packard. In an essay devoted to the study of relative social value ascribed to various occupations, Packard noted a trend that indicated that 'as professions become less "practical" (and are less likely to be in demand by profit-making organizations)' they actually enjoy less social value, partly because their practitioners 'command less and less money' for their services (1964: 88). This trend has if anything continued, as is suggested by the greater salaries commanded by professional faculty in many universities as a result of 'market differential'. Meanwhile, the prestige associated with purely academic positions has declined in relative measure.[25] Formally recognizing the extent to which our scholarly activity is infused with pragmatic application, and achieving institutional acceptance of the professional practice in which many rhetoricians already engage may actually turn out to be a timely response to the erosion of practical value and the corresponding loss of status currently being felt in other humanities disciplines. It may also provide a suitable counterweight to the 'insignia of mental work ranking higher than the insignia of physical work' observed by James Chesebro (1996: 12) as permeating our academic institutions.

In the end, I have few illusions as to settling the debate over whether rhetoric *should* be considered a professional activity; however, whether it *can* be so considered seems clear. Rhetorical knowledge can plainly be seen to conform to the three principles of professionalism outlined by Schein. The first, a theoretical foundation or 'underlying discipline', is primarily concerned with the 'whys' of human communication behaviour, what it means, and what it's for. This element can be understood as the body of rhetorical theory from Aristotle to Burke to McGee and beyond. The second, an 'applied science' of rhetoric, is concerned with the 'hows' of human communication—how meanings are actually communicated and the effects achieved when they are. This component considers the critical or analytical process of rhetorical criticism applied to a diversity of rhetorical acts and artifacts, whether broadly or narrowly defined, and also describes the 'reflection-in-action' of the rhetor who carefully considers the elements of the rhetorical scene and creates an appropriate response. The third element, a 'skills' component, is also concerned with the 'doing' itself—with the service of the discipline to its constituency or to society generally. This final aspect may be considered to be the use of rhetorical knowledge and skill in hands-on application such as public debate, speech-making, teaching, consulting, debate coaching, and the like, and in the larger and more visible role of social responsibility that will make us 'a disciplinary leader in this 21st Century movement' toward greater engagement with society (NCA, 1999: 325).

The conversation about the function and scope of rhetorical study and about its role in the pursuit of human knowledge has been going on for a long time, and there is no doubt that it will continue. Nevertheless, despite a consistent thread of discussion in which speech communication has been conceptualized as a profession, we are far from having achieved true professional status, either on university campuses or in the eyes of the public. The use of the term 'profession' to describe speech communication has not typically been consistent with its meaning as applied to fields like medicine, law, engineering, or business. Indeed, it lacks even the connotation of the 'minor' professions such as education, social work, public relations, or journalism (Moore, 1976: 141). As well, although we clearly recognize that graduates of rhetoric and communication programs are receiving education suited to a relatively well-defined *range* of career choices (Berki, et al., 1995), we have failed to fully embrace the notion of 'professionalism', perhaps partly because we have conceived of the designation as requiring students to be prepared for a *specific* career path.

That such a requirement is not necessarily so is evident in the fact that, despite their apparently career-driven education, not all those with law degrees take the bar, not all engineering graduates pursue careers in technical fields, and not all education majors end up as teachers. A career in the profession in which one has trained is an option rather than an imperative, just as it is for those who currently 'train' as communication specialists. Formalizing our academic specialty into a professional program need not narrow the choices available to our students. Indeed, doing so might well increase their options, since their academic education would be understood to be enhanced by a level of professional skill, making these 'scholar-practitioners' desirable to a variety of potential employers whose interest in communication skill has been repeatedly emphasized.[26]

Seeing rhetoric as a professional undertaking and not just an academic pursuit may well have the power to enlarge rather than to restrict or reduce our richness as a discipline. If fully embraced, it also has the potential to increase our understanding of the relationship between theory and practice, teaching and scholarship, that many of us have found elusive in an arts-and-science context, where practical application is often frowned upon or dismissed as, in the words of my former department chair, 'a petty diversion'. Most important, understanding our own professionalism can help us to foster an environment where an integrated program of research, teaching, and service is not only possible but expected and welcomed, and in so doing, can bring us to a fuller understanding of rhetorical communication as both theoretical process and human practice. Perhaps it is time to suggest that when the rhetoricians withdrew from literature departments in 1914 to form their own scholarly organizations and departments, they acted appropriately, but incompletely: perhaps they should have kept going, right out of Arts and Science into a professional faculty of their own.

Notes

1. See Enrique D Rigsby, 'African American Rhetoric and the "Profession"', *Western Journal of Communication* 57 (1993): 191–9; Cal W. Downs, Paul Harper, and Gary Hunt, 'Internships in Speech Communication', *The Speech Teacher* 25 (1976): 276–82; Melbourne S. Cummings, 'The Profession and How It Relates to the Black Experience', *Bulletin of the Association of Departments and Administrators in Speech Communication* 13 (1975): 27–8; D. Thomas Porter, 'Affirmative Action: The SCA and the Profession', *Bulletin of the Association of Departments and Administrators in Speech Communication* 6 (1974): 30–46; D. Thomas Porter, 'Affirmative Action: The SCA and the Profession', *Bulletin of the Association of Departments and Administrators in Speech Communication* 6 (1974): 30–46; Robert T. Oliver, 'A View Ahead: The Speech Profession in 1984', *Today's Speech* 20 (1972): 9–13; Theodore Jr Clevenger, 'Column Two: A Survival Manual for the Speech Profession', *Today's Speech* 20 (1972): 3–10; Waldo Braden, 'An Uncommon Profession', *Southern Speech Journal* 36 (1970): 1–10; Marceline Erickson, 'The Required Speech Courses and the Speech Profession', *The Speech Teacher* 12 (1963): 26; Brooks Quimby, 'Is Directing Forensics a Profession?', *The Speech Teacher* 12 (1963): 41–2; Elbert W. Harrington, 'The Academic and the Rhetorical Modes of Thought', *Quarterly Journal of Speech* 72 (February 1956): 25–30; F.L. Whan, 'The Speech Profession Jilts Radio', *Quarterly Journal of Speech* 30 (1944): 439–44.
2. For example, Keith V. Erickson, T. Richard Cheatham, and Carroll R. Haggard, 'A Survey of Police Communication Training', *The Speech Teacher* 25 (1976): 299–306; Kathleen M. Jamieson and Andrew D. Wolvin, 'Non-Teaching Careers in Communication: Implications for the Speech Communication Curriculum', *The Speech Teacher* 25 (1976): 283–9; Gomer Pound, 'Dual Responsibility: Career Training and General Education', *Bulletin of the Association of Departments and Administrators in Speech Communication* 6 (1974): 27–9; Darrell T. Piersol, 'Responsibility for Career Training', *Bulletin of the Association of Departments and Administrators in Speech Communication* 6 (1974): 22–4; Vernon A. Stone, 'Broadcast News Education and the Profession', *Journalism Quarterly* 47

(1970): 162–5; D. Hugh Gillis, 'Broadcasting as Profession: A Socio-Economic Approach', *Journal of Broadcasting* 11 (1966): 73–82; James H. McBath, 'Speech and the Legal Profession', *The Speech Teacher* 10 (1961): 44–7; Glenn Frank, 'The Profession of Journalism', *Journalism Bulletin* 2 (1926): 5–8.

3. Although the discipline traditionally known as speech, or more recently speech communication, is a familiar one in American universities, it is all but unknown in Canada. With its roots in the ancient discipline of rhetoric, the modern department of speech communication offers courses in a variety of areas of communication study, including rhetoric, instructional communication, communication theory, philosophy of communication, organizational communication, and so on. For a more detailed discussion of the discipline's nature, see MacLennan (1999 and 1998), National Communication Association (1996), and Smith (1954). A brief overview is also available online at the National [that is, American] Communication Association.
4. Garry Wacker, 'Governance of Professions: What is the Best Way to Serve Society?', *Proceedings of the Canadian Conference on Engineering Education* (Ottawa, ON: Canadian Council of Professional Engineers, 1996); Garry Wacker, 'The Engineering Profession in Canada: Challenge, Renewal, and Opportunity'. Keynote address to the 10th Canadian Conference on Engineering Education. *Proceedings of the Canadian Conference on Engineering Education* (Ottawa, ON: Canadian Council of Professional Engineers, 1996); Donald A. Schon, *The Reflective Practitioner: How Professionals Think in Action* (New York: Basic Books, 1983), 23. See also S.C. Florman, *The Existential Pleasures of Engineering*, 2nd ed. (New York: St Martin's, 1994); Donald A. Schon and Chris Argyris, *Theory in Practice: Increasing Professional Effectiveness* (San Francisco: Jossey-Bass Publishers, 1974).
5. For example, Canadian Engineering Qualifications Board, *Admission to the Practice of Engineering in Canada* (Ottawa: Canadian Council of Professional Engineers, 1992).
6. Lloyd F. Bitzer, 'The Rhetorical Situation', *Philosophy and Rhetoric 1* (Winter 1968): 3–4. See also 'Functional Communication: A Situational Perspective', *Rhetoric in Transition: Studies in the Nature and Uses of Rhetoric*, ed. Eugene E. White (University Park, PA: Pennsylvania State University Press, 1980), 21–38; Kenneth Burke, 'The Rhetorical Situation', *Communication: Ethical and Moral Issues*, ed. Lee Thayer (London: Gordon and Breach Science Publishers, 1973); Richard A. Vatz, 'The Myth of the Rhetorical Situation', *Philosophy and Rhetoric* 6 (1973): 154–61.
7. The systematic rigour of his treatment is further suggested by Rhys Roberts's note in his translation of Aristotle that the word 'art' as used in the *Rhetoric* 'stands for methodical treatment of a subject'. *The Rhetoric and Poetics*, 19.
8. For an overview, see Ann Gill, *Rhetoric and Human Understanding* (Prospect Heights, IL: Waveland, 1994); James McCroskey, *An Introducton to Rhetorical Communication* (Englewood Cliffs, NJ: 1993); Patricia Bizzell and Bruce Herzberg, *The Rhetorical Tradition: Readings from Classical Times to the Present* (Boston: St Martin's, 1990); James L. Golden, Goodwin F. Berquist, and William E. Coleman, *The Rhetoric of Western Thought*, 4th ed. (Dubuque, Iowa: Kendall/ Hunt, 1989); Brian Vickers, *In Defence of Rhetoric* (New York: Oxford University Press, 1989).
9. Practical rhetorics for both speakers and critics still invoke Aristotelian principles. Typical examples currently available are David Zarefsky and Jennifer MacLennan, *Public Speaking: Strategies for Success*, Canadian ed. (Scarborough, ON: Prentice Hall Canada, 1997) and Sonja Foss, *Rhetorical Criticism: Exploration and Practice* (Prospect Heights, IL: Waveland, 1996).
10. See, for instance, John Douglas and Mark Olshaker, *Journey into Darkness* (New York: Scribner, 1997); Ruth P. Rubinstein, *Dress Codes: Meanings and Messages in American Culture* (Boulder, CO: Westview, 1995); Margaret Visser, *Much Depends on Dinner* (Toronto: HarperPerennial, 1992); William Lutz, *Doublespeak* (New York: HarperCollins, 1989); John Fiske and John Hartley, *Reading Television* (London: Methuen, 1978); Pierre Berton, *The Big Sell* (Toronto: McClelland and Stewart, 1963).
11. For example, few works are more celebrated (or more practical) than Dale Carnegie's *The Quick*

and Easy Way to Effective Speaking (New York: Dale Carnegie and Associates, 1962) or *How to Win Friends and Influence People* (New York: Simon & Shuster, 1981 [orig 1936]), both of which feature treatments of rhetoric that owe much to Aristotle.

12. Kenneth Burke, *A Grammar of Motives*. (Berkeley: University of California Press [rpt 1969], 1945), and *Counter Statement* (Berkeley: University of California [rpt 1968], 1931).
13. Walter R. Fisher, 'Clarifying the Narrative Paradigm', *Communication Monographs* 56 (1989): 55–8; 'The Narrative Paradigm: An Elaboration'. *Communication Monographs* 52 (1985): 347–67; 'Narration as a Human Communication Paradigm: The Case of Public Moral Argument', *Communication Monographs* 51 (1984): 1–22.
14. Northrop Frye, *The Educated Imagination* (Bloomington: Indiana University Press, 1964); *The Well-tempered Critic* (Bloomington: Indiana University Press, 1963); *By Liberal Things* (Toronto: Clarke Irwin & Co., 1959); and *Anatomy of Criticism* (Princeton, NJ: Princeton University Press, 1957).
15. Ernest G. Bormann, 'Coloquy I. Fantasy and Rhetorical Vision: Ten Years Later', *Quarterly Journal of Speech* 68 (1982): 288–305, and 'Fantasy and Rhetorical Vision: The Rhetorical Criticism of Social Reality', *Quarterly Journal of Speech* 58 (1972): 396–407.
16. For an overview of these approaches, see Sonja K. Foss, *Rhetorical Criticism*, 1996, Sonja K. Foss, Karen A. Foss, and Robert Trapp, *Contemporary Perspectives on Rhetoric*, 2nd ed. (Prospect Heights IL: Waveland Press, 1991); and Roderick Hart, *Modern Rhetorical Criticism*, 2nd ed. (Glenview IL: Scott Foresman, 1997).
17. As they are in Barry Brummett, *Rhetoric in Popular Culture* (New York: St Martin's Press, 1994) and *Rhetorical Dimensions of Popular Culture (Studies in Rhetoric and Communication* (Tuscaloosa: University of Alabama Press, 1991). See also William A. Covino and David A. Joliffe, 'An Introduction to Rhetoric', *Rhetoric: Concepts, Definitions, Boundaries* (Boston: Allyn & Bacon, 1995), 1–26.
18. Accounts are provided by, among others, Scott Consigny, 'Dialectical, Rhetorical, and Aristotelian Rhetoric', *Philosophy and Rhetoric* 22 (1989): 281–7; Robert S. Cathcart, 'McLuhan vs. Aristotle: Media Dominance in the 1980s', *Journal of the Association for Communication Administration* 27 (1979): 52–5; Wayne E. Brockriede, 'Toward a Contemporary Aristotelian Theory of Rhetoric', *The Quarterly Journal of Speech* 52 (1966): 33–40; Edwin Black, 'Aristotle and Rhetorical Criticism', *Rhetorical Criticism: A Study in Method* (New York: Macmillan, 1965), 36–90; Dilip Gaonkar, 'Object and Method in Rhetorical Criticism: from Wichelns to Leff and Mcgee', *Western Journal of Speech Communication* 54 (Summer 1990) 290–316; James T. Boulton, 'The Criticism of Rhetoric and the Act of Communication', *Essays on Rhetorical Criticism*, ed. Thomas R. Nilsen (New York: Random House, 1968); Robert D. Clark, 'Literary and Rhetorical Criticism' also in Nilsen; and Lane Cooper, 'The Rhetoric of Aristotle', *The Quarterly Journal of Speech* 21 (1935): 10–19.
19. Other relevant articles include William Eadie, 'Reitzel's Selection as Outstanding Professor Honors Both Teaching and Service Learning', *Spectra* 35 (April 1999); Judith A. Rolls, 'Experiential Learning as an Adjunct to the Basic Course: An Analysis of Student Responses to a Pedagogical Model', *Basic Communication Course Annual* 5 (1993): 182–99; L.B. Specht and P.K. Sandlin, 'The Differential Effects of Experiential Learning Activities and Traditional Lecture Classes in Accounting', *Simulation and Gaming* 22 (1991): 196–210.
20. Jennifer MacLennan, 'What Do We Have to Take this Stuff For? Integrating Theory and Practice Through the Group Independent Study'. Paper presented at the Conference of the Northwest Communication Association, Coeur D'Alene, ID, April 1997; G.R. Bringle and J.A. Hatcher, 'Implementing Service Learning in Higher Education', *Journal of Higher Education* 67, 2 (1996): 221–3; I. Harkavay, 'Service Learning as a Vehicle for Revitalization of Education Institutions and Urban Communities'. Paper presented at the American Psychological Association Annual Meeting, Toronto, ON, August 1996; J.S. Berson, 'A Marriage Made in Heaven: Community Colleges and Service Learning' [Online].

Available URL: <http://www.broward.cc.fl.us/bcc/st_affairs/judity2.html> (1995); S. Cohen and C.A. Sovet, 'Human Service Education, Experiential Learning, and Student Development', *College Student Journal* 23 (1989): 117–22.

21. 'History of the Communication Discipline'. Online: National Communication Association <www.natcom.org/nca/Template2.asp?bid=398>. Accessed 22 March 2007.
22. Noam Chomsky and E.S. Herman, *Manufacturing Consent* (New York: Pantheon, 1988); G.R Funkhouser, *The Power of Persuasion* (New York: Times Books, 1986); A. Ries and J. Trout, *Positioning: The Battle for Your Mind* (New York: McGraw Hill, 1981); Vance Packard, *The People Shapers* (New York: Little, Brown & Co., 1977) and *The Hidden Persuaders* (New York: Pocket Books, 1961); Chaim H. Perelman and L. Olbrechts-Tyteca (1971, op. cit.); David Ogilvy, *Confessions of an Advertising Man* (New York: Dell, 1963).
23. E.L. Boyer, *College: The Undergraduate Experience in America* (New York, Harper & Row, 1987); R. Grantz and M. Thanos, 'Internships: Academic Learning Outcomes', *National Society for Experiential Education Quarterly* 22 (1996): 10–27;
24. See, among others, Wayne C. Booth, *The Rhetoric of Fiction*; Deanne Bogdan, 'Is It Relevant and Does It Work? Reconsidering Literature Taught as Rhetoric', *Journal of Aesthetic Education* 16 (Winter 1982): 27–38; Carroll C. Arnold, 'Oral Rhetoric, Rhetoric, and Literature', *Contemporary Rhetoric*, ed., W. Ross Winterowd (New York: Harcourt Brace Jovanovich, 1975), 60–73; Kenneth Burke, 'Rhetoric and Poetic', *Language as Symbolic Action* (Berkeley: University of California Press, 1966), 295–307; Reed Whittemore, 'Literature as Persuasion', *The Fascination of the Abomination* (New York: Macmillan, 1963), 240–51; Donald K. Smith, 'Origin and Development of Departments of Speech', *History of Speech Education in America*, ed., Karl R. Wallace (New York: Appleton-Century-Crofts, 1954), 447–70; Herbert A. Wichelns, 'The Literary Criticism of Oratory', *Studies in Rhetoric and Public Speaking in Honor of James A. Winans*, ed., A.M. Drummond (New York: Century, 1925: 181–216).
25. In my former university, for example, beginning Assistant Professors in Arts and Science started at under $40,000. Assistant Professors in the Faculty of Management (the only professional faculty in the institution) received starting salaries of $60,000 or better. See University of Lethbridge, *The University of Lethbridge Faculty Handbook* (Lethbridge: University of Lethbridge, 1 July 1999).
26. College of Engineering, *University of Saskatchewan: Terms of Reference for the D.K. Seaman Chair in Technical and Professional Communication* (Saskatoon: University of Saskatchewan, 1998); Joseph Hoey, *Employer Satisfaction with Alumni Professional Preparation* (North Carolina State University, University Planning and Analysis, 1997) Online. Available URL: <http://www2.acs.ncsu.edu/UPA/survey/reports/employer/employ.htm>; DiSalvo, D.C. Larsen, and W.J. Seiler, 'Communication Skills Needed by Persons in Business Organizations', *The Speech Teacher* 25 (1976): 269–75; Elton L. Francis, 'When the Engineer Speaks', *Today's Speech* 8 (1960): 6–9; Arthur Eisenstadt, 'The Employer and 'The Speech Teacher', *Journal of Communication* 3 (1953): 105–9.

References

Applegate, J., and Sherry Morreale. 2001. 'Creating Engaged Disciplines', in V. Hendley, ed., *AAHE Bulletin* 53, 9.

Aristotle. 1954. *The Rhetoric and Poetics*, W. Rhys Roberts, trans. Fredrich Solmsen, ed. New York: The Modern Library [Random House].

Baskerville, Barnett. 1971. 'Rhetorical Criticism, 1971: Retrospect, Prospect, Introspect', *Southern Speech Communication Journal* 37 (Winter): 122.

Berko, Roy, Megan Brooks, and J. Christian Spielvogel. 1995. *Pathways to Careers in Communication*. Annandale, VA: Speech Communication Association.

Bitzer, Lloyd F. 1968. 'The Rhetorical Situation', *Philosophy and Rhetoric* 1 (Winter): 3–4.

Booth, Wayne C. 1983. *The Rhetoric of Fiction*, 2nd ed. Chicago: University of Chicago Press.

Brock, Bernard L., and Robert L. Scott. 1980. *Methods of Rhetorical Criticism*. Detroit: Wayne State University Press.
Bryant, Donald C. 1963. 'Rhetoric: Its Functions and Its Scope', *Quarterly Journal of Speech* 39 (December): 401–4.
Burke, Kenneth. 1968 [1931]. *Counter Statement*. Berkeley: University of California.
———. 1969 [1945]. *A Grammar of Motives*. Berkeley: University of California Press.
———. 1969 [1962]. *A Rhetoric of Motives*. Berkeley: University of California Press.
Chesebro, James. 1996. 'Communication Vistas: Futures from a 1996 Perspective', *Spectra* 32 (October): 12.
Cialdini, Robert. 1993. *Influence: Science and Practice*. New York: Scott Foresman.
———. 1998. *Influence: The Psychology of Persuasion*. New York: William Morrow.
Corbett, Edward P.J. 1971. *Classical Rhetoric for the Modern Student*. New York: Oxford University Press.
Cosnett, Gary. 1996. 'Ancient Wisdom for the Modern Professional'. Available at: http:www.salesdoctors.com/cosnett/index.htm.
Dillon, George. 1986. *Rhetoric as Social Imagination*. Bloomington, Indiana: Indiana University Press.
Glazer, Nathan. 1974. 'Schools of the Minor Professions', *Minerva*.
Katula, Richard A., and E. Threnhauser. 1999. 'Experiential Education in the Undergraduate Curriculum', *Communication Education* 48 (July): 248.
Kuhn, Thomas. 1970. *The Structure of Scientific Revolutions*, 2nd ed. Chicago: University of Chicago Press.
Moore, Wilbert E. 1976. *The Professions: Roles and Rules*. New York: Russell Sage.
National Communication Association. 1999. 'Call for Papers for the 2000 Convention of the National Communication Association'. *NCA Convention Program 1999*. Annandale, VA: National Communication Association.
Packard, Vance1964a. 'Totem Poles of Job Prestige', *The Status Seekers*. New York: Cardinal.
———. 1964b. 'The Trouble with People', *The Hidden Persuaders*. New York: Cardinal.
Perelman, Chaim, and L. Olbrechts-Tyteca. 1971. *The New Rhetoric: A Treatise on Argumentation*, John Wilkinson and Purcell Weaver, trans. Notre Dame: University of Notre Dame Press.
Schein, Edgar H. 1992. *Organizational Culture and Leadership*, 2nd ed. San Francisco: Jossey-Bass Publishers.
Schon, Donald A. 1983. *The Reflective Practitioner: How Professionals Think in Action*. New York: Basic Books.
Wichelns, Herbert A. 1925. 'The Literary Criticism of Oratory', *Studies in Rhetoric and Public Speaking in Honor of James A. Winans*, A.M. Drummond, ed. New York: Century.

QUESTIONS FOR CRITICAL THOUGHT

1. Why does MacLennan open with a comparison of her experience in the English department and her experience in an engineering college? What does this personal anecdote have to do with the substance of her essay?
2. Why does MacLennan refer to so many other works? Why is the essay so heavily endnoted?
3. The essay is a research report. How do its diction, organization, logical development, and pacing show this? For whom do you think it was originally written? How do you know?
4. What is the central feature of professional activity, according to MacLennan?
5. What two features distinguish a professional education from other areas of university study?

6. In what sense is rhetoric a 'practical, instrumental art'? How does MacLennan establish this?
7. Compare MacLennan's essay with Lloyd Bitzer's 'Functional Communication'. In what ways are their arguments similar?
8. Why does MacLennan survey so many definitions of rhetoric? Does doing so strengthen her argument?
9. What are the three features of professional knowledge?
10. In what sense is this essay about language? Could this essay have been placed in any of the other categories in this book? If so, which one(s)? Should it have been? Why?
11. How personal is this essay? How formal is it? If you are familiar with MacLennan's voice from her other contributions to this volume, to what extent does this essay also 'sound' like her? What are the elements that distinguish her style from that of the other writers in this book? What are the elements that distinguish any writer's style from any other?
12. Outline the major points in MacLennan's argument. How difficult is this to do? How is the paper structured?
13. MacLennan draws support for her argument from at least two separate fields. Given that she is working with so many quotations from others, what is original about this argument?
14. How does MacLennan establish ethos with her intended readers?
15. Like Speed's 'What Do You Mean I Can't Call Myself a Software Engineer?', MacLennan's essay is an extended definition of a single contested term. Why is MacLennan's piece so much longer than Speed's? What elements of context, purpose, and audience may have shaped each writer's rhetorical choices?
16. Why does a 'full understanding of rhetoric rely on direct hands-on experience'?
17. How difficult was it for you to read this essay? To what extent might your difficulty be attributed to a lack of context for the argument? Do you think MacLennan's original readers might have had a similar difficulty?
18. How provocative is MacLennan's ending? How provocative might it have been for her original readers? Why?
19. How is the word 'engineering' used in this essay? Does it meet the criteria set out by Speed ('What Do You Mean I Can't Call Myself a Software Engineer?')? Does it need to do so? Why or why not?
20. To what extent is this essay about professional writing?

PART VII

Ethical and Political Constraints

LIKE other human actions, our communication is a product of the choices we make as we go about our business. These choices are shaped by a number of factors, and are rarely pure. In addition to considerations of scientific validity, professional honour, public welfare, or the pursuit of truth, human behaviour is also affected by such issues as power, status, self-interest, security, greed, pride, face loss, and fear. The way we behave, and the way we communicate, are not isolated phenomena; they are affected by our inner states as well as by outer factors, and our actions and our messages frequently can have a powerful impact on the experience and behaviour of others, influencing them in sometimes profound ways.

It is this potential to influence, shape, and constrain the choices and beliefs of other people that introduces a political and ethical dimension to the study and practice of communication. Principles of ethics—the ability to choose between right and wrong, between competing 'goods', between self-preservation and the possibility of harm to others—have evolved to help us to make decisions in keeping with this fundamental principle of respect for others as human persons. However, our capacity to choose an ethical course of action can sometimes be constrained by forces that can best be understood as political. Politics involves the management of power, status, and authority—not only in government, but in corporate, organizational, and social environments as well.

Power, as we know, both seduces and corrupts, sometimes even at the lowest of organizational and personal levels; as a result, making ethical choices in what and how we communicate is not always as straightforward as it might initially seem. As our selections suggest, professional communication is far more politically charged than we are ready to acknowledge, and ethical communication is more difficult than simply presenting the 'facts' as we know them. We are all familiar with the abuses of power that sometimes occur in bureaucracies and in governments; scandals having to do with these forms of corruption are reported in the press so often and in such detail that they that they have almost lost their power to shock and surprise us. But we may not be as aware of the extent to which issues of power shape our everyday choices as we communicate with others, or the extent to which communicating effectively means making sometimes difficult ethical choices.

Corporate politics, office politics, social and sexual politics, and interpersonal politics are all ultimately matters of power or self-interest, and most of these attempts to influence us are carried out exclusively through communication. Our political survival—both in the public arena and in private—frequently depends on our ability to understand as much about what a message is hiding as about what it reveals, and our ethical challenge is to make choices that preserve the fundamental respect and trust on which our society depends.

CHAPTER 39

Communicating Ethically

Jennifer M. MacLennan

Communication, like other forms of human choice and action, can be ethical or unethical; it can be honest or dishonest; it can be helpful or harmful. But like other human actions, communication is ethical insofar as we accord to others at least the same consideration that we hope to receive in return. This essay explores the nature of ethics and its relationship to communication, and articulates a code of ethics for communicating more responsibly.

What is Ethics?

The word 'ethics' is derived from the Greek word 'ethos', or 'public character',[1] and for many of us, the word still carries some of the flavour of its original meaning; for example, we still recognize ethical behaviour as a key component of good character (Kilpatrick, 1992). The word as it's used now has at least two distinct, though related, meanings, both of which retain a clear relationship to the word's Greek origins.

First, 'ethics' refers to a branch of academic study which concentrates on how people make decisions between right and wrong, how they choose between two conflicting values, and on how deeply-held principles actually guide our choices and our actions. Although it's usually thought of as a specialization within philosophy or theology, the study of ethics may also take place in professional contexts, such as law, medicine, engineering, or business. In these contexts, ethicists not only study values and their sources, but also conduct analyses of various cases to identify the pertinent issues that shape decision-making in specific situations.[2]

The word 'ethics' is also used in a second, more practical or applied sense. In addition to referring to the formal study of the way people reason and choose between right and wrong, 'ethics' also refers to the systems of values or codes of behaviour that we have evolved to help us make reasonable and effective choices as we live, work, and socialize with other people. To act ethically is to act responsibly and with good judgement, taking into account not only our own wants, but also the needs and the autonomy of others. To act ethically is to respect and interact with other people as beings, not as objects to be manipulated for our own benefit, and our ethical systems have evolved to help us to make decisions in keeping with this fundamental principle of respect for others as human persons.

It is important to understand that ethical systems do not originate with individuals. Although each of us is free to evaluate and respond to situations using our own judgement and values, ethics is not simply a matter of doing what you feel like, nor of expediency or convenience or self-gratification.[3] These are all motives that, from time to time, drive our behaviour, but they are not effective interpersonal or leadership practice, and they are certainly not the result of ethical decision-making, which is a thoughtful, deliberative process and not an automatic, accidental, or unintentional outcome.

Individuals do not invent their own ethical codes. Instead, codes of ethics typically find their justification in some authority beyond the individual: the authority of religious belief, the sanction of society, the collective experience of a profession, the traditions of a culture, the wisdom of a distinguished individual. Whether based on religious, cultural, familial, or professional codes of behaviour, all ethical systems have a strong collective component; they are based on interests broader than those of the individual, and are not simply a matter of personal whim.[4] Thus, our interactions with others at all levels have an ethical dimension that is given its particular flavour by religious beliefs and cultural norms, by social custom, by church doctrine, by family tradition, by law, by the code of a profession, even by standards of conduct within a given workplace.

Some Universal Ethical Principles

Interestingly, although there are variations in the specific precepts of different ethical systems, there are some universal principles that they all seem to share. At the heart of all ethical codes is some form of what we might call respect: regard and honour for others with whom we interact. The philosopher Martin Buber characterized this as an 'I–thou' orientation for its recognition of the other as a human self, in order to distinguish it from what he described as an 'I–It' orientation, in which others are regarded and treated essentially as objects (Buber, 1993).

This principle permeates all ethical codes, and can be seen in the ancient admonition of Confucius that has come to be known as the golden rule, which is most frequently stated as 'do unto others as you would have them do unto you' (Stein, 1975). This fundamental principle translates into a recognition of, and respect for, what might be called the 'selfhood' of other people.

A sense of right and wrong, of appropriateness, of ethical conduct, permeates all human interaction, including the professions. Most have developed explicit codes of conduct to guide their members in carrying out their responsibilities.[5] The best-known and longest-standing code of professional ethics, the Hippocratic oath of physicians, implicitly embodies this same principle in its admonition to 'first, do no harm.'[6] Many organizations, including universities, have developed codes of conduct for their members that implicitly assume an ethical standard.[7]

A code of ethics does not typically itemize rules of behaviour or set out specific regulations such as the 'pool rules' posted in the hot tub room at the neighbourhood condo. Its purpose is not to directly govern behaviour, but to provide a foundation for thoughtfully evaluating and judging the situations in which we must act. A true code of ethics (as opposed to a code of conduct) is meant to assist, and not replace, human assessment and choice, and thus offers general principles rather than a template for how to behave. The ultimate responsibility of choosing how to act remains with the individual.

As the ethicist Andrew Olson explains, the purpose of a code of ethics is not to tell us what to do, but to give us a foundation for deciding what we must or should do. According to Olson, 'a code of ethics increases ethical sensitivity and judgement, strengthens support for individuals' moral courage, and helps to hone an organization's sense of identity' (Olson, nd). Because the situations in which we have to make judgements are not always clear-cut, because they are not entirely predictable, and because each is unique, we must draw on our understanding to make choices that will best fulfill our obligations to our listeners, our messages, and ourselves. A code of ethics is intended to help us do that.

If you study the ethical codes of a number of professions and organizations, you will be initially struck by the range of variation in the specific details each addresses. Some are closer to a code of conduct than to a code of ethics, since they address specific actions that must be performed or avoided, while others provide general principles to guide decision-making. But despite their outward variations, the majority of such codes share some striking similarities. Two features in particular stand out: first, ethical codes of all professions focus not on the benefits to the individual practitioner, but on our obligations and responsibilities to those we serve, to society, to the profession or organization of which we are a part, or to some other higher good. The language of the codes frequently emphasizes service, focusing on values like duty, responsibility, or obligation.

The codes may also speak of such personal qualities as honour, respect, honesty, and integrity, all in the context of obligation and service. As an example, consider these four 'core values' that inform the character-building focus of the Boy Scouts: 'personal honesty, fairness in one's dealings, respect for others, and the maintenance of a healthy self' ('Four Core Boy Scout Values', nd). While most codes of ethics are not so explicitly aimed at character-building, they all, at least implicitly, encode the notion of Aristotle's 'good character'. Thus, one feature of all ethical practice is a pervasive sense of responsibility or obligation to others, to the profession, or even to a higher power.

Second, in addition to their emphasis on service and social obligation, ethical codes always presume a fundamental regard for the sanctity of human life. They are marked, both explicitly and implicitly, by respect for the integrity and autonomy of other people, for their safety and well being, for their individual liberty, for their essential value as human beings.

A Communication Code of Ethics

Ethical communication involves genuine respect for the other person, a sincere commitment to what you are saying, and respect for your own integrity. We communicate ethically when we choose a course of action that limits the unnecessary harm done to others, when we take responsibility for our choices and actions, and when we do not knowingly deceive or misrepresent facts, beliefs, or actions. The ethical principles of communication—credibility, sincerity, and integrity—are anchored in the participatory process of building connections with other people. Let's consider the three principles of our ethical code in greater detail, along with examples of how they might translate into specific behaviour.

I. **I will take responsibility for my words and my actions**. I will acknowledge and honour my obligation to stand behind my words. I will do my best to ensure the validity of my interpretations. I will not present interpretations and judgements as if they are simple reports. I will keep my word. I will try to recognize and defuse defensive reactions before conflict escalates. If I do not know the answer to a question, I will acknowledge it rather than trying to bluff my way through. I will focus on accommodating others' need for understanding, information, and support before indulging my own need for self-expression. I will hold myself responsible for how well someone else understands my message. When I am wrong, I will admit it. I will not divulge things said to me in confidence. I will offer advice only when it is asked for. I will do my best to listen genuinely in the spirit of understanding. I will keep in mind that my assessments are provisional and can be altered by subsequent information.

II. **I will take care not to misrepresent myself or my message**. I will strive to communicate clearly, simply, and tactfully. I will not pretend to be something I am not or claim expertise that I do not have. I will not assume authority that I do not legitimately merit. I will use language with care and precision. I will label inferences and judgements as such, and not present them as fact. I will not distort or misrepresent issues in order to win a point. I will present my thoughts in my own language, and will not present the ideas or thoughts of others as though they're my own. I will pay attention to what I may be communicating nonverbally. I will know the source of information that I use, and be able to explain how and why it's credible. I will not claim expertise I do not possess. I will recommend to others only actions that I have myself taken, or genuinely would take. I will not make promises I cannot keep.

III. **I will avoid unnecessary hurt to others by my words or tone.** I will treat others as I would like to be treated in their place. I will assume good will on the part of others, and act with good will in return. I will listen to others as carefully as I expect them to listen to me. I will respect others as autonomous human beings and treat them with courtesy. I will follow through on my promises to others. I will give face when necessary, and will avoid causing face loss to others. I will make reasonable requests. If I must criticize others, I will speak with tact and generosity of attitude. Where an apology is due, I will apologize to others genuinely and sincerely. I will avoid repeating unfounded gossip. If a conflict arises, I will do my best to avoid escalating it. I will try not to say things in anger that I may later regret.

Practices that violate the fundamental ethic of respect and deny the humanity of others, even when they are entrenched cultural practices or sanctioned by government, law, or religious custom, are at their root unethical. Apartheid, 'ethnic cleansing', slavery, head taxes, expulsion of whole groups of people from their homes, stripping others of their ethnic identity—all have at one time or another been in force somewhere on the globe, many of them even in Canada. The fact that such policies were sanctioned by various political or cultural regimes does not make them ethical, nor does the fact that hundreds, even thousands, of people subscribe to whatever system of belief supports such acts of dehumanization.[8] Thus, when we deny the humanity of others, we behave unethically.[9]

Even though we cannot always fully accommodate the needs of others, we nevertheless have a fundamental obligation to behave in ways that respect their humanity and their autonomy. As the sociologist Erving Goffman explains, 'participation in any contact with others is a commitment' (1967: 5) to others and to the social structure to which we belong.

However, because people have freedom of choice in their dealings with others, they sometimes behave in ways that violate their ethical obligations to those around them. This mistreatment of others may arise unintentionally, such as when an attempt to save face for ourselves inadvertently causes someone else to lose face, or it may be the product of deliberate acts of disrespect, such as the desecrating of the local mosque in the wake of 9/11. It may be a minor infraction, as when the clerk in the registrar's office places policy above the human interest, or it may be an action bordering on the criminal, as when someone is made the victim of a hate crime.

However, whatever the circumstances, we need to be very clear that whenever we choose to disregard the welfare and face of others, when we treat others simply as things that can be used for our own purposes and then brushed aside, when we serve ourselves at the expense of others' welfare, when we attempt to influence others without regard to their well being, we are choosing to behave unethically, just as much as when we lie, cheat, steal, or betray others. As individuals, we should regard any policy or practice that treats human beings as things to be manipulated for profit or other benefit as unethical to its core.[10] Indeed, disregard for the safety or welfare of others is in certain circumstances a criminal act.[11]

S.I. Hayakawa, the celebrated linguist, wrote that as a result of our participation in human society we reap many rewards, including a sense of belonging and the social cooperation that ensures our very survival. But our membership in human society also entails, along with these significant benefits, an equally significant obligation to those with whom we share its benefits. We are bound by our humanity to honour and respect the human dignity and worth of others. To do less than this, no matter who might sanction it or what short-term benefits it might bestow, is to behave unethically.

Notes

1. S. Michael Halloran, 'Aristotle's Concept of Ethos, or if not His, Somebody Else's', *Rhetoric Review* 1 (September 1982): 58–63; William M. Sattler, 'Conceptions of Ethos in Ancient Rhetoric', *Speech Monographs* 14 (1941): 55–65.
2. For example, see Margaret Sommerville, 'Ethical Dilemmas in Medicine', *McGill Faculty of Medicine Mini-Med Study Corner*. Available at: http://www.medicine.mcgill.ca/minimed/archive/Sommerville2002.htm.
3. Such behaviour actually signals a disregard of others' well being, a failure of socialization, and even possible psychopathology. See Paul Babiak and Robert Hare, *Snakes in Suits* (Toronto: HarperCollins, 2006); Sgt Matt Logan, 'The Psychopathic Offender: How Identifying the Traits Can Solve Cases', *Royal Canadian Mounted Police Gazette* 66 (2004), available at: http://www.gazette.rcmp.gc.ca/article-en.html?&article_id=39. See also the classic work Harvey Cleckley, *The Mask of Sanity*, 5th ed. (Augusta, GA: Emily Cleckley, 1988), available at http://www.cassiopaea.org/cass/sanity_1.PDF.
4. For a more extensive and very readable discussion of the root and nature of ethics, see Peter Singer, *Practical Ethics*, 2nd ed. (Cambridge, UK: Cambridge University Press, 1993).
5. Many of these can be found at the Center for the Study of Ethics in the Professions. Available at http://ethics.iit.edu/codes/.
6. See, for example, Illinois Institute of Technology, 'Codes of Ethics Online', available at http://ethics.iit.edu/codes/.
7. Including, for example, those posted on line by Queen's University, (http://www.queensu.ca/secretariat/senate/policies/codecond.html); the University of Toronto (http://www.utoronto.ca/govcncl/pap/policies/studentc.html); Athabasca University (http://www.athabascau.ca/calendar/page11.html); or McMaster University (http://www.mcmaster.ca/univsec/policy/StudentCode.pdf), to mention only a few.
8. For some interesting discussions of policies of dehumanization, the work of Haig A. Bosmajian is informative. See, for instance, 'Defining the "American Indian": A Case Study in the Language of Suppression', *Speech Teacher* 22 (March 1973): 89–99; 'The Language of Sexism', *ETC: A Review of General Semantics* 29 (September 1972): 305–13; 'The Language of White Racism', *College English* 31 (December 1969): 263–72.
9. As was, of course, underlined by the famous Nuremberg War Crimes trials of 1945–9. You can find information about them at the 'Famous World Trials' website. See Doug Linder, 'The Nuremberg Trials' (2000), http://www.law.umkc.edu/faculty/projects/ftrials/nuremberg/nurembergACCOUNT.html.
10. Corporations and bureaucracies get a lot of heat on this score. See, for instance, Joel Bakan, 'The Externalizing Machine', *The Corporation: The Pathological Pursuit of Profit and Power* (New

York: Simon and Schuster, 2004), 60–84 and Noam Chomsky, *Profit over People* (New York: Seven Stories Press, 1999).

11. *Criminal Code of Canada*, Article 219 (1): 'Every one is criminally negligent who in doing anything, or (b) in omitting to do anything that it is his duty to do, shows wanton or reckless disregard for the lives or safety of other persons.' Available at: http://laws.justice.gc.ca/en/c-46/267361.html.

References

Buber, Martin. 1993 [1923]. *I–Thou*. New York: Touchstone.

'Four Core Boy Scout Values'. nd. *DELTA: An Ethics in Action Program For Boy Scouts*. Available at: http://pinetreeweb.com/delta-1.htm.

Goffman, Erving. 1967. 'On Face-Work', *Interaction Ritual: Essays on Face-to-Face Behaviour*. New York: Anchor Books.

Kilpatrick, William. 1992. *Why Johnny Can't Tell Right from Wrong: Moral Illiteracy and the Case for Character Education*. New York: Simon and Schuster

Olson, Andrew. N.d. 'Authoring a Code of Ethics: Observations on Process and Organization', Illinois Institute of Technology Centre for the Study of Ethics in the Professions. Available at: http://ethics.iit.edu/codes/Writing_A_Code.html.

Stein, Jess, ed. 1975. *The Random House College Dictionary*, rev. ed. New York: Random House.

QUESTIONS FOR CRITICAL THOUGHT

1. What is the origin of the word 'ethics'? Why does its etymology matter?
2. What two definitions of 'ethics' does MacLennan offer? With which is she more concerned in this essay?
3. What, according to MacLennan, does it mean to behave ethically?
4. What is the difference between a code of ethics and a code of conduct? Why does MacLennan draw this distinction?
5. What is an I–thou orientation?
6. What is the connection between ethics and professional communication?
7. MacLennan articulates a code of ethics for communicators. What are the three principles she identifies?
8. How personal is MacLennan's essay? How formal is its style? If you have read other selections by her in this book, compare them with this one. What differences, if any, do you see? How might context and purpose have shaped each piece?
9. Compare MacLennan's discussion of ethics with the others in this chapter. What commonalities do you see?
10. To what extent is this selection an extended definition?

CHAPTER 40

Ethos: Character and Ethics in Technical Writing

Charles P. Campbell

In *Zen and the Art of Motorcycle Maintenance*, Robert Pirsig characterizes thus the technical description of a motorcycle:

> The first thing to be observed about (it) is so obvious you have to hold it down or it will drown out every other observation. This is: It is just duller than ditchwater. Yahda, yah-da. yah-da, yah-da, yah, carburetor, gear ratio, compression, yah-dah-yah, piston, plugs, intake, yah-dayah, on and on and on. That is the romantic face of the classic mode. Dull, awkward, and ugly. Few romantics get beyond that point. (1975: 71)

A romantic, in Pirsig's terms, does not look at the underlying form of things, as expressed in specifications, flow diagrams, and equations. Rather, a romantic looks at surfaces—the curve of a parabolic antenna dish, the sleekness of a console. Pirsig's classicist, on the other hand, hardly sees surfaces at all, being preoccupied with the underlying principles that make things function.

Though Pirsig uses 'classic' and 'romantic' idiosyncratically, his comment about technical prose resonates even for us who read it and write it constantly. Technical prose, we might say, just lacks character.

Pirsig ascribes the classic–romantic split to a falling away from Quality. This prelapsarian state supposedly existed before Athenian philosophers wielded their analytic knives, creating such divisions as truth/probability, mind/body, philosophy/rhetoric—divisions so deeply ingrained in Western culture that they appear 'natural'. One consequence of the resulting dualism, as Daniel R. Jones (1989) points out in an insightful interpretation of Pirsig's book, is a 'spectator attitude' toward technology. Spectators are alienated from technology, whether they consume its products or even work with it.

Pirsig notes, early in the book, that the technical manuals he worked on as writer and editor were spectator manuals:

> It was built into the format of them. Implicit in every line is the idea that 'Here is the machine, isolated in time and space from everything else in the universe. It has no relationship to you, you have no relationship to it. . . . We were all spectators. And it occurred to me that there is no manual that deals with the real business of motorcycle maintenance, the most important aspect of all. Caring about what you are doing is considered either unimportant or taken for granted. (Pirsig, 1975: 27)

This is the character of the objective narrative stance, which is calculatedly depersonalized.

This stance, according to my colleague Lynn Deming,

> comes from the difficulties writers in technical fields seem to have inserting themselves into their scientific and technical documents. . . . They feel the presence of an actor—an 'I' or 'we' or even a 'researcher' or 'scientist' or 'engineer'—dilutes or contaminates the objectivity and authenticity of the data or distracts the reader from the real subject—the chemical

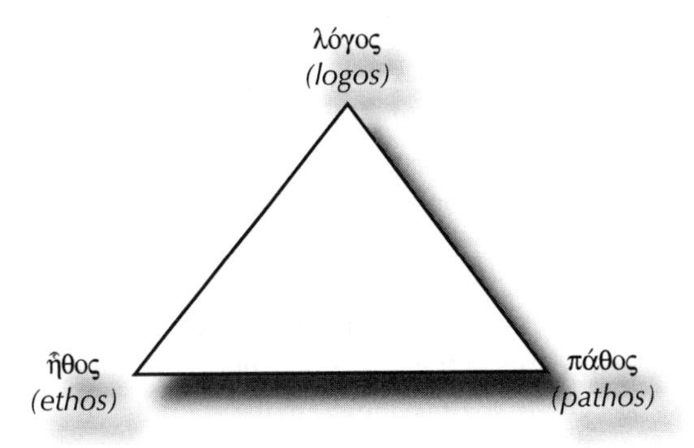

Figure 40.1 Aristotle's 'Communication Triangle.'

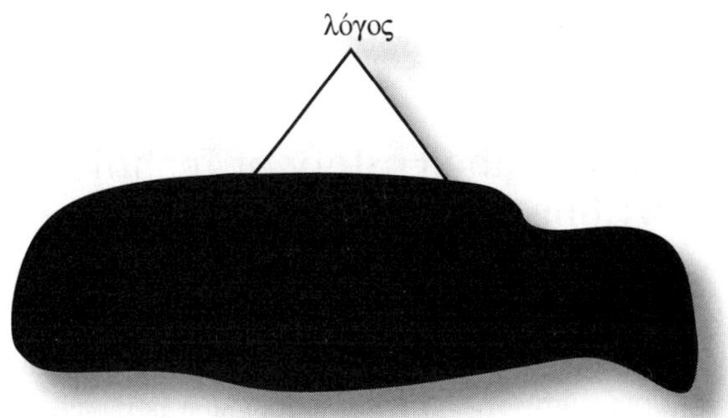

Figure 40.2 Early Twentieth Century: 'Content is all that matters.'

> reaction or the fabrication. This is just not true. (1993: 154)

In Pirsig's terms, they are spectators too—even as writers.

The problem that Pirsig ascribes to the absence of Quality and that Deming suggests stems from a problematical narrative stance looks somewhat different from another perspective, that of Aristotelian rhetoric. The Aristotelian perspective has been used forcefully by Arthur E. Walzer and Alan Gross (1994) in their analysis of accounts of the Challenger disaster. They found these accounts conforming to one of two perspectives: positivist or postmodernist. Positivist accounts emphasize a failure in communication—the disaster occurred because engineers and managers failed to transmit or receive all the facts. Postmodernist accounts stress differences in interpretive frameworks that cause engineers and managers to draw different conclusions from the same data. 'The positivistic perspective . . . attempts, in effect, to minimize deliberation and circumvent ethics . . . . Since the postmodernists have been forceful critics of the positivists, it seems even more surprising that their analyses of the Challenger case all but dismiss ethics as irrelevant and virtually deny the possibility of rhetorical deliberation reaching consensus' (Walzer and Gross, 1994: 431).

From a rhetorical perspective, there are two sorts of deliberation that could have taken place: those contributing to technical knowledge for making engineering decisions, and those contributing to normative understanding for making ethical decisions. When the technical arguments cannot produce a clear consensus but a decision must be made anyway, deliberations need to become normative—to ask, as Walzer and Gross do for Challenger, 'What rules ought to guide our decision in the absence of knowledge of how the o-rings will perform?' (1994: 427).

In such circumstances, neither the spectator status conferred by a positivist focus on 'objective facts' nor the insulation afforded by immersion in a particular profession's interpretive framework serves the public interest. With Walzer and Gross, I regard Aristotelian rhetoric as offering a method for conducting public discourse, one that could improve the quality of public decision-making. Yet a part of Aristotelian rhetoric, the artistic proofs, appears to support the radical individualism that is part of the positivistic culture of professionalism. I think in fact that such support is only the result of a fallacy of translation. Aristotle's *ethos* is usually translated as 'character', a quality we conventionally ascribe to individuals. But *ethos* is a richer concept, involving the individual in a deliberative community and thus having, as the

Greek root implies, an ethical dimension as well as a transactional one.

Ethos in the History of Technical Writing

The idea that technical writing involves rhetorical transaction at all has only become utterable within about the last 20 years. It was initially resisted by a prevailing view, still common, that language is merely a package for information—a transparent medium in which clarity is the highest value. Yet, as Mary B. Coney (1992) has pointed out, as the history of experimental science has come to be understood in terms of how communities construct knowledge, 'No longer can one assume that meaning is something developed by independent researchers, encoded into messages, packed into containers, and sent off to readers who are isolated from these processes.'

Information as message packets is another manifestation of spectatorhood, which stands in the way of effective communication so long as documents focus solely on the technical. Fortunately, the emphasis of technical writing has been shifting late in this century from 'technical' to 'writing'. This history can be explained through the paradigm of the 'communication triangle' implicit in Aristotle's artistic proofs (Fig. 40.1), which regard not only reasoning (*logos*) as a persuasive element, but also the condition of the audience (*pathos*) and the character of the speaker (*ethos*).

Before there was a professional field called technical communication, technical writing was done by technical people. Like Pirsig's classicists, they cared mostly about such matters as design and function. In terns of Aristotle's triangle, these folks thought only of *logos*: content and reasoning.

Did their writing therefore escape the claims of pathos and *ethos?* No: often their prose created such obstacles to understanding, through its jargon and density, that it created the *ethos* of *the expert*: one whose esoteric knowledge makes him—and it still was mostly him—a member of a priesthood, a wizard whose mysteries aren't supposed to be understood by the uninitiated. The *pathos* dimension could be summed up in the sentence, 'If you're smart enough to understand this, fine; if not, too bad.' Call this dominance of *logos* the objectivist model (Fig. 40.2).

The perils of the objectivist model are summed up well in a 1946 book written for technical specialists, ironically with a point of view that objectivizes its readers:

> The greatest mistake in the preparing of reports is that the technician does not put himself in the place of the audience or the readers, and does not give them what they want. He sees the subject too much from his own point of view, not enough from *their* point of view. This comes from being too close to the subject and from lack of imagination. This criticism can scarcely be repeated too often. It is exceedingly important. (MacDonald, 1946: 66–7)

In reaction to this 'objectivist' neglect by specialists of the needs of readers, managers, and bureaucrats, who needed to understand technical material well enough to make decisions about dam sites and weapons systems, sought to find *technical writers* who could explain technical matters. Thus began technical communication as a discipline. Sometimes these writers were technical people themselves, perhaps engineers who did some reading outside their own field and who had a knack (but not a heuristic) for writing pretty well. Others were liberal arts majors who had the interest and patience to understand and explain technical matters, but who also lacked a heuristic since rhetoric had largely disappeared from college curricula around 1900. I was one of the latter: thirty years ago, I parlayed a master's in literature and some Navy electronics background into a technical writing job.

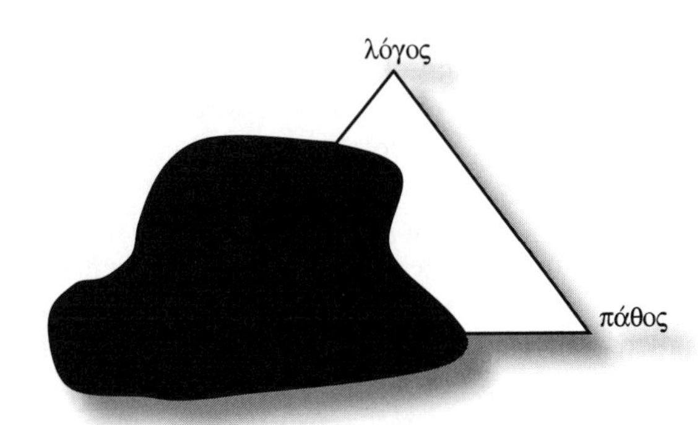

Figure 40.3 Late Twentieth Century: 'Think about the audience.'

As the field gained momentum, two complementary things happened, both on the pathos corner of the triangle. First, books on technical writing, such as Houp and Pearsall's (1995) and Mathes and Stevenson's (1991), began stressing audience needs and ways to analyze them; such books, used in college technical writing courses, began to chip away at the 'objectivist' model. Second, with the growing use of computer software, the field of usability testing developed. Bringing the audience into the design phase brought much new and productive thinking about how documents should be organized, text chunked for easy access, and graphics used for information and access.

Ethos and the Tyranny of Audience

Yet the empowerment of audience still leaves out something crucial. Last October at the annual meeting of the Council of Programs in Technical and Scientific Communication, one of the discussions surfaced some feelings that there is by now a 'tyranny of audience' brought about by too much attention to what the audience wants and too little to what the audience needs, the latter question involving the *ethos* dimension. This is the situation described in Fig. 40.3.

The tyranny of audience may be seen most clearly in the current fashion of considering it as 'users'. Users are implicitly conceived of as being somewhat infantile: they want specific bits of pragmatic knowledge (e.g., 'How can I get my system to stop crashing when I have three programs open at the same time?') and they want it *now*. How this tyranny might be undermined has been shown humorously by Marilyn Cooper (1994), who put together a short manual conflating cartoons from Michael Paul McLester's *Beset by Demons*—a bloated spider singing, 'Every blade of grass you see, every flower and every tree. . . . Everything belongs to ME!'—and parts of a chain saw manual.

Cooper's 'manual' makes it hard for the user to be a spectator: 'Yes, indeed, your new chain saw is not only a precious commodity but is also instrumental in the effort to transform the natural world for the market. In purchasing and using this chain saw, you have chosen to participate in this effort.' Cooper also forces the user to be aware of being part of a community with experiences in common: 'Perhaps because of its awkward position, the spider has cut off part of its left foot. Never allow any part of your body to touch the rotating chain. You know chain saws are dangerous. You've seen chain saw massacre movies.'

The dimension Cooper brings in is the *ethos* still missing in Fig. 40.3. It's an *ethos* situated within the cultural context of the late twentieth century in America, not floating disembodied outside time and place.

Ethos and Persona

Ethos is sometimes treated as a synonym for persona. Cicero sometimes used persona (in Latin, 'mask') to translate *ethos* because Latin has no exact synonym. Thus, according to Sharon Crowley, 'Roman rhetoricians who relied on Greek rhetorical theory sometimes confused ethos with pathos . . . . This lack of a technical term is not surprising, because the requirement of having a respectable character was built into the very fabric of Roman oratory' (1994: 85). That is, character was conceived in social terms—who you were, and thus your built-in credibility as a speaker, had to do with your station in society as well as your living up to the expectations of that role. This is a point to which I shall return shortly.

In late Roman times, and again after the rhetorical reforms of Peter Ramus in the sixteenth century, when rhetoric became reduced mostly to questions of style, style became the means of creating persona—of giving the prose personality. Merrill Whitbum wrote an article almost twenty years ago (1976) that discussed personality in technical writing. Like T.A. Rickard in his 1908 *Guide to Technical Writing* (1908), Whitbum was mainly interested in improving technical documents by having them exhibit better prose style.

Style and authorial voice are strongly linked, but strong persona is not necessarily a plus in technical writing. For example, humour might not be appreciated by the harried computer user as she consults 'help' to figure out why a program function isn't working, and it seems almost unimaginable in an operating manual for a nuclear power plant.

Ethos is not, however, at all the same thing as persona, described thus in the last two editions of the venerable *Reporting Technical Information* by Houp, Pearsall, and Tebeaux:

> Writers make important decisions about content and style based upon consideration of the audience and the persona the writer wants to project. *Persona* refers to the role the writer has or assumes when writing. It relates to, among other things, the position of the writer and his or her relationship to the audience and the situation. (1995: 18)

This view acknowledges that persona and audience are in some way fungible—interchangeable, in the sense that the speaker seems at one with the audience and the audience is drawn into the speaker's perspective. This fungibility is also implicit in the egocentric method of audience analysis used by Mathes and Stevenson (1991): when the writer positions herself at the centre of a series of circles representing various levels of audience, she is implicitly positioning herself to take a narrative stance appropriate for her several audiences. The audience as analyzed has individuals with particular 'operational', 'objective', and 'personal' characteristics, and the writer has a purpose: to authorize, propose, recommend, request, instruct. But the fungibility remains implicit, because the writer remains a black box, an unknown—a spectator. And from fields like advertising and public relations, we're all too aware of the possibility of using persona in ways that are ethically dubious.

Aristotle too treats *ethos* and pathos as fungible, and character as something that can be crafted for particular audiences and situations:

> Since all people receive favorably speeches spoken in their own character and by persons like themselves, it is not unclear how both speakers and speeches may seem to be of this sort through use of words. (cited in Kennedy, 1991: 168)

In the part of Book II of *Rhetoric* where this passage appears, Aristotle first gives a catalog of human emotions, then a catalog of characteristics appropriate to persons of certain ages and stations in life. Here, however, the fungibility of *ethos* and *pathos* comes from the fact that in Aristotle's time, *ethos* did not correspond to what we would now call personality, but rather was more like the public reputation one acquired by habitually acting in a particular societal role. In modern terms, *ethos* is the public image one acquires, say, from acting habitually as an engineer among engineers, or as a banker among bankers.

What I am describing as fungibility strongly resembles Kenneth Burke's principles of identification and consubstantiality (1969). As one works as an engineer or a banker, one identifies his or her interests to a large extent with those of the group, and through habituation acquires the group *ethos*. Consubstantiality comes into play when the engineer, while remaining an engineer, becomes 'substantially one' with a banker in creating a common sphere of interest through a business proposal. Or does in successful proposals.

To the extent, then, that one is not born a banker or an engineer, the *ethos* one has is partly acquired and partly invented. Under the circumstances shown in Figs. 40.2 and 40.3, *ethos* in technical prose was largely unconscious. Yet, as James S. Baumlin notes, 'More than an expression of individual psychology or an intersection of social forces, *ethos* is, as Aristotle himself suggests, quintessentially a linguistic phenomenon' (1994: 23). Such a view of *ethos* seems to justify Houp, Pearsall, and Tebeaux's treating, in the passage quoted earlier, the relation of persona and audience as something that can be done consciously—that is, one can invent a persona appropriate to a particular document's intended audience.

Now, there's a virtual industry that depends on persona. It provides software instruction to the uninitiated by writing in a breezy, you-oriented style with humorous authorial asides. Here is an example chosen at random: 'That 8 MB of extra memory that Quattro Pro would like could cost you as much as $400. (That's almost $3,700 in dog dollars)' (Goodkin, 1993).

So, though persona is part of *ethos*, it is not the whole. Killingsworth and Gilbertson summarize well the relatively few articles on the primary character in technical writing, which, they say, follow three general theoretical trends:

> 1. They conflate the concepts of *ethos* and *persona*.
> 2. They recommend the adoption of personae, but without being clear about the ethical responsibilities of the author or the general relation of writer to reader.
> 3. They recommend an aggressively personal approach to ethos without being clear about the technical means or possible outcomes of such an approach. (1992: 110)

Ethos as Construct

If *ethos* is not identical with persona, neither is it, because of its ancient association with reputation, identical with formal ethics. For persons, the ethics/character link is easier than for written documents, where the ethical dimension can only be inferred from the prose. In reading literature we are used to filtering 'truth' out of the utterances of unreliable, self-serving, or incompetent narrators, fully aware that the narrative voice is a fiction, a construction, and that the ethical probity of the narrator need not reflect that of the author.

But in technical writing, the narrative voice is also a construction, not just a transparent window on truth. (I suggested something of the sort in an earlier article on engineering style [Campbell, 1992].) It is even more obviously a construction, in that it is likely to be either a corporate or a generic voice. Killingsworth and Gilbertson (1992) assert that the poststructuralist notion of an author submerged in a network of intertextuality applies even better to technical writing than to literary works. That is, text is the medium by which ideas are mediated and compromises reached. The narrative voice of corporate documents, how they are developed and maintained and how they sustain a corporate *ethos*, is a subject worthy of study in itself.

Construction of *ethos* is the flip side of writers' constructing audiences. *Ethos* stands in relation to persona as Mary B. Coney's constructed audiences stand in relation to the empirical audiences of the technical writing textbooks. Coney has argued that writers usually require readers to read in a variety of roles, so that a reader-in-the-flesh has to adapt to the role theorized for her by the writer. Coney's article (1992) anatomized the ways technical readers construct the meaning of technical texts. Coney provides an alternative to regarding readers as static empirical subjects, analyzable in terms of their roles, backgrounds, and biases. Instead, she looks at the roles that readers are called upon to play as they are reading—how the text calls the reader to fill the roles it requires.

This way of looking at readers reading technical texts represents an advance: it treats reading as an active part of a rhetorical transaction that has the potential to change both the writer and the reader: 'roles are always transforming themselves in the course of the reading process: what started out as a naive user on page two of a manual is transformed *by the very act of reading* into a more sophisticated chooser of options on page 72' (1992: 61, emphasis in original).

Many people of a certain age could vouch for such transformation by a manual. John Muir's *How to Keep Your Volkswagen Alive* (1969), now available in a 25th anniversary edition, illustrates very well both Aristotle's adaptation of speaker to audience and of Coney's providing a role for the reader. Muir hypothesized an audience who, like him, wanted to be able to fix their Volkswagens in out-of-the-way places and who, like him, were willing to get their hands dirty. And he found that audience. The authorial persona was prominent but not dominant; mainly it established a dynamic with the reader. As a result, people otherwise technically innocent became technically adept, unafraid of adjusting their valves or even replacing the whole top end under a shade tree.

Ethos as a Vital Component in Technical Communication

One thing that should be apparent by now is that relations among *logos*, *pathos*, and *ethos* are not static, but dynamic, and that all are constructed in the act of writing. These dynamic relations also construct the lateral relations among the dimensions *I* (or *we*), *ethos*; *you* or *pathos*; and *it*, or *logos* to give us the typical genres of technical communication (Fig. 40.4). It's significant that two of the genres depend heavily on *ethos*. In fact, Killingsworth and Gilbertson (1992: 113–19)

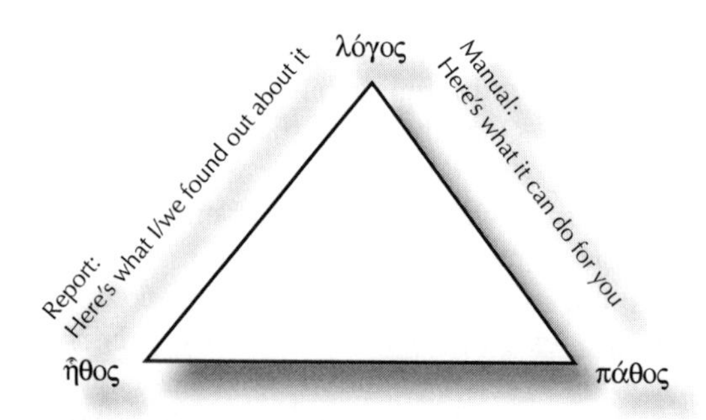

Figure 40.4 *Ethos* and the genres of technical writing.

have derived a whole series of subgenres based on the permutations of *I*, *you*, and *this*, which I adapt here:

- *The manual*: Maybe it shouldn't surprise us that pathos was revalorized at the same time there was a growing need for usable documentation. Manuals work hard along the *logos–pathos* leg of the triangle, trying to establish a strong bond between product and user: 'Here's what it can do for you.'

  Traditionally, the *ethos* in manuals is Platonic, or seemingly transparent—the reader is unaware of a narrative voice. But as the Cooper and Muir examples suggest, *ethos* could mediate powerfully between the technical material and the needs of the reader—particularly in answering the questions so often neglected in manuals, 'Why would I want to use this feature in the first place? How could it relate to my needs? Would using it affect the dominant practices of my professional community?'
- *The proposal*: The emphasis here is along the *ethos–pathos* axis, the primary relation being distillable to 'Here's what I/we can do for you.' *Logos* comes into play because without content, a proposal goes nowhere. *Ethos* traditionally results from the way proposers bring ideas to bear upon prospective clients'

needs. And though the *ethos* of a proposal is crucial to its success, many proposals seem to rely primarily on abstract discussions of methodology and secondarily on 'quals', canned blurbs, describing key personnel and previous projects. (Granted, it is hard to develop a consistent *ethos* when a proposal is being written by 16 people all following faithfully the demands of a poorly written RFP.)

- *The report*: The main relation is 'Here's what I/we found out about it.' Reports' credibility is largely gained through the handling of *logos*. Good ones, of course, also acknowledge the needs of readers through focus and f organization. But in foregrounding 'it', technical reports often sidestep the ethical dimension.

Ethos as Structural Element of Argument

The ethical dimension in argument could be described as a dynamic relation among a set of external conditions (situations, facts, laws, common places) and a narrative voice and reader as roles constructed in the text by the writer. That's a fairly abstract description, but the relations are shown very well by Scott Sanders (1988) in an article that illustrates how a model of argument derived from the psychologist Carl Rogers employs the constructs of *logos*, *ethos*, and *pathos* to 'place' writer and reader in relation to a set of facts. For his demonstration, Sanders uses a poem of William Carlos Williams and an oil-company memo.

In this Rogerian form of argument (derived from Young, Becker, and Pike, 1970), *logos* comes first as a neutral statement of the facts of the case, then comes *pathos* as an acknowledgment of the reader's perspective. Finally comes *ethos*, which conveys not only the desire of the writer but the relation in which reader and writer may stand to one another after the reading. This method of argument is particularly useful when establishing a relationship is as important as conveying information, as is the case in proposals, job-application letters, and customer-relations letters.

For example, suppose that a customer has ordered a specific hydraulic valve from a wholesaler. The wholesaler has sold all of them, and the manufacturer has substituted a design that is functionally the same but physically different. A faxed letter giving the customer the bad news might go like this:

> Dear Customer,
> Last week you ordered from us an XYZ Co. 127336 solenoid-operated valve. This valve is no longer available. The manufacturer recommends substituting a 14-7339 valve, which has the same throughput capacity but is physically 3" shorter and 1" larger in outside diameter, and requires a 24-V actuating pulse instead of 12 V.
>
> We realize that this substitution may cause you some problems. You may have to redesign parts of your piping and electrical systems. That's a real pain, particularly if you're on a short schedule. If you chose a supplier who could get you a valve with the right specs on short notice, we'd understand.
>
> But we do have the 14-7339 in stock, and we can have it to you tomorrow by 10 AM so your engineers can quickly find out how much modification you'd have to do. If it works for you, we'll bill you as usual. If not, send it back. Either way, we'll work with you. We value your continued business.

The structure of this letter creates an *ethos*, one hopes, of a company that is reliable (it doesn't shy away from unpleasant facts), understanding (it has empathy for the reader's situation), and accommodating (it will do all it can to keep its customer happy). It is also ethical in giving the customer options rather than recommending another valve that would less clearly meet the customer's needs, or, worse yet, sending a substitute valve with no explanation.

Ethos and Ethics

From the generic relations discussed earlier, it seems that *ethos* does have a real ethical component. Whether the constructed *ethos* originates in a person or in a committee, if those responsible for the writing don't consider their own stance vis à vis the audience and the material presented, they may wind up being ethically remiss. In the interpretation of the Challenger disaster by Walzer and Gross (1994), the ethical remission was not in a failure to consider or convey all the facts but in the lack of an appropriate deliberative model for proceeding in the case of indeterminate technical data.

Part of the ethics problem is the very codes that are supposed to describe ethical behaviour in professional settings. Like the manuals Pirsig complained about, codes of ethics objectify behaviour, alienating it from the people who are behaving through their writing. This problem is discussed well by Brown (1994), who notes that the STC Code for Communicators' 'vague and abstract language . . . makes adherence a problem' and that the Code's 'metaphor of "bridge" turns the writer into a conveyance, one implicitly neutral'. Brown worries that writers in institutional settings 'implicitly assume the privatized sphere of an organizational *ethos* rather than that derived from personal ethics or the public good'.

If writers do implicitly assume an organizational *ethos* (which would be like trying to adhere to an objectified ethical code), it may be because they and their organizations still operate under a model in which organization is hierarchical and power flows down from the top. Yet even in such organizations, documents that convey the organizational *ethos* emerge from a process of negotiation—the organization can speak to its customers or the public with one voice only after it has resolved differences among many internal voices. That may account for the character of much organizational prose: it represents a lowest common denominator, and those who are supposed to turn out the prose—writers and editors—are often the people with the least influence in shaping the message.

Those with more clout, the scientists and engineers and managers, often insist on writing in the same way no matter what the audience. Scientists may think it unethical to present their work in anything less than the full complexity it has for them, or to tailor presentations to nonspecialist audiences. Engineers often want to drag the audience through all the details of a technical analysis. Managers may think it undignified to use ordinary language and instead prefer an emptily sonorous, abstract prose.

One might well ask, though, how ethical it is *not* to be aware of, nor to adapt to, the audience's prior understandings—to risk either snowing them or talking over their heads. Here, we might take a hint from Coney and look at organizational prose in terms of the kind of audience it seems to imply. Then we could compare the description of that audience with the characteristics of the audience we're actually trying to reach.

Another dimension of this exercise is to look at organizational prose in terms of its implicit attitude toward its audience. Attitude is the aspect of *ethos* defined partly by purpose or motive and partly by choices of content and style.

Purpose is the guise of *ethos* in composition books in this century, borrowing Alexander Bain's nineteenth-century ideas of purpose as narration, description, exposition, and argument. Purpose is usually stated quite explicitly in technical prose. Motive, however, is another matter.

> As Kenneth Burke suggests, motive may be conveyed by tonality, described as 'a barely detectable inflection . . . which unmistakably implies, 'This is the slant you have too, if you have the proper slant.' If [the teacher] explicitly mustered the arguments for that position, he would risk *freeing* the students of his

> limitations, by enabling them to become critically aware of those limitations'. (1969: 98)

The situation Burke describes is an underhanded *ethos*, conveyed as he says not on the editorial page but in the headlines. It is the conveying of attitude by code words, as in promotion of 'neighborhood schools' in Boston in the 1960s as a way of opposing school desegregation, or in Rush Limbaugh's current demonizing of 'liberals'—or for that matter, in dismissing arguments by labelling them 'positivist'.

Attitude may show through content and style in several ways. Prose may pander to the audience by telling it only what it seems to want to hear. It may ignore audience by paying no attention to its level of understanding or its needs. It may convey an attitude about the audience through the choice of words and information, like online help that, by repeatedly belabouring the obvious, evidently considers its audience dimwitted.

Summary

In summary, it is time we explicitly theorized the *ethos* component of technical writing. We need to recognize that ethos is a construction, not a function of personality, and that writers can ethically construct a variety of *ethoi*, just as readers, according to Mary Coney, customarily read in a variety of roles. Until we pay the kind of explicit attention to *ethos* that we've given pathos and logos, our technical rhetoric will continue to be a spectator's rhetoric.

References

Baumlin, J.S. 1994. 'Introduction', in *Ethos: New Essays in Rhetorical and Critical Theory*, p. 23. Dallas: SMU Press.

Brown, S.C. 1994. 'Rhetoric, Ethical Codes, and the Revival of *Ethos* in Publications Management', in *Publications Management: Essays for Professional Communicators*. Amityville, NY: Baywood.

Burke, K. 1969. *A Rhetoric of Motives*. Berkeley: University of California Press.

Campbell, C.P. 1992. 'Engineering Style: Striving for Efficiency', *IEEE Trans. Professional Communication* 35: 130–7.

Coney, M.B. 1992. 'Technical Readers and their Rhetorical Roles', *IEEE Trans. Professional Communication* 35: 58–63.

Cooper, M. 1994. Handout from a Talk Presented at New Mexico Tech, Feb.

Crowley, S. 1994. *Ancient Rhetorics for Contemporary Students*. New York: Macmillan.

Deming, L.H. 1993. 'The Nature of the Narrator in Technical Writing', in *Proceedings of the 40th Annual Conference, Dallas, TX, June 1993*, pp. 154–7. Arlington, VA: Society for Technical Communication.

Goodkin, D. 1993. *MS-DOS 6.2 Upgrade for Dummies*. San Mateo, CA: IDG.

Houp, K.W., T.E. Pearsall, and E. Tebeaux. 1995. *Reporting Technical Information*, 8th ed. Boston: Allyn and Bacon.

Jones, D.R. 1989. 'Exploring Quality: What Robert Pirsig' s *Zen and the Art of Motorcycle Maintenance* Can Teach Us about Technical Communication', *IEEE Trans. Professional Communication* 32: 154–8.

Kennedy, Jr, O.A. 1991. *Aristotle on Rhetoric: A Theory of Civil Discourse*. New York: Oxford.

Killingsworth, M.J., and M. Gilbertson. 1992. *Signs, Genres, and Communities in Technical Communication*. Amityville, NY: Baywood.

MacDonald, P.B. 1946. *Personality and English in Technical Personnel*. New York: Van Nostrand.

Mathes, J.C., and D.W. Stevenson. 1991. *Designing Technical Reports*, 2nd ed. New York: Macmillan.

Muir, J. 1969. *How to Keep Your Volkswagen Alive: A Manual of Step by Step Procedures for the Compleat Idiot*. Santa Fe, NM: Muir.

Pirsig, R.M. 1975. *Zen and the Art of Motorcycle Maintenance*. New York: Bantam.

Rickard, T.A. 1908. *A Guide to Technical Writing*, 2nd ed. San Francisco: Mining and Scientific.

Sanders, S.P. 1988. 'How Can Technical Writing

be Persuasive?', in *Solving Problems In Technical Writing*. New York: Oxford.

Walzer, A.B., and A. Gross. 1994. 'Positivists, Postmodernists. Aristotelians, and the *Challenger* Disaster', *College English* 56: 420–33.

Whitbum, M.D. 1976. 'Personality in Scientific and Technical Writing', *Journal of Technical Writing and Communication* 6: 299–306.

Young, R.E., A.L. Becker, and K.L. Pike. 1970. *Rhetoric: Discovery and Change*. New York: HBJ.

QUESTIONS FOR CRITICAL THOUGHT

1. 'Ethos' is a rhetorical term. What does it mean? What does it have to do with technical writing?
2. Campbell characterizes technical prose as 'lacking character'. Buehler makes a similar charge in 'Situational Editing', as does Glaser in 'Voices to Shun'. Is 'depersonalized' style desirable? Defensible?
3. Campbell quotes extensively from Robert Pirsig's *Zen and the Art of Motorcycle Maintenance*. Why?
4. Campbell introduces rhetoric as an antidote to what he seems to consider a mistaken notion of objectivity. Why do so many theorists of technical writing challenge the objective narrative stance? Why do so many see rhetoric as the solution to this problem?
5. According to Campbell, Aristotelian rhetoric is about public decision-making. Why then is it of benefit to technical writers?
6. What is the difference between what the audience wants and what the audience needs? What does Campbell mean by the 'tyranny of audience'? Why is it detrimental to the practice of professional writing?
7. What is ethos as Campbell defines it? Is his treatment of the concept consistent with that of other contributors to this volume?
8. Campbell speaks of the 'fungibility' of ethos and pathos. What does this mean? What is the relationship between the two concepts?
9. For whom was Campbell's article written? How do you know?
10. Campbell identifies three genres of technical writing: the report, the user manual, and the proposal. Each, he argues, has a different relationship to its audience. Explain.
11. Campbell suggests that technical writing has employed a 'spectator's rhetoric'. What does this mean? What reasons does Campbell provide for this state of affairs?

CHAPTER 41

Between Efficiency and Politics: Rhetoric and Ethics in Technical Writing

Cezar M. Ornatowski

Traditional textbook rationales for the technical writing course locate the essence of technical writing in objectivity, clarity, and neutrality, and the need for teaching it in its usefulness to employers. Such rationales, however, are unable to accommodate a notion of ethics and responsibility: if the writer merely serves the interests that employ her by reporting facts in an objective way, how can she exercise choice when ethical problems arise? An alternative view is to see technical writing as always rhetorical and involved with potentially conflicting agendas and interests, with objectivity, clarity, and neutrality serving merely as stylistic devices in the writer's rhetorical toolbox. Technical writers are rhetoricians who continually make ethical choices in serving diverse interests and negotiating between conflicting demands. The recognition of the fundamental rhetoricity of technical writing is the first step towards accommodating a meaningful notion of ethics into the technical writing curriculum.

Every semester, I begin the first hour of my basic technical writing class with a brief introduction to 'What Is Technical Writing and Why Do We Need It?' I say a few things about what I think technical writing is all about and tell anecdotes from my consulting practice to illustrate some major principles that we are going to cover during the semester. Then I ask students to read Chapter 1 from their textbook, a chapter that is usually called something like 'Writing in the Workplace' or 'Why You Need This Course'.

Here is what my students have found out by reading Chapter 1 in the textbooks I have used in recent years. In the second edition of his *Technical Writing*, Paul Anderson tells the student, 'From the perspective of your professional career, one of the most valuable subjects you will study in college is Writing' (1989: 4). The major argument for the value of writing is implicit in the subheading that follows, 'Writing Will Be Critical to Your Success', and the chief premise is usefulness of good communicators to employers. 'Besides enabling you to perform your job,' Anderson tells the student, 'writing well can bring you many personal benefits, including recognition in the form of praise, raises, and promotions' (1989: 5). In a similar vein, Steven Pauley and Daniel Riordan, in their *Technical Report Writing Today*, tell the student, 'In industry and business today, technical writing is an important part of everyone's career' (1990: 4). After citing the usual statistics to show that people at work write a great deal, Pauley and Riordan conclude that 'writing is extremely important for moving ahead in any profession' (1990: 4). And, they add, when 'done well, technical writing is an exciting, fulfilling experience', but when it is 'done poorly, it is frustrating, even harmful to career advancement' (1990: 3). Finally, in his *Technical Writing*, John Lannon devotes one page to a section entitled 'Writing Skills in Your Career'. After citing two corporate executives on the necessity of effective communication on the job, Lannon concludes that 'good writing gives you and your ideas *visibility* and *authority* within your organization,' while 'bad writing . . . is not only useless to readers and politically damaging to the writer; it is also expensive' (1988: 10–11). Lannon cites the figure of $75 billion spent annually in American business and industry on communication, with roughly 60 per cent of the writing produced being 'inefficient: unclear, misleading, irrelevant, or otherwise wasteful of time and money' (1988: 11).

Behind the brevity of such rationales seems to lie the assumption that no serious effort—on either the author's or the reader's part—is needed to justify the enterprise, perhaps because the need for 'good writing skills' appears self-evident and common-sensical in the terms in which it is presented. These terms are usually, as we saw, those of efficiency, effectiveness, and usefulness. The underlying catechism runs something like this: to work in a social collective (a business organization, a government institution, a cooperative, etc.) you need to communicate. To work effectively, you need to communicate effectively. The principles of effective communication can be derived from the general characteristics (structural, professional, cultural, personal) and needs of the appropriate social collectives and of the people who work in them. More effective communication means a more effective collective. More effective collectives make for a better society. More effective collectives also make more money. An incidental, but not unimportant, benefit is that if you communicate well and make the collective more effective, you advance (and make more money). Ergo: effective communication is useful all around.

Substantively, the major contentions that underlie this catechism are of course true, although the two middle terms, the contentions that the principles of effective communication may be derived from organizations and that more effective organizations make for a better society, have been convincingly criticized by, respectively, David Dobrin and Carolyn Miller. For the rest, however, one must admit that people at work do indeed write a lot, a lot more than most students realize. Communication, and especially written communication, is indeed one of the major modes of operating in any organization or institution, just as the production, circulation, and consumption of information is one of the major activities of modem organizations. Finally, communication does cost money (I was recently told that the average 'cost' of a technical report in an aerospace firm to which I was a consultant was around $5,000), and poor communication may seriously impede the functioning and effectiveness of an organization.

Recently, however, I have been increasingly uncomfortable with this whole rationale as a way of introducing a course in technical writing. It does not seem to account for what we know about the complex nature of the relationship between language and its social contexts or to provide any meaningful and practical way to talk about the ethical dilemmas faced by writers in organizational contexts. Effectiveness and efficiency, understood in terms of usefulness to employers, as the basic premises for communicative action appear to leave the communicator no provision, at least in theory, for action that does not 'efficiently' further the goals of the institution or interests she serves. The clearest illustration of the problem comes from Anderson. In the introduction to his text, Anderson tells the student, first, that 'writing well enables you to make a personal impact' (1989: 6), while a little later he tells the student that 'as an employee, you will communicate for instrumental purposes. That is, most of your communications will be designed to help your employer achieve practical, business objectives, such as improving a product, increasing efficiency, and the like' (1989: 7). Not much scope is left in the end of the 'personal impact' Anderson promises students.

Several well-publicized cases of communicative 'irresponsibility', notably the Challenger disaster and the Three Mile Island nuclear accident, have been widely discussed in the literature (see, for example, Farrell and Goodnight; Winsor).

Yet little seems to have changed in the wake of these discussions in terms of how we conceive of the nature of technical writing or how we see its teaching. Could it be that 'irresponsibility' is built into our present thinking about the nature of technical writing and into the very foundation of how we go about teaching it?

I shall argue that it may be so and try to articulate the problem by redefining technical writing in rhetorical terms.

The Problem

In their study of the public rhetoric surrounding the nuclear accident at Three Mile Island, Farrell and Goodnight conclude that 'the inadequacies of accidental rhetoric at Three Mile Island point to a failure larger than the technical breakdown of 1979: the failure of technical reason itself to offer communication practices capable of mastering the problems of our age' (1981: 271). The 'rhetorical crisis' that followed the technical crisis, and that contributed to a sense of helplessness, paralysis, and loss of public confidence in the ability of appropriate agencies to act in the public interest, was caused, according to Farrell and Goodnight, by 'systematic failures, emanating from deep-seated contradictions in contemporary theories and practices of communication' (1981: 273).

The ethical dilemma that I have articulated at the beginning of this discussion is, I think, but a surface manifestation of one such 'deep seated contradiction'. The contradiction goes beyond the difficulty of reconciling usefulness to employers with a sense of personal and social responsibility. It is, I think, ultimately the contradiction between two incompatible claims we make at once about the nature of technical writing and two incompatible conceptions of language that these claims imply.

One claim has to do with what technical writing does: it is effective and useful to employers because it accomplishes practical goals. I will use Pauley and Riordan's definition of technical writing as an example. (I realize, of course, that textbooks do not represent the profession and that to cite them as representative of what 'we' hold is presumptuous, if not wrong. However, textbooks may be taken, I think, to represent the standard 'paradigm' of current classroom practice.) Technical writing, Pauley and Riordan say,

> is the practical writing people do as part of their jobs . . . . [P]eople generate documents as an expected part of their responsibilities. These documents enable businesses, corporations; and public agencies, including governmental units, schools, and hospitals, to achieve their goals and maintain their operations. (1990: 7)

The second claim has to do with what technical writing is: it is objective, plain, neutral, clear, and so on. Pauley and Riordan, for example, characterize technical writing as

> written in plain, objective language. Since its purpose is to inform or persuade a reader about a specific matter, technical writing focuses the reader's attention on the relevant facts . . . . As much as possible, the words should not cause the readers to add their own personal interpretations to the subject. (1990: 8)

The essence of the contradiction is that two incompatible goals are held out for the technical writer: to serve the interests that employ her effectively and efficiently while being objective, plain, factual, and so on. What she finds in practice is that serving specific interests (any interests, even the most public-spirited) requires at least a degree of rhetorical savvy and that doing so is incompatible with 'objectivity', 'plainness', or 'clarity'. The latter, in fact, are not a writer's goals or purposes but stylistic devices that the technical writer employs in serving whatever interests it is she serves. Objectivity may also, of course, describe the writer's attitude (however difficult such an attitude may be to maintain in practice), but the problem is precisely the confusion between attitudes and rhetorical devices. As long as we do not distinguish, and train writers to distinguish, between rhetorical effects and human attitudes, as long as we promulgate the view that there is a language that does not involve 'personal interpretation', we have no way of talking about responsibility or ethics that does not appear self-contradictory.

I confront this contradiction every day in my technical communication class. Like probably most of us, I have students with different professional backgrounds. Some are humanities majors, some are engineering majors, and a few are working engineers or other professionals. I cannot help noticing that the working students make adjustments to the stylistic principles we discuss, adjustments that are often as enlightening as they are exasperating: 'I can't tell my boss about problems so directly'; 'We can't just tell the customer clearly the product failed the test'; 'How do I use active voice without blaming my boss?' The honesty and, let's face it, soundness of such comments contrasts with what I can't help feeling is the naïveté of the nonworking students, who take at face value—often with a vengeance—the stylistic injunctions to be clear and direct, to state the problem and to make messages explicit. The result may be well-formed reports, but I am forced to admit that they would never 'fly' in the complex game of trade-off's and judgment calls that defines the communicative dynamics of real organizations. The student who writes 'The engine failed the test and shipping it may endanger the lives of innocent passengers' in a technical report may be ethically well-meaning, but the report would never pass the review cycle. The problem, of course, is how do I explain this to the student without making her feel that this ethics stuff is just school baloney?

Rhetoric at Work: An Example

Let me offer an example of the kind of daily ethical and rhetorical decisions writers in organizational contexts typically (in my experience) face. Consider one engineer's reflections on the ethical dilemmas involved in writing a single sentence of an aircraft engine test report. The engineer (whom we shall call Stephen) works as a test engineer in a major US aerospace firm. By education and training, Stephen is a mechanical engineer. He has worked in the firm for over eight years. He is in his mid-30s, intelligent, quick, and articulate. He has no formal training in technical writing. What he knows about writing technical reports he learned on the job, mostly from reading other reports, from company manuals, as well as from having his reports 'redlined' and discussed by his supervisors. This practical training has made him well acquainted with the rhetorical and political subtleties of functioning effectively in the organization. I observed Stephen (and many other engineers in the firm) for nine months as part of a study of the rhetoric of technical information in the firm. I also conducted numerous interviews with Stephen and with other engineers.

The excerpts below are taken from one of the taped interviews with Stephen, conducted while he was engaged, over a period of several weeks, in testing an electronic controller (which I will call ELCON) for an auxiliary aircraft engine. The tests in question were environmental tests, in which ELCON was subjected to a variety of extreme environmental conditions: cold, heat, water, humidity, sand, and others. At the conclusion of the tests, Stephen began working on a formal engineering test report that would go through an appropriate approval cycle within the company and then be released to the customer (a major airplane manufacturer) and to the Federal Aviation Administration. It is this report, and specifically one sentence in it, that is the subject of the remarks quoted below.

Stephen is trying to express the fruitlessness of repeated attempts to start the engine through the ELCON during the low temperature test. He is trying to express the 'hopelessness' of trying to start the engine. He does not want to 'lie' (his word) and make it seem that ELCON worked much better than it did, yet he is aware that he cannot simply say 'hopeless', because that puts a very important and expensive product in an unfavourable light, and he is sure his supervisors, and especially the program office, would object to his choice of words. The sentence he ended up writing in the official test report reads: 'There was a problem encountered during the low temperature test, in which the power-up of the ELCON after cold soak was *fruitless initially*' (emphasis added). Here is how Stephen explained his choice of the word 'fruitless'. (The quotation is long, but I think worth citing in its entirety so as not to lose the sense of the full dimensions of the rhetorical choices Stephen is making here.)

> Here is a political sentence. 'There was a problem encountered during the low temperature test, in which the power-up of the ELCON . . . ' After it sits in a box at minus 67 degrees for six hours, then you have to turn it on. Well, when it came time to turn it on, the ELCON, the power supply inside, didn't respond. Kind of like when you turn your car on and the battery's dead and you just hear the click. So, how do you write that up? So, . . . I don't think I actually put the agonizing I did over this sentence in here. But I agonized over this sentence because I didn't want to say, let's see, it still says 'fruitless initially', and that's essentially what . . . that's what I ended up saying. But you don't want to say 'hopeless', whereas there was no way this thing was going to start, and that's really what it was. We tried for about a half hour and that thing wasn't going to start, the power wouldn't get to it. It had to do with all the circuitry being soaked to that temperature. So we tried for a half hour, which is eons in terms of electricity. You know, you flip a switch the light goes on, that's how electricity is, so half an hour is . . . So how do I write this up? The ELCON failed in power-up. That does not describe it, because we tried again and again and again. Do I really want to say that we tried, you know, forty-five times in thirty minutes to power this thing up, you know? So I came up with 'fruitless' only because it tells the story without saying that I did it again and again and again. To me it was hopeless, but 'fruitless' says it like, 'Yeah, we didn't get the power up,' and it doesn't say 'We tried ninety billion times,' and it doesn't say 'We tried once,' it just says 'We didn't get the power up.' And that is as vanilla as I can make that, that's as benign a statement as I can make without tipping my hand, so to speak . . . . It's not so difficult to try and document what happened but there is . . . there are points at which you don't want to hurt yourself, you don't want to . . . you don't want to harp on anything in particular. This, the only way I could document what happened without making it sound like the horror show it was, was to just say 'Initially we had trouble,' Which is true. It's a bit of an understatement, but it is true.

Stephen's dilemma is how to tell the technical 'truth' while at the same time telling it within the boundaries of outcomes and implications of events acceptable to his supervisors and to the report's customers. Note, however, that even the 'technical fact' is largely a matter of personal perception and interpretation: is the attempt hopeless or just fruitless? It may depend on how impatient one feels that day or how long it is until lunch time. Elsewhere, Stephen described the rhetorical strategy he uses in writing his reports as 'selective emphasis'. In the statements below (taken from the same interview), he explains what he means by 'selective emphasis'.

> You can change the message that you give with a report by not highlighting certain data. It's probably most apparent in something like a performance report. If I'm documenting engine performance there is [sic] a lot of

> things that I could look at, to say that it's a good performer or a bad performer. You ask can I change the slant of the report by changing the data. I would say no, the data is what the data is: sacrosanct. What I can do is highlight data, withhold data, present it in a different fashion, like, I could tell you this engine was a great performer. It's delivered its shaft power; it's delivered bleed. And you might nor know that this engine is high on specific fuel consumption . . . . So, yes, there are ways you can change how a report is perceived; or how a test event is perceived. Actually, if you present everything, it becomes a matter how you highlight it with the prose. And that's the final thing I would say to you: *You can't change the data, but you can certainly change the positive/negative of a given issue*. I even think *it is incumbent* on *everybody to do just that*, particularly when you are writing for the customer. If you ask anybody in this [test] department, I'm sure you'll get some hard-ass answers about, you know, the data is what the data is. I don't think anybody here [in the test department] has ever cut a corner in that sense. On the other hand, I think between us test [department] and the project [project management, responsible for liaison with customers], the stuff that gets out of here is reasonably, I wouldn't say 'censored' but its . . . it's, maybe I will say 'censored', yeah, it's been reviewed to the point that as little damaging statement as possible is included.

Stephen is considered a 'good writer' among the people for whom he writes: the program managers and the customers. His success appears to be due, in large part, to his rhetorical acumen, to his ability to make the delicate trade-offs and decisions alluded to in the passages quoted above. By contrast, I have talked to, and seen reports written by, engineers who do not enjoy the reputation of being 'good' (read 'effective') writers and whose reports are always heavily reviewed and often need to be revised by the supervisors themselves before being officially released. The major problem that these writers have (as I gathered from studying their reports and from discussing their reports with their supervisors and co-workers) is not that they do not know what an engineering report should look like (they have learned that on the job and the format is prescribed in departmental Policies and Procedures) or that they cannot write 'good' sentences (their sentences are no better or worse than Stephen's). Rather, they are heavy-handed and naïve in reporting facts, not politically astute or willing to be astute, or not linguistically accomplished enough to be able to be astute, to negotiate the subtle lines between political prudence and necessary disclosure.

From my research in that firm I concluded that a good organizational writer is one who can negotiate successfully the subtle boundary between, on the one hand, the stylistic and formal demands of clarity, objectivity, neutrality, format, and effective use of visual devices, and, on the other hand, the institutional, social, and situational (read: political) demands placed on the text. In my experience, all organizational communication, including technical documentation, bridges the gap between the physicality of technical phenomena on the one hand and their organizational and social meanings on the other hand, between data as technical and as social facts. It is this inevitable gap—the gap that arises, for instance, when unruly events and the idiosyncrasies of gas turbine thermodynamics are at odds with the political and economic exigencies and stakes—that defines the rhetorical 'space' that most technical documents written in real environments must bridge. As Edwin Layton has observed, 'organizations like the federal government or a modern corporation have other ends in view than the best and most efficient engineering' (3). It is the difference between these 'other ends' and the theoretical ideal of the 'best and most efficient engineering' that define the major arena of rhetorical choice and ethical conflict for writers of technical documents in organizations. The ethos of engineering report, I noted in my research, is always constructed out of uneasy compromise and accommodations, even though many engineers are troubled by the fact that their reports are 'censored'—as Stephen hesitantly put it.

The Dilemma Revisited

My problem as a teacher (and theorist) comes down to this: I know that technical writing in real-world contexts is much more political and rhetorical than the textbooks make it out to be and that it is in this political/rhetorical sphere that ethical problems arise. The problem is that the standard textbooks and the standard curriculum offer no way to deal with this knowledge without seeming to throw away the whole shop. If writing documents is only a matter of clearly marshalling objectives facts and designing readable texts, 'ethical' problems should not arise. Something would be a fact or not be a fact, be clearly relevant or be clearly irrelevant. My students, the ones who do not work, feel very smug about being honest and get impatient when I talk about ethics. How could one be so stupid or such a crook, they ask when we discuss the Challenger tragedy. They are right. If technical writing is just a matter of designing clear documents, the only way ethical problems could arise is through ignorance of facts or sheer stupidity (the former being, interestingly, the favourite defense of those who are accused of 'irresponsible' practices). In fact, the way ethics are alluded to in most textbooks, almost incidentally and parenthetically and usually in the form of a simple injunction to be 'honest' or 'ethical', shows the inability of the standard rationale to develop a more sophisticated notion of ethics than simple avoidance of outright dishonesty. The mechanisms for 'audience analysis', for determining the reader's needs, for analyzing the situation, offer no tools for discerning or discussing the more subtle, and essentially political, choices, pressures, and agendas that one encounters in real environments.

As a teacher, one could agree with Sherry Southard that 'being professional means knowing the proper protocol for the corporate world' (1990: 90) and that it is such protocol we should be teaching. While such a position is no doubt realistic, it amounts to throwing in the towel on ethics, because it contains few provisions for action outside bureaucratic control and outside simply serving corporate interests that pay the bill. One could go to the other extreme and advocate rebellion, a rhetoric of 'dissention', as proposed by John Trimbur, who suggested that we should replace 'the "real world" authority of consensus with a rhetoric of dissensus' (1989: 615). The problem with this attitude as a premise for teaching and practicing technical writing is that assuming a perennially contestatory stance puts the communicator in danger of becoming, at best, substantially irrelevant (like those proverbial 'nyet' men in the Soviet Politburo) or, at worst, unemployed. It is also questionable whether programmatic contestation really leads to a better understanding of the nature of language, of the relationship between language and its social contexts, and of the nature and role of technical communication. I have a problem seeing it, for instance, as a coherent rationale for a technical writing curriculum.

Carolyn Miller has suggested that technical writing ought to be reconceived as 'a matter of conduct rather than of production, as a matter of arguing in a prudent way toward the good of the community rather than of constructing texts' (23). Although emotionally I am in agreement with this sentiment, I do not see how it solves the practical and pedagogical problem. Employees, after all, are paid to render services to employers and to further their goals 'effectively' and 'efficiently'. Institutional readers 'need' a document that does a certain job, and they won't accept one that does not do it to their satisfaction, no matter how lofty the social sentiments of the writer.

Efficiency is Politics, Technical Writing is Rhetoric

To begin to find a way towards dealing with the problem of ethics realistically and effectively, we must, I think, accept two premises. The first premise is that *organizational efficiency is always involved with politics.* Aaron Wildavsky has offered a definition of politics that, I think, is very appropriate in this context. Politics, Wildavsky suggests, is an essentially social process by which an institution 'mobilizes resources to meet pressing needs' (1964: 94), The second premise is that *technical writers are always inevitably rhetoricians* precisely because they are 'useful' to employers and their writing is 'effective'. In fact, technical writing is a form of rhetoric developed in, and uniquely suited to, the social contexts in which it typically takes place. Let me explain.

The most common contexts for technical writing are business/industrial organizations and government institutions. The dominant purpose of such organizations (as well as the major criterion by which the people who work in them are judged as effective) is the maximization of advantage and increased control of resources in the major operative spheres: economic, political, intellectual. It is the maximization of the symbolic or real 'capital'.

Communication serves this goal as much as do other institutional activities. Technical documents (engineering reports, model specifications, proposals, annual reports, financial reports, business plans, product brochures, manuals, and other documents) play an important role in the dramaturgy of organizational life through which the organization and, within the organization, the various departments, interests, or individuals, garner resources and advance their causes (for an example, see Barbara Mirel's discussion of the political functions of a user manual). In this dramaturgy, reason (or the appearance of reason) is the fundamental condition of effectiveness, and it is reason, or its appearance, that well-designed technical documentation provides. Richard Ritti has argued, following Max Weber, that the dominant 'world-view' (by which Ritti understands 'the framework of beliefs, attitudes, and values that provides the basis for classifying, interpreting, and understanding the events of everyday life') of most contemporary organizations, whether private, public, or governmental, is bureaucracy (1986: 104). Bureaucracy is characterized by legality, impersonality, and rationality, subsumable under the major functions of planning, organizing, leading, and controlling. Within the compass of bureaucratic rationality, all problems must be (or must appear to be) amenable to technical, rational (note the common collusion of the two terms) solutions comprised of a step-by-step cycle of setting goals, developing a plan, accumulating relevant facts, deciding on the best course of action, and so on. Because of its implicit and traditional association with science and technology and its 'rational' pretensions, embodied in the stylistic attributes of clarity, directness, factuality, objectivity, and neutrality, technical communication is well suited to serve as the rhetorical instrument of organizational–bureaucratic rationality. It is also uniquely, and conveniently, congruent with the cultural self-perceptions of the people who dominate business and industrial organizations. Ritti describes American corporate culture as a culture of '"decision-makers"; of crisp, hard-hitting, two-fisted decision-makers; of bottom-liners; of players of hardball. The metaphors are redolent of *efficiency and effectiveness.* Facts exist, can be weighed, and objective outcomes maximized' (1986: 104, emphasis added). Such people want their reports to be factual, objective, and effective—as long as the facts 'objectively' and efficiently support their agendas and objectives.

It is largely in this sense that technical writing is 'useful' to employers and that is implicitly and ultimately why we teach it.

The traditional catechism has got it wrong—it is not that technical writing is important because communication costs money and may have an impact on one's career, but the other way round—organizations and institutions are willing to spend big money on communication, and communication may have an impact on one's career because it is useful—and that's why it is in demand and that is why we teach it. Here comes the rest of my catechism. It is useful because it is a vital element in furthering the goals (business, political, economic, and other goals) of organizations and institutions. It furthers those goals because it is fundamentally rhetorical, just like any other use of language, and its rhetoric is uniquely suited to organizational and institutional rationality. The essence of its usefulness lies in the fact that it is capable of giving a rational, technical, 'scientific' mantle to agendas that are always in some measure political and that it gives the illusion of objectivity to what are always and inevitably interpretations. The fact that we do not commonly think that way about it is its greatest rhetorical asset. Realizing this is the first step towards having the capacity to analyze the trade-offs and bargains that one makes. And that is the first step to talking sensibly about ethics.

Of course, in saying this, I have not solved my initial problem. But I have articulated the problem in a way that lets me talk abut it as an integral part of what I see technical writing to be, and that articulation is in keeping with some other things I know about language, writing, and social contexts. That way of articulating it also engages technical writing with theories of human action and its institutionalizations. Paul Anderson is right in telling students that writing is one of the most valuable courses they will take in college. But I think the reasons we usually give for this value miss the real point.

References

Anderson, Paul V. 1989. *Technical Writing: A Reader-Centered Approach*, 2nd ed. New York: Harcourt.

Dobrin, David N. 1985. 'What's the Purpose of Teaching Technical Communication', *The Technical Writing Teacher* 7: 146–76.

Farrell, Thomas B., and G. Thomas Goodnight. 1981. 'Accidental Rhetoric: The Root Metaphors of Three Mile Island', *Communication Monographs* 48: 271–300.

Lannon, John M. 1988. *Technical Writing*, 4th ed. Glenview: Scott.

Layton, Edwin, Jr. 1971. *The Revolt of the Engineers: Social Responsibility and the American Engineering Profession*. Baltimore: Johns Hopkins University Press.

Miller, Carolyn R. 1989. 'What's Practical About Technical Writing?', in Bertie E. Fearing and W. Keats Sparrow, eds, *Technical Writing: Theory and Practice*, pp. 14–26. New York: MLA.

Mirel, Barbara. 1988. 'The Politics of Usability: The Organizational Functions of an In House Manual', in Stephen Doheny-Farina, ed., *Effective Documentation: What We Have Learned from Research*, pp. 277–98. Cambridge: MIT Press

Pauley, Steven E., and Daniel G. Riordan. 1990. *Technical Report Writing Today*, 4th ed. Boston: Houghton.

Ritti, R. Richard. 1986. 'The Social Basis of Organizational Knowledge', *Organization Communication: Emerging Perspectives. Vol. 1*. Lee Thayer, ed. Norwood: Ablex.

Southard, Sherry G. 1990. 'Interacting Successfully in Corporate Culture', *Journal of Business and Technical Communication* 4: 79–90.

Trimbur, John. 1989. 'Consensus and Difference in Collaborative Writing', *College English* 51: 602–16.

Wildavsky, Aaron. 1964. *The Politics of the Budgetary Process*. Berkeley and Los Angeles: University of California Press.

Winsor, Dorothy. 1990. 'The Construction of Knowledge in Organizations: Asking the Right Questions About the Challenger', *Journal of Business and Technical Communication* 4: 7–20.

QUESTIONS FOR CRITICAL THOUGHT

1. Ornatowski observes that traditional rationales for technical writing courses tend to emphasize 'efficiency, effectiveness, and usefulness', but they are incomplete. Why? What is missing from these justifications?
2. Ornatowski suggests that an assumption of 'irresponsibility' is built into the values that mark technical writing—objectivity, efficiency, and usefulness to an employer—because they leave so little scope for personal and social responsibility. Why is this a flaw in the teaching of technical writing?
3. What is the difference that Ornatowski wishes to draw 'between rhetorical effects and human attitudes'? Why is this distinction important in the teaching and practice of technical writing?
4. Ornatowski warns of the 'complex game of trade-offs and judgement calls that defines the communicative dynamics of real organizations'. In what sense is he simply describing the dynamics of a rhetorical situation?
5. Why does Ornatowski provide the details of the rhetorical choices made by Stephen? What effect do these details have on his argument?
6. How does Ornatowski define a 'good organizational writer'?
7. What is the 'rhetorical space' in which most professionals must write? Is this the same as rhetorical situation? Why or why not?
8. Ornatowski observes that 'technical writing in real-world contexts is much more political and rhetorical than the textbooks make it out to be' and that it is this 'political/rhetorical' pressure that gives rise to ethical challenges in workplace writing. What challenges does such a fact present for the *teacher* of technical writing?
9. Why does Ornatowski conclude that organizational efficiency and political interests are always inextricably linked? What does he mean by 'political' in drawing this conclusion?
10. Technical communication, by Ornatowski's reckoning, is always rhetorical. In what sense is he using the word 'rhetorical'? Does his definition correspond to the other definitions of rhetoric offered in this book? Why or why not?
11. Ornatowski concludes by reversing the usual justification for teaching communication. Is his re-statement of the standard argument significantly different, or is it essentially the same argument with additional detail? What is the 'real point' that most textbooks miss?
12. How personal is Ornatowski's style? How readable is his article? Does it have a distinctive 'voice'? Is it appropriate that it have one? Why or why not?

CHAPTER 42

Developing Ethical Decision-Making Skills: How Textbooks Fail Students

Jim Gough and Anne Price

Introduction

Business communications textbooks are often sufficiently wrong, vague, misleading, or confusing as to make the problem of effective ethical decision-making more difficult.

The Enron scandal and other recent newsworthy events relating to business practices have brought issues of business ethics into sharp focus. While business students may be taught business ethics in a variety of courses in their programs, the discreditable actions of some business leaders suggest that more could and should be done to develop in business students an understanding of the theories of ethics. This would enable them to make good ethical decisions. Perhaps the vague and confusing texts to which we refer have contributed to the inability of some business people to recognize (or at least acknowledge) unethical behaviours and/or participate in good ethical judgments.

This paper will critically review a number of texts that are taken to be suitable for second year business students and evaluate those parts that address ethics to determine their acceptability at this latter task. While being a random selection of texts, these are nevertheless typical of the current texts available in North America. Most of these texts provide a summary section on ethics in the first chapter and a section in each chapter providing more specific suggestions for ethical behaviour. The findings are organized here into four categories. They are:

- texts that provide problem scenarios for evaluation, but no method for decision-making.
- texts that conflate issues of ethics with other issues such as propriety and legality and provide no philosophical approach.
- texts that suggest that there are (or specifically identify) several philosophies of ethics and allow that any one of these may be acceptable. The implication in these texts is that it is possible to switch between philosophies according to the circumstances.
- texts that identify several theories of ethics and indicate that one must be consistent in the application of one of these theories.

The Conditions for Ethics

Before testing the purported ethical decisions or advice made by oneself or others, it is necessary to identify, in schematic terms, the set of conditions and circumstances that go into making ethical decisions or giving good ethical advice. If any decision or advice fails to take account of these conditions and circumstances, then we can say either:

1. it is not an ethical decision at all but rather a different kind of decision, and/or;
2. it is a weak or poor ethical decision or piece of ethical advice because it fails to employ to any significant degree a sufficient number of these conditions.

So, first, some conditions for making an ethical decision or giving ethical advice, as opposed to other kinds of decisions or other kinds of advice.

- Anyone making an ethical decision or giving ethical advice must be aware of the persistent and on-going need to be knowledgeable, self-conscious or self-aware, self-reflective, and self-critical of all such ethical decisions or ethical advice. This may not be a requirement in other areas of choice, such as in commerce or the arts. It is possible to make an acceptable aesthetic or artistic choice that is based on the authority of the muses or intuition or gut-feeling but not possible to use this unreflective process in making an ethical decision.
- Anyone giving ethical advice must be aware of the difference between ethical considerations and considerations of propriety or expediency, the latter being that which is done for the sake of appearances, reputation, social acceptance or conformity but not necessarily done because it is considered to be the right or ethically correct thing to do.
- Anyone giving ethical advice must be aware of the difference between a descriptive claim and a prescriptive claim, with ethical advice focused on the latter and not the former. To describe a situation or state of affairs or practice is not to give prescriptive advice that it ought to be done or should be done. It is necessary to give reasons for supposing that what is the case, should be the case. Alternatively, it is necessary to give reasons why what is the case, and unacceptable, should not be the case but changed somehow. This means that traditional, conventional, customary, or majority practices are not uncritically or automatically accorded any authority for being ethically acceptable based on a description of what is a practice or what has been a practice in the past. Ethical decisions and advice must be deliberative and not simply reactive, nor accidental, nor unintentional.
- There is a significant sense in which someone who makes an ethical decision or gives ethical advice makes the decision and the advice his or her own. Thus, he or she becomes, in a strong sense, responsible for the decision and the advice advocated. That is, the ethical actor or advisor is accountable and not detached, distanced or aloof from the results or consequences of his or her choices and/or advice. This means that if the advice given, for example, exhibited unethical bias or prejudice, then the person making the decision or giving the advice could be held accountable for this ethical fault.
- Anyone making an ethical decision or giving ethical advice should be aware of the necessity to employ some acceptable standard or other in the judgement to arrive at the decision. This standard must be decided independently from the subjective interests of the individual making the decision and independent of the situation in which the decision is made. It also needs to be consistent over similar cases so that a principle of: (a) reciprocity: assuming others will act in the same way you act towards them, since your choices imply that you approve of these choices for others, or (b) equity: treating like cases alike, when there are no relevant ethical differences to distinguish them, is employed.
- Anyone making an ethical decision or giving ethical advice should be aware of the difference between what is ethical and what is legal, such that any standard test for the former may not be identical to a test of the latter. What is legal, for example certain kinds of discrimination (against blacks in several countries during the era of slavery) may not be ethical. Conversely, what is ethical, for example, the freedom to choose a certain kind of lifestyle (homosexual life-

styles were not legal in England for many years) may not be legal.

- Anyone making an ethical decision or giving ethical advice should be aware that, while:
  - information used in making an ethical judgement may come from a particular profession or practice,
  - the need for an ethical judgement might occur within some profession or practice, like medicine, nursing, business, and
  - the consequences of an ethical judgement may be felt within a distinct profession,
  - the standard for the ethical judgement and the ethical decision is independent of the ethos or practices of the particular profession.

It is important not to confuse the interests of a profession with ethical interests. Sometimes, for example, professional codes of ethics are not clearly understood to provide standards of professional practice, violations of which might reflect badly on the profession, and ethical standards, violations of which might simply be wrong or ethically unacceptable.

Circumstances In Which Ethical Considerations Occur

There are certain specifiable circumstances in which it is reasonable to claim that certain ethical considerations could occur, under the above-specified conditions for making ethical decisions and giving ethical advice. Some of these are the following:

- Ethical considerations occur only in a community or social setting where there is possible conflict between competing interests for scarce resources or the satisfaction of more than one individual's interests or set of interests. It is, then, unlikely that ethical considerations would occur in a situation where only one's own interests at stake. For example, it is relatively easy to construct plausible desert island scenarios where no ethical choices are made because there is no possible inter-subjective conflict of interests since only one person's interests are at stake. However rare such cases are, they serve to highlight the essential part played by a community in the determination of ethical considerations.
- Ethical considerations require a human interest as part of the identification or attempted resolution of any conflicts. For example, while it may be possible for one animal to harm another, it is not possible for one animal to be held ethically responsible for harming another. However, a human being in these circumstances could be held accountable. If, for example, a corporation were not a human being, then it might not be responsible in the same way human beings are held accountable, while those humans who controlled the organization might he considered responsible for ethical decisions and their outcomes.
- Ethical deliberations can occur in any situation where there is an actual or a perceived conflict of values. Ethical deliberations rarely occur where there is a difference in factual accounts unless this difference in interpretations about the facts has some implications for values. So, conflicts of values can occur between the interests and the values of business or commerce practitioners or participants and religious, political or ethical practitioners even where there are no factual disputes at issue. All the facts may be known about the effects of acid rain, but these facts do not resolve conflicts about what values should predominate in changing the situation—economic values, political values, or wellness values.
- To resolve a conflict requires minimally that there be a means of resolving the conflict.

This means is one which itself satisfies certain ethical conditions. The resolution should be intended to persuade or convince someone in the situation and others connected to the situation on the basis of reasons, not emotions or simple self-interest.

- To resolve an ethical conflict could on many, if not most, occasions require the sacrifice of something of value in order for it to be possible to achieve a resolution of the conflict. The false belief that nothing need be sacrificed can lead to the pessimistic view that there are no resolutions possible to ethical problems or issues. So, anyone who suggests that we need not sacrifice, for example, any values accrued in the practice of business in order to resolve an ethical issue is simply not considering the situation accurately or clearly and potentially giving false advice.
- In a non-legal, informal sense, ethical decisions based on ethical judgements establish a possible precedent for others to follow, whether these others employ non-precedent reasoning or not. So, there is a public, example-setting component to ethical decisions, with one decision lending initially plausible support to another decision in a similar situation. This isn't simply a matter of how someone appears to others or a matter of a person's reputation in some public context but rather an obligation to insure that our decisions are considered carefully as they may be used to support other similar choices.
- Ethical decisions should follow from the individual's desire or intention to do the right thing or achieve the ethically best result without external coercion. That is, an ethical decision should be one that each of us makes our own. If someone is forced (public pressure, constraints of the position within an organization or company, fear of repercussions, and so on) to make an ethical decision, then it is the force that carries the authority and responsibility for the decision, not the individual decision maker. External compulsions are also often temporary and only as effective as the force or threat of sanction can be maintained, rendering them generally ineffective in the creation of good ethical character.
- Both the ethical decisions someone makes and how this person makes these decisions tells the individual and others something important about this person's ethical character. It also tells them the kind of life s/he values and promotes in a more or less (most of the times, it not all) consistent portrait or characterization. That is, it may or may not be true that you could tell something of my character as a result of the car I choose or the clothes I wear, but it is always ethically significant that my ethical character is reflected in my ethical choices. Ethical standards must be internalized, whether we identify such internalization with the fabled 'conscience' or not, since internalization is the best way of insuring that intention matches action and actions remain consistent over a protracted period of time.
- Ethical advice must be consistent with both ethical judgements and ethical decisions made by the person giving the advice and that promoted of others in similar situations making decisions under similar circumstances. This avoids the serious logical fault of inconsistency or the negative ethical charge of hypocrisy. If we were to suggest (mistakenly) that all ethical decisions could vary from one individual to another and from one circumstance to another, then consistent ethical advice would be difficult if not impossible. The mistake here is confusing propriety (which may change by circumstance) with sound ethical standards (which are invariant to circumstances).
- Ethical decisions must be based on ethical judgments which require some minimal

rational ability to choose among alternative decisions, which involves knowing the short and long term positive and negative consequences of the ethical choices made. Hence, good ethical decision-making requires a degree of rational maturity that should be shared by those giving sound ethical advice.

Using the above considerations, we can evaluate the contributions to ethical decision-making and the adequacy of ethical advice given in some contemporary business textbooks. These considerations are reflective but not exhaustive of the entire set of such texts. Parenthetically, it is not our intent to moralize about the texts but rather to offer advice that is open to reflective criticism and change, if appropriate. We are not claiming to have the best word nor the last word in this area.

The Findings

TEXTS PROVIDING PROBLEM SCENARIOS WITHOUT METHODS FOR DECISION-MAKING

In trying to simplify this very complex issue, some texts address it so superficially as to diminish its significance. One text (Fox, 1994) provides students with an 'Ethics Test' (18–19) in multiple-choice format. For example, one question asks:

1. If a fellow employee was a victim of racial discrimination, I would:
   a. Create a file documenting the problem.
   b. Tell the employee I would provide support if he or she complained.
   c. Complain to a superior likely to be sympathetic
   d. Advise the person that he or she might be happier elsewhere.

There are nine such questions and 'A Guide to How you Did' at the end of the chapter. It tells the reader the strength of their ethics based on whether they answered predominantly a, b, c, or d to the questions. Fox says, 'If you answered "a" most often, you have a strong sense of ethics, but tend to be rigid. You will run into ethical conflicts in your career unless you find a very like-minded company' (24).

This text is accompanied by a commentary that claims that 'Because people have always disagreed about what constitutes good and bad or right and wrong, ethical dilemmas are seldom "solved"' (18). This neatly avoids providing any philosophical basis for decision-making and, indeed, none is attempted. It is misleading on several accounts. First, most ethical decision-making is not about ethical dilemmas and most get solved. Second, since the inception of the free-market there have been disagreements about what 'forces' do or do not control the market outcomes, yet this has not led to a situation where there are no proposed or workable solutions. So, why must this be the case with ethics? The pseudo advice given in this case may be an attempt at a self-fulfilling prophecy. Fox does, however, provide some suggestions for ethical writing practices in specific circumstances such as persuasion (don't distort or misrepresent), and in writing headings that truthfully represent the contents of a section of a document.

Anderson (1995: 12) dodges the difficulties of this issue by stating 'personal values differ at least some what from person to person. Consequently, this book won't tell you what your values ought to be.' This may provide the reader with a sense of empowerment, or it may leave them floundering in a sea of conflicting perceptions and ideas. There is a difference between resolving conflicts or attempting to do so, and washing your hands of them. The latter is an unacceptable case of diversion, pulling the covers over your head to deny the existence of six-foot snowdrifts outside your window. It is also a feigned and inappropriate application of tolerance. We ought not to be tolerant of the sexist and racist values of others, yet, according to this view; I simply might not be able to condemn your racist and sexist values. Similarly, Ober (2003: 22) says that 'each of us has a personal code of ethics,' and goes on to focus on the legal problems that may arise in business communication. The principal guide given by Anderson is to 'think of yourself not only as an *employee* but also as a *citizen* of the organization for which you work.' This is somewhat baffling since it seems to imply that citizens are somehow more ethical than employees, but the explanation seems to be that we must think of ourselves as individuals but also as members of various groups. It also seems to convey an attitude of expediency, do what works to your advantage in the situation or role you happen to be in at the time, rather than an ethical attitude, which suggests that whether something is right or wrong is not relative to one's situation.

In a more detailed section on ethical decision-making, Lannon (2000) provides a similar approach to Anderson. He begins by pointing out the need to recognize unethical communication, which is an important first step. He also identifies some causes of unethical communication: social pressure, groupthink, and a personal bias that influences judgment (65–7). But when Lannon discusses the critical thinking required for ethical decisions, the reader is simply provided with a list of the varying obligations one has: to oneself, one's clients, one's company, one's coworkers, one's community, and one's society. Beyond saying that 'ethical questions often resist easy answers' (71), no ethical system of decision-making is put forward. Platitudes, instead of process, provide an insubstantial basis for providing good ethical advice.

Even less practical advice is provided in Locker's *Business and Administrative Communication* (2000). However, it is interesting to note that the section on legal and ethical concerns begins with the observation that 'legal fees cost US businesses hundreds of thousands of dollars. . . . Ethical concerns don't carry the same clear dollar cost as legal fees. But over the last 25 years [various high profile ethical cases] have left many consumers with a deep distrust of both business and government.' It is because of this (and the immediately preceding reference to costs) that she suggests business and government must act ethically. When we carefully scrutinize the reasons for acting 'ethically', the response, in this case, seems to be because of expediency (it is in my best interests to do so in this case) or propriety (I will look good in the eyes of others if I do so). Neither of these factors has any demonstrable good effect on the internalization of ethical standards so crucial to creating good ethical characters. So, neither of these should be the basis for good ethical advice.

A figure in Locker's text identifies some issues that arise in business communication that have ethical components, such as the use of emotional appeals and the manipulation of audiences through careful organization of messages. However, no means of identifying or addressing ethical issues and resolving them are put forward. Providing a description of the best means used to argue about something does not help us identify what it is we should be arguing about. This is a serious problem since recognizing that there is an ethical issue may be the most significant step in the process of confronting it and attempting to resolve it.

TEXTS THAT CONFLATE ISSUES OF ETHICS WITH OTHER ISSUES

The relationship between ethics and the 'moral judgment accepted by society' is the path taken by Boone, Kurtz, et al. (1999: 27). One margin note is a quotation from a public relations executive, Chester Burger, who says 'Ask yourself how your action would look if it were reported tomorrow on the front page of the local newspaper' (28). This is a fallacious appeal to the irrelevant authority of a mythical entity, 'society', which is too vague. If this refers to public opinion, then the conflicting reactions of black and white populations in the United States to the outcome of the O.J. Simpson trial is enough to convince us that 'society' is by no means uniform in its judgements. The real basis of this appeal may be personal 'reputation', namely how someone is perceived by others. Such an appeal is superficial at best, since it is not at all about the 'real' you or what kind of ethical character you are, once the persona or public image is removed. There is no ethical or even logical connection between reputation (good or bad) and any corresponding ethical character. Ethics is lived, not worn like a coat for convenience or appearance.

These authors also provide a convenient tool, in the form of a checklist, to enable readers to evaluate ethical decisions as they relate to writing. For example, the writer should ask him or herself 'If I accomplish my purpose, will it be in the best interest of my audience?' and 'Did I reveal relevant information in documents and conversation or hold key details back to protect my interests?' (derived from Brownell and Fitzgerald, 1992).

The question here is, 'What constitutes an ethical interest?' Clearly there are a number of interests involved in any ethical judgement: business interests, personal interests, legal interests, and so on. These are not necessarily ethical interests and should not be confused with them. Each of these interests could have different goals, purposes, aims, and tests to determine whether the goals have been satisfied.

This text also equates ethical business practices with profitability by saying 'According to one university study [not named], corporations that have paid dividends for 100 years or more tend to be companies that place a high priority on ethics' (28). To attempt to justify an ethical judgement or ethical advice on the basis of business interests is to commit a logical category mistake. This is to suggest that ethics depends for its justifications on business or the interests of business, when it should be the other way around. While it may be the case that 'ethics pays dividends', this is not the justification for doing what is right or avoiding what is wrong. This claim is, in itself, a little troubling in light of the checklist question 'Are my references concrete and specific instead of ambiguous and abstract?' (29).

Individual chapter sections on ethics in the Boone, Kurtz, et al. text include questions such as that of whether or not email messages on company computers are private messages. They point out that the laws relating to the rights of the company and the expectation of privacy for the employee are in conflict on this issue. Again, readers are provided with a series of questions, but no clear mechanism for the resolution of the issue. As such, the implication is that what one 'feels' is ethical is, in fact, ethical. Also, this is to confuse the use of laws as the authority for ethical advice or decisions, when laws themselves have been and continue to be evaluated on the basis of whether or not they exemplify ethical standards of non-discrimination, fairness, or equality, for example. Laws, for example, which exhibit ethical failures, cannot themselves be used to support sound ethical decisions or advice.

Two other texts (Thill, Bovee, and Cross, and Guffey, Rhodes, and Rogin) provide a similar relationship between ethics and the moral judgements of society. Both suggest that the person looking for guidance ask themselves a series of questions. Thill, et al. (13) say that a person *might* ask:

- Is the decision legal? (Does it violate civil law or company policy?)
- Is it balanced? (Does it do the most good and the least harm?)
- Is it a decision you can live with? (Does it make you feel good about yourself?)
- Is it feasible? (Can it work in the real world?)

In a similar vein, Guffey, et al. (29) suggest that a person *may* ask:

- Is the action you are considering legal?
- How would you see the problem if you were on the opposite side?
- What are alternative solutions?
- Can you discuss the problem with someone whose advice you trust?
- How would you feel if your family, friends, employer, or coworkers learned of your action?

Both of these texts seriously *confuse* a legal justification with an ethical justification, falsely suggesting that legality can be successfully used to provide rational support for the ethical, when it cannot. If this were the case, then all civil disobedience would be unethical, when such actions are often more ethical than the laws they condemn.

Both of these texts fall back on the golden rule: Do unto others as you would have them do unto you. This is something that students readily agree with but which is not very helpful when faced with two conflicting alternatives that are both ethical and fair; or between two alternatives that are both unkind to others; or between two alternatives that both fall into grey areas between what is and what is not clearly ethical or unethical.

Ethical behaviour, according to Guffey, et al. (26), 'involves four principles: honesty, integrity, fairness, and concern for others'. Thill, et al. (12) define ethical people as 'generally trustworthy, fair, and impartial, respecting the rights of others and concerned about the impact of their actions on society'. The vagueness and essentially meaningless homily since it provides no specific process, direction, or advice to anyone reading it. This similarity of perception echoes the ways in which business communications texts seem to reflect one another in context, organization, and tone (see also Kolin, 2001). There are also many similarities between the appearance, style, and content of the textbooks currently available. Inasmuch as the advice offered in regard to ethics is inadequate, the fact that the same kind of advice is often repeated in slightly different forms in other texts means that they are few opportunities for instructors or students to find more acceptable or less vague advice elsewhere within the genre.

TEXTS THAT SUGGEST THAT SEVERAL PHILOSOPHIES OF ETHICS MAY OFFER ACCEPTABLE GUIDES

A pleasing exception to the notion that most textbooks are vague about ethics is *How to Write for the World of Work* by Pearsall, Cunningham, and Smith (2000). What distinguishes this text is, first of all, that it clearly references and distinguishes three ethical systems: utility-based, rule-based, and rights-based. Perhaps the texts previously mentioned felt that making reference to ethical systems would unnecessarily complicate the discussion for business students, but the Pearsall, et al. work makes it much easier to see that these ideas are not necessarily always conflated.

This text also provides a case for analysis and shows how each of the ethical systems identified will evaluate the ethics of the situation, in order to provide plausible ethical advice. Further, it points at that this is only the first step. The individual involved must then make a decision as to how to respond under the circumstances. The authors don't go so far as to say that making an ethical choice may be personally costly or damaging, but they do say that 'sometimes, it takes enormous courage to be ethical.' This identifies the important component of character referred to earlier in this paper since it suggests that the individual has to have a certain deliberate determination, courage to stick to a principled response over expediency or propriety, sacrificing something important to 'do the right thing'.

Lehman, Himstreet, and Baty (1996) also identify some ethical theories. They begin by suggesting that the Golden Rule, if applied by everyone, would result in a much-improved working environment. They go on to suggest that the Pagano Model (1996: 157) is a more sophisticated model for determining if a proposed action is ethical. This states that one must answer honestly a series of six questions:

1. Is the proposed action legal—the core starting point?
2. What are the benefits and costs to the people involved?
3. Would you want this action to be a universal standard, appropriate for everyone?
4. Does the action pass the light-of-day test? That is, if your action appeared on

television or others learned about it, would you be proud?

5. Does the action pass the Golden Rule test? That is, would you want the same to happen to you?
6. Does the action pass the ventilation test? Ask the opinion of a wise friend with no investment in the outcome. Does this friend believe that the action is ethical?

Beyond that, though, Lehman, et al. identify and explain utilitarian theory, and the doctrine of *prima facie* duties. These become a part of a framework for analyzing ethical issues (154). In this framework the individual is instructed to:

1. Identify the legal implications of the alternative.
2. Determine whether the alternative is consistent with company or professional code of ethics.
3. Use ethical principles and theories to assess whether the alternative is ethical.
4. Implement the alternative.
5. Communicate ethical decisions to appropriate individuals inside or outside the organization.

The implication here is that any one of (or group of) the identified ethical theories and principles is as good as any other. It is significant, though, that among the ethical systems offered the focus is on fulfilling one's duties and optimizing the satisfaction of the greatest number of stakeholders. Thus, the needs of the business world, while not being identified as being more significant than others' needs, would still benefit from the framework provided here.

TEXTS THAT IDENTIFY SEVERAL PHILOSOPHIES OF ETHICS BUT POINT OUT THAT ONE MUST BE CONSISTENT

As indicated earlier, this is not an exhaustive study of business communications texts, but it is significant that none of the texts available suggests that one must be consistent in the use of a particular ethical system or philosophy. It would not be appropriate for an author to identify one system as having merit over others, of course, although some texts—as we have seen—come quite close to that. At the same time, though, the reader is left with the impression that either there is no system for ethical decision-making, or that whatever system one chooses on any given occasion will suffice.

QUESTIONS FOR CRITICAL THOUGHT

1. This is a research report. What is the problem that Gough and Price are researching?
2. What is the purpose of the report? For whom do you think it was originally written?
3. Consider the style of writing: how would you describe it? How personal is it? How readable? How formal? What is its tone?
4. Why do Gough and Price employ so many enumerated lists?
5. Why do they spend so long describing the conditions under which ethical decisions might be made?

6. How is the passage organized? How many segments does it contain, and what are they?
7. What is the difference, according to Gough and Price, between propriety and ethical standards?
8. Gough and Price condemn what they call the 'platitudes, not process' approach. What are their main complaints about the way ethics is handled in the books they surveyed?
9. Where, according to Gough and Price, does ethical responsibility ultimately reside? To what degree is ethical responsibility personal? To what degree does ethics transcend the individual?
10. Gough and Price are concerned with how ethics is treated by college textbooks, and they identify a number of shortcomings in these treatments. What are these shortcomings?
11. MacLennan's 'Communicating Ethically' originally appeared in one of her textbooks for beginning technical writers. Does that essay suffer from the same weaknesses identified by Gough and Price? Why or why not?
12. Compare this article with those by Dombrowski ('Can Ethics Be Technologized?') and Shay ('Aristotle's *Rhetoric* as a Handbook of Leadership'). What connection is there between ethics and leadership? Between ethics and persuasion?

CHAPTER 43

Can Ethics Be Technologized? Lessons from Challenger, Philosophy, and Rhetoric

Paul M. Dombrowski

Abstract

Technology informs many aspects of our lives. Many critics perceive technology as a system of values, seeing it as an incomplete 'ethic'. I explore the converse. Using Ellul's technique (translated as 'technologism'), I ask, 'Can ethics be technologized?' I show how the Challenger disaster delimits the range of technologism with regard to ethics. Collecting additional technical information cannot of itself prevent ethical lapses. Furthermore, the investigations implicitly show the assumption that technologism can apply to ethics in their call for additional procedures. The recognition that procedures already in place were adequate, however, shows this assumption to be fallacious. I also show that trying to technologize ethics is a recent instance of an old reductive fallacy. The ancient sophists were criticized for trying to technologize both rhetoric and ethics. In recent philosophy, too, many critics insist that ethics cannot be reduced, systematized, or technologized. Ethics then is innately problematic, so ethical choices must always be continually deliberated among people in an indeterminate way.

Technolgoy dominates our age. This dominance alarms ethicists because technology can often seem on another plane from ethics. These critics, in order to situate technology within ethical discourse, continually raise the question, 'Is technology an implicit system of values, an "ethic"?'

Cultural critics such as J. Ellul (1990), S. Monsma (1986), L. Winner (1986), and L. Marx (1988) critique the autonomy too readily accorded technology, as do feminist critics such as S. Harding (1986) and E.F. Keller (1989). Within technical and professional communication studies, D. Sullivan (1990), C. Miller (1989), and S. Katz (1992) all warn against the narrow pursuit of a single technological value (whether named excellence, expedience, or utility), while M. Markel (1991) explores the deep complexity of real ethical problems. R.L. Johannesen (1990) offers a comprehensive review of the many approaches to ethics in communication generally. My own *Humanistic Aspects of Technical Communication* (1994) overviews both feminist critiques of science and ethics in technical and professional communication. G. Clark (1987) and E. Garver (1985) argue that ethics cannot be reduced to rules and urge instead a holistic rhetorical approach to ethics attuned to situational exigencies.

I would like to raise the converse question: 'Can ethics be technologized?' (I am rendering Ellul's French 'technique' as 'technologism'.) That is, can ethics be systematized into a sort of technology, say in the form of a system of procedures, which can substitute for personal or social ethical responsibly? Can technologism carry over even to ethics?

Such technologizing could have many forms, such as attempting to reduce ethics to a code of conduct or to operationalize ethics in a decision algorithm. Codes of conduct will not be dealt with the ethics here due to space limitations; see Markel (1991) and Buchholz (1989) for ethical appraisals of codes of conduct. D. Sturges offers an example of a procedural or algorithmic approach to ethics. Sturges's Ethical Dilemma Decision Model 'provides a systematic framework for processing decisions and their ethical implications consistently', a framework that aims at 'greater control and less uncertainty' in such decisions (1992: 50). Though the impulse to systematize and concretize is practically the wellspring of technology, I argue that ethics is one area of human conduct that cannot be technologized. I am not saying that any attempt to clarify, operationalize, or codify ethical responsibilities is always futile, only that we should not expect too much of such attempts. We should not expect such technologies to substitute for personal judgment and responsibility.

In this article, I conclude that the ethical burden is just that: a weight from which we can never fully be relieved, try as we might. I will show that the fallacious technologistic assumption was operative in the Challenger disaster in two ways. I also will draw from two fields closely connected to ethics. From rhetoric, I show that the assumption that ethics can be technologized parallels the mistaken assumption by the sophists of classical Greece that rhetoric could be technologized. From philosophy, I show how two recent philosophers affirm the inescapability of the ethical burden. Therefore the question of technologizing human conduct is both very old and very current.

The Challenger Disaster

The Challenger disaster itself illustrates the question 'Can ethics be technologized?' My research into the disaster convinces me that we cannot legitimately expect that increasing amounts of objective technical information will always resolve ethical questions. The disaster could not have been prevented by collecting more technical information, because technical information does not determine fully either its own meaning or its own ethicality. Ethics, then, presents a limiting case for technologism.

For those unfamiliar with the Challenger disaster, let me review it. I will focus on the issue of '*anomalousness*' and on the issue of '*flightworthiness*' involved in the L–l (read as 'L minus one') preflight review meeting the night before the launch. My studies and those of many others reveal that even in a highly technical activity such as the Shuttle program, raw technical information does not signify its own ethicality. Instead, the very human act of conceptualizing and interpreting data is the crux of ethical responsibility. Though technologism assumes that more and more technical information will resolve dilemmas, Challenger shows that this assumption can be fallacious and can lead to disaster (Pace, 1988; Winsor, 1990; Dombrowski, 1992; Moore, 1992).

From early Shuttle missions, it was apparent that something was seriously wrong. Inspection of the spent boosters revealed that the O-rings seals frequently were charred to varying degrees, some almost halfway through. Initially this charring was said to be 'anomalous' because it was not supposed to happen. These early observations of charring were noted with alarm by technicians and reported to higher authorities because the seals were vital to the safety of the flight. However, little was done to change things.

The fact that successive flights were launched and boosters were recovered intact, even with some charring of the O-rings, was taken as evidence that charring should be understood in a new light: the charring of the O-rings was not a serious concern and should be tolerated.

According to the Congressional committee investigating the disaster, language about the charring took a startling turn. The charring began to be referred to as 'acceptable erosion', 'allowable erosion', and 'acceptable risk' (US Congo House, 1987: 53, 55, 56). This was a complete reversal of the original interpretation of the charring: 'But rather than identify this condition as a joint that didn't seal, that is, a joint that had already failed, NASA elected to regard a certain degree of erosion or blow-by as "acceptable"' (1987: 62). In this way, the anomalous was no longer anomalous because it happened all the time, and what was cause for alarm became grounds for reassurance.

This reversal shows how technical information may be interpreted in light of many social considerations, some ethical and relevant, some not. Therefore we cannot expect technical data to tell their own story as though from their own autonomous authority. And so collecting more and more data, as technologism seems to direct us to do, cannot be expected always to guide us out of ethical dilemmas.

The flightworthiness assumption operative in the L–1 teleconference between managers from NASA and engineers and managers from Morton Thiokol Corp (MTI) further illustrates that technical information cannot of itself guide ethical conduct. During this teleconference, management and engineers at MTI repeatedly raised questions about the safety of the O-rings (some engineers refused to accept the reconceptualization of the anomaly). The MTI engineering group was 'very adamant about their concerns . . . ' (Rogers Commission, 1986: 66–9). NASA management was 'appalled' (Hardy at Marshall SFC) at these objections because of the possibility of having to postpone a flight that was already far behind schedule. MTI management called for an off-line caucus to discuss among themselves the actions to be taken.

In the off-line caucus, management at MTI and NASA pressed engineers to prove absolutely that the mission was doomed. Engineers were confused by the argumentative assumption behind this questioning. The standard assumption by management is that the vehicle *will not* fly;

engineers then have to prove that the vehicle *will* fly safely. This standard assumption makes safety paramount by saddling engineers with the burden of proof; without convincing proof, the flight is cancelled. When engineers could not prove that the flight would certainly end in disaster, management took their implicit but abnormal assumption as confirmed.

R. Boisjoly (Member, Seal Task Force, MTI) describes this reversal of assumption: 'This was a meeting where the determination was to launch, and it was up to us to prove beyond a shadow of a doubt that it was not safe to do so. This is in total reverse to what the position usually is in a preflight conversation or flight readiness review. It is usually exactly opposite to that' (1986: 93). When it became clear that management would not be persuaded, the engineers gave up. Thus the decision to launch was determined less by technical information itself than by the prior interpretive framework that management chose to adopt.

In both these instances, the collection of more technical information would have been pointless, as several engineers testified. In one of the most objective of technologies, then, the technical facts did not suffice to guide their ethical use. Therefore technologism of itself could not have prevented the ethical lapses that resulted in tragedy.

The Challenger Investigations

The subsequent investigations of the disaster by the Presidential Commission (known as the Rogers Commission) and by the Congressional Committee also reveal the problems with trying to technologize ethics. The call for more procedures by the Rogers Commission assumes that instituting additional impersonal procedures can substitute for personal ethical responsibility and prevent ethical lapses—this I call the technologizing of ethics. This assumption is shown to be fallacious, however, by testimony and evidence within the Rogers investigation indicating that the procedures already in place were adequate. It is also shown to be fallacious by the findings of the Congressional committee, which explicitly insisted that impersonal procedures cannot substitute for personal judgment.

The Rogers Commission report did not explicitly grapple with the question of the locus of responsibility—whether it was primarily people, or primarily procedures, or a mixture of the two. The Conclusions and Recommendations sections of the Rogers report reflect this lack of clarification. At the end of the volume in which causes are discussed, paragraph one states, 'the decision was flawed'; paragraph two states, 'the decision-making process was flawed'; and paragraph three states that the cause was 'failures in communication' (1986, V: 82). These expressions are confusing, ambiguous, perhaps even contradictory. Were particular decisions flawed, suggesting personal responsibility; were the procedural systems themselves flawed, indicating procedural responsibility; or did someone fail to do something required by procedures, indicating personal responsibility?

Elsewhere I have reported that my analysis of the evidence and testimony before both the commission and the committee shows that personal judgment, rather than procedural shortcomings, accounts for the loss of Challenger. The Conclusions and Recommendations sections of both investigations, however, make little mention of personal judgment or responsibility. That the conclusions do not follow logically comes to light only when key decisions are examined to assess whether personal judgment or procedural requirements determined important decisions (Dombrowski, 1991). Repeatedly, key decisions show that personal decision-making was much more important than impersonal procedural decision-making. Two examples follow.

First, Lawrence Mulloy (Manager, SRB Project, Marshall Space Flight Center or MSFC) testified that he had told 'everyone' about the problem with the O-ring seals, yet there is absolutely no mention of it in the flight-readiness reviews (1986, V: 85). Thus the all-important decision whether to put a verbalized reservation into print was a personal judgment. This decision shows that procedures operate only derivatively on the basis of written inputs that might not reflect the whole decision-making picture.

Second, there existed alternative, independent paths for reporting problems such as O-ring charring, but their existence did not prevent the disaster (1986, V: 84). Thus deliberate procedural redundancy was rendered ineffective by decisions that were erroneous or misleading (e.g., that charring was not 'anomalous'). This circumventing of procedural safeguards shows the futility of expecting too much of procedures themselves.

Thus, many crucial decisions were made not through procedural algorithms but personally and separately from the system of procedures. Procedures were involved only after personal decision-making, to effect decisions already made. As Boisjoly summarized the L–1 meeting: 'This was a meeting where the determination was to launch . . . ', regardless of the technical facts presented. The absolutely vital decision had already been made prior to the teleconference required by procedures; even then an off-line caucus made the discussion off-the-record. It therefore does not make sense to recommend the implementation of more and more impersonal procedures to prevent poor judgment and ethical lapses.

The importance of assumptions about the roots of ethicality is even more apparent in comparing the Rogers and Congressional reports. The two bodies came to different conclusions, despite the Congressional investigation's heavy reliance on the very evidence and testimony gathered by the Rogers Commission. Thus the two bodies looked at the same basic information but saw different causes and responsibilities.

For example, the Rogers Commission frequently mentions 'flawed procedures' and difficulties with the 'communication system'. On the other hand, the Congressional committee arrived at conclusions focusing more on personal responsibility and judgment, even to the point of highlighting this difference in its introductory remarks. They found, for example, that 'the failure was not the problem of technical communication, but rather of technical decision making' (1986: 30). The committee also is quite clear about the limitations of procedural safeguards: 'It seems clear that the process [i.e., procedures] cannot compensate for faulty engineering judgment' (1986: 70).

The Congressional committee even more clearly explained these limitations:

> [T]he committee finds no basis for concluding that the Flight Readiness Review procedure is flawed; on the contrary, the procedure appears to be exceptionally thorough and the scope of the issues that are addressed at the FRRs is sufficient . . . . However, the Flight Readiness Reviews are not intended to replace engineering analysis, and therefore, they cannot be expected to prevent a flight because of a design flaw that management had already determined represented an acceptable risk . . . . [A] process is only as effective as the responsible *individuals make it*. (1957: 150, emphasis added)

The most important lesson we can learn from this analysis is that impersonal procedures cannot substitute for personal ethical responsibility—ethics cannot be *technologized*. Though the basic impulse to institute additional procedures is understandable, we should not expect too much of it. As the Challenger disaster shows, we cannot responsibly expect that additional procedures of themselves can substitute for, or prevent, ethical lapses.

Classical Philosophy and Rhetoric

To show that the technologist frame of mind is quite old and to suggest a reasoned response to it, let me shift to rhetoric in classic Greece. Rhetoric from its beginning has had to grapple with attempts to technologize it. In fact, the entire history of rhetoric shows repeated alterations between, on the one hand, an open-ended, personalistic, and holistic view of rhetoric, and, on the other hand, a systematic, impersonal, 'skills' view—basically, a technology of rhetoric.

Rhetoric was a vitally important topic in classical Athens because persuasion, the function of rhetoric, was intimately linked to the conduct of civic life, to the advancement of culture, and to ethics. As to what the term 'rhetoric' meant, however, there were many camps.

One camp was the sophists. Traditionally, the sophists have been understood as charlatans who taught a base, self-serving, unethical sort of persuasion that did not merit being called 'true rhetoric' (Guthrie, 1971; Kerferd, 1981). One of the chief criticisms of the sophists was that they taught only a bag of tricks, a collection of *technai* [the same root as 'technology'], a set of procedures by which the audience could be led in the direction the rhetor wished. They were accused, that is, of technologizing rhetoric.

Another criticism was that the sophists claimed to teach virtue—not that they taught *about* virtue but that their course of study would *make* a student into an ethical citizen. Utilizing their *technai*, the sophists contended, would necessarily yield an ethical citizen. In effect, for the sophists, their technologic system was ethics (or at least functionally constituted ethics). Thus, supposedly, ethical conduct could be communicated clearly and mechanically in a closed set of principles, then handed out like a course syllabus—cut-and-dried, bought-and-paid for.

Socrates, Plato, Aristotle, and other philosophical rhetors held a very different view, however. They taught that rhetoric was much nobler, much more complicated, and much more weighty in its personal responsibility. Rhetoric was the means by which one soul could lead another to recollect its prior association with the divine. Two people would use noble rhetoric, then, to urge each other mutually in the love and pursuit of wisdom—philosophia. The dialogic interchange between these two people was not only very close but very organically specific to the two particular people. As a result, nothing but the individual particulars of their coming together could determine the rhetoric by which they enlightened and exhorted each other. Rhetoric for Socrates could not be reduced to a set of rules—it could not be technologized.

Socrates also held that virtue—ethical conduct—could not be taught because it too could not be reduced to a set of rules. Ethics could not be reduced to *technai* because of the uniqueness of every situation. Instead, the right thing to do had to be determined in each particular situation, a determination that required the engagement of real individuals earnestly arguing according to their own enlightenment. Rhetoric thus is the vital communicative means by which ethics is both revealed and practiced. Ethics could also not be given to another, like a commodity, for it was not a thing but an attitude and a process.

Ethics for Socrates also required hard work. Socrates, the perpetual social critic, was so annoying precisely because he refused to let people take the easy, mindless road in life. He would not allow people to forsake their ethical responsibility for a set of rules that could make decisions for them. Socrates's scrupulously ethical life caused him, however, a good deal of hardship, ultimately costing him his life.

These very ancient ideas—rhetoric cannot be technologized, ethics cannot be technologized, and ethics is an inescapable task—have not died off. Miller, Katz, Sullivan, Garver, and others continue to remind us of the same basic issue: the fallacy of assuming that complex human activities such as rhetoric and ethics can be reduced to fixed sets of skills. We can also hear echoes within contemporary philosophy.

Contemporary Philosophy

M. Grene (1984) was one of the great American scholars, teachers, and humanists of recent times. She was admired for her keen mind and lucid writing, but also for her sense of humane responsibility and for her ethical criticism. In her famous introduction to existentialism, Grene admitted that although she disagreed with important features of Jean-Paul Sartre's existentialism, she highly admired Sartre because of what was implicit in his thinking. The very basis of Sartre's existentialism was good faith between people—in effect, ethical responsibility. Good faith required constant engagement in the immediate here-and-now through communication with others that was open to the unexpected. Grene says that Sartre thus rejected any suggestion that philosophy could be reduced to a set system of principles that, once established, defined philosophy forever. For Sartre, philosophy is what philosophers *do*. It is not a lifeless system of principles but a *task*, and one which is never complete. It is a task with which we are always saddled—a burden, yes, but also the root of our very humanity.

Grene and Sartre thus had the same antitechnologizing understanding of philosophy: it is not a thing but a human activity and one that never (at least in this life) attains its end. In like manner, presumably Grene and Sartre would agree that ethics, as part of philosophy, is the same never-ending activity that cannot be reduced to a technology.

Continuing in philosophy in this century, the continental ethicist Emmanuel Levinas (1990) also emphasized the inescapability of the ethical burden. Levinas has received a good deal of attention since the late 1970s from postmodem literary theorists such as Jacques Derrida. He was strongly opposed to objectifying, systematizing, and technologizing ethics, while he strongly advocated the vital importance of particularity, uniqueness, and situational specificity.

Levinas held that morality begins with our awareness of, and response to, what he calls 'the face of the other person'. This 'other' is a particular, real, other person (rather than a generic abstraction) whose otherness is his or her fundamental difference from ourselves. This uniqueness and difference is embodied most strikingly in 'the face', the usual locus of our distinguishing someone from ourselves and from all others. The essential particularity and uniqueness of this distinctly other person opposes any inclination we might have to generalize about 'others'—to establish, and act out of, general rules of conduct—because the effect of our actions on any one person cannot be predicted certainly. Thus the other's capacity to respond, to criticize, and to disagree all elementally oppose our inclination to objectivize him or her. It also opposes any inclination to systematize our behavior with respect to him or her, which would be to technologize our ethics.

Let me try to make concrete Levinas's theorizing and to illustrate its rhetorical, emotional, and moral force. Consider the famous Vietnam Memorial by Maya Lin in Washington. It is a simple, long, low wall of polished marble set into a trench in the earth. On it are engraved the names of most of the US personnel who died in the Vietnam Conflict. Like the real bodies of veterans buried in a grave, the monument is buried in the ground and the names are engraved into the stone. It is widely recognized as an architectural masterpiece and a monumental moral statement. Directly behind Lin's memorial is a more conventional monument to the same conflict, a group of bronze statues with generic faces, set at ground level. These generic statues are much less famous precisely because they lack the ethical force of Lin's personalized memorial.

Lin's Vietnam Memorial, rather than presenting generic faces, instead presents each individual name before the viewer, who in effect faces the death of each one of these unique individuals. At the same time, the viewer morally faces himself or herself while seeing his or her own unique human face optically reflected in the polished marble and superimposed on the other's name. The effect is to feel attached to, and somehow responsible to, that other person.

Let me show how a Levinasian perspective could shed light on the investigations of the

Challenger disaster. One of the findings of the Presidential commission and the Congressional committee was that the shuttle astronauts, those real people most directly affected by decisions about the significance of O-ring charring or about flightworthiness, had not been included in the decision-making process. Thus the impersonal process proceeded without having to *face* directly the people affected by those decisions. Following the recommendations of both investigative bodies, however, astronauts now are included in such decisions—real faces before decision makers.

Levinas explains that we must continually face (Levin as deliberately uses 'face' in several senses) the unremitting task of ethical examination. This characteristically human task is our unending burden precisely because we can never fully know another; we can never fully predict what he or she thinks, feels, or believes; and we can never be certain how our actions will be perceived by the other. Compounding this difficulty is our interaction, the reciprocal evolution of ourselves stemming from our social communications. This interaction not only makes the other unique and continuously changing, but ourselves too as we are changed through the interaction. For Levinas as for Sartre as for Socrates, this is our burden: to have to weigh continually the rightness of our actions. This burden simply is part of being human.

Conclusion

Ethics *is* what is problematic. It is not a fixed set of rules but an ongoing human activity that must continually be thrashed out for particular circumstances and people. This is not to say that nothing can be said about ethics in general, only that the difficult business of arguing between competing values cannot entirely be circumvented. Any attempt to treat ethics as entirely operationalizable, to technologize ethics, is self-deluding and ineffective.

References

Buchholz, W.J. 1989. 'Deciphering Professional Codes of Ethics', *IEEE Trans. Professional Communication* 32: 62–158.

Clark, G. 1987. 'Ethics in Technical Communication: A Rhetorical Perspective', *IEEE Trans. Professional Communication* 30: 190–5.

Dombrowski, P.M. 1991. 'The Lessons of the Challenger Disaster', *IEEE Trans. Professional Communication* 34, 4: 211–16.

———. 1992. 'Challenger and the Social Contingency of Meaning: Two Lessons for the Technical Communication Classroom', *Technical Communication Quarterly* 1, 3: 73–86.

———, ed. 1994. *Humanistic Aspects of Technical Communication*. Amityville, NY: Baywood.

Ellul, J. 1990. *The Technological Bluff*, G. Bromiley, trans. Grand Rapids, MI: Eerdmans.

Garver, E. 1985. 'Teaching Writing and Teaching Virtue', *Journal of Business Communication* 22, 1: 51–73.

Grene, M. 1984. *Introduction to Existentialism*. Chicago, IL: University of Chicago Press (Midway Reprint).

Guthrie, W.K.C. 1971. *The Sophists*. Cambridge: Cambridge University Press.

Harding, S. 1986. *The Science Question in Feminism*. Ithaca, NY: Cornell University Press.

Johannesen, R.L. 1990. *Ethics In Human Communication*, 3rd ed. Prospect Heights, IL: Waveland.

Katz, S.B. 1992. 'The Ethic of Expediency: Classical Rhetoric, Technology, and the Holocaust', *College English* 54, 3: 255–75.

Keller, E.F. 1989. 'The Gender/Science System: or, Is Sex to Gender as Nature is to Science', in *Feminism & Science*. Bloomington, IN: Indiana University Press.

Kerferd, G.B. 1981. *The Sophistic Movement*. Cambridge: Cambridge University Press.

Levinas, E. 1990. *Collected Philosophical Papers*, A. Lingis, trans. Dordrecht, The Netherlands: Martinus Nijhoff.

Markel, M. 1991. 'A Basic Unit on Ethics for Technical Communicators', *Journal of Technology, Writing and Communication* 21, 4: 327–50.

Marx, L. 1988. *The Pilot and the Passenger: Essays on Literature, Technology, and Culture in the United States*. New York: Oxford University Press.
Miller, C.R. 1989. 'What's Practical about Technical Writing', in *Technical Writing: Theory and Practice*, B.E. Fearing and W.K. Sparrow, eds. New York: Modern Language Association.
Monsma, S.V. 1986. *Responsible Technology: A Christian Perspective*. Grand Rapids, MI: Eerdmans.
Moore, P. 1992. 'When Politeness is Fatal: Technical Communication and the Challenger Accident', *Journal of Business and Technical Communication* 6, 3: 269–92.
Morgenstern, J. 1995. 'The Fifty-Nine-Story Crisis', *The New Yorker* 29 May: 49–53.
Pace, R.C. 1988. 'Technical Communication, Group Differentiation, and the Decision to Launch the Space Shuttle Challenger', *Journal of Technology, Writing and Communication* 18, 3: 207–20.
Rogers Commission [United States, Presidential Commission on the Space Shuttle Challenger Accident]. 1986. *Report to the President by the Presidential Commission on the Space Shuttle Challenger Accident*. 86-16083. GPO: Washington, DC.
Sturges, D.L. 1992. 'Overcoming the Ethical Dilemma: Communication Decisions in the Ethic Ecosystem', *IEEE Trans. Professional Communication* 35: 44–50.
Sullivan, D.L. 1990. 'Political–Ethical Implications of Defining Technical Communication as a Practice', *Journal of Advanced Composition* 10: 375–86.
United States, Congo House. 1987. *Investigation of the Challenger Accident: Report of the Committee on Science and Technology*. House of Reps., 99th Congress, 874033. GPO: Washington, DC.
Winner, L. 1986. *The Whale and the Reactor: A Search for Limits in an Age of High Technology*. Chicago. IL: University of Chicago Press.
Winsor, D.A. 1990. 'The Construction of Knowledge in Organizations: Asking the Right Questions about the Challenger', *Journal of Business and Technical Communication* 4, 2: 7–20.

QUESTIONS FOR CRITICAL THOUGHT

1. Dombrowski argues that ethics cannot be technologized. What does he mean by this term?
2. According to Dombrowski, ethics are a matter of personal judgement and responsibility, for which 'impersonal procedures cannot substitute'. Why is this so?
3. Why does Dombrowski use the example of the Challenger disaster to illustrate his case?
4. What kind of evidence does Dombrowski offer to support his assertion that ethics cannot be reduced to a cookie-cutter heuristic?
5. What is 'technologism'?
6. According to Dombrowski, 'raw technical information does not signify its own ethicality.' What does this statement mean?
7. Dombrowski recounts significant changes in the language used to describe the O-ring failure on the space shuttle. In what way did this change in 'naming' contribute to the disaster, and why? What does this fact tell us about the way we use language, and about its effects?
8. Dombrowski describes the Challenger disaster as an ethical lapse rather than a

technological failure, even though the failed O-rings were the cause of the explosion. Why does he make this claim?

9. Compare Dombrowski's discussion of rhetoric with those provided by Smith ('What Connection does Rhetorical Theory Have to Technical and Professional Communication?'), Campbell ('Ethos: Character and Ethics in Technical Writing'), and Buehler ('Situational Editing').
10. What relationship exists, according to Dombrowski, between ethics and situation? Does he contradict Gough and Price ('Developing Ethical Decision-Making Skills') on this point?
11. Dombrowski's style appears relatively personal, while still being formal. Can you point to some features of the article that help to create this effect?
12. To what extent does Dombrowski's article conform to the narrative style called for by Carol Forbes in 'Getting the Story, Telling the Story: The Science of Narrative, the Narrative of Science'?

CHAPTER 44

The Moral Un-Neutrality of Science

Charles P. Snow

Scientists are the most important occupational group in the world today. At this moment, what they do is of passionate concern to the whole of human society. At this moment, the scientists have little influence on the world effect of what they do. Yet, potentially, they can have great influence. The rest of the world is frightened both of what they do—that is, of the intellectual discoveries of science—and of its effect. The rest of the world, transferring its fears, is frightened of the scientists themselves and tends to think of them as radically different from other men.

As an ex-scientist, if I may call myself so, I know that is nonsense. I have even tried to express in fiction some kinds of scientific temperament and scientific experience. I know well enough that scientists are very much like other men. After all, we are all human, even if some of us don't give that appearance. I think I would be prepared to risk a generalization. The scientists I have known (and because of my official life I have known as many as anyone in the world) have been in certain respects just perceptibly more morally admirable than most other groups of intelligent men.

That is a sweeping statement, and I mean it only in a statistical sense. But I think there is just a little in it. The moral qualities I admire in scientists are quite simple ones, but I am very suspicious of attempts to oversubtilize moral qualities. It is nearly always a sign, not of true sophistication, but of a specific kind of triviality. So I admire in scientist very simple virtues—like courage, truth-telling, kindness—in which, judged by the low standards which the rest of us manage to achieve, the scientists are not deficient. I think on the whole the scientists make slightly better husbands and fathers than most of us, and I admire them for it. I don't know the figures, and I should be curious to have them sorted out, but I am prepared to bet that the proportion of divorces among scientists is slightly but significantly less than that among other groups of similar education and income. I do not apologize for considering that a good thing.

A close friend of mine is a very distinguished scientist. He is also one of the few scientists I know who has lived what we used to call a Bohemian life. When we were both younger, he thought he would undertake historical research to see how many great scientists had been as fond of women as he was. I think he would have felt mildly supported if he could have found a precedent. I remember his reporting to me that his researches hadn't had any luck. The really great scientists seemed to vary from a few neutral characters to a large number who were depressingly 'normal'. The only gleam of comfort was to be found in the life of Jerome Cardan; and Cardan wasn't anything like enough to outweigh all the others.

So scientists are not much different from other men. They are certainly no worse than other men. But they do differ from other men in one thing. That is the point I started with. Whether they like it or not, what they do is of critical importance for the human race. Intellectually, it has transformed the climate of our time. Socially, it will decide whether we live or die, and how we live or die. It holds decisive powers for good and evil. *That* is the situation in which the scientists find themselves. They may not have asked for it, or may only have asked for it in part, but they cannot escape it. They think, many of the more sensitive of them, that they don't deserve to have this weight of responsibility heaved upon them. All they want to do is to get on with their work. I sympathize. But the scientists can't escape the responsibility—any more than they, or the rest of us, can escape the gravity of the moment in which we stand.

Doctrine of Ethical Neutrality

There is of course one way to contract out. It has been a favourite way for intellectual persons caught in the midst of water too rough for them.

It consists of the inventions of categories—or, if you like, of the division of moral labour. That is, the scientists who want to contract out say, *we* produce the tools. *We* stop there. It is for *you*—the rest of the world, the politicians—to say how the tools are used. The tools may be used for purposes which most of us would regard as bad. If so, we are sorry. But as scientists, that is no concern of ours.

This is the doctrine of the ethical neutrality of science. I can't accept it for an instant. I don't believe any scientist of serious feeling can accept it. It is hard, some think, to find the precise statements which will prove it wrong. Yet we nearly all feel intuitively that the invention of comfortable categories is a moral trap. It is one of the easier methods of letting the conscience rust. It is exactly what the early nineteenth century economists, such as Ricardo, did in the face of the facts of the first industrial revolution. We wonder now how men, intelligent men, can have been so morally blind. We realize how the exposure of the moral blindness gave Marxism its apocalyptic force. We are now, in the middle of the scientific or second industrial revolution, in something like the same position as Ricardo. Are we going to let our consciences rust? Can we ignore that intimation we nearly all have, that scientists have a unique responsibility? Can we believe it, that science is morally neutral?

To me—it would be dishonest to pretend otherwise—there is only one answer to those questions. Yet I have been brought up in the presence of the same intellectual categories as most western scientists. It would also be dishonest to pretend that I find it easy to construct a rationale which expresses what I now believe. The best I can hope for is to fire a few sighting shots. Perhaps someone who sees more clearly than I can will come along and make a real job of it.

The Beauty of Science

Let me begin with a remark which seems some way off the point. Anyone who has ever worked in any science knows how much esthetic joy he has obtained. That is, in the actual *activity* of science, in the process of making a discovery, however humble it is, one can't help feeling an awareness of beauty. The subjective experience, the esthetic satisfaction, seems exactly the same as the satisfaction one gets from writing a poem or a novel, or composing a piece of music. I don't think anyone has succeeded in distinguishing between them. The literature of scientific discovery is full of this esthetic joy. The very best communication of it that I know comes in G.H. Hardy's book *A Mathematician's Apology*. Graham Greene once said that he thought that, along with Henry James's prefaces, this was the best account of the artistic experience ever written. But one meets the same thing throughout the history of science. Bolyai's great yell of triumph when he saw he could construct a self-consistent, non-Euclidean geometry; Rutherford's revelation to his colleagues that he knew what the atom was like; Darwin's slow, patient, timorous certainty that at last he had got there—all these are voices, different voices, of esthetic ecstasy.

That is not the end of it. The *result* of the activity of science, the actual finished piece of scientific work, has an esthetic value in itself. The judgments passed on it by other scientists will more often than not be expressed in esthetics terms: 'That's beautiful!' or 'That really is very pretty!' (as the understating English tend to say). The esthetics of scientific constructs, like the esthetics of works of art, are variegated. We think some of the great syntheses, like Newton's, beautiful because of their classical simplicity, but we see a different kind of beauty in the relativistic extension of the wave equation or the interpretation of the structure of deoxyribonucleic acid, perhaps because of the touch of unexpectedness. Scientists know their kinds of beauty when they see them. They are suspicious, and scientific history shows they have always been right to have been so, when a subject is in an 'ugly' state. For example, most physicists feel in their bones that the present bizarre assembly of nuclear particles, as grotesque as a stamp collection, can't possibly be, in the long run, the last word.

We should not restrict the esthetic values to what we call 'pure' science. Applied science has its beauties, which are, in my view, identical in nature. The magnetron has been a marvelously useful device, but it was a beautiful device, not exactly apart from its utility but because it did, with such supreme economy, precisely what it was designed to do. Right down in the field of development, the esthetic experience is as real to engineers. When they forget it, when they begin to design heavy-power equipment about twice as heavy as it needs to be, engineers are the first to know that they are lacking virtue.

There is no doubt, then, about the esthetic content of science, both in activity and the result. But esthetics has no connection with morals, say the categorizers. I don't want to waste time on peripheral issues—but are you quite sure of that? Or is it possible that these categories are inventions to make us evade the human and social conditions in which we now exist? But let us move straight on to something else, which is right in the grain of the activity of science and which is at the same time quintessentially moral. I mean, the desire to find the truth.

The Search for Truth

By *truth*, I don't intend anything complicated, once again. I am using the word as a scientist uses it. We all know that the philosophical examination of the concept of empirical truth gets us into some curious complexities, but most scientists really don't care. They know that the truth, as they use the word and as the rest of us use it in the language of common speech, is what makes science work. That is good enough for them. On it rests the whole great edifice of modern science. They have a sneaking sympathy for Rutherford, who, when asked to examine the philosophical bases of science, was include to reply, as he did to the metaphysician Samuel Alexander: 'Well, what have you been talking all your life, Alexander? Just hot air! Nothing but hot air!'

Anyway, truth in their own straight-forward sense is what the scientists are trying to find. They want to find what is *there*. Without that desire, there is no science. It is the driving force of the whole activity. It compels the scientist to have an overriding respect for truth, every stretch of the way. That is, if you're going to find what is there, you mustn't deceive yourself or anyone else. You mustn't lie to yourself. At the crudest level, you mustn't fake your experiments.

Curiously enough, scientists do try to behave like that. A short time ago, I wrote a novel in which the story hinged on a case of scientific fraud. But I made one of my characters, who was himself a very good scientist, say that, considering the opportunities and temptations, it is astonishing how few such cases there are. We have all heard of perhaps half a dozen open and notorious ones, which are on the record for anyone to read—ranging from the 'discovery' of the L radiation to the singular episode of the Piltdown man.

We have all, if we have lived any time in the scientific world, heard private talk of something like another dozen cases which for various reasons are not yet public property. In some cases, we know the motives for the cheating—sometimes, but not always, sheer personal advantage, such as getting money or a job. But not always. A special kind of vanity has led more than one man into scientific faking. At a lower level of research, there are presumably some more cases. There must have been occasional PhD students who scraped by with the help of a bit of fraud.

But the total number of all these men is vanishingly small by the side of the total number of scientists. Incidentally, the effect on science of such frauds is also vanishingly small. Science is a self-correcting system. That is, no fraud (or honest mistake) is going to stay undetected for long. There is no need for an extrinsic scientific criticism, because criticism is inherent in the process itself. So that all that a fraud can do is waste the time of the scientists who have to clear it up.

The remarkable thing is not the handful of scientists who deviate from the search for truth but the overwhelming numbers who keep to it. That is a demonstration, absolutely clear for anyone to see, of moral behaviour on a very large scale.

We take it for granted. Yet it is very important. It differentiates science in its widest sense (which includes scholarship) from all other intellectual activities. There is a built-in moral component right in the core of the scientific activity itself. The desire to find the truth is itself a moral impulse, or at least contains a moral impulse. The way in which a scientists tries to find the truth imposes on him a constant moral discipline. We say a scientific conclusion—such as the contradiction of parity by Lee and Yang, is 'true' in the limited sense of scientific truth, just as we say that it is 'beautiful' according to the criteria of scientific esthetics. We also know that to reach this conclusion took a set of actions which would have been useless without the moral nature. That is, all through the marvelous experiments of Wu and her colleagues, there was the constant moral exercise of seeking and telling the truth. To scientists, who are brought up in this climate, this seems as natural as breathing. Yet it is a wonderful thing. Even if the scientific activity contained only this one moral component, that alone would be enough to let us say that it was morally un-neutral.

But is this the only moral component? All scientists would agree about the beauty and the truth. In the western world, they wouldn't agree on much more. Some will feel with me in what I am going to say. Some will not. That doesn't affect me much, except that I am worried by the growth of an attitude I think very dangerous, a kind of technological conformity disguised as cynicism. I shall say a little more about that later. As for disagreement, G.H. Hardy used to comment that a serious man ought not to waste his time stating a majority opinion—there are plenty of others to do that. That was the voice of classical scientific nonconformity. I wish that we heard it more often.

Science in the Twenties

Let me cite some grounds for hope. Any of us who were working in science before 1933 can remember what the atmosphere was like. It is a terrible bore when aging men in their fifties speak about the charms of their youth. Yet I am going to irritate you—just as Talleyrand irritated his juniors—by saying that unless one was on the scene before 1933, one hasn't known the sweetness of the scientific life. The scientific world of the twenties was as near to being a full-fledged international community as we are likely to get. Don't think I'm saying that the men involved were superhuman or free from the ordinary frailties. That wouldn't come well from me, who have spent a fraction of my writing life pointing out that scientists are, first and foremost, men. But the atmosphere of the twenties in science was filled with an air of benevolence and magnanimity which transcended the people who lived in it.

Anyone who ever spent a week in Cambridge or Göttingen or Copenhagen felt it all round him. Rutherford had very human faults, but he was a great man with abounding human generosity. For him the world of science was a world that lived on a plane above the nation-state, and lived there with joy. That was at least as true of those two other great men, Niels Bohr and Franck, and some of that spirit rubbed off on to the pupils round them. The same was true of the Roman school of physics.

The personal links within this international world were very close. It is worth remembering that Peter Kapitza, who was a loyal Soviet citizen, honoured my country by working in Rutherford's laboratory for many years. He became a fellow of the Royal Society, a fellow of Trinity College, Cambridge, and the founder and kingpin of the best physics club Cambridge has known. He never gave up his Soviet citizenship and is now director of the Institute of Physical Problems in Moscow. Through him a generation of English scientists came to have personal knowledge of their Russian colleagues. These exchanges were then, and have remained, more valuable than all the diplomatic exchanges ever invented.

The Kapitza phenomenon couldn't take place now. I hope to live to see the day when a young Kapitza can once more work for 16 years in Berkeley or Cambridge and then go back to an eminent place in his own country. When that can happen, we are all right. But after the idyllic years of world science, we passed into a tempest of history, and, by an unfortunate coincidence, we passed into a technological tempest too.

The discovery of atomic fission broke up the world of international physics. 'This has killed a beautiful subject,' said Mark Oliphant, the father figure of Australian physics, in 1945, after the bombs had dropped. In intellectual terms, he has not turned out to be right. In spiritual and moral terms, I sometimes think he has.

A good deal of the international community of science remains in other fields—in great areas of biology, for example. Many biologists are feeling the identical liberation, the identical joy at taking part in a magnanimous enterprise, that physicists felt in the twenties. It is more than likely that the moral and intellectual leadership of science will pass to biologists, and it is among them that we shall find the Rutherfords, Bohrs, and Francks of the next generation.

The Physicist, a Military Resource

Physicists have had a bitterer task. With the discovery of fission, and with some technical breakthroughs in electronics, physicists became, almost overnight, the most important military resource a nation-state could call on. A large number of physicists became soldiers not in uniform. So they have remained, in the advanced societies, ever since.

It is very difficult to see what else they could have done. All this began in the Hitler war. Most scientists thought then that Nazism was as near absolute evil as a human society can manage. I myself thought so. I still think so, without qualification. That being so, Nazism had to be fought, and since the Nazis might make fission bombs—which we thought possible until 1944, and which was a continual nightmare if one was remotely in the know—well, then, we had to make them too. Unless one was an unlimited pacifist, there was nothing else to do. And unlimited pacificism is a position which most of us cannot sustain.

Therefore I respect, and to a large extent share, the moral attitudes of those scientists who devoted themselves to making the bomb. But the trouble is, when you get onto any kind of moral escalator, to know whether you're ever going to be able to get off. When scientists became soldiers they gave up something, so imperceptibly that they didn't realize it, of the full scientific life. Not intellectually. I see no evidence that scientific work on weapons of maximum destruction has been in any intellectual respect different from other scientific work. But there is a moral difference.

It may be—scientists who are better men than I am often take this attitude, and I have tried to represent it faithfully in one of my books—that this is a moral price which, in certain circumstances, has to be paid. Nevertheless, it is no good pretending that there is not a moral price. Soldiers have to obey. That is the foundation of their morality. It is not the foundation of the scientific morality. Scientists have to question and if necessary to rebel. I don't want to be misunderstood. I am no anarchist. I am not suggesting that loyalty is not a prime virtue. I am not saying that all rebellion is good. But I am saying that loyalty can easily turn into conformity, and that conformity can often be a cloak for the timid and self-seeking. So can obedience, carried to the limit. When you think of the long and gloomy history of man, you will find that far more, and far more hideous, crimes have been committed in the name of obedience than have ever been committed in the name of rebellion. If you doubt that, read William Shirer's *Rise and Fall of the Third Reich*. The German officer corps were brought up in the most rigorous code of obedience. To them, no more honorable and God-fearing body of men could conceivably exist. Yet in the name of obedience, they were party to, and assisted in, the most wicked large-scale actions in the history of the world.

Scientists must not go that way. Yet the duty to question is not much of a support when you are living in the middle of an organized society. I speak with feeling here. I was an official for 20 years. I went into official life at the beginning of the war, for the reasons that prompted my scientific friends to begin to make weapons. I stayed in that life until a year ago, for the same reason that made my scientific friends turn into civilian soldiers. The official's life in England is not quite so disciplined as a soldier's, but it is very nearly so. I think I know the virtues, which are very great, of the men who live that disciplined life. I also know what for me was the moral trap. I, too, had got onto an escalator. I can put the result into a sentence: I was coming to hide behind the institution; I was losing the power to say no.

A Spur to Moral Action

Only a very bold man, when he is a member of an organized society, can keep the power to say no. I tell you that, not being a very bold man, or one who finds it congenial to stand alone, away from his colleagues. We can't expect many scientists to do it. Is there any tougher ground for them to stand on? I suggest to you that there is. I believe that there is a spring of moral action in the scientific activity which is at least as strong as the search for truth. The name of this spring is *knowledge*. Scientists *know* certain things in a fashion more immediate and more certain than those who don't comprehend what science is. Unless we are abnormally weak or abnormally wicked men, this knowledge is bound to shape our actions. Most of us are timid, but to an extent, knowledge gives us guts. Perhaps it can give us guts strong enough for the jobs in hand.

I had better take the most obvious example. All physical scientists *know* that it is relatively easy to make plutonium. We know this, not as a journalistic fact at second hand, but as a fact in our own experience. We can work out the number of scientific and engineering personnel needed for a nation-state to equip itself with fission and fusion bombs. We *know* that, for a dozen or more states, it will only take perhaps six years, perhaps less. Even the best informed of us always exaggerate these periods.

This we know, with the certainty of—what shall I call it?—engineering truth. We also—most of us—are familiar with statistics and the nature of odds. We know, with the certainty of statistical truth, that if enough of these weapons are made, by enough different states, some of them are going to blow up, through accident, or folly, or madness—the motives don't matter. What does matter is the nature of the statistical fact.

All this we *know*. We know it in a more direct sense than any politician because it comes from our direct experience. It is part of our minds. Are we going to let it happen?

All this we *know*. It throws upon scientists a direct and personal responsibility. It is not enough to say that scientists have a responsibility as citizens. They have a much greater one than that, and one different in kind. For scientists have a moral imperative to say what they know. It is going to make them unpopular in their own nation-states. It may do worse than make them unpopular. That doesn't matter. Or at least, it does matter to you and me, but it must not count in the face of the risks.

Alternatives

For we genuinely know the risks. We are faced with an either-or, and we haven't much time. The *either* is acceptance of a restriction of nuclear armaments. This is going to begin, just as a token, with an agreement on the stopping of nuclear tests. The United States is not going to get the 99.9 per cent 'security' that it has been asking for. This is unobtainable, though there are other bargains that the United States could probably secure. I am not going to conceal from you that this course involves certain risks. They are quite obvious, and no honest man is going to blink them. That is the *either*. The *or* is not a risk but a certainty. It is this. There is no agreement on tests. The nuclear arms race between the United States and the USSR not only continues but accelerates. Other countries join in. Within, at the most, six years, China and several other states have a stock of nuclear bombs. Within, at the most, ten years, some of those bombs are going off. I am saying this as responsibly as I can. *That* is the certainty. On the one side, therefore, we have a finite risk. On the other side we have a certainty of disaster. Between a risk and a certainty, a sane man does not hesitate.

It is the plain duty of scientists to explain this either-or. It is a duty which seems tome to come from the moral nature of the scientific activity itself.

The same duty, though in a much more pleasant form, arises with respect to the benevolent powers of science. For scientists know, and again with the certainty of scientific knowledge, that we possess every scientific fact we need to transform the physical life of half the world. And transform it within the span of people now living. I mean, we have all the resources to help half the world live as long as we do and eat enough. All that is missing is the will. We *know* that. Just as we know that you in the United States, and to a slightly lesser extent we in the United Kingdom, have been almost unimaginably lucky. We are sitting like people in a smart and cozy restaurant and we are eating comfortably, looking out of the window into the streets. Down on the pavement are people who are looking up at us, people who by chance have different coloured skins from ours, and are rather hungry. Do you wonder that they don't like us all that much? Do you wonder that we sometimes feel ashamed of ourselves, as we look out through that plate glass?

Well, it is within out power to get started on that problem. We are morally impelled to. We all know that, if the human species does solve that one, there will be consequences which are themselves problems. For instance, the population of the world will become embarrassingly large. But that is another challenge. There are going to be challenges to our intelligence and to our moral nature as long as man remains man. After all, a challenge is not, as the word is coming to be used, an excuse for slinking off and doing nothing. A challenge is something to be picked up.

For all these reasons, I believe the world community of scientists has a final responsibility upon it—a greater responsibility than is pressing on any other body of men. I do not pretend to know how they will bear this responsibility. These may be famous last words, but I have an inextinguishable hope. For, as I have said, there is no doubt that the scientific activity is both beautiful and truthful. I cannot prove it, but I believe that, simply because scientists cannot escape their own knowledge, they also won't be able to avoid showing themselves disposed to good.

QUESTIONS FOR CRITICAL THOUGHT

1. What is the purpose of this article?
2. Snow's article was originally written as a speech. Are there any elements in its style that reflect this fact?
3. Why is the world frightened of scientists, according to Snow?
4. Writing in 1961, Snow refers to scientists almost exclusively as 'men'. Is he justified in doing so? How does this usage strike the ear of a reader 45 years later?
5. What makes scientists the most important occupational group in the world, according to Snow? Is it the same quality that impels them to make moral choices?
6. What was going on politically and internationally at the time Snow delivered this speech? To what extent did context inform his ideas? To what extent are they still relevant today?
7. Why, exactly, does Snow reject the 'doctrine of the neutrality of science'? Can you reconstruct his argument?
8. How common is the argument that technology is a neutral tool? Why do so few of the contributors to this volume believe that argument?

9. Snow speaks of the aesthetic satisfaction, even ecstasy, of scientific activity and scientific results. He compares the experience to the experience of art. Does he mean to suggest, then, that science is a kind of art?
10. Why does he lead into his point about morality by talking about aesthetics? What have the two to do with each other?
11. Snow argues that science is a 'self-correcting system' that represents 'moral behaviour on a grand scale'. Explain what this means.
12. In what sense is the search for truth, which Snow says is the key activity of science, an essentially moral activity?
13. Snow laments the international community that he says distinguished the field of physics in the 1920s. What is the connection between community and ethics? He doesn't make this connection explicit, but he implies that it's there. Can you make it explicit?
14. Snow argues that by working on weapons of mass destruction, scientists traded away an essential component of scientific morality. What is this component?

PART VIII

Communication in a Technological Society

WHAT we communicate depends in part on what our technology allows us to communicate, or so argued Canada's celebrated theorist of the electronic age, Marshall McLuhan. McLuhan established that, although its effects are often invisible to us, technology is not neutral. It shapes not only the form of our messages, but also the values and assumptions that drive them, and as a result our entire way of interacting with others. If it is true that we are what we communicate, then to a great degree we are being re-made by the technological advancements that increase the range and frequency of our interactions.

However, as McLuhan warned, every extension of our ability to communicate involves a corresponding loss of some dimension of the interaction: through email, for instance, we can instantly reach the furthest corner of the world and simultaneously communicate with hundreds or even thousands of other people, but at the same time that we connect with so many over such a distance, we lose something else: the interpersonal and nonverbal dimensions that we have relied upon to make our contact with others meaningful.

To observe the profound impact of our technology on the way we interact does not make us anti-technology or opposed to its influence. Instead, to ask questions about how our machines shape our communication is to probe the role and meaning of communication in our professional and personal lives. For example, if we communicate with others primarily through words on a screen, beyond reach of the sound of a human voice, the sight of a human face, or the touch of a human hand, what is the impact on our sense of others as beings worthy of our respect and concern? What happens to human intimacy and connectedness? Are these important features of communication lost to us permanently, or are they being remade and redefined? In either case, what will be the impact of such developments on the way we live our lives?

To study the impact of technology on human interaction is to open the door to a fuller understanding of what our machines mean to us and how they are helping to shape our view of what it means to be human. The essays in this section of the book were selected because each explores some aspect of the impact of communication technology on the way we see ourselves and each other.

CHAPTER 45

Thinking about Technology

George Grant

A computer scientist recently made the following statement about the machines he helps to invent: 'The computer does not impose on us the ways it should be used.' Obviously the statement is made by someone who is aware that computers can be used for purposes of which he does not approve—for example, the tyrannous control of human beings. This is given in the word 'should'. He makes a statement in terms of his intimate knowledge of computers which transcends that intimacy, in that it is more than a description of any given computer or of what is technically common to all such machines. Because he wishes to state something about the possible good or evil purposes for which computers can be used, he expresses, albeit in negative form, what computers are, in a way which is more than their technical description. They are instruments, made by human skill for the purpose of achieving certain human goals. They are neutral instruments in the sense that the morality of the goals for which they are used is determined outside them.

Many people who have never seen a computer, and only slightly understand the capacity of computers, have the sense from their daily life that they are being managed by them, and have perhaps an undifferentiated fear about the potential extent of this management. This man, who knows about the invention and use of these machines, states what they are in order to put our sense of anxiety into a perspective freed from the terrors of such fantasies as the myth of Doctor Frankenstein. His perspective assumes that the machines are instruments, because their capacities have been built into them by human beings, and it is human beings who operate those machines for purposes they have determined. All instruments can obviously be used for bad purposes, and the more complex the capacities of the instrument, the more complex can be its possible bad uses. But if we apprehend these machines for what they are, neutral instruments which we in our freedom are called upon to control, we are better able to come to terms rationally with their potential dangers. The first step in coping with these dangers is to see that they are related to the potential decisions of human beings about how to use computers, not to the inherent capacities of the machines themselves. Indeed the statement about the computer gives the prevalent 'liberal' view of the modem situation which is so rooted in us that it seems to be common sense itself, even rationality itself. We have certain technological capacities; it is up to use to use those capacities for decent human purposes.

Yet despite the seeming common sense of the statement, when we try to think the sentence 'the computer does not impose on us the ways it should be used,' it becomes clear that we are not allowing computers to appear before us for what they are. Indeed the statement (like many similar) obscures for us what computers are. To begin at the surface: the words 'the computer does not impose' are concerned with the capacities of these machines, and these capacities are brought before us as if they existed in abstraction from the events which have made possible their existence. Obviously the machines have been made from a vast variety of materials, consummately fashioned by a vast apparatus of fashioners. Their existence has required generations of sustained effort by chemists, metallurgists, and workers in mines and factories. Beyond these obvious facts, computers have been made within the new science and its mathematics. That science is a particular paradigm of knowledge and, as any paradigm of knowledge, is to be understood as the relation between an aspiration of human thought and the effective conditions for its realization.

It is not my purpose here to describe that paradigm in detail; nor would it be within my ability to show its interrelation with mathematics conceived as algebra. Suffice it to say that what is given in the modern use of the word 'science' is the project of reason to gain ' objective' knowledge. And modern 'reason' is the summoning of anything before a subject and putting it to the question, so that it gives us its reasons for being the way it is as an object. A paradigm of knowledge is not something reserved for scientists and scholars. Anybody who is awake in any part of our educational system knows that this paradigm of knowledge stamps the institutions of that system, their curricula, in their very heart, in what the young are required to know and to be able to do if they are to be called 'qualified'. That paradigm of knowledge is central to our civilizational destiny and has made possible the existence of computers. I mean by 'civilizational destiny' above all the fundamental presuppositions that the majority of human beings inherit in a civilization, and which are so taken for granted as the way things are that they are given an almost absolute status. To describe a destiny is not to judge it. It may indeed be, as many believe, that the development of that paradigm is a great step in the ascent of man, that it is the essence of human liberation, even that its development justifies the human experiment itself. Whatever the truth of these beliefs, the only point here is that without this destiny computers would not exist. And like all destinies, they 'impose'.

What has been said about the computer's existence depending upon the paradigm of knowledge is of course equally true of the earlier machines of industrialism. The Western paradigm of knowledge has not been static, but has been realized in a dynamic unfolding, and one aspect of that realization has been a great extension of what is given in the conception of 'machine'. We all know that computers are machines for the transmitting of information, not the transformation of energy. They require software as well as hardware. They have required the development of mathematics as algebra, and of algebra as almost identical with logic. Their existence has required a fuller realization of the Western paradigm of knowledge beyond its origins, in this context the extension of the conception of machine. It may well be said that where the steel press may be taken as the image of Newtonian physics and mathematics, the computer can be taken as the image of contemporary physics and mathematics. Yet in making that distinction, it must also be said that contemporary science and Newtonian science are equally moments in the realization of the same paradigm.

The phrase 'the computer does not impose' misleads, because it abstracts the computer from the destiny that was required for its making. Common sense may tell us that the computer is an instrument, but it is an instrument from within the destiny which does 'impose' itself upon us, and therefore the computer does impose.

To go further: How are we being asked to take the word 'ways' in the assertion that 'the computer does not impose the ways'? Even if the purposes for which the computer's capacities should be used are determined outside itself, do not these capacities limit the kind of ways for which it can be used? To take a simple example from the modern institutions of learning and training: in most jurisdictions there are cards on which children are assessed as to their 'skills' and 'behaviour', and this information is retained by computers. It may be granted that such information adds little to the homogenizing vision inculcated throughout society by such means as centrally controlled curricula or teacher training. It may also be granted that as computers and their programming become more sophisticated the information stored therein may be able to take more account of differences. Nevertheless, it is clear that the ways that computers can be used for storing and transmitting information can only be ways that increase the tempo of the homogenizing processes. Abstracting facts so that they can be stored as information is achieved by classification, and it is the very nature of any classifying to homogenize. Where classification rules, identities and differences can appear only in its terms. Indeed the word 'information' is itself perfectly attuned to the account of knowledge which is homogenizing in its very nature. 'Information' is about objects, and comes forth as part of that science which summons objects to give us their reasons.

It is not my purpose at this point to discuss the complex issues of good and evil involved in the modern movement towards homogeneity, nor to discuss the good of heterogeneity, which in its most profound past form was an expression of autochthony. Some modern thinkers state that beyond the rootlessness characteristic of the present early stages of technological society, human beings are now called to new ways of being rooted which will have passed through modem rootlessness, and will be able at one and the same time to accept the benefits of modern homogenization while living out a new form of heterogeneity. These statements are not at issue here. Rather my purpose is to point out that the sentence about computers hides the fact that their ways are always homogenizing. Because this is hidden, questioning homogenization is closed down in the sentence.

To illustrate the matter from another aspect of technological development: Canadians wanted the most efficient car for geographic circumstances and social purposes similar to those of the people who first developed the mass-produced automobile. Our desire for and use of such cars has been a central cause of our political and economic integration and our social homogenization with the people of the imperial heartland. This was not only because of the vast corporate structures necessary for building and keeping in motion such automobiles, and the direct and indirect political power of such corporations, but also because any society with such vehicles tends to become like any other society with the same. Seventy-five years ago somebody might have said 'The automobile does not impose on us the ways it should be used,' and who would have quarelled with that? Yet this would have been a deluded representation of the automobile.

Obviously, human beings may still be able to control, by strict administrative measures, the ways that cars are used. They may prevent the pollution of the atmosphere or prevent freeways from destroying central city life. It is to be hoped that cities such as Toronto will maintain themselves as communities by winning popular victories over expressways and airports. Whatever efforts may be made, they will not allow us to represent the automobile to ourselves as a neutral instrument.

Obviously the 'ways' that automobiles and computers can be used are dependent on their being investment-heavy machines which require large institutions for their production. The potential size of such corporations can he imagined in the statement of a reliable economist: if the present growth of IBM is extrapolated, that corporation will in the next 30 years be a larger unit than the economy of any presently constituted national state, including that of its homeland. At the simplest factual level, computers can be built only in societies in which there are large corporations. This will be the case whatever ways these institutions are related to the states in which they are incorporated, be that relation some form of capitalism or some form of socialism. Also those machines have been and will continue to be instruments with effect beyond the confines of particular nation states. They will be the instruments of the imperialism of certain communities towards other communities. They are instruments in the struggle between competing empires, as the present desire of the Soviet Union for American computers illustrates. It might be that 'in the long run of progress', humanity will come to the universal and homogeneous state in which individual empires and nations have disappeared. That in itself would be an even larger corporation. To express the obvious: whatever conceivable political and economic alternatives there may be, computers can only exist in societies in which there are large corporate institutions. The ways they can be used are limited to those situations. In this sense computers are not neutral instruments, but instruments which exclude certain forms community and permit others.

QUESTIONS FOR CRITICAL THOUGHT

1. What is the purpose of this article?
2. For whom was it written? Do you think it was intended for the audience of a book like this one? If not, why might it have been included in this book?
3. Grant's article is an early consideration of the constraints placed upon our choices by the technology we use. Grant suggests that the computer will, like the auto, 'exclude certain forms of community and permit others'. To what extent has his prediction come true with the advent of the Internet?
4. In arguing that our technology does impose some constraints that determine how it might be used and for what, Grant echoes the work of the communication theorist Marshall McLuhan, who argued that the media we use themselves impose meaning. Even before McLuhan, Winston Churchill famously observed that 'We shape our buildings; thereafter they shape us.' Compare Grant's observations about the shaping power of technology with McLuhan's 'Mechanical Bride'. What correspondences do you see?
5. Grant contends, like McLuhan, that technology is not neutral. Can you think of any other technologies that carry their own behavioural or social imperatives? Has the technology available to you made your life different from your grandparents' lives? From your parents' lives? How?
6. Compare Grant's view of technology as a social force with the views articulated in the essays by by Lorinc ('Driven to Distraction'), Talbott ('The Deceiving Virtues of Technology'), McKenzie ('First Flight'), and Kelsey and Pankl ('Verbal Text').
7. How personal is the essay? How formal is it? How do these two variables interact in Grant's work?

CHAPTER 46

Motorcar: The Mechanical Bride

Marshall McLuhan

Here is a news item that captures a good deal of the meaning of the automobile in relation to social life:

> I was terrific. There I was in my white Continental, and I was wearing a pure-silk, pure-white, embroidered cowboy shirt, and black gabardine trousers. Beside me in the car was my jet-black Great Dane imported from Europe, named Dana von Krupp. You just can't do any better than that.

Although it may be true to say that an American is a creature of four wheels, and to point out that American youth attributes much more importance to arriving at driver's-license age than at voting age, it is also true that the car has become an article of dress without which we feel uncertain, unclad, and incomplete in the urban compound. Some observers insist that, as a status symbol, the house has, of late, supplanted the car. If so, this shift from the mobile open road to the manicured roots of suburbia may signify a real change in American orientation. There is a growing uneasiness about the degree to which cars have become the real population of our cities, with a resulting loss of human scale, both in power and in distance. The town planners are plotting ways and means to buy back our cities for the pedestrian from the big transportation interests.

Lynn White tells the story of the stirrup and the heavy-armored knight in his *Medieval Technology and Social Change*. So expensive yet so mandatory was the armored rider for shock combat, that the cooperative feudal system came into existence to pay for his equipment. Renaissance gunpowder and ordnance ended the military role of the knight and returned the city to the pedestrian burgess.

If the motorist is technologically and economically far superior to the armored knight, it may be that electric changes in technology are about to dismount him and return us to the pedestrian scale. 'Going to work' may be only a transitory phase, like 'going shopping'. The grocery interests have long foreseen the possibility of shopping by two-way TV, or video-telephone. William M. Freeman, writing for *The New York Times* Service (Tuesday, 15 October 1963), reports that there will certainly be 'a decided transition from today's distribution vehicles. . . . Mrs Customer will be able to tune in on various stores. Her credit identification will be picked up automatically via television. Items in full and faithful coloring will be viewed. Distance will hold no problem, since by the end of the century the consumer will be able to make direct television connections regardless of how many miles are involved.'

What is wrong with all such prophecies is that they assume a stable framework of fact—in this case, the house and the store—which is usually the first to disappear. The changing relation between customer and shopkeeper is as nothing compared to the changing pattern of work itself, in an age of automation. It is true that going-to and come-from work are almost certain to lose all of their present character. The horse has lost its role in transportation but has made a strong comeback in entertainment. So with the motorcar. Its future does not belong in the area of transportation. Had the infant automotive industry, in 1910, seen fit to call a conference to consider the future of the horse, the discussion would have been concerned to discover new jobs for the horse and new kinds of training to extend the usefulness of the horse. The complete revolution in transportation and in housing and city arrangement would have been ignored. The turn of our economy to making and servicing motorcars, and the devotion of much leisure time to their use on a vast new highway system, would not even have been thought of. In other words, it is the framework itself that changes with new technology, and not just the picture within the frame. Instead of thinking of doing our shopping by television, we should become aware that TV intercom means the end of shopping itself, and the end of work as we know it at present. The same fallacy besets our thinking about TV and education. We think of TV as an incidental aid, whereas in fact it had already transformed the learning process of the young, quite independently of home and school alike.

In the 1930s when millions of comic books were inundating the young with gore, nobody seemed to notice that emotionally the violence of millions of cars in our streets was incomparably more hysterical than anything that could ever be printed. All the rhinos and hippos and elephants in the world, if gathered in one city, could not begin to create the menace and explosive intensity of the hourly and daily experience of the internal-combustion engine. Are people really expected to internalize—live with—all this power and explosive violence, without processing and siphoning it off into some form of fantasy for compensation and balance?

In the silent picture of the 1920s a great many of the sequences involved the motorcar and policemen. Since the film was then accepted as an optical illusion, the cop was the principal reminder of the existence of ground rules in the game of fantasy. As such, he took an endless beating. The motorcars of the 1920s look to our eyes like ingenious contraptions hastily assembled in a tool shop. Their link with the buggy was still strong and clear. Then come the balloon tires, the massive interior, and the bulging fenders. Some people see the big car as a sort of bloated middle age, following the gawky period of the first love-affair between America and the car. But funny as the Viennese analysts have been able to get about the car as sex object, they have at last, in so doing, drawn attention to the fact that, like the bees in the plant world, men have always been the sex organs of the technological world. The car is no more and no less a sex object than the wheel or the hammer. What the motivation researchers have missed entirely is the fact that the American sense of spatial form has changed much since radio, and drastically since TV. It is misleading, though harmless, to try to grasp this change as middle-age reaching out for the sylph Lolita.

Certainly there have been some strenuous slimming programs for the car in recent years. But if one were to ask, 'Will the car last?' or 'Is the motorcar here to stay?' there would be confusion and doubt at once. Strangely, in so progressive an age, when change has become the only constant in our lives, we never ask, 'Is the car here to stay?' The answer, of course, is 'No.' In the electric age, the wheel itself is obsolescent. At the heart of the car industry there are men who know that the car is passing, as certainly as the cuspidor was doomed when the lady typist arrived on the business scene. What arrangements have they made to ease the automobile industry off the centre of the stage? The mere obsolescence of the wheel does not mean its disappearance. It means only that, like penmanship or typography, the wheel will move into a subsidiary role in the culture.

In the middle of the nineteenth century great success was achieved with steam-engined cars on the open road. Only the heavy toll-taxes levied by local road authorities discouraged steam engines on the highways. Pneumatic tires were fitted to a steam car in France in 1887. The American Stanley Steamer began to flourish in 1899. Ford had already built his first car in 1896, and the Ford Motor Company was founded in 1903. It was the electric spark that enabled the gasoline engine to take over from the steam engine. The crossing of electricity, the biological form, with the mechanical form was never to release a greater force.

It is TV that has dealt the heavy blow to the American car. The car and the assembly line had become the ultimate expression of Gutenberg technology; that is, of uniform and repeatable processes applied to all aspects of work and living. TV brought a questioning of all mechanical assumptions about uniformity and standardization, as of all consumer values. TV brought also obsession with depth study and analysis. Motivation research, offering to hook the ad and the *id*, became immediately acceptable to the frantic executive world that felt the same way about the new American tastes as Al Capp did about his 50,000,000 audience when TV struck. Something had happened. America was not the same.

For 40 years the car had been the great leveller of physical space and of social distance as well. The talk about the American car as a status symbol has always overlooked the basic fact that it is the *power* of the motorcar that levels all social differences, and makes the pedestrian a second-class citizen. Many people have observed how the real integrator or leveller of white and Negro in the South was the private car and the truck, not the expression of moral points of view. The simple and obvious fact about the car is that, more than any horse, it is an extension of man that turns the rider into a superman. It is a hot, explosive medium of social communication. And TV, by cooling off the American public tastes and creating new needs for unique wrap-around space, which the European car promptly provided, practically unhorsed the American auto-cavalier. The small European cars reduce him to near-pedestrian status once more. Some people manage to drive them on the sidewalk.

The car did its social levelling by horsepower alone. In turn, the car created highways and resorts that were not only very much alike in all parts of the land, but equally available to all. Since TV, there is naturally frequent complaint about this uniformity of vehicle and vacation scene. As John Keats put it in his attack on the car and the industry in *The Insolent Chariots*, where one automobile can go, all other automobiles do go, and wherever the automobile goes, the automobile version of civilization surely follows. Now this is a TV-oriented sentiment that is not only anti-car and anti-standardization, but anti-Gutenberg, and therefore anti-American as well. Of course, I know that John Keats doesn't *mean* this. He had never thought about media or the way in which Gutenberg created Henry Ford and the assembly line and standardized culture. All he knew was that it was popular to decry the uniform, the standardized, and the hot forms of communication, in general. For that reason, Vance Packard could make hay with *The Hidden Persuaders*. He hooted at the old salesmen and the hot media, just as *MAD* does. Before TV, such gestures would have been meaningless. It wouldn't have paid off. Now, it pays to laugh at the mechanical and the merely standardized. John Keats could question the central glory of classless American society by saying, 'If you've seen one part of America, you've seen it all,' and that the car gave the American the opportunity, not to travel and experience adventure, but 'to make himself more and more common'. Since TV, it has become popular to regard the more and more uniform and repeatable products of industry with the same contempt that a Brahmin like Henry James might have felt for a chamber-pot dynasty in 1890. It is true that automation is about to produce the unique and custom-built at assembly-line speed and cheapness. Automation can manage to bespoke car or coat with less fuss than we ever produced the standardized ones. But the unique product cannot circulate in our market or distribution setups. As a result, we are moving into a most revolutionary period in marketing, as in everything else.

When Europeans used to visit America before the Second War they would say, 'But you have communism here!' What they meant was that we not only *had* standardized goods, but *everybody* had them. Our millionaires not only ate cornflakes and hot dogs, but really thought of themselves as middle-class people. What else? How could a millionaire be anything but 'middle-class' in America unless he had the creative imagination of an artist to make a unique life for himself? Is it strange that Europeans should associate uniformity of environment and commodities with communism? And that Lloyd Warner and his associates, in their studies of American cities, should speak of the American class system in terms of income? The highest income cannot liberate a North American from his 'middle-class' life. The lowest income gives everybody a considerable piece of the same middle-class existence. That is, we really have homogenized our schools and factories and cities and entertainment to a great extent, just because we are literate and do accept the logic of uniformity and homogeneity that is inherent in Gutenberg technology. This logic, which had never been accepted in Europe until very recently, has suddenly been questioned in America, since the tactile mesh of the TV mosaic has begun to permeate the American sensorium. When a popular writer can, with confidence, decry the use of the car for travel as making the driver 'more and more common', the fabric of American life has been questioned.

Only a few years back Cadillac announced its 'El Dorado Brougham' as having anti-dive control, outriggers, pillarless styling, projectile-shaped gull-wing bumpers, outboard exhaust pores, and various other exotic features borrowed from the non-motorcar world. We were invited to associate it with Hawaiian surf riders, with gulls soaring like sixteen-inch shells, and with the boudoir of Madame de Pompadour. Could *MAD* magazine do any better? In the TV age, any of these tales from the Vienna woods, dreamed up by motivational researchers, could be relied upon to be an ideal comic script for *MAD*. The script was always there, in fact, but not until TV was the audience conditioned to enjoy it.

To mistake the car for a status symbol, just because it is asked to be taken as anything but a car, is to mistake the whole meaning of this very late product of the mechanical age that is now yielding its form to electric technology. The car is a superb piece of uniform, standardized mechanism that is of a piece with the Gutenberg technology and literacy which created the first classless society in the world. The car gave to the democratic cavalier his horse and armor and haughty insolence in one package, transmogrifying the knight into a misguided missile. In fact, the American car did not level downward, but upward, toward the aristocratic idea. Enormous increase and distribution of power had also been the equalizing force of literacy and various other forms of mechanization. The willingness to accept the car as a status symbol, restricting its more expansive form to the use of higher executives, is not a mark of the car and mechanical age, but of the electric forces that are now ending this mechanical age of uniformity and standardization, and recreating the norms of status and role.

When the motorcar was new, it exercised the typical mechanical pressure of explosion and separation of functions. It broke up family life, or so it seemed, in the 1920s. It separated work and domicile, as never before. It exploded each city into a dozen suburbs, and then extended many of the forms of urban life along the highways until the open road seemed to become nonstop cities. It created the asphalt jungles, and caused 40,000 square miles of green and pleasant land to be cemented over. With the arrival of plane travel, the motorcar and truck teamed up together to wreck the railways. Today small children plead for a train ride as if it were a stagecoach or horse and cutter: 'Before they're *gone*, Daddy!'

The motorcar ended the countryside and substituted a new landscape in which the car was a sort of steeplechaser. At the same time, the motor destroyed the city as a casual environment in which families could be reared. Streets, and even sidewalks, became too intense a scene for the casual interplay of growing up. As the city filled with mobile strangers, even next-door neighbours became strangers. This is the story of the motorcar, and it has not much longer to run. The tide of taste and tolerance has turned, since TV, to make the hot-car medium increasingly tiresome. Witness the portent of the crosswalk, where the small child has power to stop a cement truck. The same change has rendered the big city unbearable to many who would no more have felt that way ten years ago than they could have enjoyed reading *MAD*.

The continuing power of the car medium to transform the patterns of settlement appears fully in the way in which the new urban kitchen has taken on the same central and multiple social character as the old farm kitchen. The farm kitchen had been the key point of entry to the farmhouse, and had become the social centre, as well. The new suburban home again makes the kitchen the centre and, ideally, is localized for access to and from the car. The car has become the carapace, the protective and aggressive shell, of urban and suburban man. Even before the Volkswagen, observers above street level have often noticed the near-resemblance of cars to shiny-backed insects. In the age of the tactile-oriented skin-diver, this hard shiny carapace is one of the blackest marks against the motorcar. It is for motorized man that the shopping plazas have emerged. They are strange islands that make the pedestrian feel friendless and disembodied. The car bugs him.

The car, in a word, has quite refashioned all of the spaces that unite and separate men, and it will continue to do so for a decade more, by which time the electronic successors to the car will be manifest.

QUESTIONS FOR CRITICAL THOUGHT

1. What is McLuhan's central thesis?
2. This essay was originally written in 1965, and yet its observations seem remarkably contemporary. Why?
3. Gutenberg was the inventor of the printing press, a technology that made the mass production of printed materials possible and inexpensive. What does McLuhan mean by 'Gutenberg technology'?
4. McLuhan argues that 'Gutenberg technology' created the first classless society in the world. Explain what he means.
5. List some of the changes to the way humans live and work that were 'caused' by the advent of the car.
6. McLuhan asserts that 'the car . . . has quite refashioned all of the spaces that unite and separate' people, and predicts that it will eventually be supplanted by its electronic successors. What evidence is there currently that this might be taking place?
7. Is there any evidence of the 'return to the pedestrian scale' that McLuhan suggests could happen as electronic technology displaces the mechanical? You may wish to do a bit of research on recent changes to urban planning theories.

8. McLuhan argues that most people who try to imagine future interactions as shaped by technology fail because they 'assume a stable framework of fact' which will not persist. McLuhan predicts, for example, that the advent of two-way electronic communication would not simply change activities like shopping, but actually eliminate the entire notion of the store as we know it. To what extent has his prediction already come true?
9. McLuhan was one of the first to observe how our interaction is shaped by technology, often without our knowing it. These ideas have been extremely influential, and not only among those who study media. How many of the essays in this collection take for granted this principle that the media we use shape our interactions, our assumptions, and our culture?
10. McLuhan uses several examples to support his argument. What are they? How convincing are they?
11. Does McLuhan's writing have an audible 'voice'? If so, what are its characteristics?
12. How personal or impersonal is McLuhan's style? For whom is the article likely written? It's over forty years old; in what ways is it dated?

CHAPTER 47

Verbal Text: Electronic Communication in the Information Age

Sigrid Kelsey and Elisabeth Pankl

Introduction

A vital tool for the contemporary workplace, email replaces a great deal of communication that formerly transpired orally or through non-electronic, written correspondence. The American Management Association's 2003 *E-Mail Rules, Policies and Practices Survey* found that in the United States, the average employee spends about an hour and forty-five minutes daily using email. Other research has estimated that number to be even higher. A 1999 study reports that in the United States, 2.8 billion email messages, sent by 130 million workers, go out each day (Ciocchetti, 1999: 1). Thus, the effectiveness of email is often in question—not only from a time-management perspective, but also from an interpersonal one. If computer mediated communication (CMC) is the preferred method of communication, both personally and professionally, how are continuity and meaning established for those involved? In fact, the proliferation of CMC in and out of the workplace can be viewed as a very visible climax of our electronic era. This discussion is not a call for a return to the golden days of carbon copies and memos composed on a typewriter, but an investigation of how and why CMC, particularly the cultural phenomenon of email, has both altered our conceptions of written communication and reinforced our early reliance on oral communication.

Historical Background: How Communication Has Repeated Itself

CMC, while providing us with the ability to communicate with anyone, anywhere, and at anytime, has also brought formal communication back to an emphasis on the oral, an emphasis that the West hasn't experienced since ancient times. The earliest beginnings of what we now consider the cradle of Western civilization were located in the oral tradition. The development of our logocentric (or word-centred) culture grew out of an established method of communication that utilized the spoken word. Yet, speaking and writing have also been inseparable throughout history. Since the Sumerians carved Cuneiform symbols into clay about 3500 BC to present day CMC, the spoken and written language have interacted with each other in countless ways. Egyptian hieroglyphics and other forms of early writing used signs to represent objects. The first alphabet, created by the Phoenicians, used symbols to represent sounds rather than pictures or ideas. As a tool to record oral speech in writing, an alphabet presents early evidence of the effect of spoken language on writing. Yet, the West developed into a logocentric culture, so much so that we now assume the superiority of the written word over the spoken. However, the age-old discipline of rhetoric began in the oral tradition. The act of speaking was prized far beyond that of the new technology of writing. In fact, Plato, a philosopher who is still very much at the heart of the university in the West, was very suspicious of the written word. Throughout history, advances in technology have continued to change language and the way people communicate. The printing press first made it possible to distribute the written word to the masses, and later typewriters, telegraphs, and telephones facilitated the ease of written and spoken communication. Computers advanced communication further with word processing, email, the World Wide Web, instant messaging, and other forms of CMC. Both the linguistic content and the sociolinguistic context of CMC have brought spoken and written forms of languages ever closer.

Speaking the Written Word

The social factors of the language, or sociolinguistics, of email more often resemble speech than writing. Traditional features of writing, for example, the characteristic that it is 'planned', are not normally associated with email messages, generally sent with spontaneity. Little thought is given to composing and sending email messages, even in the workplace. Like informal conversation, most messages are sent rapidly with little or no rereading and editing. Each technological evolution of written language, from pen to typewriter to computer, has made it faster, easier, and more economical to edit the written word; thus, the lack of editing common in email messages is ironic, especially considering that many email editors even have spell checking capabilities. Examining the social context of the medium, however, provides an explanation for this paradox. Perhaps the reason email messages are not often edited carefully, despite the ease of doing so, is because people treat writing an email the same as engaging in casual speech. It is quite easy to see how one could make this connection or leap. In fact, studies show that the faster technology allows people to engage in written communication, the closer the writing resembles spoken communication. Rather than a separate form of language, people are likely to treat writing in general as a written conversation, often applying little or no editing (Baron, 1998: 48).

In addition to the fast pace of CMC, the dialogue form that most email correspondence takes presents another sociolinguistic feature of spoken language that other forms of written discourse usually do not exhibit. Email most often is a conversation tool, replacing telephone conversations to some degree in most work places. Email 'threads' refer to strings of messages that can be pieced into a conversation, often unintelligible when taken out of context. It is used to ask a quick question and receive an answer, and like spoken dialog, an email message is expected to be answered much more quickly than a paper correspondence.

The interpersonal social dynamics of email correspondence also bring it closer to speech than writing. The context of most email messages is on an interpersonal level, meant for one person or to be placed in a certain context, most likely irrelevant to others. John Locke, in *The De-Voicing of Society*, contends that it is only the sound of our voice that can transmit 'who we are' (1998: 163) and that a reliance on technologies to communicate encourages us to excommunicate ourselves. Yet, because the majority of professional communication is reliant on technology, users must learn to how to communicate personal thoughts and opinions effectively through the use of CMC. As technologies change, so do notions of identity and habits of interpersonal interaction.

Equally important as the sociolinguistic commonalities between speech and email are the social differences. For example, CMC facilitates the ease with which people can engage in casual conversation with people they would not otherwise have contact with. Communicating to a CEO or company president is almost as easy as clicking 'send', drastically different than the social context of other forms of discourse, both written and spoken. CMC removes the hesitation that might come from interrupting a person with a phone call and reduces the time and thought it takes to send a traditional written communication.

Email lists, commonly referred to as listservs, display another sociolinguistic attribute of CMC not found in traditional speech or writing. Lists enable subscribers to exchange information with others worldwide. Users give little thought to sending out a message to an electronic mailing list with hundreds of subscribers, some or all of whom may be strangers to the sender. These unique characteristics of CMC make communication with a large and wide audience easier than ever before, but the lack of planning associated with what is communicated can create problems. Most people would not wish to speak in front of the same audience with that spontaneity, or send a paper letter to hundreds of recipients with a trite request or improper joke. Effective users of listservs should follow the same guidelines when sending a message to the list.

Understanding the sociolinguistics and linguistics of the medium give email users a basis for effective communication. In writing email, as in all forms of communication, it is not only what is communicated that is important, but how it is communicated. Many people fail to realize that that within CMC there exists a delicate balance of power. And, unfortunately, many of those who do grasp the implications of this cyclical power structure fail to utilize the medium to their benefit, and, often as not, they employ CMC to their detriment. For example, most college-level students would never consider being rude directly to their instructor's or professor's face; however, the impertinence of students via email is legendary among campus faculty. Literature concerning the CMC behaviours of students contains numerous examples of these incidents; everything from outrageous flaming from disgruntled students to merely clueless rudeness has been documented. Not only is flaming inappropriate when addressing an individual who exerts authority over the sender, but the use of CMC to barter for a better grade is another bad choice made by students. IM proves especially dangerous in this area. While most students would know when to stop when lobbying for a better grade in person, when this lobby is conducted over IM facial and other clues of body language are lost on an already thoughtless student. In addition, the urgency of IM encourages students to post and respond at a tempo even faster than standard email communication.

On the opposite end of the spectrum, students who 'gush' to their teachers via email, stemming either from insecurity or calculated attempts to win favour, prove equally annoying. Repeated phrasings of 'thank-you' and 'please' do not necessarily endear the sender to the receiver. Additionally, misusing titles (i.e., addressing someone with a master's degree as Dr or employing Mrs for a single woman) only point to the insensitivity and ignorance of the sender. And, as first names increasingly become non-gendered, for example Taylor and Alexis, from influences of various cultures (including consumer culture), it is absolutely necessary not to assume the gender of the receiver. Gender is no longer—if it ever was—a stable category and one-time clues such as first names should no longer be assumed as signifiers of gender.

Why, some may ask, be so concerned about what one does as a student? Student rudeness to faculty via CMC prepares them to be unfit workplace employees. If a student's 'training' has included rash and highly inappropriate emails, IMs, and blog messages to a supervisor (i.e., instructor or professor) then she has set herself up to continue that behaviour and, unfortunately, reap the rewards which might include loss of employment and/or a negative reputation within a given industry.

The consequences of rash emailing are evident and the durability of an email message is a perhaps its most dangerous feature. Email messages can easily be saved, forwarded, and printed. Email listservs are commonly archived on the Internet and searchable through Internet search engines. Sending a private email to a family member from work is not necessarily safe, either. Employer-issued email accounts have little constitutional protection of privacy, Ciocchetti (1999) reports, advising that employees should be aware of their rights regarding email messages. In fact, email messages are allowable as evidence in most American states. Enron's demise, for example, was caused in part by email evidence. The consequences of the durability and reproducibility of email can be far reaching.

Not surprisingly, in addition to the shared sociolinguistics attributes of email and oral communication, the linguistics of email closely resembles speech. Studies have shown that written language, compared to oral, usually contains shorter texts, longer words, more attributives, and more varied vocabulary. Written language is also found to have more difficult words, but simpler sentences. Spoken language usually has fewer nouns and adjectives, but more verbs and adverbs, and tends to have more repetition and elaborations, more self references, and is found to be more readable and interesting, with simpler vocabulary. Email messages tend to have the linguistic characteristics or spoken, rather than written, communication.

Understanding the linguistic differences between email and spoken communication is an important step to learning to communicate effectively via email. The lack of paralinguistic cues, such as facial expressions and vocal changes in pitch, in CMC represents a major difference in the two forms of conversation. Irony, for example, is more difficult to detect in written communication, especially with someone who is not a close acquaintance. The development of 'emoticons', like smileys, is an interesting linguistic device in CMC that helps make up for, in some part, this lack of cues. Emoticons, however, can be interpreted as unprofessional and immature in a professional setting, and should be used infrequently, if at all, when communicating with professional colleagues, especially superiors.

Despite the implications of a more durable form of conversation, studies have shown that people tend to be less reserved regarding whom they communicate with, and what they communicate, when email is used as the form of communication. While the linguistic structure of email resembles speech, the content of messages and the receivers of those messages differ greatly. People sometimes send an email message out on a list with hundreds of names, with little fore- or afterthought. Additionally, people often send an email message to a person they would normally hesitate to 'bother', in addition to strangers. While the ease of contacting and communicating with people over email can greatly enhance and facilitate communication in a profession, it can present pitfalls as well when abused. These 'warnings' on the uses and misuses of email should not scare users from the useful technology. In fact, it would be unfortunate if users were to resist the development of a unique voice in their presentation of their online selves or ethos (Lenard, 2005).

Misspellings are nonexistent in oral communication, yet represent a consequence of treating written communication as if it were spoken. While bad grammar and mispronunciations happen in casual spoken communication, they are perhaps overlooked more easily as a normal part of the discourse. With email, however, misspellings and bad grammar stand out, because of the differences between the two formats.

Indeed, linguistically and many times sociolinguistically, email is almost synonymous with oral communication. Used correctly, email can advance professional communication, making it more efficient and inexpensive to communicate internally and externally. Email facilitates various forms of professional communication, among them point-to-point, broadcast announcements, internal, external, public, and private. Wireless devices that enable instant messaging anytime, anywhere, broaden the spatial boundaries and time constraints of email even further. But the features of email that facilitate communication can also make it a professional hazard when used without forethought.

Conclusion

Marshall McLuhan, a communications and media theorist, once noted that 'societies have always been shaped more by the nature of the media by which men communicate than by the content of the communication.' CMC is shaping language and communication in society, and a successful communicator will use the beneficial changes brought about by email while attempting to contain the detrimental. Understanding possible implications of using CMC in the workplace is key to successful communication in all professions.

In addition to understanding how evolving forms of media and discourse are shaping professional communication, writing style guides are helpful for writing effectively. Email etiquette and style guides and rules abound, many of them offering similar tips for using email to communicate. Everything from short guides on the Internet to entire books on writing effective email are available to professionals. Besides understanding the broad implications of the email media on communication discussed in this paper, reading some email style guides will provide practical tips for communicating effectively. Common tips from style guides include using proper grammar and spelling, writing concisely, avoiding the use of all uppercase letters (it conveys shouting), avoiding 'urgent', 'please read', and other such tags, writing useful subject headings, and other etiquette and style tips.

Email has brought about a faster paced, more impatient work world. The ability to communicate quickly and inexpensively with coworkers and colleagues across the world is a great advantage to many professions, but it is prudent to use the speed and ease of email wisely. Email memos composed with the thought taken in the days of typewriters and pens will likely result in greater efficiency and more effective communication in the long run.

References

Baron, Naomi S. 1998. 'Writing in the Age of Email: The Impact of Ideology versus Technology', *Visible Language* 32, 1: 35–53.

Lenard, Mary. 2005. 'Dealing with Online Selves: Ethos Issues in Computer-Assisted Teaching and Learning', *Pedagogy: Critical Approaches to Teaching Literature, Language, Composition, and Culture* 5, 1: 77–95.

Locke, John L. 1998. *The De-Voicing of Society*. New York: Simon and Schuster.

QUESTIONS FOR CRITICAL THOUGHT

1. What is the purpose of this essay?
2. Kelsey and Pankl argue that electronic communication, and in particular email, shapes interaction so that 'users must learn how to communicate interpersonal thoughts and opinions effectively' through these new media. Do they therefore support George Grant's contention that media are not neutral instruments?
3. The authors make some pretty large assertions about human interaction that are passed over without much comment, including the following: 'As technology changes, so do notions of identity and habits of interpersonal interaction.' What would you have to believe to accept this comment without objection, as Kelsey and Pankl expect you to do? Unearth as many as you can of the assumptions that underlie this statement.
4. Kelsey and Pankl argue that email communication in some respects represents a return to the conventions of orality. What are these conventions? Are they advantageous to communication or disadvantageous?
5. Why do Kelsey and Pankl specifically target student uses of email that they deem to be inappropriate? What exactly are these practices? Why are they problematic?
6. The authors argue that email communication changes communication practices on a socio-linguistic level, and that users must learn to adapt their communication to these new expectations. Yet they also admonish their readers about using electronic communication carelessly. The standards they advocate are clearly standards derived from other forms of written communication. Does this mean that they contradict their own thesis? Why or why not?
7. Throughout the essay Kelsey and Pankl use the acronym 'CMC'. What is the effect of this usage? Why do they not simply write out the words?
8. How personal is this essay? How formal is it? How clearly does it project a human voice? To help you answer this question, compare it, for example, to Peter Elbow's 'Three Tricky Relationships to an Audience' or David Ingham's 'These Minutes Took 22 Hours'.

9. Kelsey and Pankl give a brief history of communication. What is its purpose? Compare their treatment to that provided by Buehler ('Situational Editing'). What differences do you see? What might account for those differences?
10. For what audience is this passage written? Support your argument with evidence from the essay itself.

CHAPTER 48

The Perils of PowerPoint

Thomas R. McDaniel and Kathryn N. McDaniel

College professors everywhere are incorporating PowerPoint presentations into their classroom lectures. Faculty often pressure their deans to make every classroom a 'smart classroom', and those fuddy-duddy faculty too slow to embrace this quickly-emerging technology are considered Luddites, resisters to change, out of step with modern student expectations. Technology can be a boon to pedagogy, but it is not without its perils. Before jumping headlong into the rushing tide of PowerPoint presentations, consider these cautions and criticisms about this popular teaching tool.

It's Inflexible

When you use PowerPoint to convey information and ideas, it limits not only the content you can convey, but also the pace at which you present. If a student has a question (which the format of PowerPoint discourages anyway), the presenter may lose the flow of the PowerPoint in trying to answer it. If the student's question requires a quick jump ahead to a later point, the presenter will (if the program will allow it) have to scroll through upcoming points to address it. This can lead to confusion and a sense of disorder for both the presenter and the students. If the presenter has included too much information on the slides, students may delay the presentation by insisting that they 'haven't finished' copying everything down. If you find out that your audience has a different level of knowledge than you expected (for example, if they didn't do the reading they were supposed to), the presentation cannot easily be adapted to fit the new situation. What all of these 'ifs' demonstrate is that there's insufficient flexibility in the presentation form to allow for any surprises—those wonderful 'teachable moments' that energize a lesson.

It's Risky

How many times have you seen a PowerPoint presentation where some technical difficulty

a. made it impossible to start the presentation on time?
b. caused the presenter to lose the presentation entirely and end up fumbling halfheartedly through the presentation?
c. made it difficult to change the 'slides', making every transition a long or clumsy process?
d. created a problem with the sequencing of points such that the presenter lost his or her place?
e. all of the above?

Technology is a wonderful thing, but its use also opens up all kinds of possible delays and technical difficulties. The real trouble with PowerPoint technology is that the presenter becomes too dependent on it and often cannot simply abandon the technological 'enhancement' to perform the lesson anyway when technical difficulties arise, as they invariably do.

It's a Crutch

PowerPoint often serves as a crutch that prevents academics from developing real teaching skills. This is particularly a problem for academics who have spent most of their training in relatively isolated activities (researching in labs and libraries and then writing up their research) and who often have introverted tendencies. Instead of having to develop a pedagogy that engages the class at some level, instead of having to learn to communicate ideas to the individuals within the class, the professor can spend hours laying out a PowerPoint presentation that resembles a scholarly publication more than a lesson and that presents the information in a way that stifles communication between teacher and students. This is 'presentation', not teaching.

It's Boring

One of our students talks about PowerPoint classes as a 'Zone-Out Zone'. Not only is it easy for students to zone out during a presentation, it's often actually difficult for them to stay focused and attentive. This occurs for several reasons. First, a PowerPoint presentation seems to signal to students that they will not be necessary for the next 50 minutes or so, that their presence is purely as an audience, and as a result many students automatically disengage even at the very outset. (Having the lights out provides a cue.)

Second, presenters often put all of the salient points of the presentation on the slides, bullet-pointed for clarity. Sometimes they even distribute a handout of the information on each slide. Why does a student need to listen to the presenter read through each of these, even if there is a longer explanation? The pacing seems to slow down painfully; the students never have to figure out for themselves what the key ideas or points are; they have become merely scribes, copying down information. No matter how many 'cool graphics' you have, if they don't relate to the material (and are just 'frills'), students will tune out everything of substance.

It's Style Without Substance

The stylish presentation that PowerPoint offers often occurs at the expense of substance. Instead of spending time researching and studying the content, the presenter spends hours figuring out how to have the bullet points 'fly' in. Examples end up watered down because of technological limitations or the lack of an appropriate graphic. Complex ideas are reduced to bullet points and clever images that don't allow for nuance, multiple perspectives or definitions, or points of contention. Excessive stylish features slow the pace of the lesson and reduce the amount of material that can be conveyed effectively.

Even the best PowerPoint presentation is impressive not because of the insights and ideas conveyed, but because of the skilled use of technology it represents. In thinking about whether or not a PowerPoint presentation was effective, people will often focus on the technologies used, the frills and graphics, the smoothness with which the technology functioned. This is the last thing you want students to be getting out of your lesson—that you, the teacher, are good with technology.

Like Any Tool . . .

While PowerPoint can be a great addition to a teacher's pedagogical repertoire, it is no magic bullet guaranteed to make professors better (and more impressive) teachers. Like any tool, it can be misused or abused, and when that happens teaching effectiveness may be undermined instead of enriched. Effective pedagogy means knowing the benefits of any given teaching tool. Those who know the 'perils of PowerPoint' are most likely to avoid its pitfalls.

QUESTIONS FOR CRITICAL THOUGHT

1. McDaniel and McDaniel offer several 'cautions and criticisms' of PowerPoint as a teaching tool. To what extent does their article confirm the observations of other contributors about the power of technology to shape (or to limit and constrain) our communication?
2. In what sense does PowerPoint or similar technology 'prevent academics from developing real teaching skills'? What are these skills?
3. In what sense does the technology 'signal to students that they will not be necessary for the next 50 minutes or so'? Have you had such an experience in any of your own classes? Why is this so?
4. According to the authors, 'Effective pedagogy means knowing the benefits of any given teaching tool.' What are the benefits of PowerPoint as a teaching tool? Do the McDaniels discuss them? Why or why not?
5. The McDaniels identify several 'perils' that cause PowerPoint-based teaching to fail; do they offer any advice for overcoming these drawbacks? Why or why not?
6. Is the assertion in the first paragraph that 'technology can be a boon to pedagogy' warranted? Is it supported by the discussion? If not, why is it included?
7. How personal is this essay? How formal is its style? How engaging is it?
8. Describe the structure of the passage. How many parts does it have, and how are they arranged? To what extent did the original context (the essay was a contribution to a newsletter on teaching and learning) constrain the shape and content of this article?

CHAPTER 49

Rewind, Pause, Play, Fast-Forward

Ibrahim Khider

In this age of MP3 players, CDs, DVDs, and whatever-else medium is out there, I grow an audiocassette collection. Nobody wants tapes any more; garage sales and public libraries unload them for a song, friends hand over bags full (happy for the freed-up space); co-workers have literally dumped boxes of them on my desk. The result? I now have access to a library of sound that not even a file-sharing site could afford.

But tape is outdated and unlike its predecessor, the vinyl record, it will probably never go through a nostalgic Renaissance—except with eccentrics such as myself. As our culture becomes more predisposed to dispose of them, I become the incidental archivist. Stacks upon stacks of audiocassettes line my shelves, company-recorded and individually compiled. Cassettes are not so much a medium as a mentality about how sound can be experienced.

Part of this is because tapes are too cumbersome to navigate between tracks, one must fast-forward or rewind to favoured moments as opposed to getting to those moments instantaneously, as with a CD. Choosing between digital or analog is where a mentality toward listening comes into play: the reflective 'analog' mindset versus the digital 'instant gratification' one. It is with the former where one can best appreciate, because of its constraints, hearing the content gradually unfurl like flower petals after rain.

The main objection to tape, however, would be the analog hiss found in playback, especially pronounced between tracks or in silent moments of a recording such as classical music. It is in this field of white noise, caused by the movement of magnetic media over playback heads, the listener hears minute details—a field virtually unnoticeable while a song plays.

It is also this field that gives tape an edge over CDs; studio engineers for CD recordings pare down extraneous noises, particularly from analog-source recordings, to ensure maximum background silence, not unlike trimming a lawn right down to the soil—rendering a recording bare. Sound on tape is like a lawn growing lushly, yielding frequencies digital is too stingy to afford. Identical recordings on different mediums (one analog, the other digital) reveal a liveliness of the former that the latter has yet to capture.

Home-made mixed compilation tapes are not unlike being handed handwritten letters (remember those?); the tape compiler arranges songs to convey subtle messages to the recipient. There is only so much space to record, so the compiler must carefully select her/his message. In addition, the compiler (even recording studios) had to record content in 'real time', bestowing an organic quality to the final result.

Then there is the packaging. I enjoy reading (sometimes with amusement) often-handwritten liner notes with heartfelt or snide remarks; I marvel at tape covers made of cut-outs from magazines, photocopies, or hand-drawn art. The cassette itself often bears inscriptions from the compiler, similarly filled with remarks.

The medium of music is important because it is the difference between a lovingly prepared home-cooked meal and something slid off the conveyor belt of an assembly-line restaurant.

Like fast food, MP3s are disposable. It is simple to delete files from an MP3 player, and once it is gone, it has seemingly vanished, without a trace of it ever having existed. The same cannot be said of tape as it is much more cumbersome to erase and record over—and even then vestiges of past material remain.

The products of the digital medium, by contrast, are compiled near-instantaneously, with little thought afforded to track selection because space constraints are not there. Consequently, like a sprawling buffet, it is possible to eventually consume without tasting. Sending a zip file jammed with MP3 selections does not have the personal touch of a mixed tape, which is more like a specialty restaurant. If you send a mixed tape you either have to give it to a recipient or have it delivered—like sending flowers or a box of chocolates.

Computers date fast; software dates faster. With new computers and software continually redesigned and redeveloped, instead of perfecting what already exists, errors (the digital version of decay) pervade.

Consequently, large hunks of data go missing when moving material from the old to the new medium and, as in deletion, it seems there is no evidence of it ever having existed. Because of this approach to technology and the way information is stored, the choice our society faces is what information will be preserved for the future and what will be allowed to decay into oblivion. Ancient stone and parchment scripts seem to have outlasted attempts to capture them on a more contemporary medium, such as a computer.

I have had the eerie experience of listening to my voice on cassette, recorded when I was a child a quarter-century ago. My then high-pitched voice is still crisp, piercing through the sheen of analog hiss. That cassette still has a way to go before it is rendered useless by time. My tapes will probably kick around long after my computer, long after an MP3 player, and probably long after I am gone; they will be destined for a museum or landfill.

Whatever alien civilization excavates our extravagant culture will probably have an easier time playing back tapes (which will probably still work) than figuring out MP3 players (which probably won't).

QUESTIONS FOR CRITICAL THOUGHT

1. What is the purpose of Khider's essay?
2. The essay originally appeared in the *Globe and Mail*; who, then, is his likely audience? In what sense can the article be said to 'select' its audience? Is this true of all writing?
3. The piece is essentially an extended comparison. How, then, is it organized?
4. What is the function of the food analogy that Khider employs throughout the essay? Is it effective? Why?
5. According to Khider, what choice faces society with regard to preserving and retrieving information? Why?
6. How personal is this essay? Is it mostly formal or mostly informal? What are the elements that establish these qualities of footing?
7. How does this essay link to the others in this chapter? In this book?
8. Khider advances his argument in a series of oppositions. What are they? How do they work?
9. Consider Khider's language: 'growing lushly', 'too stingy', 'organic', 'liveliness', 'flower petals after a rain'. What is the effect of such usage on the overall essay?
10. Khider notes that his tape collection will likely outlive him, after which it will be destined to end up in 'a museum or a landfill'. Does the yoking of these two possibilities strike you as odd? Why or why not? Is this statement ironic? Is it intentional? Is it a comment on museums? On landfills? On the society that contains both? How do you know? In what sense are museums and landfills similar?
11. The great Canadian media theorist Marshall McLuhan noted that our increased dependence on technology pressures the culture toward greater instantaneity and

greater simultaneity. In other words, we are pushed to want things immediately, and we are offered multiple stimuli at once. What effect does this 'all-at-onceness' of the culture have, according to Khider? What is lost in the process?

12. What does Khider mean when he says we are driven to 'consume without tasting'?
13. Why is the opposition between 'a home-cooked meal' and 'an assembly-line restaurant' particularly apt?

CHAPTER 50

Driven to Distraction: How Our Multi-channel, Multi-tasking Society is Making it Harder for Us to Think

John Lorinc

In the late 1920s, a Russian-born psychologist named Bluma Zeigarnik found herself sitting in a crowded Viennese coffee house, wondering how the waiters could accurately recall the minute details of numerous orders without committing them to paper. That casual observation proved to be the basis of a paper published in 1927, in which she laid out what came to be known as the Zeigarnik Effect. Her now-famous thesis was that the human mind is better at remembering incomplete tasks, an insight that proved useful to generations of marketers and managers, who devised ways of leveraging our mental response to interruptions and manipulating individuals into buying products or completing tasks more efficiently.

I first learned about the Zeigarnik Effect on a fascinating website titled 'Interruptions in Human-Computer Interaction' (interruptions.net), which gathers the work of a diverse collection of researchers—social scientists, software engineers, psychologists, and neuroscientists—who study what may well be the defining condition of the information age: chronic distraction.

During the past three or four years, quantum leaps in wireless digital technology have brought us to the point where high-powered portable devices permit us to be in constant contact with one another, to access vast storehouses of digitized entertainment, and to plug into the Internet virtually anytime, anywhere. The unveiling earlier this year of Apple's new iPhone anticipates an era dominated by a gadget that effortlessly functions as a cellphone, a personal digital assistant, and a camera; holds hundreds of hours of digital music; streams high-resolution digital video; receives digital satellite radio and maybe even television; and offers full Internet access regardless of the time of day and where the user happens to be.

Even before this all-in-one technology makes its grand debut, we are revelling in the miracle of nearly ubiquitous connectivity. But all this access has not come without a psychological cost that is ultimately rooted in the way our brains function. If we now find ourselves adrift in an ocean of information, our mental state increasingly resembles the slivered surface of a melting glacier. As the dozens of studies at interruptions.net attest, we have created a technological miasma that inundates us with an inexhaustible supply of electronic distractions. Rather than providing necessary interruptions to assist us in focusing on the incomplete task at hand, as Zeigarnik proposed, the deluge of multi-channel signals has produced an array of concentration-related problems, including lost productivity, cognitive overload, and a wearying diminishment in our ability to retain the very information we consume with such voraciousness. It may be that our hyper-connected world has quite simply made it difficult for us to think.

The irony is that one of the fundamental promises of information technology—the radical improvement in the efficiency of our interactions with one another—is being undermined by the technology's enormous capacity to overwhelm us with information and thus short-circuit our need to concentrate. Cognitive psychologists are beginning to understand why the human brain isn't well suited to the sort of communications environment we've built for ourselves. Yet in post-industrial urban societies, few of us are willing or able to disengage, because going offline in a wireless world is no longer an option. This raises a pair of tough questions: Do we control this technology or has it come to control us? And have we arrived at a point, fifteen years after the advent of the web, where we need to rethink our relationship with a technology that may well be altering the way our minds function?

This isn't going to be a Luddite rant. Like many people, I spend much of my working day in front of a computer screen. I have instant access to information I could never have obtained even a decade ago. At the same time, I find myself asking why phrases like 'train of thought' and 'undivided attention' are part of our linguistic geography, and what's become of the underlying mental states they refer to. It often seems as though the sheer glut of data itself has supplanted the kind of focused, reflective attention that might make this information useful in the first place.

The dysfunction of our information environment is an outgrowth of its extraordinary fecundity. Digital communications technology has demonstrated a striking capacity to subdivide our attention into smaller and smaller increments; increasingly, it seems as if the day's work has become a matter of interrupting the interruptions. Email for many people has become an oppressive feature of work life. MySpace, YouTube, chat rooms, and the blogosphere, for all their virtues as new mediums of political debate and cultural activity, have an amazing ability to suck up time. During this decade, executives and the political class became literally addicted to BlackBerrys and these devices are now being taken up by consumers who obsessively check them while waiting for coffee or minding the kids at the playground. Information workers spend their days pursuing multiple projects that involve serpentine email threads, thousands of files, and endless web searches. Paradoxically, the abundance of information begets a craving for even more, a compulsion that results in diminishing returns and is often remarkably undiscerning. Scientists at the Xerox Palo Alto Research Center have labelled this kind of online behaviour as 'information foraging' and coined the pithy term 'informavores' to describe this new species.

The technology, in theory, has the ability to emancipate individuals from tedious minutiae: we no longer need to memorize vast amounts of quotidian information (phone numbers, addresses, trivia of any sort) because a digital version is always retrievable. So, in principle, we should have more mental space to focus on the things that are important to us. And yet, the seductive nature of the technology allows us to sample almost anything and, when addicted to this foraging of bits and bytes, focus on nothing. The resulting cognitive overload has become the occupational hazard for the technorati.

Some years ago, researchers surveyed managers in the United States, the United Kingdom, Australia, Singapore, and Hong Kong. Two-thirds reported stress, tension, and loss of job satisfaction because of cognitive overload. 'Information is relentlessly pushed at us, and no matter how much we get we feel we need more, and of better quality and focus,' wrote David Kirsh, a Toronto-born expert in cognitive science who runs the Interactive Cognition Lab at the University of California, San Diego. When I spoke to him recently, he said his basic concerns about information fragmentation hadn't changed, and the problem may accelerate with the new portable technologies that have turbocharged the data environment.

The mechanics of cognitive overload are similar to the problem of insufficient RAM. 'In most models of working memory and attention, everything has to go through a central executive processor before being passed into long-term memory,' explains Frank Russo, an assistant professor of psychology at Ryerson University. Our built-in CPUs are found in the brain's frontal lobe. These centres need time to 'rehearse' or 'scaffold' incoming information by building the neural circuits on which the data will eventually be stored. 'If it is not rehearsed enough or elaborated upon,' Russo told me, 'the information never makes it to the long-term store.' When someone is bombarded by data, the executive processor doesn't have the time or the resources to encode everything and starts to show signs of fatigue.

While 'memory' is a word that has been appropriated by the information technology world, human and digital memory function very differently. Absent corrupted documents or bugs, the act of saving a file means saving it in its entirety, with an understanding that when it is retrieved the file will be in the state the user left it. Human memory is much more error-prone and subjective. Our memories decay and reshape themselves over time. To appreciate the contrast, imagine if you saved a document and when you reopened it the text contained only the parts that pleased you.

Russo illustrates the point with an experiment he does for his students. He shows them a video of a group of young men and women tossing two basketballs among themselves quite rapidly. The students are asked to count the number of passes. At a certain point in the video, a man in a gorilla suit walks through the frame. After the video, when the students are asked if they noticed anything odd, about a third say they didn't see this absurd disruption. 'I use this experiment to demonstrate the point that perception and memory are not like running a tape,' Russo says. 'We do have selective attention and we miss things, especially if we're very focused on a particular task.' As Jeffery Jones, an assistant professor of psychology at Wilfrid Laurier University, puts it, 'There seems to be a limit to the amount of information we can process at one time.'

It's not just a matter of quantity either. Kathy Sierra is a Boulder, Colorado-based educator who designed and created the bestselling 'Head First' software-development guides, which are based on neuroscience research about cognition and human memory. In developing her approach, she pored over evidence that revealed how the human brain, from an evolutionary point of view, remains a machine programmed primarily to look out for its owner's survival, like the threat of an approaching predator. 'Our brain cares about things that are very different than the conscious mind wants to learn,' she says. It is geared to respond to novel, surprising, or terrifying emotional and sensory stimuli. Her conclusion: the fast-paced, visually arousing hit of video games is intensely captivating for the human brain, whereas the vast amount of text found on websites, blogs, and databases tends to wash over us.

The human mind, well-suited as it is for language, has always adapted to new information technologies such as the printing press and the telephone, so why should the latest generation be any different? It may be partly a matter of the quantity of information at our disposal, and the speed and frequency with which it comes at us. The research on cognitive overload and multi-tasking reveals that our brains are ill equipped to function effectively in an information-saturated digital environment characterized by constant disruptions. While there's much hype about how young people weaned on the Internet and video games develop neural circuits that allow them to concentrate on many tasks at once, the science of interruptions suggests our brains aren't nearly that plastic.

Russo cites epidemiological studies showing that drivers who are talking on a cell phone are four times more likely to be involved in an accident than those who remain focused on the road. Aviation experts have understood this phenomenon for years. A large proportion of plane crashes involving pilot error can be traced to cockpit interruptions and distractions. A 1998 study pointed out that when people are engaged in highly familiar or routine tasks—the things we say we can do in our sleep—they become vulnerable to distraction-related errors because the brain is, essentially, on autopilot and doesn't recover well when it is called on to respond to information that is unpredictable, even casual conversation. 'Cognitive research indicates that people are able to perform two tasks concurrently only in limited circumstances, even if they are skillful in performing each task separately,' concluded a recent NASA study on cockpit distractions. That's why pilots are required to keep banter to a minimum.

Multi-tasking, however, is the signature behaviour of the wired world. We spend our days ricocheting between websites, blogs, our own files, and the various communications devices demanding our attention. Ironically, humans have misappropriated the nomenclature of digital technology to describe this phenomenon. The phrase 'multi-tasking', David Kirsh observes, was invented to describe a computer's capabilities, not a person's.

Yet wireless devices encourage ill-advised multi-tasking: driving and checking BlackBerrys; talking on the phone and reading email; working on two or more complex projects at once. In corporate meetings, participants discreetly text one another or check email while the boss is talking. University classrooms are now filled with students tapping away at their wireless laptops. They may be focused on a document or a website related to the lecture or they may not. Digital technologies invite disruption and pose a daunting challenge to the possibility of a group of individuals applying their collective attention to a particular chore.

Not surprisingly, a growing body of scientific literature has demonstrated that multi-tasking in an office setting is a recipe for lost productivity—a message that runs directly counter to the way many companies want their employees to work. When someone is bouncing between complex tasks, he loses time as the brain is forced to refocus. An American Psychological Association study has found that those 'time costs increased with the complexity of the tasks, so it took significantly longer to switch between more complex tasks.' When multi-tasking, the brain's executive processor performs a two-stage operation: the first is 'goal shifting' (e.g., shifting from editing a text file to checking email), and the second is 'rule activation' (turning off the learned rules for editing on a word processing program and turning on the rules for managing the email program that's being used). According to the APA, Joshua Rubinstein, a psychologist with the US Federal Aviation Administration, determined that 'rule activation itself takes significant amounts of time, several tenths of a second—which can add up when people switch back and forth repeatedly between tasks. Thus, multi-tasking may seem more efficient on the surface, but may actually take more time in the end.'

Uncontrolled interruptions create a similar cognitive response. You're working on your computer and the cell rings, the BlackBerry buzzes, or the incoming email notification pings. Out of a sense of urgency, curiosity, or simply a craving for a distraction from an arduous task, you break away to deal with the interruption, which may be something very simple (a quick cell phone exchange) or something quite complex (a detailed email from a co-worker that's been marked urgent). In other cases, the interruption leads you off on an entirely new tangent and you may not end up returning to the original project for hours. By that point, you have forgotten where you were or you may have closed windows that now need to be found and reactivated. It's like putting a novel down for days and then discovering you need to reread the last chapter in order to figure out what is happening.

When a large British company evaluated the emails sent by its employees, it discovered that almost 30 per cent were unnecessarily copied, irrelevant, or difficult to understand. The annual cost in lost productivity was estimated to be about £3,400 per person, or almost £10 million across the firm. Those numbers don't include the time lost as employees try to get back on task.

The annual Computer–Human Interaction (CHI) conference, held last year at Montreal's cavernous Palais des congress, was in many ways a classic nerdapalooza—hundreds of grad students, post-docs, professors, and software industry types networking and swapping business cards.

The Palais had set up hotspots throughout the building, which virtually guaranteed that the sessions would be a study in multi-tasking and fragmented attention. The presenters plugged their laptops into digital projectors and fired up their PowerPoint presentations, while the conference delegates whipped out their PowerBooks and promptly went online. Almost everyone fiddled with some kind of portable device—laptops, BlackBerrys, Palms, or camera phones. One young woman played with an Etch-A-Sketch key fob.

Interestingly, many of the most popular sessions dealt with finding technological solutions to the daily problems precipitated by the combination of too much communication and too little time. Similar preoccupations have turned up at other information technology conferences in the past year or two, according to Sierra, who attends many techie gatherings. At one conference she went to last year, the dominant topic of discussion had to do with ways of filtering out unwanted information. 'I've never seen that before and I've been going to these tech conferences for fifteen years.'

Some groups debated the failings of recommender systems—software on movie, music, and book sites that purport to provide tips based on user profiles but typically generate unwieldy lists instead. Others deliberated on why it had become so difficult to electronically set up face-to-face meetings in the age of crushingly crowded schedules. At a session about text messaging, a young post-doc presented the results of a study on the costs of interruptions among MSN users.

But it was during an esoteric debate on 'information architecture' that one participant drove right to the heart of the issue that seemed to be on everyone's mind. 'If information is like the sea,' this delegate asked, 'what is seamanship?' The question seemed to me to be about as profound an observation as anything I've come across in all the discussions about the geography of the digital universe. 'We don't talk about "human–wind interactions",' he continued. 'We talk about sailing. We don't talk about "human–saw interactions". We talk about woodworking.'

His point was that we don't have a relationship with a toaster because it is nothing more or less than an object we use to perform a discrete task. But information/communications technology is unlike any other human invention because it performs data processing tasks more adroitly than the human brain. What's more, the wireless advances of the past decade have created portable devices that purport to augment our minds. Whether or not they do, we have become more and more dependent on these fabricated cortexes. We have complex relationships with such gadgets because, increasingly, we can't really function without them. I could get along without a car but I can no longer earn a living without my browser.

When these technologies create an unintended consequence—i.e., a Google search that produces millions of hits that may not be ordered according to the user's needs and is therefore a self-defeating solution to the problem of finding information on the Net—we seek to engineer our way out of the box. At the CHI conference, technical people debated technical solutions to the failings of a techno-culture that throws up too much information and too many distractions. While the participants were clearly preoccupied with this Catch-22, they largely believed that technology must deliver the solutions.

This orientation was glaringly obvious during a seminar entitled 'Because I Carry My Cell Phone Anyway,' by Pam Ludford, a PhD candidate at the University of Minnesota. She developed a prototype of a 'place-based reminder system', dubbed 'PlaceMail'. Every day, she began, Americans spend about two-and-a-half hours doing chores at different places—the mall, the dry cleaners, the supermarket. But, she said, 'people have imperfect practices for managing these tasks'. We make lists, then misplace or forget to check them. By way of a solution to such common imperfections, she has devised a 'location-based reminder system'. In broad strokes, you key your to-do list into a web-interface feature on a cell phone or BlackBerry equipped with a global positioning system chip or other location-sensitive technology. Next, you input the locations of the places where said chores can be accomplished. Then, as you're driving around, the GPS chip will detect if you are close to the supermarket, say, whereupon the phone rings and an electronic message appears, reminding you to pick up eggs and toilet paper.

After the session, Victoria Bellotti, a principal scientist and area manager at the Palo Alto Research Center, told me that such aides may be 'the next big thing'. But she also seemed dubious. For most people, a mnemonic scribble—'Mother's Day' or 'Beth blah blah'—is more than enough of a trigger to retrieve the memory necessary for an intended task, especially if it is stored in some kind of chronological context, such as an appointment book. 'Your brain,' Bellotti said, 'is basically a pattern-matching instrument'. Such memory prostheses may prove to be overkill, she added. But then she quickly noted that she herself has a dreadful memory. 'Maybe we could stop worrying about certain things and focus on other things if we had that prosthetic device on us.' On the other hand, it might simply prove to be yet another dispenser of interruptions that further atomize our capacity to concentrate.

PlaceMail, in fact, is evidence of the feedback mechanism in our over-connected culture. In our relentless drive for more data-friendly wireless communications, we have produced a surfeit of communication and information, the combination of which has a tendency to clog up our schedules, splinter our attention spans, and overwhelm our short-term memories. Given the way our brains actually function, it may turn out that what we need is more time and fewer distractions, even if that means less information.

Not all of us are looking to key our way out of this box. We are now witnessing the emergence of a non-technological response to the symptoms of an accelerated info-culture. In San Francisco, a writer and consultant named Merlin Mann runs a blog called '43 Folders', which is about 'personal productivity, life hacks, and simple ways to make your life a little better'. The popular site has become a focal point of debate about ways to manage the downside of too much digital communication, but from the perspective of users rather than technophobes.

The name of Mann's blog comes from an idea in *Getting Things Done: The Art of Stress-Free Productivity*, the best-selling 2001 time-management guide by David Allen, who has become a guiding light for people in BlackBerry twelve-step recovery programs. A hippie in dress pants, Allen is an Ojai, California-based consultant who was an educator and jack-of-all-trades until the mid-1980s, when he set himself up as a productivity consultant. *Getting Things Done*—in the sturdy tradition of American marketing, he has trademarked the phrase and the acronym *GTD*—offers a smorgasbord of ideas about how to take back your life using a mixture of common sense, mind-clearing techniques, and self-discipline.

He's big on creating paper to-do lists and eliminating the minor sources of frustration that pollute the typical workday. The forty-three folders idea involves setting up a system of forty-three ordinary manila folders in one's office—one for each day of the month, and another dozen for the months—in which you place reminders of tasks that need to be completed and when. One of Allen's premises is that much of the stress associated with an information-saturated workplace is that we end up over-committing ourselves without quite knowing how much we're on the hook for. You have a vague sense of emails that have gone unanswered and interrupted projects dangling in digital limbo, going nowhere but nonetheless giving you grief.

By taking up the ideas in *GTD*, he contends, one can compile 'a complete and current inventory of all your commitments, organized and reviewed in a systematic way [in which] you can focus clearly, view your world from optimal angles, and make trusted choices about what to do (and not do) at any moment'. One of Allen's most popular ideas is the 'Hipster PDA'—a little notebook or a sheath of index cards, which you keep on your person so you can make notations as they come to you, rather than committing them to some digital black hole or, worse, forgetting these fleeting thoughts as other sources of distraction muscle their way into your consciousness. It doesn't get any more low-tech than that.

The growing interest in such 'solutions'—to borrow a favourite techie buzzword—indicates the way portable information technologies have unwittingly created new problems while solving old ones. The BlackBerry that doesn't stop pinging, the tsunami of email, the relentless subdividing and cross-posting of online data—these features of our daily information diet hint that something's gone awry. If we are to establish balance in our relationship with the digital information that envelops us, we must reconsider our understanding of the inner workings of our pre-existing mental machinery and the limits of its capacity to adapt to the electronic environment.

One approach is to recognize the futility of the compulsion to inundate ourselves with information in the hope of meaningfully processing everything that comes over the digital transom. Kathy Sierra says one of the most widely read and copied posts on her blog was a *cri de coeur* in which she confessed that she had stopped trying to keep up with all the technical reading she was supposed to be doing. The post brought an enormous sigh of relief in response from thousands of distracted bloggers who, she says, were grateful to be released from that treadmill of surfing, reading, forgetting, repeating.

The chronic memory loss prompted by such online behaviour is, in fact, the canary in the coalmine. Our information technologies have created an epidemic of engineered forgetfulness—a symptom of the massive quantity of data we attempt to cram into our minds each day. We inevitably fail, yet our social biases about forgetting are thoroughly negative. A great memory is still considered to be a sign of mental acuity while we associate forgetfulness with aging and decline. But, as Sierra points out, a healthy brain actively rejects much of the information we're trying to stuff into it; the brain is designed to be selective. 'There's a lot of chemistry in the brain devoted to stopping us from remembering things,' she told me. 'That means forgetting is pretty darn important. We feel guilty about it. But we should have a great deal of respect for that [mechanism].' Discarding information that is not urgent or relevant is crucial to our ability to think in ways that are efficient and creative.

The point is that we must acknowledge the self-inflicted memory lapses triggered by information overload, chronic interruptions, and relentless electronic multi-tasking. The need to be much more conscious of our information diet, in turn, is a reflection of the imbalance between our technical capacity to record information digitally and our neural capacity to remember it chemically. After

15 years of web access, we haven't really tried to reconcile these unevenly matched features of our mental geography. Moreover, amid all the transformations, we have been devaluing those very neurological capabilities that technology has not been able to mimic, and none more thoroughly than the biological need to concentrate as a way of allowing long-term memory to transform into thought and, when necessary, action.

As a consequence, our perennially distracted Net culture seems programmed to eliminate time for thinking, which is not the same thing as time for finding and saving data. We have unleashed an explosion of digital media but, paradoxically, we have less and less opportunity to digest it, and then to allow all the information to, well, inform. Being online has become a state of being, while going off-line increasingly represents either an act of will or a tiny gesture of rebellion against the status quo. 'We're at a point when we can't be alone,' Wilfrid Laurier psychology professor Jeffery Jones told me as we talked about his research on technology and interruptions. We were sitting in his small office: the husk of an old computer sat on the floor, and there was a new wide-screen terminal on his desk, along with a joystick and his cell phone. Toward the end of the interview, someone knocked, but he ignored it. He said he now knows that if he wants to focus, he must make a point of not picking up the phone or answering his email—even though that failure to connect leaves him feeling vaguely guilty. 'But I've learned,' Jones reflected, 'that you have to have some time when you are unavailable.'

QUESTIONS FOR CRITICAL THOUGHT

1. What, according to Lorinc, is the 'defining condition of the electronic age'?
2. Human beings need to feel some connection with others in order to fulfill their most basic interpersonal needs. Is this need for intimacy being fulfilled by our 'nearly ubiquitous connectivity'? Why or why not? What is the cumulative impact of this phenomenon?
3. Albert Einstein once famously remarked that 'information is not knowledge.' To what extent could Lorinc's essay be considered an extended exploration of, or an answer to, Einstein's observation?
4. Our culture has long prized efficiency as a virtue, but according to Lorinc, there are costs. What are they?
5. Consider the words Lorinc uses to describe the so-called 'informavores': 'addicted', 'obsessive', 'craving', 'compulsive' people who experience high levels of 'stress', 'tension', 'loss of job satisfaction', and 'fatigue'. What kind of mental state do such words suggest? Do you think this impression is deliberate on Lorinc's part?
6. Why does Lorinc use the 'insufficient rain' metaphor? What is its rhetorical effect?
7. What is 'multi-tasking'? Why is it so widely embraced? What does it say about our culture and its values? What are the dangers, according to Lorinc?

8. Lorinc intimates that information technology is fundamentally different in its impact on human experience than other kinds of technology. Why?
9. Marshall McLuhan argued that our technologies offer us the capacity to extend our abilities and our understanding, but he also warned that each technology has its own imperatives, and that each extension of ourselves comes at the cost of some other human capability. What is it that we are losing for the sake of our 'ubiquitous connectivity'?
10. Lorinc quotes a study that found as many as 30 per cent of emails to be 'irrelevant, unnecessarily copied, or difficult to understand'. Does this figure surprise you? Why or why not? What is the cost, in monetary terms, of this proliferation of messages? What is its cost in other terms?
11. What is the meaning of the metaphor 'the canary in the mine'? Lorinc suggests that chronic memory loss is just such an early indicator; what, exactly, is it a harbinger of?
12. How would you characterize Lorinc's style in this essay? Is it a personal exploration, or an impersonal analysis? Is Lorinc's approach mostly formal or mostly informal? How audible is the voice of the essay? What constraints shape the choices that a writer makes?

CHAPTER 51

The Deceiving Virtues of Technology

Stephen L. Talbott

The following is the text of an invited address I gave at the Cognitive Technology 2001 conference at the University of Warwick in England, held 6–9 August.

This morning I would like to take a long view of technology. A very long view. It begins with Odysseus and his beleaguered companions penned up in the cave of Polyphemus, the great, one-eyed, Cyclopean giant, and offspring of Poseidon. Polyphemus had already twice brained a couple of the men by smashing their heads against the earth, then devouring them whole for a day's meal.

Odysseus, of course, was desperate and, as he later told the story, 'I was left there, devising evil in the deep of my heart, if in any way I might take vengeance on him, and Athena grant me glory.' So he hit upon a plan. Finding a huge beam in the cave, he and his companions sharpened it, hardened the point in the fire, and hid it beneath one of the dung heaps littering the place. When Polyphemus returned from pasturing his flocks, and after he had dined on a third pair of the companions, Odysseus offered him a wondrously potent wine the Greeks had brought with them.

The Cyclops drank without reserve, draining three bowls and then falling into a drunken stupor. But before passing out, he asked Odysseus for his name, and the warrior answered, 'Nobody is my name, Nobody do they call me.'

As the giant then lay senseless, dribbling wine and bits of human flesh from his gullet, Odysseus and his comrades heated the end of the beam in the coals of the fire and then, throwing all their weight onto it, thrust it into the eye of Polyphemus. Roaring mightily, the blinded Cyclops extracted the beam from his bloodied eye, groped to remove the huge stone blocking the mouth of the cave, and bellowed his outrage to the other Cyclopes living nearby. But when they came and asked who was causing his distress, his answer that 'Nobody' was the culprit left them perplexed. 'If nobody is tormenting you, then you must be ill. Pray to Poseidon for deliverance.' And so they left him to his troubles.

At this, said Odysseus, 'my heart laughed within me that my name and cunning device had so beguiled' the Cyclops. Danger remained, however. Polyphemus stationed himself at the cave mouth to make sure no man escaped. So again Odysseus devised a plan. He used willow branches to tie his men beneath the bellies of the giant's huge sheep. Polyphemus, feeling only the backs of the sheep as they filed out of the cave to pasture, failed to note the deception.

The escape, it appeared, was made good. But the Greek captain's bravado would yet endanger the lives of all his comrades. As they silently fled to their ship and plied their oars to distance themselves from the frightful abode of the Cyclops, Odysseus was loath to remain an anonymous 'Nobody'. In his pride, he could not resist the temptation to call ashore to Polyphemus, taunting him and naming himself the author of the successful strategem: 'O Cyclops,' he shouted, 'Odysseus, the sacker of cities, blinded thine eye.'

Infuriated, Polyphemus broke off a huge piece of a mountain and hurled it in the direction of the taunt, nearly demolishing the ship. Then he prayed to his father, Poseidon, asking that Odysseus should endure many trials and that all the company, if not Odysseus himself, should perish before arriving home. Poseidon honored the prayer; Odysseus alone, after long wandering and many sufferings, returned to his beloved Ithaca.

Devices of the Mind

Now, jumping ahead to our own day, I'd like you to think for a moment of the various words we use to designate technological products. You will notice that a number of these words have a curious double aspect: they, or their cognate forms, can refer either to external objects we make, or to certain inner activities of the maker. A 'device', for example, can be an objective, invented *thing*, but it can also be some sort of scheming or contriving of the mind, as when a defendant uses every device he can think of to escape the charges against him. The word 'contrivance' shows the same two-sidedness, embracing both mechanical appliances and the carefully devised plans and schemes we concoct in thought. As for 'mechanisms' and 'machines,' we produce them as visible objects out there in the world even as we conceal our own machinations within ourselves. Likewise, an 'artifice' is a manufactured device, or else it is trickery, ingenuity, or inventiveness. 'Craft' can refer to manual dexterity in making things or to a ship or aircraft, but a 'crafty' person is adept at deceiving others.

So we find this interesting link between technological products and inner cleverness. Hardly surprising when you think about it. But we don't in fact think much about it. If we did, we might wake up to some of the distortions in our current relations to technology. To begin with, we might wonder why the element of guile or deceit figures so strongly in the various meanings I've just reviewed.

This odd association between technology and deceit occurs not only in our own language, but even more so in Homer's Greek, where it is much harder to separate the inner and outer meanings, and the deceit often reads like an admired virtue. The Greek *techne,* from which our own word 'technology' derives, meant 'craft, skill, cunning, art, or device'—all referring without discrimination to what *we* would call either an objective construction or a subjective capacity or maneuver.

Techne was what enabled the lame craftsman god, Hephaestus, to trap his wife, Aphrodite, in a promiscuous alliance with warlike Ares. He accomplished the feat by draping over his bed a wondrously forged snare whose invisible bonds were finer than a spider's silken threads. The unsuspecting couple blundered straightway into the trap. As the other gods gathered around the now artless couple so artfully imprisoned, a gale of unquenchable laughter celebrated the guile of Hephaestus. 'Lame though he is is,' they declared, 'he has caught Ares by craft (*techne*).' Here *techne* refers indistinguishably to the blacksmith's sly trickery and his skillful materialization of the trick at his forge.

Likewise, the Greek *mechane*, the source of our 'machine', 'mechanism', and 'machination', designates with equal ease a machine or engine of war, on the one hand, or a contrivance, trick, or cunning wile, on the other. The celebrated ruse of the Trojan Horse was said to be a *mechane*, and it was admired at least as much for the devious and unexpected turn of mind behind its invention as for the considerable achievement of its physical construction.

The Man of Many Devices

We come back, then, to Odysseus, the trickster par excellence, introduced in the first line of the *Odyssey* as 'crafty-shifty'—a man of many turnings, or devices. One of his standard epithets is *polumechanos*—'much-contriving, full of devices, ever-ready'. It was he, in fact, who conceived the Trojan Horse, one of the earliest and most successfully deceitful engines of war. Listen to how Athena compliments Odysseus:

> Only a master thief, a real con artist,
> Could match your tricks—even a god
> Might come up short. You wily bastard,
> You cunning, elusive, habitual liar!
> (Stanley Lombardo, trans.)

These traits, any psychologist will tell you, are closely associated with the birth of the self-conscious individual. The ability to harbour secrets—the discovery and preservation of a private place within oneself where one can concoct schemes, deceive others, contrive plans, and invent devices—is an inescapable part of every child's growing up. The child is at first transparent to those around him, with no distinct boundaries. If he is to stand apart from the world as an individual, he must enter a place of his own, a private place from which he can learn to manipulate the world through his own devisings.

Granted, such manipulative powers may be exercised for ill as well as good, and the Greeks sometimes appear to us remarkably casual about the distinction. But, in any case, the gaining of such multivalent power is inseparable from growing up; to give people greater capacity for good is also to give them greater capacity for evil. In what follows, it *is* the conscious *capacity* that I will speak of as having been necessary for our development, not its employment in a negative or destructive manner.

What I want to suggest is that, to begin with, technology was a prime instrument for the historical birth of the individual self. And the *Odyssey* is almost a kind of technical manual for this birth—for the coming home, the coming to himself, of the individual. When you realize this, you begin to appreciate how the 'My name is Nobody' story, which seems so childish and implausible to us, might have entranced Homer's audiences through one telling after another. You can imagine them wondering at Odysseus's presence of mind, his self-possession, his ability to wrest for himself a private, inner vantage point, which he could then shift at will in order to conceal his intentions from others—something no one lacking a well-developed ego, or self, can pull off. And they doubtless wondered also at his self-control, as when he refused his immediate warrior's impulse to respond in kind to the Cyclops' aggressions—an impulse that would have proven disastrous. Instead he pulled back, stood apart within himself, and devised a trick. In re-living Odysseus's machinations, the hearers were invited into that place within themselves where they, too, might discover the possibilities of invention and craft. It requires a separate, individual self to calculate a deceit.

The classicist, George Dimock, has remarked that Homer makes us feel Odysseus's yearning for home as 'a yearning for definition'. The episode with Polyphemus is symbolic of the entire journey. In the dark, womb-like cave, Odysseus is as yet Nobody. Horner intimates childbirth by speaking of Polyphemus 'travailing with pains' as his captive is about to escape the cave. Only upon being delivered into freedom, as we have seen, can Odysseus declare who he is, proclaiming his true name (Dimock, 1990: 15, 111). Further, every birth of the new entails a loss—a destruction of the old—and the thrusting of the sharpened beam into the great Cyclopean eye suggests the power of the focused, penetrating, individual intellect in overcoming an older, perhaps more innocent vision of the world (Holdrege, 2001).

To grow up is to explore a wider world, and Dimock points out that, first and last, Odysseus 'got into trouble with Polyphemus because he showed nautical enterprise and the spirit of discovery'—not because of recklessness or impiety. 'In Homer's world, not to sail the sea is finally unthinkable.' Perhaps we could say, at great risk of shallowness: in those days, to set sail was to embark upon the information highway. There were risks, but they were risks essential to human development.

Homer certainly does not downplay the risks. Having been warned of the fatally entrancing song of the Sirens, Odysseus plugged his sailors' ears with wax, but not his own. Instead, he had the others lash him to the ship's mast, sternly instructing them not to loose him no matter how violent his begging. And so he heard those ravishing voices calling him to destruction. His desire was inflamed, and he pleaded for release, but his men only bound him tighter.

You may wonder what the Sirens offered so irresistibly. It was to celebrate in song the great sufferings and achievements of Odysseus and his followers, and to bestow upon them what we might be tempted to call the 'gift of global information'. In the Sirens' own words:

> Never yet has any man rowed past this isle in his black ship
> until he has heard the sweet voice from our lips.
> Nay, he has joy of it and goes his way a wiser man.
> For we know all the toils that in wide Troy
> the Argives and Trojans endured through the will of the gods,
> and we know all things that come to pass upon the fruitful earth.
> (xii:186-91, A.T. Murray, trans.)

'We know all things.' The rotting bones of those who had heeded this overpowering invitation to universal knowledge lay in heaps upon the shores of the isle of the Sirens. Only the well-calculated balance of Odysseus's *techne*—only the developing self-awareness with which he countered the excessive and deceitful offer—enabled him to survive the temptation. As Dimock observes about Odysseus lashed to the mast:

> Could a more powerful example of the resisted impulse be imagined? . . . Odysseus has chosen to feel the temptation and be thwarted rather than not to feel it at all.

Here we see the perfect balance between the open-hearted embrace of life with all its challenges, and artful resistance to the ambitions of hubris. The temptation of knowledge leads only to those rotting bones unless it is countered by the kind of self-possession that enables us to resist our own impulses. The external gifts of *techne* come, in the end, only through the strengthening of the *techne* of our own consciousness. When you look today at the mesmerized gaze of web surfers as they hypnotically respond to the sweetly sung promises of online information and glory, you realize that our own culture honours the Sirens far more than it does the healthy respect for risk, the self-discipline, and the inner cunning of Odysseus, man of many devices.

Balance and Separation

If my first point, then, was that technology can serve as midwife to the birth of the individual, the second is that this midwifery requires a well-calculated balance between the challenges we take on and our self possession, our wide-awake, conscious resourcefulness. This sensible calculation is part of what it means to be grown up, notwithstanding the widespread, if impossibly foolish, notion today that whatever *can* be attempted *ought* to be attempted.

There's a third point here. The Cyclopes, unlike Odysseus, lived in a kind of state of nature, and they spurned all advanced technologies. Never faring upon the open sea, they refused voyages of discovery. Odysseus describes them this way:

> To the land of the Cyclopes, violent, innocent
> of laws,
> we came; leaving it all to the gods
> they put hand to no planting or plowing;
> their food grows unsown and uncultivated,
> wheat, barley, vines which produce
> grapes for their wine; Zeus' rain makes it grow
> for them.
>
> For the Cyclopes have no red-cheeked ships,
> no craftsmen among them, who could build
> ships with their rowing benches, all that is
> needful
> to reach the towns of the rest of the world as
> is common—
> that men cross the sea in their ships to meet
> one another;
> craftsmen would have built them handsome
> buildings as well.
> (ix.106–30, George Dimock, trans.)

If 'nature is good to the Cyclopes', observes Dimock, it is 'not because they are virtuous. Rather, the kindness of nature has deprived them of the stimulus to develop human institutions.' To venture out—to separate themselves from the womb of nature—would have brought risk and pain, but it could also have brought self-development. Technology, I would add, is an instrument, a kind of lever, for this necessary detachment of the individual self from a nurturing surround that otherwise can become stifling, as when an infant remains too long in the womb.

My third point, then, is this: technology assists the birth of the individual in part by separating him from the natural world. To begin with, this separation, this loss of paradise, reconstitutes the world as an alien, threatening place, continually encroaching upon the safe habitations fortified by human *techne*.

Reversals

But here things begin to get interesting, for if you look at technology and society today only through the lens of my argument so far, you will be badly misled. After all, nearly three millennia—most of recorded history—lies between Homer's day and our own. Things have changed. What we see, in fact, almost looks like a reversal.

There is, to begin with, the 'inversion' of nature and culture that philosopher Albert Borgmann talks about. Early technological man carved out his civilized enclosures as hard-won, vulnerable enclaves, protected places within an enveloping wilderness full of ravening beasts and natural catastrophes. We, by contrast, live within a thoroughly technologized and domesticated landscape where it is the remaining enclaves of wildness that appear painfully delicate and vulnerable (Borgmann, 1984: 190 ff). Today, if we would set bounds to the wild and lawless, it is the ravening beast of technology we must restrain. If nature still threatens us, the threat is that it will finally and disastrously succumb to our aggressions.

A second reversal is closely related to this. You will recall that the *Odyssey* opens with its shipwrecked hero on the isle of Ogygia, where the beautiful goddess, Kalypso, has kept him as her consort for seven years while urging him to marry her. She would have made him a god and given him a good life, free of care. The name 'Kalypso', of course, means 'the Concealer', and her offer of an endless paradise would in effect have kept Odysseus unborn and nameless, concealed within an immortal cocoon. But he chose instead to pursue the painful path to his own home so as to realize his mortal destiny as a man.

The contrivings and devisings of *techne*, as we have seen, served Odysseus well in his striving toward self-realization and escape from anonymity. But now note the reversal: as Neil Postman has famously elaborated in *Amusing Ourselves to Death* and other works, today it is technology that cocoons us and promises us endless entertainment, distraction, and freedom from cares. I'm sure I don't need to elaborate this point for you. Just watch the advertisements on television for half an hour.

I remarked earlier that when Odysseus set sail on his perilous journey over the high seas, he was, in a sense, embarking upon the information highway of his day. But I added that the comparison might be a shallow one. Why shallow?

Well, look at the differences. Odysseus's journey was a continual risking of life and happiness. It was a journey of horrific loss as well as gain, so that preventing the ultimate loss required every ounce of strength, every bit of cunning he could muster, every crafty art he could set against the temptation to abandon his mission and therefore also himself. He wrestled not only with the foolishness of his companions and the armed might of his opponents, but also with the enticements and hostilities of the gods and the despair of the shades in Hades. Faced with the Sirens' promise of boundless knowledge, he could not simply lean back and choose among the knowledge-management systems offered by *high-techne* solution providers. Any lapse of will or attention on his part, any succumbing to temptation, would have been fatal.

When, by contrast, I venture onto the information highway today, I put almost nothing of myself on the line. I know, we hear much talk about transformation—about the coming Great Singularity, the Omega Point, the emergence of a new global consciousness. But, to judge from this talk, we need only wire things up and the transformation will occur—automatically. Complexity theorist Ralph Abraham says that 'when you increase the connectivity, new intelligence emerges.' Our hope, he adds, is for 'a global increase in the collective intelligence of the human species . . . a quantum leap in consciousness'. And computer designer Danny Hillis tells us that 'now evolution takes place in microseconds. We're taking off . . . . There's something coming after us, and I imagine it is something wonderful.'

Call this, if you will, 'Evolution for Dummies' or 'Plug-and-Play Evolution'—just add connections and—presto!—a quantum leap in consciousness. What easy excitements we revel in! But our excitement is not for the potentials of our own growth; what we anticipate, rather, is our sudden rapture by the god of technology. No blood and sweat for us, no inner work, no nearly hopeless perils of the hero's quest. If, through our own folly, we face the end of the natural world, no problem: we will be spared the Tribulation because technology, in a singular salutation, will translate us into altogether new and better conditions of life.

Victory of the Contrivance

Personally, I see a rather different promise in all the machinery of the digital age. The *techne* we invest in outward machinery always gains its character and meaning from the *techne* of our inner devisings. What we objectify in the hard stuff of the world must, after all, first be *conceived*. Look at the technologies heralded by people like Abraham and Hillis, and you will notice that the conceiving has a distinctive and limited character. We have invested only certain automatic, mechanical, and computational aspects of our intelligence in the equipment of the digital age, and it is these aspects of ourselves that are in turn reinforced by the external apparatus. In other words, you see here what engineers will insist on calling a 'positive feedback loop', a loop almost guaranteeing one-sidedness in our intelligent functioning. This one-sidedness is nicely pictured in the lameness of Hephaestus, the craftsman god.

You can see, then, why it is not really such a great paradox to say, as I have often told audiences, 'technology is our hope if we can accept it as our enemy, but as our friend, it will destroy us.' *Of course* technology threatens us, and *of course* it calls for a certain resistance on our part, since it expresses our dominant tendencies, our prevailing lameness or one-sidedness. The only way we can become entire, whole, and healthy is to struggle against whatever reinforces our existing imbalance. Our primary task is to discover the potentials within ourselves that are *not* merely mechanical, *not* merely automatic, *not* reducible to computation. And the machine is a gift to us precisely because the peril in its siding with our one-sidedness forces us to strengthen the opposite side—at least it does if we recognize the peril and accept its challenge.

Unfortunately, there does not seem to be much recognition yet. In fact, in many quarters there is nothing but an exhilarated embrace of one-sidedness. Where, for the Greeks, *techne* always had two complementary but never completely separable aspects—the increasingly self-aware inner originating and the outer result—our technology has become so much gadgetry and wiring and abstract protocols and transistors in one physical state or another. We have forgotten the crafty inner origin and essence of the *techne* that once served our ancestors so well.

And so we reconceive the interior space within which Odysseus hatched his plots and secured his name, telling ourselves that it is merely filled with mindless brain mechanisms, more gadgets, not coincidentally, exactly like the external ones we are so adept at making. In other words, the *techne* that devises is being co-opted by its own devices. Odysseus was on his way to being a true contriver; we seem content to be mere contrivances.

Compare Homer's man of many devices with Silicon Valley's man of many gadgets, and you will immediately recognize a reversal of emphasis within *techne*. Where the individual's consciousness of self once became more vivid through the experience of his own capacity to objectify his inner contrivings in the outer world, today the objects as such have engulfed us, threatening the originating self with oblivion.

Rousing Ourselves

All this suggests to me that if we are to escape the smothering technological cocoon, our *techne* today must, in a sense, be directed against itself. Our trickery must be aimed at overcoming the constraints imposed by our previous tricks. What we must outwit is our own glib, technical wit.

Or, putting it a little differently: we are engaged in a continual conversation between what you might call the frozen *techne* already embodied in the vast array of our external devices, and the conscious, living *techne* we can summon from within ourselves in the current moment. It is always disastrous for the future of the self when we abdicate the living half of this conversation, as when we yield ourselves uncritically to what we consider the purely objective promise of technology.

In Odysseus's day, *techne* was a conscious resourcefulness that had scarcely begun to project itself into the material apparatus of life. What apparatus existed was an enticement for further creative expression of the nascent human self. While the technology of the Greeks may seem hopelessly primitive to us, it is worth remembering that the balanced awakening heralded by Homer culminated in a flowering of thought and art that many believe has never been surpassed for profundity or beauty anywhere on earth.

Today, that balance seems a thing of the past. The powers of our minds crystallize almost immediately, and before we are aware of them, into gadgetry, without any mediating, self-possessed reflection, so that we live within a kind of crystal palace that is sometimes hard to distinguish from a prison. The question is no longer whether we can use the enticement of clever devices as a means to summon the energies of dawning selfhood; rather, it is whether we can preserve what live energies we once had, in the face of the deadening effect of the now inert cleverness bound into the ubiquitous external machinery of our existence.

This machinery, this inert cleverness, is the greatest threat to our future. We require all our highest powers of contriving to overcome our contrivances. In the end, the contriving—not the contrivance—is the only thing that counts. There is a law of human development traditionally stated this way:

> Whoever has, to him shall be given, and he shall have an abundance; but whoever does not have, even what he has shall be taken away from him.
>
> (Matt. 13:12)

It is a hard saying because it makes no sense in regard to our external possessions, where it would be pure injustice. But when you realize that it is a natural law of our inner life, the meaning becomes clear: we either grow and develop, reaping inner riches upon inner riches, or else we lose whatever we started with. For the self is a conscious power of originating; there are no external gains for the self, and there *is* no remaining in one place. We cannot *be* static selves; the only life available to us consists of self-realizings or self-abdicatings. The image of the semi-comatose, automatically responding figure in front of a screen *is* the image of the self extinguishing itself—and in some ways I suppose it recalls the image of Nobody in the dark cave of the one-eyed Cyclops. Odysseus managed to rouse himself. Our own choice is not yet clear.

Reckoning with the Scoundrel

Before proceeding to a conclusion, I would like to make one matter fully explicit. To admire Odysseus for his self-arousal is not to deny that he was, in many ways and by our lights, a scoundrel. On their way home after the fall of Troy, he and his men sacked the city of Ismarus simply because it was there. Likewise, as Helen Luke reminds us, they came to the land of the Cyclopes seeking plunder, so it is hard to blame Polyphemus for responding in the same spirit. The Cyclopes themselves were a pastoral folk who kept peaceably to themselves, and the crude Polyphemus was able to speak quite tenderly to his sheep (Luke, 1987: 13–15; but compare Dimock, 1990: 110–15).

Nothing requires us to repress our own judgments about Odysseus's behaviour. But it is always problematic when such judgments are not tempered by a sense for historical and individual development. None of us would like to be judged solely by what we have been, as opposed to what we are becoming. And all human becoming is marked by certain tragic necessities, partly reflecting the progress of the race to date.

This is clear enough when we look at the developing child. 'Blessed are the little children'—profoundly true, for they have a wonderful openness to everything that is noble, beautiful, healing. But children have also been characterized as beastly little devils, casually inflicting horrible pain upon each other. This, too, has its truth. The point is that neither judgment makes a lot of sense when taken in the way we would assess the well-developed character of a fully self-conscious adult. The child is only on the way to becoming an adult self, and much of what we see in his early years is less the expression of the individual to come than it is the raw material—both noble and diabolical—from which the individual must eventually shape himself.

History Matters

In light of all I have *said*, perhaps you will not be surprised when I make a fourth and final point, concerning history. Today the computer gives us the reigning image of the human mind, and it seems to me highly curious that those researchers aiming to formalize this image and make it more rigorous almost completely ignore the history of what they are trying to understand. One hundred and fifty years after Darwin, when we have learned to explore almost everything from bacteria to galaxies in a developmental context, how can we blithely set about explaining the human mind—and even trying to implement it in software—without having made the slightest effort to see what it is in historical terms? Can we expect to be any more successful than those biologists who sought to understand what a species was without any sense for biological evolution (Barfield, 1981)?

Look at it this way. When we try to create an artificial mind, are we trying to program an Odysseus or a Danny Hillis? It makes a difference!—and, if I may say so, it is vastly easier to capture aspects of Hillis's intelligence in a computer than it would be to capture much of Odysseus's intelligence. We have, after all, spent the last several hundred years learning to think computationally, to formulate and obey rules, to crystallize our thoughts into evident structures of logic. It was on this path that we felt compelled to develop computers in the first place, and it is hardly surprising that these computers turn out to be well designed for representing the kind of Hillisian thinking embodied in their design. But to glory in this fact as if it were the solution of age-old puzzles about the mind—well, as many have recognized in recent years, this is a bit premature.

History can help us to counter our preoccupation with external devices. When Odysseus's heart laughed within him at the success of his cunning device in beguiling the Cyclops, he was rejoicing first of all in the developing awareness of his inner capacities as a centred and conscious self. He reveled in his devices because they arose from an intensifying experience of *his* own powers, not because he saw in them a wholly independent promise. Our crisis today is a crisis of conviction about the primacy of our conscious powers of devising. What Odysseus was gaining, we are at risk of giving up. The evidence of our self-doubt is on every hand:

- Media gurus such as Ray Kurzweil and Hans Moravec are telling us flat out that our devices will soon render us obsolete by taking over all devising for us.
- The discipline of cognitive science, compulsively outward-oriented as it is toward devices rather than toward the self-aware potentials of the deviser, has all but declared the problem of consciousness to be intractable. It seems easier to some simply to deny consciousness as a significant fact—or, rather, as the stage upon which alone significant facts can manifest themselves—than to accept that we are more than devices.
- Futurologists lead us in an orgy of prognostication about what sort of life our gadgets will bring us, instead of facilitating a societal conversation about what sort of future we might want to *choose*. The human being as devising agent vanishes from the discussion.
- The corporations driving our future run more and more like machines merely calculating the bottom line. Especially in high-tech, the next generation product is cranked out based purely on technical and market feasibility. Employees cannot comfortably

ask the one question most urgently dictated by responsible selfhood: Toward what end are we making this product?

- International capital flows are becoming mere data flows, so far as our participation in them is concerned. The automatisms governing these flows—which are the flip side of our abdication of selfhood and responsibility—leave us little room to concern ourselves with the concrete effects of our capital upon the world's communities.

In sum, what the global picture reveals is a radical displacement of the devising self by its own devices—not because of any necessity, but because the devising self has hesitated, become unsure of itself. And at this moment of crisis, the Cyclopes in their might and the Sirens with their enticements confront us from everyone-eyed screen, every newspaper, magazine, and billboard, every mechanism for social transaction, persuading us that we are powerless to affect the technological future and inviting us to dull the pain of consciousness and responsibility by partaking of the delights and wonders that await us. We are, in other words, being asked to become Nobodies again. But the invitation toward self-dissolution is at the same time an opportunity to seize ourselves at a higher level than ever before. Everything depends upon our response. In contemplating our choices, it would not be a bad idea to look back to the Greeks and to Odysseus, man of many devices, for some wily insight into our current predicament.

References

Barfield, Owen. 1981. 'The Nature of Meaning', *Seven* 2: 32–43.

Borgmann, Albert. 1984. *Technology and the Character of Contemporary Life: A Philosophical Inquiry*. Chicago: University of Chicago Press.

Dimock, George E. 1990. *The Unity of the Odyssey*. Amherst, MA: University of Massachusetts Press.

Holdrege, Craig. 2001. Personal communication.

Luke, Helen M. 1987. *Old Age: Journey into Simplicity*. New York: Parabola Books.

QUESTIONS FOR CRITICAL THOUGHT

1. The essay is about the effects of contemporary technology. Why, then, does Talbott begin with the reference to the classical legend of Odysseus?
2. In defining the word *techne*, Talbott provides a history of its meaning. What is the purpose of an etymological definition? What does it contribute to his argument?
3. What meanings, or implications, does Talbott tease out of the etymology of words like device, machine, and technique? How important are these to his argument?
4. 'To give people a greater capacity for good is also to give them a greater capacity for evil.' What does this mean? What does it mean when applied to technology?
5. As part of his celebrated definition of what it means to be human, the great rhetorician and communication theorist Kenneth Burke argued that we are separated from ourselves 'by instruments of [our] own making'. Talbott makes a similar argument.

What role does technology play in creating (or compromising) our sense of individual selves?

6. In what sense is this an essay about technology? In what sense is it about language? About ethics? About human identity? Should it have been placed in another section of this book? Why or why not?
7. The essay is one of the lengthiest in this book. Why does it take Talbott so long to make his argument?
8. What is the connection between technology and the self?
9. In his essay 'The Moral Un-neutrality of Science', Charles Snow argues that scientific and technological advances carry their own moral imperatives. How does Talbott see the same issue?
10. Outline the development of Talbott's argument. How is it structured? How many phases are there? In what order does he arrange his points? What kind of evidence does he provide?
11. What does Talbott mean when he says 'technology is our hope if we can accept it as our enemy, but as our friend, it will destroy us'?
12. What is the tragedy in the loss of the 'human being as devising agent'? What is the reason for this loss?
13. What does Talbott's argument have in common with that offered by John Lorinc in 'Driven to Distraction'? Why are the two articles so different?
14. The article originated as an oral address. What features does it still retain of its origins?
15. What kind of background knowledge do you need to make sense of Talbott's argument? Does he provide sufficient context?
16. How personal is Talbott's article? How formal is it? Why might Talbott have made the rhetorical choices he did?

PART IX

Trouble in the Office: A Communication Case Study

As the essays in this book have established, communication in a professional setting is far from being as straightforward as most university texts suggest. Instead, it is complicated by ethical, political, social, cultural, and interpersonal dynamics.

Understanding the motivations that drive us is a complex business. We are admonished by nearly every one of the contributors to this book that we must become adept at reading situations, at discerning human motives, at accommodating the needs and expectations of our various audiences. But recognizing motives and understanding the dynamics of interaction is difficult and often confounding. Human communication, in the workplace or elsewhere, is filled with risk and interpretation, with judgement and choice, with ethical responsibility and self-interest. It involves the management of power, status, security, and authority. Each of us makes choices that reflect any number of combinations of these issues, and we do the best we can in a sometimes confusing and frequently challenging world. Because we must rely on our own perceptions and judgements, we can sometimes be profoundly mistaken.

This chapter offers a complex case study intended to demonstrate the challenges of making reliable assessments of others based on what we observe in the workplace. This collection is the result of an assignment given to students in an MEng degree programme, as part of their professional communication course. The case study depicts the behaviour of one Bob Eaglestone, hired on a probationary contract at his new company. Eaglestone's actions have been causing his co-workers some distress, and the seminar participants were asked to analyze the scenario using the theoretical models and insights provided in the course.

Just as might happen in real life, Paul Zepf and Joe Azzopardi present very different readings of the same set of events. Both men read the scenario in light of some course material about so-called 'office psychopaths', those 'difficult' personalities who are all too frequently encountered in the workplace.

Since the scenario is largely fictional, neither of these writers is absolutely 'correct' in his interpretation; instead, their analyses reveal important insights into the way in which we interpret and assess the behaviour of others, and show how what appear to be reasonable judgements could actually be mistaken. Their responses also demonstrate how labels like 'psychopath' can be potentially damaging when wielded so casually by those who are not experts in making such a diagnosis or whose understanding of the theory may be imperfect. No matter how sure we may be about how to interpret what we have witnessed, we need to recognize that our judgements of others based on what we see in a specific context must always be provisional.

CHAPTER 52

Trouble in the Office: The Case of Bob Eaglestone

Jennifer M. MacLennan

Probationary contracts in the firm you work for are 12 months in duration, and are not automatically extended. Typically, when the first contract expires at the end of the 12-month period, promising candidates may apply for a second probationary contract. Those who successfully complete the second contract may then be hired into a permanent position.

Bob Eaglestone is completing his first probationary contract in your small department. Although his resume was impressive and his interview was strong, his on-the-job performance seems to have raised the ire of others in the department. Though he has never overtly said so, he seems to consider routine tasks to be beneath him, and he has developed a reputation of pushing off tasks onto secretarial staff, administrative assistants, and interns, who have begun to resent 'taking up the slack' for him. Worst of all, Eaglestone seems eager to take credit when things go well, but he is perceived as slow to pitch in when something needs doing.

On more than one occasion, Eaglestone has even failed to follow through on critical tasks, calling in sick at the last minute or claiming that he's too overloaded with something more important (some of his more cynical co-workers have suggested that 'more important' usually just means something that he perceives to be higher profile or more prestigious). He talks a lot around the lunchroom and water cooler about his prowess as an engineer and about the big money he made in his previous positions, but his accomplishments in this office seem modest by comparison to these reports. People have begun to view him as something of a blowhard, and they're tired of his failure to shoulder what they view as his share of the work; in fact, some of your best people have begun to avoid working with him. Everyone assumes he will not be offered a second contract when the current one expires.

Today is the monthly department meeting. Because the department is small, meetings have always been fairly casual. You have never had need of a formal structure or a chair, but by mutual consent the meetings have always been kept short and on track. The agenda for today's meeting is the new policy manual that has been sent around for review. Everyone has been asked to bring comments and feedback to the meeting, which is expected to take an hour.

Discussion moves along as usual until near the end of the meeting, when it is Eaglestone's turn to speak. He makes a brief reference to the policy manual, but soon veers off onto other topics. People have come to expect a certain amount of self-promotion from Eaglestone, but it takes a few minutes before you realize that what he's actually doing is more than his usual bragging. He appears to have sensed that he will not be granted an interview when the new contracts are announced, and he seems to initially intend to make a case for being re-hired. However, as he continues to speak, he works himself into a bit of a lather. He accuses his colleagues of jealousy; he says they are holding him back and trying to damage his professional reputation. He says that his problems in the workplace have been the result of their unwillingness to cooperate with him. Nobody can stand it, he says, that he's better and more skilled than they are. Nevertheless, he says, though he is constantly being undermined—even sabotaged—he holds no grudges against anyone and he has tried to get along with everyone. He concludes by saying that it's obvious that this is a third-rate place where only mediocre people are wanted. He himself deserves to be somewhere much better.

This monologue continues for nearly half an hour, as the group sits in stunned silence. Everyone appears acutely embarrassed; nobody makes eye contact with each other or with the speaker. Nothing remotely like this has ever happened in your unit before; it's always been a collegial environment and your meetings have focused on the agenda at hand. It has not occurred to any of you that the meetings could be, or would be, used as a platform of this sort. The meeting has no official chair, and nobody has the presence of mind to call a halt to this unexpected harangue.

When Eaglestone finishes speaking, you collect yourself enough to suggest that the meeting adjourn for now. Nobody responds to Eaglestone's outburst; everyone files out in embarrassed silence.

QUESTIONS FOR CRITICAL THOUGHT

1. Many of the articles in this book have emphasized the importance of understanding both audience and context. In what ways does this case study illustrate or dramatize this important principle?
2. Take a look at the way the narrative is presented; does MacLennan 'slant' the description in any way toward a particular result? Provide some specific examples.
3. Compare the scenario to other pieces by MacLennan in this book; to what extent can you 'hear' her voice? To what extent is it muted? What elements of the context might affect how audible this text is?
4. According to several of our contributors, such as Dombrowski ('Can Ethics Be Technologized?'), Buehler ('Situational Editing'), MacLennan, and others, professional communication involves elements of ethics, politics, interpersonal dynamics, and face risk. How do those factors influence the unfolding of events in this scenario?
5. Take a few minutes to jot down your own impressions and interpretations of Eaglestone. Compare your reactions with the analyses by Zepf and Azzopardi. What assumptions did you bring that might have conditioned your response? Did their

interpretations change your mind at all? The instructions to the group for this examination assignment were as follows:

> Discuss this situation as fully as you can, from a communication perspective. Your analysis should include some discussion of the following:
> i) In this scenario, who is saying what, to whom, and for what purpose?
> ii) What, if anything, could have been done to prevent Eaglestone from hijacking the meeting for his own agenda? Should something have been done?
> iii) What role, if any, did issues of face and footing play in the way this event unfolded?
> iv) What does this event reveal about Eaglestone's attitudes and motives? What does it reveal about this workplace?
> v) What effect might this event have on future meetings in the department? What effect, if any, should it have?

CHAPTER 53

Reading Eaglestone: A Corporate Psychopath?

Paul J. Zepf

Who is Saying What, To Whom, and For What Purpose?

The problem described in the scenario, from Eaglestone's point of view, seems to be centred on his desire to be granted a second interview to discuss a second probationary contract, and ultimately to obtain a permanent position. However, he is getting impatient and upset, because his contract is running out, and he has gotten the impression that he will not get the cooperation and acknowledgement represented by an interview. If he's correct, he will definitely be out of the company at the end of his present contract.

Although a second interview may not be a guarantee of getting the new contract, the probability appears to be very high. Eaglestone appears to believe—at least he claims—that he is doing a great job despite the reactions of the people around him. He declares that they are beneath him and in his way, but it appears nevertheless that his co-workers have some influence with higher-ups as to whether he will be granted a second interview. In particular, the key to his successful bid to continue his job appears to be the workers in his department, especially the better workers who have recently begun avoiding working with him because of his poor overall performance and loud mouth.

The situation is intensified by the fast-approaching end of his current contract, the time deadline for the second interview, the shrinking opportunities to persuade people to support his second interview, and the growing worker resistance to his continued presence. He probably sees this monthly meeting as his last attempt to solve his exigence.

This regular monthly meeting is perfect for his ends. All the workers are present and trapped. The format is loose enough to ensure success in sliding the meeting over to his agenda. Since the others are trapped and he can direct the meeting, he sees himself in control. At this point, his objective appears to be to bully the group to insure that they support him. Barring that, he may think he can intimidate them sufficiently that they will not prevent him from getting the second interview.

For the rest of the members of the department, the problem is somewhat different. They want to remove Eaglestone, not only from the department but also from the company, as a result of his arrogance, his questionable ability, and his lack of reliability and team support. The people in the department know that they have to present a united front and solid evidence to upper management if they are to insure Eaglestone is not considered for a second interview. They will need to try to prevent any misguided support from any of them or any other influential source within the company. They must be careful not to make any mistakes that would cause upper management to see Eaglestone in a different light or compromise's management's ability to deny him a second interview. If they are successful, Eaglestone's contract will expire and he will be out of their lives.

In the meeting, the co-workers find themselves trapped by Eaglestone, and it is a face-losing confrontation. Everyone has become frustrated with his poor attitude and productivity over the past year, but this is the last straw, and any remaining sympathy or support for him has dried up. Obviously, the workers have not anticipated his outburst and are ill-prepared for it, so no plan is in place to put an end to it. Nevertheless, if Eaglestone succeeds in provoking a confrontation, he may create an opportunity to cry legitimately cry foul to upper management. They respond to the outburst with confused silence.

What could they have done? When Eaglestone initially attempted to go off topic, someone should politely but firmly have called the meeting to order and re-focused on discussing the new policy manual, which after all is the purpose of the meeting. Since an astute senior member of the department did not handle the situation immediately, then to stop it a little later could result in Eaglestone's claiming that he has been treated with prejudice by being prevented from speaking. If he is unscrupulous enough—and he appears to be, based on his charges that his colleagues are conspiring to do him out of a job—Eaglestone might even force the hand of management by laying a formal complaint of unfair dismissal. Then any attempt by members of the department to prevent his being granted a second interview would appear to confirm his complaint, making it more difficult to refuse him an interview and a permanent position. Thus, once he has begun his tirade and gotten away with it, it may be best for members of the department to ride out the outburst and adjourn, as was done. This move at least will enable them to avoid an ugly scene that in turn might allow Eaglestone to win the interview that will land him the extension of probation.

It is possible, even likely, that members of the department have overlooked earlier indications of trouble from Eaglestone that might have alerted them to potential problems at the meeting. For example, Eaglestone's actions during normal working hours perhaps should have suggested that he would bring this behaviour into the meeting at one time or another. The impending expiration of his first contract and the increasing likelihood that he would be refused a second interview would certainly have provided him with further provocation, and perhaps should have prepared the group for a possible disruption of the normal flow of the meeting. In light of the Eaglestone incident, perhaps a plan should be developed to prevent people like him from turning the meeting into a forum to continue to disrupt the department. Had such a plan existed, it could have been be implemented as soon as Eaglestone moved off topic.

Assuming that there is no substance to his complaints, Eaglestone's pattern of behaviour points to the possibility that he may be what Robert Hare has called a 'corporate psychopath', a person who, while nonviolent, is nevertheless prone to the 'selfish, callous, and remorseless use of others'. Business psychopaths can wreak havoc in the workplace, dragging others into their destructive web of deceit and exploitation, and Eaglestone appears to exhibit many of the sixteen traits that red-flag this type of behaviour. These include unusual insecurity combined with arrogance (exemplified, for instance, in his excessive talking about prowess and big money); untrustworthiness and unethicality (shown in his taking credit for the achievements of others and his exaggeration of his previous accomplishments); manipulation of others (ambushing people at the meeting or tricking others into doing his work), insensitivity, remorselessness, and blaming of others (his insistence, for example, that he is being sabotaged by jealous co-workers); erratic behaviour, lack of focus, unreliability (his inability to finish or take responsibility for critical tasks assigned to him); selfishness and impatience (shown, for example, in his sense of entitlement to a second interview that he has not earned); dramatic self-display (he seizes opportunities for self-aggrandizement); and bullying. In keeping with this pattern, Eaglestone knows that his first contract is up soon and senses that he may not get the second contract. He obviously is anxious about his future, and his self-centred focus will cause him to zero in on an opportunity to manipulate and bully people into giving him want he wants. Eaglestone needs to dominate others and control events.

If Eaglestone indeed has psychopathic tendencies, he will not, or cannot, feel any loss of face since he has little sensitivity to or concern for other people's feelings. He will therefore feel little or no guilt or hesitation in launching an attack to press his agenda and goals, since his sense of urgency is driven by his self-centred focus on the contract expiry date. His outburst will be out of proportion, and he is unlikely to be constrained by a sense of appropriateness. Quite simply, Eaglestone's use of the monthly meeting to gain control and to bully his co-workers into supporting a second interview is exactly as expected.

There is potential for an additional loss of face by the people in the department, if Eaglestone's antics during this meeting become public company gossip. This could cause discomfort and embarrassment to the department, even though they are victims, since reports of the incident could reach upper management and even provoke some sympathy for Eaglestone, resulting in investigations and meetings that could involve placating Eaglestone.

How did this company end up with an Eaglestone in the first place? The workplace described in the scenario, though functional and healthy when the department is made up of those with good will, has weaknesses that unscrupulous people could exploit. The best solution is not to hire them in the first place. It might be possible to improve hiring practices so as to identify troublemakers before they are hired. However, it is notoriously difficult to detect psychopathic tendencies during the constrained situation of an interview, since such people are superficially charming and likeable, and very adept at concealing their true nature in the short term. They learn quickly how to stay submerged until they are in a position of strength, and often present such a good front that those hiring them don't carefully check references, even where they are supplied. In this case, Eaglestone provided an impressive resume and gave a strong interview; nevertheless, his references should have been requested and verified in detail as to attitude, behaviour, and work performance.

Furthermore, in this workplace, though an agenda was set for each monthly meeting, no chair was assigned. As well, the meetings were small in size (just the department) and used a casual approach. This lack of structure had worked well in the past due to mutual understanding and good will among co-workers, and the resulting casual atmosphere had been productive in encouraging input from all. This situation is a prime target for a person looking to be disruptive and manipulative. Once the meeting structure had been compromised, control transferred to Eaglestone. The impact of the attack will cause co-workers to be apprehensive and insecure regarding future meetings. The fear is of grandstanding and surprises, especially as new employees come into your department. Unfortunately for the department, the Eaglestone experience might compromise some of the good will that the department has enjoyed in the past by imposing more structure in an attempt to control or minimize the damage that an Eaglestone can cause. While it might be prudent to prepare for the worst, too many procedural rules can stifle creativity and trust.

The moral of the story is that manipulative, destructive people must be dealt with when exposed and not allowed to continue taking advantage of the next victim. The ethical thing to do is nail him hard and fast.

QUESTIONS FOR CRITICAL THOUGHT

1. What is a corporate psychopath? How effectively does Zepf support his assertion that Eaglestone is such a person?
2. How systematic is Zepf's analysis? Point to some specific examples to support your argument.
3. Zepf wrote his response as a final examination for his graduate communication course. What are the constraints of such a situation? Is there anything about the essay that marks it as an examination response?
4. How audible is Zepf's article? How personal is the style? To what extent are Zepf's rhetorical choices conditioned by the situation in which he is writing?
5. How does Zepf build his ethos? On what does his credibility depend?
6. To what extent is Zepf's analysis another example of argument by definition? Point to some examples from the text to support your answer.

CHAPTER 54

Defending Eaglestone: Bad Fit or Wrongful Hire?

Joe Azzopardi

Bob Eaglestone's outburst is indefensible, in any context. Irrespective of any underlying issues, he should have maintained his composure and left this position with his ethos intact, and saved face for future opportunities (Moore, 2002).

However, there may be more to the events than meets the eye. Except for a brief description of his half-hour rant, little is provided from Eaglestone's perspective, and this perspective might shed some different light on the events that took place in his workplace. The analysis that follows is based on a scenario re-creation. The re-creation adds context to subtle cues in the scenario that are, in my opinion, open to interpretation; indirectly, it explores some of the relational issues involved.

TWELVE MONTHS AGO

Bob Eaglestone is a knowledgeable engineer with a track record of successful projects and is a force to be reckoned with at his current employer, Digi-X Inc., a large multinational firm. With an innate sense of mechanical systems, like many engineers, he feels this skill, along with the credibililty he has developed over a decade with Digi-X, are sufficient for a successful career. As a typical engineer, he doesn't place much value on audience appeal. In fact, he is upset that several colleagues have been promoted on audience appeal *alone* and he considers this to be the corporate equivalent of pandering; he will never stoop to that level.

What motivates an accomplished engineer like Bob to seek opportunities outside of his current firm? Perhaps, despite his accomplishments, he feels frustrated at the lack of career progression, or maybe a co-worker with questionable motives has manoeuvred himself into a position that forces Bob to re-evaluate the wisdom of his firm's management. Bob applies to the following advertisement, unaware of the challenges ahead, not only to clear the hurdles of the recruitment process, but more importantly, to re-establish himself as a figure of authority, in a new firm.

> **Senior Engineering Position Available**
>
> An exciting opportunity has been created within a dynamic organization for an accomplished senior engineer with management potential. The company is going through a rapid growth phase and you will be expected to effect change; you must demonstrate an ability to juggle multiple priorities and delegate effectively, increasing staff utilization rates of junior engineering, administrative, and secretarial employees. You have a portfolio of successful projects that speaks to your technical abilities and are a leader by example. As some resistance to change may be encountered, you have a strong character and are adept at managing conflict. You will have the full support of senior management in this newly created role. Compensation will be excellent and a significant component of the remuneration will be a bonus tied to achievement of performance metrics.[1]

As part of the selection process, Bob undergoes a rigorous battery of interviews and favourably impresses the senior management of DB Ltd., a medium-sized Canadian firm. After a spirited negotiation, he receives an offer of employment that includes immediate participation in the company's seemingly lucrative bonus plan. The offer contains what appears to be an unconventional probation period, two separate terms, one-year in length each, before the position becomes permanent, but Bob is inclined to take a risk. The management of Digi-X attempts to keep him, but he considers this effort too little, too late, and he accepts the offer with DB Ltd., despite his concerns about the lengthy probationary periods. Little does Bob know: he is merely a trophy hire, a trinket to impress DB's financial institution in upcoming negotiations.

Within his first few days at DB Ltd., it becomes apparent to him that all is not well. Profitability has been overstated; senior management has misrepresented the true situation. The relationship between senior management and employees appears to be one of mistrust; the lucrative bonus plan is an unattainable carrot for everyone, dangling hopelessly out of reach.

Undaunted by this unexpected setback, Bob has a depth of character that allows him to see beyond the barriers, but what are his chances of success? The answer depends on how success is measured; if the measure is whether Bob can exit with his good reputation intact, he is doomed to fail.

BACK TO THE RECENT PAST

Faced with what appeared to be an insurmountable problem, Bob has spent the past 11 months trying to build relationships with his new co-workers, but senior management has thwarted him at every turn. Even his attempts to delegate work to junior staffers, to expose them to new experiences, have been blocked, so he has given up trying. Bob has seen, in his new manager, Pat, psychopathic characteristics that eclipse those of his past colleagues by an incalculable amount; how could he have missed the signs during his interviews? His co-workers are oblivious, though, as they have each developed a coping mechanism that amounts to an impenetrable shell. Senior management appears to be quite pleased with this state of affairs, and actually encourages this interpersonal disconnect, as the shell each employee has built around him or herself minimizes personal interactions. Meetings are short and focused on the tasks.

Whenever Bob appeared to achieve some success at penetrating a colleague's shell, it seemed Pat would assign a new, 'mission-critical' task, with an impossible, artificial deadline, and with insufficient resources to complete the task. At last count, Bob had 11 active projects assigned to him, all urgent, while his colleagues, coincidentally 11 in total, typically handled one or two projects each. The stress of the situation began to take its toll, particularly when artificial deadlines loomed. After months of such frustrations, there were days when Bob was simply unable to get out of bed, knowing that he would be facing a confrontation with Pat for missing yet another deadline.

Unknown to Bob at first, and partially to save his own face,[2] Pat would encourage Bob's colleagues to pick up the slack when an 'important' task had to be completed; doing so would result in a token reward. Like a group of trained seals, the employees of DB Ltd played into Pat's bizarre power game. One by one, despite the token perks offered by Pat, they have each come to loathe Bob and his deadline-avoidance tactics. Bob, on the other hand, could not understand why they would play into Pat's hand so easily; he felt that he was being undermined, not only by Pat, but also by his colleagues.

When he first began to feel the effects of his dysfunctional workplace, Bob would stay at home, in bed, watching *Oprah*. As time passed, he had the presence of mind to seek professional help, but decided not to disclose this to DB Ltd, even though he felt legitimately harassed. Bob could not be certain that the information would be kept confidential at DB. Armed with this information, Pat would come in for the kill; Pat might even leak the information to Bob's colleagues outside of the firm. Bob knew that he had to rebuild his strength if he was to have any chance of maintaining any dignity in this situation, so he continued to bear the cost of his absence and pay for medical treatment out of his own pocket.

During his counselling sessions, Bob discussed DB's 'management by policy' approach, amongst other things. To Bob, the policy manual was an incoherent collection of 250 seemingly disjointed policies. Any time a situation arose, a new policy letter was issued, and would simply be added to the manual without any amendments. Policy Letter #77, for example, deals with theft of toilet paper from the ladies' washroom and points out that if the theft continues, toilet paper will no longer be supplied free of charge; the letter is dated 2 August 1987.

A token review of the policy manual takes place annually; all employees are invited to comment on the policy manual, but it is widely understood by all, except Bob, that employees are expected to make only superficial comments, and that the policy manual will be affirmed with only minor corrections to punctuation and grammar. When it is Bob's turn to speak, he goes postal.[3] . . .

ANALYSIS OF SITUATION

Back at Digi-X, Bob had developed an effective form of what the communication theorist Erving Goffman calls a 'face', which he typically displayed at meetings and other encounters in his workplace

(Goffman, 1967). In general, Bob presented a calm, non-confrontational demeanour, even when expressing disapproval to his staff, and took care to provide only thoughtful and deeply introspective comments. Bob had appeared out-of-face only once, when he had raised his voice enough to be heard above a trivial argument and put two junior engineers in their respective places. Under these conditions, those who overheard the exchange approved of Bob's actions; even his young protégés acknowledged that they were deserving of such attention. In the right context, an occasional *minor* outburst from an otherwise well-mannered associate is acceptable.

However, after almost a year of trying to make the best of a bad situation, Bob Eaglestone had still not developed any similar level of authority and face at DB Ltd. Opportunities for interaction were few and far between, and lacked any element of personal connection. While he may not have been free to establish appropriate credibility amongst his new peer group, Bob failed to recognize that he *was* free to move on at any time. Instead, he chose instead to stay on, hoping to alter the situation. His reluctance to move on may have been in part an attempt to save face with his peer group at Digi-X: to leave DB Ltd so soon after joining them would have been an admission that he was wrong to leave Digi-X.

Bob should have chosen the lesser face loss; it would have been in his best interest to do so, as it is not safe to assume that events that happen inside an organization will stay within that organization. Despite the proliferation of non-disclosure agreements, industry information is a commodity that is routinely used as a negotiating tool. Executive recruiters work with many firms and human resource specialists may switch firms frequently; knowledge about practitioners in their respective industries goes along with them.

While recognizing that under these circumstances it was wise to keep sensitive personal information out of the public domain, Bob has inadvertently created a situation where his actions may precede him at future job interviews, and he will have to work harder to establish credibility in the future, both at interviews and at any new firms where he may be hired. Bob should have guarded his reputation with the same amount of care he took with his personal private information.

At the meeting, Bob challenged his colleagues' inability to recognize his 'engineering prowess'; unfortunately, his overt message is self-centred, not audience-centred, and is doomed to fail. The real motive for his message, which is embedded indirectly in his outburst, is his frustrations with the blindness of his co-workers, who are unable to see the effect that Pat has had, and is having, on the group.[4] Unfortunately, Bob's attempt to awaken them to the situation is doomed to fail. His senior job title by itself does not automatically grant him the footing to convince this group of the severity of the situation, especially when paired with his recent pattern of behaviour. Without having first established his own authority to speak, he simply cannot effectively influence his audience.

Bob failed to recognize that all communication takes place 'within a context of persons, objects, events, and relations' (MacLennan, 2003: 9).

In his former role at Digi-X, his past contributions had granted him automatic recognition and given him a strong basis for establishing credibility and authority. While he made no explicit effort to manage interpersonal relation, his managers and peers knew that he could be relied upon to meet or exceed his customer expectations. This same reputation was what had got him in the door at DB Ltd, but once he was inside, his former achievements were considered irrelevant in the context of the new organization. Continuing to rely on past performance as an indicator of future performance, Bob failed to recognize that the reputation and credibility he had been relying on for the past several years no longer existed in his new workplace.

In the meeting, the group is shocked by Bob Eaglestone's outburst and apparently has no mechanism to deal with incidents of this nature.

That past meetings have followed an informal protocol, without assigned responsibilities, and have stayed on track, is highly unusual. Typically, the only meetings that stay on topic and are short in duration are meetings that are held under duress; emergency meetings called at 5:00 pm tend to be short, especially if same-day actions are required.

'The group sits in stunned silence,' avoiding 'eye contact with each other' and with Eaglestone. In focusing on the content of his complaints instead of on the context and likely impact, he is merely venting his frustrations, and with the wrong audience. Bob fails to recognize the fact that he is not in a situation where his words can make a difference; while a problem certainly exists, and may need to be addressed in some form, in the meeting Bob is simply preaching to a group that cannot acknowledge his message or do anything to change the situation. The group is unable to grasp what needs to change and even if its members could understand, they have been rendered impotent by Pat's domination.

Certainly, someone should have stepped in to stop Bob, to save him from himself, but he appears not to have forged a single alliance in his new peer group. If not to save face for Bob, one leader within the group should have stepped in to save face for the rest of the group. As this did not happen, it appears that there is no social structure within the group; active participation is unlikely, and information flows in only one direction: down, from management. This group needs a more formal protocol for conducting meetings, with distinct roles and responsibilities, if for no other reason than to ensure that every participant feels free to contribute without fear of retribution.

AFTER THE MEETING

After the meeting, the group reconvenes and discusses Bob's outburst. How dare Bob rock the boat like that? Is he trying to get everyone fired? What gives him the right to stand up and hijack the meeting like that? Good thing his contract is nearly up; he'll never be re-hired *now*. What was he thinking—*was* he thinking? This happens every time *they* go and hire someone from outside. I'll be glad when he's gone and things return to *normal* around here.

Nothing like this has ever happened before, and now that Bob is unlikely to stay much longer, it will never happen again. The other group members know better, so no change to the protocol is required, or so the 'team' thinks.

After the meeting, Bob is exhausted. He returns to his desk, collects his personal belongings and leaves through a back door without speaking to anyone. The following morning, the group reconvenes; the policy manual is marked up with the usual meaningless drivel and passed along to Pat for corrections and publication.

Pat is satisfied that order has been restored now that Eaglestone is out of the picture. He and the senior management team acknowledge their victory with a celebratory lunch, oblivious to the fact that DB Ltd is squandering its most valuable resource, its human resource.

Notes

1. From my experience, recruitment, selection and hiring practices are random events, despite trends that are embraced as panaceas, but are really nothing more than flavour-of-the-month solutions. Uncertainty exists for employers and employees alike, but employers are usually in control of the situation. An oft-cited issue, by employers, is resume embellishment and yet, promises of enhanced responsibilities, generous annual raises, and bonus plans to lure candidates are commonplace. While legal recourse exists, wrongfully hired employees are generally not in a position, neither financially nor emotionally, to launch legal action, even if they are aware of their options.
2. Pat could not simply allow these deadlines to pass unnoticed; to do so would amount to admitting to their artificial nature, giving Bob a

legitimate grievance, one he could use against Pat.

3. Colloquially, 'to become extremely angry', (http://encarta.msn.com/dictionary/postal.html, accessed 16 October 2005).
4. And by its choice, conscious or otherwise, to do nothing to curtail Pat's anti-social behaviour, the effect that the whole senior management team is having.

References

Goffman, Erving. 1967. 'On Facework', ***Interaction Ritual***. Garden City, NY: Anchor Books.

MacLennan, Jennifer. 2003. ***Effective Communication for the Technical Professions***. Toronto, ON: Pearson Education Canada.

Moore, D. 2002. 'Last Impression Counts When Leaving a Job', ***The Toronto Star***, 21 August 2002. Available at http://www.workopolis.com/servlet/Content/torontostar/20020821/quit?gateway=work, (accessed 16 October 2005).

QUESTIONS FOR CRITICAL THOUGHT

1. Zepf and Azzopardi offer completely different interpretations of the events described in the scenario. Which do you find more compelling? Why?
2. Compare Azzopardi's analysis of the Eaglestone scenario with Zepf's analysis. What do these two different interpretations imply about communicating in a professional context?
3. How 'legitimate' is Azzopardi's analysis? Why? Does it fulfill the expectations of an exam response?
4. How audible is Azzopardi's style? How personal is it? Why might Azzopardi have chosen to write in this manner?
5. How does Azzopardi build his ethos? To what extent is ethos dependent on situation?
6. Comment on the styles of the three writers in this case study. Do they follow the 'objective narrative' style advocated by many technical style manuals? Which of the three is the most 'personal'? Why? To what extent does context shape the style of each writer? To what extent is the style shaped by the writer's own character?

Contributor Biographies

Joe Azzopardi is a professional engineer and Project Manager for Wardrop Engineering in Mississauga, Ontario. He also serves as Vice President of Academic Programs for the Ontario Product Development and Management Association. A gifted writer, Azzopardi is currently completing an MEng in Design and Manufacturing.

Brian Bauld is a distinguished former high school English teacher from Amherst, Nova Scotia. He is the creator of the rich online resource www.mrbauld.com, a collection of articles, essays, links, and observations about language, literature, and teaching that has been praised by educators world-wide. Now retired from Amherst Regional High, Bauld is the owner of b-line books, where he continues to indulge his passion for language and letters through the on-line sale of used and hard-to-find titles.

Lloyd F. Bitzer taught rhetoric and communication at the University of Wisconsin, until his retirement in the 1990s. Bitzer is best known for his influential theories about how context shapes human communication.

Mary Fran Buehler was one of the pioneers in the academic study of technical communication. During her lifetime, she worked at the Jet Propulsion Laboratory in Los Alamos, and was a prolific speaker at Society for Technical Communication, Professional Communication Society, and IEEE conferences. Buehler is best known as the author, with Robert Van Buren, of *The Levels of Edit* (1980) an influential primer on technical editing that outlined eight distinct tasks that make up the editing process (including not only proofreading for grammar and correctness, but also considering the document's conformity to company policy, its formatting, and its overall substance).

Richard T. Burton is Professor of Mechanical Engineering in the College of Engineering at the University of Saskatchewan. A thoughtful teacher and distinguished researcher, Burton has served as Assistant Dean, Undergraduate Programs for the College of Engineering, where he has also been involved in teaching the required Oral and Written Communication course for undergraduates. Burton, a Fellow of the American Society of Mechanical Engineers, has provided leadership in the establishment of the ASME Fluid Power System and Technology Division and in the development of international collaborations in fluid power education and research.

Charles P. Campbell is Professor Emeritus and former Chair of the Humanities Department at the New Mexico Institute of Mining and Technology, teaching in its Technical Communication program. He is co-editor of *The Cultural Context in Business Communication* and co-author of *How to Edit Technical Documents* (1995).

Bill Casselman is one of the country's leading etymologists and word hounds, as is shown by his numerous books on Canadian language, including the celebrated *Canadian Sayings* (1999) and its two sequels *Canadian Sayings 2* (2002) and *Canadian Sayings 3* (2004). He is also the author of *As the Canoe Tips: Comic Scenes from Canadian Life* (2006), *Canadian Food Words* (1998), *A Dictionary of Medical Derivations: The Real Meaning of Medical Terms* (1998), *Canadian Garden Words* (1997), *Casselmania: More Wacky Canadian Words and Sayings* (1996), and *Casselman's Canadian Words: A Comic Browse through Words and Folk Sayings Invented by Canadians* (1995). Casselman maintains an entertaining website devoted to Canadian language at www.billcasselman.com, where samples from his books are also available for browsing. Bill Casselman has also had a distinguished career in broadcasting, most notably as the founding senior producer of *This Country in The Morning* with Peter Gzowski.

Paul M. Dombrowski, a Professor of English at University of Central Florida, holds doctoral degrees in both rhetorical communication and the history and philosophy of science, as well as an undergraduate degree in mathematics. Among his current research and teaching interests are texts and technology, ethics in technical communication, rhetoric and philosophy of technical communication.

Peter Elbow, by his own admission, loves writing. A well-known theorist and distinguished composition scholar, he is the author of six books on the subject, including *Writing with Power: Techniques for Mastering the Writing Process*. Until his retirement in 2000, Peter Elbow served as the Director of the Writing Program at the University of Massachussetts, where he continues to offer graduate seminars focussed on issues of voice, dialect, and standard English to students in the English Department's Rhetoric and Composition Program.

Richard M. Felder is the Hoechst Celanese Professor Emeritus of Chemical Engineering at North Carolina State University. He is co-author of *Elementary Principles of Chemical Processes*, an introductory chemical engineering text now in its third edition. He has contributed over 200 publications to the fields of science and engineering education and chemical process engineering, and writes 'Random Thoughts', a column on educational methods and issues for the quarterly journal *Chemical Engineering Education*. With his wife and colleague, Rebecca Brent, he co-directs the National Effective Teaching Institute (NETI) and regularly offers teaching effectiveness workshops on campuses and at conferences around the world.

Cheryl Forbes is the Chair of the Writing and Rhetoric program at Hobart and William Smith Colleges in Geneva, New York. Forbes received her BA and MA from the University of Maryland and her PhD from Michigan State University. She previously taught English at Grand Valley State University and at Calvin College, and has experience in book and magazine publishing. Forbes has published numerous articles, has given a number of conference presentations, and directed a three-year grant awarded to Hobart and William Smith Colleges from the Fund for the Improvement of Postsecondary Education (FIPSE). Her most recent book, *Women of Devotion through the Centuries*, studies late nineteenth- and early twentieth-century women who wrote and compiled devotional books.

Joe Glaser teaches at Western Kentucky University where, in addition to serving as the Director of Composition, he has researched and taught in a variety of areas from renaissance and seventeenth-century literature to classical literature and English Grammar. His most recent research and teaching interests have focused on classical literature and grammar. Since retiring in 2003, Glaser has been at work on translations and retellings of pre-renaissance English literature.

James Gough is on faculty in the Department of Philosophy at Red Deer College in Alberta. An executive member of the Canadian Association for the Study of Practical Ethics who has served on the editorial board of several academic journals, Gough conducts research on the teaching and practice of ethics in the professions, and on the role of tradition and values in the creation and reception of arguments.

George Grant was one of English Canada's most influential political philosophers and most thoughtful social and cultural critics. In works such as *Lament for a Nation* (1965), *Technology and Empire* (1969), and *Technology and Justice* (1986), he explored the extent to which the pre-eminence of corporate capitalism and its technologies have undermined the ability of nations and individuals to pursue self-determination in the traditional sense.

Stephen M. Halloran is Professor Emeritus in the Department of Language, Literature and Communication at Rensselaer Polytechnic Institute, where he has served as Department Chair and as Associate Dean of the School of Humanities and Social Sciences. He has written widely both in rhetorical theory and in technical and professional communication, and was named a Fellow of the Rhetoric Society of America in 2002.

George C. Harwell was the author of *Technical Communication*, from which this selection comes. Throughout his book he argues that a command of clear expression—good writing style—is an essential skill for technical work, and he gives specific advice on how to achieve a clear, readable, and effective style in technical writing.

Richard M. Holliman is a sociologist working in the Science Faculty at the Open University (OU), UK. Having completed his PhD in the Department of Sociology at the OU, where he investigated representations of science in news media, he now lectures in science communication at undergraduate and postgraduate level. He is currently leading the Informing Science Outreach and Public Engagement (ISOTOPE) Project. For more information see: www.isotope.open.ac.uk.

Marjorie Rush Hovde is Associate Professor of English and Technical Communications at Indiana University-Purdue University Indianapolis, where she also serves as Coordinator of the Technical Communications Certificate. Hovde's research in technical and professional communication focuses on teaching and training of non-specialists. She has published numerous papers on audience analysis, instructional practice, and communication program evaluation. Like many experts in the discipline, Hovde emphasizes the strategic nature of professional writing and focuses on the constraints that arise from the realities of organizational politics.

David Ingham teaches in the English Department at St Thomas University in New Brunswick. Although he currently teaches primarily literature, Ingham has also taught professional communication and served as Executive Director for the Ontario Association of Cardiologists. A skilled and engaging teacher, Ingham is also talented actor, who has appeared in a variety of theatrical roles both amateur and semi-professional, including several Shakespearean offerings.

Sigrid Kelsey is an Associate Librarian at Louisiana State University. She has authored articles on email instruction and is currently editing a book on Computer Mediated Communication. She has also taught a graduate course on information technology in the LSU School of Library and Information Science.

Ibrahim Khider is a Toronto-based freelance writer of non-fiction, who publishes both in print and online. He maintains a blog under the name of 'ikhider', where he explores timely issues and events in what he hopes will be a timeless way.

Bernadette C. Longo is Associate Professor in the Department of Rhetoric, Science, and Technical Communication at the University of Minnesota. Before returning to graduate school to earn her MA and PhD, Longo was a professional writer who specialized in the medical and poultry processing industries. She became a teacher in part in order to 'give back' some of what she has learned about writing to a new generation of writers and communicators. Her current teaching includes courses in project management and design, and she is researching the history of computing.

John Lorinc is a Toronto journalist who writes about urban affairs, culture, and business. He is the author of *The New City: How the Crisis in Canada's Large Urban Centres is Reshaping the Nation* (Penguin, 2006).

William Lutz, Professor of English at Rutgers University and a member of the Pennsylvania Bar, is the well-known author of *Doublespeak: From Revenue Enhancement to Terminal Living*, from which our selection is taken. Lutz served for 15 years as head of the Committee on Public Doublespeak and for 14 years as editor of the *Quarterly Review of Doublespeak*. He has appeared on numerous television and radio programs to publicize the dangers of doublespeak, including *Today*, *Larry King Live*, *The CBS Evening News*, *The MacNeil Lehrer NewsHour*, and National Public Radio's *Morning Edition*. He is also the author of *The Cambridge Thesaurus of American English*.

Kathryn N. McDaniel is Associate Professor of History at Marietta College in Marietta Ohio, where she is currently president of the college chapter of Phi Beta Kappa. She received her BA degree with honours from Davidson College and her MA and PhD degrees from Vanderbilt University. An active scholar, she has published in such journals as *The Teaching Professor* and *The History Teacher*.

Thomas R. McDaniel is Senior Vice President and Professor of Education at Converse College in Spartanburg, SC. Author or editor of seven books, 30 textbook chapters, and over 200 articles, he has served Converse for 35 years as a department chair, division head, Dean of the College Arts and Sciences, Dean of the School of Education and Graduate Studies, Vice President for Academic Affairs, Provost, and Interim President. He holds three graduate degrees from The Johns Hopkins University.

J.S.C. McKee is Professor Emeritus in the Department of Physics, University of Manitoba, and Fellow of the Institute of Physics. As editor of *Physics in Canada*, he has written on such diverse topics as academic freedom, 'extreme' physics, and the 'benign nature of nuclear power'.

Andrea McKenzie, Director of Writing in the Disciplines at New York University, has been a tech writer in a variety of industries since about 1985. She holds a PhD in Rhetoric and Professional Writing from the University of Waterloo, where she taught communication in the Faculty of Mathematics and also in the Faculty of Arts. She has also taught communication at MIT and has provided guidance for industry in the areas of teamwork, negotiations and formal oral presentations, as well as in writing. In her work McKenzie focuses on training others how to 'read' both written and oral clues about the specific company values, organization, and methods of negotiating.

Jennifer M. MacLennan, the editor of this book, is the first occupant of the D.K. Seaman Chair in Communication and the founding Academic Director of the Ron and Jane Graham Centre for the Study of Communication in the College of Engineering at the University of Saskatchewan. As a rhetorician trained in the liberal arts who has spent the last nine years teaching in a professional college, MacLennan is very interested in the commonalities between rhetoric and professional education. She is the author of seven textbooks on a wide range of communication topics, from technical and business writing to interpersonal communication and language to practical ethics, and regularly presents talks and workshops on communication to a variety of professional, academic, and student groups and gatherings.

Marshall McLuhan was Canada's most famous scholar and critic of mass media communication, whose conception of the mass media as 'extensions' of human senses has been extraordinarily influential. Although they have at times been controversial, McLuhan's theories are evident (though not always explicitly acknowledged) in much contemporary media criticism. McLuhan taught briefly in the United States, but spent the bulk of his career at the University of Toronto, gaining fame in the 1960s with his proposal that electronic media were creating a global village. Although McLuhan of course was referring to television, his assertions anticipate the pervasive influence of the Internet and other electronic communication media.

Carolyn R. Miller is SAS Institute Distinguished Professor of Rhetoric and Technical Communication at North Carolina State University, where she conducts research in digital rhetoric, rhetorical theory, the rhetoric of science and technology, and technical and professional writing. Her publications have appeared in journals such as *Argumentation*, *College English*, the *Journal of Business and Technical Communication*, the *Quarterly Journal of Speech*, *Rhetorica*, and *Rhetoric Society Quarterly*, as well as in edited volumes published by university and commercial presses. She is a past president of the Rhetoric Society of America and was named a Fellow of the Association of Teachers of Technical Writing in 1995.

Cezar M. Ornatowski is a Professor in the Department of Rhetoric and Writing Studies at San Diego State University, where he consults, teaches, and conducts research in technical communication, organizational rhetoric, rhetoric of science, and political discourse. Ornatowski has served as Science Editor for such publications as *Science*, *Nature*, and the *Journal of Molecular Biology*. He is co-editor of *Foundations for Teaching Technical Communication: Theory, Practice, and Program Design* (1997) and *Revisioning Democracy: Central Europe and America, Critical Perspectives* (1996).

George Orwell (the pen-name of Eric Blair) was an English author, journalist, and political and cultural commentator, and is among the most widely admired English-language essayists of the twentieth century. He is best known for two novels published towards the end of his life: *Animal Farm* and *Nineteen Eighty-Four*, and for the essay included in this volume.

Elisabeth Pankl is a Humanities Librarian at Kansas State University. She has co-authored an article on blogging in academic libraries and is currently collaborating on a book chapter that addresses the confluence of CMC and Information Commons.

Arthur Plotnik is the author of *The Elements of Editing*, a standard reference through more than 20 printings, and *The Elements of Expression: Putting Thoughts into Words*, from which our selection is taken. Plotnik's writing style has been praised for its style and wit, as well as for its practicality and accuracy. His most recent book, *Spunk & Bite: A Writer's Guide to Punchier, More Engaging Language & Style* (2005) has been called 'a must for every writer's desk' and is among the best-selling recent titles on language and writing.

Anne Price has taught Communication Studies at Red Deer College in Alberta since 1993. She has written a textbook entitled *Writing Skills for the Fine Arts* and has presented conference papers discussing the cultural implications of soccer hooliganism, commercial punk music, English 'kitchen sink' films of the 1960s, and the export of American Aboriginal cultural objects.

Anatol Rapoport is a mathematical psychologist and one of the founders of General Systems Theory. He is known for his contributions to the understanding of such diverse fields as game theory, semantics, psychological conflict, and nuclear disarmament, and for his pioneering work in the modelling of parasitism and symbiosis. Since 1970 he has been Professor (now emeritus) of Psychology and Mathematics in the University of Toronto. A virtuoso musician, Rapoport has also contributed to the advancement of the fields of mathematical biology, the mathematical modelling of social interaction, and Stochastic models of contagion. He was recognized in the 1980s for his contribution to world peace through nuclear conflict restraint by his game theoretic models of psychological conflict resolution.

Neil Ryder is a lecturer in Science and Media at Royal Holloway, University of London. His research reflects his interest in the role of metaphor in the thought and writing of scientists.

Jonathan Shay is Staff Psychiatrist for the Boston VA Outpatient Clinic. A holder of both PhD and MD degrees, Shay has been a successful researcher in the biochemistry of brain-cell death, with its relevance to strokes. While recovering from a stroke of his own, Shay passed the time by 'filling in the gaps' in his education, reading English translations of the Greek epics and the Athenian plays and philosophers. He specializes in treating the psychological damage combat inflicts on soldiers using an approach that has been described as 'part neuroscience, part evolutionary theory, part psychiatric empathy and part Homer'.

Herbert W. Simons teaches rhetoric at Temple University. A prolific and distinguished scholar of political rhetoric, he has lectured widely, acted as a media commentator, developed instructional materials, and written numerous books, among them *Persuasion in Society* (2001), *The Rhetorical Turn* (1990), *Rhetoric in the Human Sciences* (1989), *Persuasion* (1988), *After Postmodernism: Reconstructing Ideology Critique* (with Michael Billig, 1994), *The Legacy of Kenneth Burke* (1988), and *Form, Genre, and the Study of Political Discourse* (1986). Simons has also served as Fulbright Senior Specialist in Political Communication, has contributed to the *Temple Issues Forum* (TIF), and is a recipient of the National Communication Association 'Distinguished Scholar' Award.

Tania Smith is Assistant Professor of Professional Communication in the Faculty of Communication and Culture at the University of Calgary. With a background in English literature and rhetoric-composition, her courses include Rhetorical History and Theory, Professional and Technical Communication, and Written and Oral Discourse. Her current research activities focus on the ways language creates communities and identities in higher education settings, and she is especially interested in teaching and learning enhancement programs that promote experiential, interactive learning and reflection, such as community service learning, peer mentoring programs, and inquiry-based learning.

Charles P. Snow was both a scientist and novelist, well known as the author of a sequence of political novels entitled *Strangers and Brothers* depicting intellectuals in academic and government settings in the modern era. However, Snow is most noted for his lectures and books regarding his concept of 'The Two Cultures', as developed in *The Two Cultures and the Scientific Revolution* (1959).

John R. Speed is a consulting engineer specializing in design and program management for public infrastructure and a frequent lecturer on engineering ethics and professional practice. He is the former Executive Director of the Texas Board of Professional Engineers. Speed holds a BSc in civil engineering and an MA in political science, with an emphasis on public management. He is a graduate of the Texas State Governor's Executive Development Program.

Stephen Strauss is a Toronto-based freelance journalist and author who served as a science writer for the *Globe and Mail* for over 20 years. Strauss is the author of numerous articles, book chapters, and editorials, as well as three books, including *The Sizesaurus: From Hectares to Decibels to Calories, a Witty Compendium of Measurements* (1997).

Stephen L. Talbot is editor of the online *NetFuture* newsletter, which he has been producing since 1995. He is currently engaged in 'an extraordinarily radical assessment of contemporary habits of thought' that will, he believes, produce a new, qualitative science. Talbot is the author of *The Future Does Not Compute: Transcending the Machines in Our Midst* and a host of papers and articles on topics as diverse as computers in the classroom, orality, and literacy in the electronic age, and the limitations of a technological understanding of the human heart.

Debbie M. Treise is Professor in the Department of Advertising and Associate Dean of the Division of Graduate Studies and Research at the University of Florida. A former freelance copywriter, Treise's research involves a range of communication topics, including health and science communication, public service advertising and medical marketing, and advertising education. Treise is currently carrying out research in science communication in conjunction with NASA's Marshall Space Flight Center.

Burton L. Urquhart is an Instructor of Communication in the College of Engineering at the University of Saskatchewan, where he is a founding faculty member in the Ron and Jane Graham Centre for the Study of Communication. Urquhart's research—beginning with an MA thesis in rhetoric and professional communication—has focused on the influence of context and audience on the making of professional messages. Urquhart has presented numerous conference papers on the rhetorical situation and is currently at work on a study of the rhetorical appeal of 'squirm' humour.

Jean Hollis Weber has a Master of Science degree and more than 20 years of experience planning, writing, editing, indexing, and testing user manuals and online help for computer software and hardware. A technical publications consultant for the past 12 years, Jean has worked with clients, written books, taught short courses in technical writing and editing, and presented parts of graduate and undergraduate courses at several Australian universities. She maintains several websites, including one for technical editors and OpenOffice.org.

Michael F. Weigold is Professor in the Department of Advertising at the University of Florida. Weigold has written widely on the subject of advertising and persuasion, source credibility, and advertising ethics. His current research interests involve impression management, information seeking, and political communication.

Helen Wilkie is a professional keynote speaker, workshop leader, and author of 'The Hidden Profit Center' (www.mhwcom.com). Educated at the University of Strathclyde and Jordanhill College of Education, both in Glasgow, Wilkie taught commerce and economics in Scotland for two years. When she moved to Canada, she also moved into the world of business, where her career encompassed marketing, human resources, international operations, shareholder relations, and corporate development. Since 1990 she has been a partner in MHW Communications, through which she speaks, writes and consults on all aspects of communication in the workplace.

Jeanie E. Wills is Assistant Professor of Communication in the Ron and Jane Graham Centre for the Study of Communication in the College of Engineering, University of Saskatchewan. A former advertising copywriter, Wills's research interests are eclectic, ranging from the rhetoric of advertising to theories of audience. Her recent papers have addressed such diverse topics as the role of Sasquatch in the Canadian imagination and the failure of Desdemona's rhetoric in Shakespeare's play *Othello*. Wills is currently completing a dissertation on the autobiographies of several of the founders of advertising.

Paul Zepf is co-founder of Oakville, Ontario-based Zarpac Inc., an international production engineering and consulting firm specializing in packaging production systems. With over 34 years of packaging production design and integration experience, Zepf has written extensively and given many seminars relating to packaging production. He holds a BSc in Mechanical Engineering from the University of Waterloo and an MEng in Design and Manufacturing from the University of Toronto.

William Zinsser is a writer, editor, and teacher. He began his career with the New York *Herald Tribune* and has been a long-time contributor to leading magazines. His 17 books include *Writing to Learn*; *Mitchell & Ruff*; *Spring Training*; *American Places*; *Easy to Remember: The Great American Songwriters and Their Songs*; and most recently *Writing About Your Life*. During the 1970s he taught writing at Yale, where he was master of Branford College. He now teaches at the New School, in New York, his hometown, and at the Columbia University Graduate School of Journalism.

Permissions

Brian G. Bauld, 'Sense and Nonsense about Grammar'. From http://www.mrbauld.com/grambb.html. Reprinted by permission of the author.

Mary Fran Beuhler, 'Situational Editing: A Rhetorical Approach for the Technical Editor'. From *Technical Education* 50, 4 (2003): 458–64. Used with permission from *Technical Communication*, the journal of the Society for Technical Communications, Arlington, VA, USA.

Lloyd F. Bitzer, 'Functional Communication: A Situational Perspective'. From *Rhetoric in Transition*, pp. 21–38, Eugene E. White, ed. University Park, PA: Pennsylvania University Press, 1980. Reprinted by permission of the author.

Charles P. Campbell, 'Ethos: Character and Ethics in Technical Writing'. From *IEEE Transactions on Professional Communication* 38, 3 (1995): 132–8. © 1995 IEEE. Reprinted, with permission, from *IEEE Transactions on Professional Communication*.

Bill Casselman, 'Digitariat'. From *The Wording Room*, http://www.billcasselman.com/wording_room/digitariat.htm (accessed 27 June 2006). Reprinted by permission of the author.

Bill Casselman, excerpt from 'Bafflegab and Gobbledygook: How Canadians Use English to Rant, to Lie, to Cheat, to Cover up Truth, and to Peddle Bafflegab'. Reprinted by permission of the author.

Paul M. Dombrowski, 'Can Ethics Be Technologized? Lessons from Challenger, Philosophy, and Rhetoric'. From *IEEE Transactions on Professional Communication* 38, 3 (1995): 146–50. © 1995 IEEE. Reprinted, with permission, from *IEEE Transactions on Professional Communication*.

Peter Elbow, 'Three Tricky Relationships to an Audience'. From *Writing with Power*, pp. 199–215. New York: Oxford University Press, 1981. Reprinted by permission of Oxford University Press, Inc.

Richard M. Felder, 'A Whole New Mind for a Flat World'. From *Chemical Engineering Education* 40 (2006): 96–7. Reprinted by permission of the author.

Cheryl Forbes, 'Getting the Story, Telling the Story: The Science of Narrative, the Narrative of Science'. From *Narrative and Professional Communication*, pp. 79–92, Jame M. Perkins and Nancy Blyler, eds. Stamford, Connecticut: Ablex Publishing Corp., 1999. Reprinted by permission of the author.

J.S.C. McKee, 'Communicating Science'. From *Physics in Canada* 54 (1998). http://www.cap.ca/pic/archives/54.4(1998)/editorial.html. Reprinted by permission of the publisher.

Joe Glaser, 'Voices to Shun: Typical Modes of Bad Writing'. From *Understanding Style: Practical Ways to Improve Your Writing*, pp. 26–41. New York: Oxford University Press, 1998. Reprinted by permission of Oxford University Press, Inc.

George Grant, 'Thinking about Technology'. From-*Technology and Justice*. Toronto, ON: House of Anansi Press, 1986. Copyright © 1986 by George Grant. Reprinted with permission from House of Anansi Press.

Stephen M. Halloran, 'Classical Rhetoric for the Engineering Student'. From *Journal of Technical Writing* 1, 1(1971): 17–24. Copyright © 1971. Baywood Publishing Company, Inc. Reprinted by permission.

George Harwell, 'Effective Writing'. From *Technical Communications*, pp. 1–15. Copyright © 1960. Reprinted by permission of Pearson Education, Inc., Upper Saddle River, NJ.

Richard M. Holliman, 'Communicating Science in the Digital Age'. http://www.open2.net/sciencetechnologynature/worldaroundus/communicate_science4.html (accessed 14 July 2006). Reprinted by permission of the author.

Marjorie Rush Hovde, 'Negotiating Organizational Constraints: Tactics for Technical Communicators'. From *Technostyle* 18 (Fall 2002): 61–94. Reprinted by permission of the author.

Ibrahim Khider, 'Rewind, Pause, Play, Fast Forward'. From *The Globe and Mail*, 23 June 2006: A22. Reprinted by permission of the author.

Bernadette Longo, 'Communicating with Non-Technical Audiences: How Much Do They Know?'. From *Proceedings of the IEEE International Professional Communication Conference*, 1993, pp. 167–71. © 1993 IEEE. Reprinted, with permission.

William Lutz, 'The World of DoubleSpeak'. From *Doublespeak*, pp. 1–7, 18–21, 46–51. New York: Harper Perennial, 1989. Reprinted by permission of Jean V. Naggar Literary Agency, Inc. for William Lutz.

Thomas R. McDaniel and Kathryn N. McDaniel, 'The Perils of PowerPoint'. From *Tomorrow's Professor* 663. Reprinted with permission from The National Teaching and Learning Forum, a bi-monthly newsletter on college teaching published by James Rhem & Associates, LLC, 301 South Bedford Street, #3, Madison, WI 53703. Individual print subscription, $49 annually. Multiple orders discounted. Internet site licenses available (http://www.ntlf.com). Orders and additional information: 608-255-4469.

Marshall McLuhan, 'Motorcar: The Mechanical Bride'. From *Understanding Media: The Extensions of Man*, pp. 217–25. New York: McGraw-Hill, 1965. Reprinted by permission of the publisher.

Carolyn Miller, 'What's Practical About Technical Writing?'. Reprinted by permission of the Modern Language Association of America from Bertie E. Fearing and W. Keats Sparrow, eds. *Technical Writing Theory and Practice*, pp. 14–24. Copyright © 1989.

Cezar M. Ornatowski, 'Between Efficiency and Politics: Rhetoric and Ethics in Technical Writing'. From *Technical Communication Quarterly* 1, 1: 91–103. Reprinted by permission of the author and publishers.

Arthur Plotnik, 'Gasping for Words'. From *The Elements of Expression: Putting Thoughts into Words*, pp. 3–11. New York: Henry Holt, 1996. Reprinted by permission of the author.

Herbert W. Simons, 'Are Scientists Rhetors in Disguise? An Analysis of Discursive Processes within Scientific Communities'. From *Rhetoric in Transition*, pp. 115–30, Eugene F. White, ed. University Park, PA: Pennsylvania State University Press, 1980. Reprinted by permission of the author.

Charles P. Snow, 'The Moral Un-neutrality of Science'. From *Science* 133 (1961): 255–62. Reprinted with permission from AAAS.

John R. Speed, 'What Do You Mean I Can't Call Myself a Software Engineer?'. From *IEEE Software* 16 (1999): 45–50. © 1999 IEEE. Reprinted, with permission, from *IEEE Software*.

Stephen Strauss, 'Avoid the Technical Talk, Scientists Told: Use Clear Language'. From *The Globe and Mail*, 27 August 1996: A1. Reprinted with permission from *The Globe and Mail*.

Stephen L. Talbott, 'The Deceiving Virtues of Technology: From the Cave of Cyclops to Silicon Valley'. From *NetFuture: Technology and Human Responsibility* 125 (15 November 2001). http://www.netfuture.org (accessed 16 April 2004). Reprinted by permission of the author.

Debbie Treise and Michael F. Weigold, 'Advancing Science Communication'. From *Science Communication* 23: 310–22. Copyright © 2002 by Sage Publications. Reprinted by Permission of Sage Publications, Inc.

Jean Weber, 'Escape from the Grammar Trap'. From *Techwr-L Magazine*. http://www.techwr-l.com/techwhirl/magazine/writing/grammartrap.html. Reprinted by permission of the author.

Helen Wilkie, 'Communicate Well and Prosper'. This article was originally published in *The Globe and Mail* (27 February 2004: C1, C5). Helen Wilkie is a professional keynote speaker, workshop leader and author of 'The Hidden Profit Center'. www.mhwcom.com. 416-966-5023

William Zinsser, 'Clutter'. From *On Writing Well: An Informal Guide to Writing Nonfiction*, pp. 6–17. New York: Harper and Row, 1976. Copyright © 1976, 1980, 1985, 1988, 1990, 1994, 1998, 2001, 2006 by William K. Zinsser. Reprinted by permission of the author.